Albert W. Alsop

CONCEPTUAL DESIGN OF CHEMICAL PROCESSES

McGraw-Hill Chemical Engineering Series

BUILDING THE LITERATURE OF A PROFESSION

Fifteen prominent chemical engineers first met in New York more than 60 years ago to plan a continuing literature for their rapidly growing profession. From Industry came such pioneer practitioners as Leo H. Baekeland, Arthur D. Little, Charles L. Reese, John V. N. Dorr, M. C. Whitaker, and R. S. McBride. From the universities came such eminent educators as William H. Walker, Alfred H. White, D. D. Jackson, J. H. James, Warren K. Lewis, and Harry A. Curtis. H. C. Parmelee, then editor of *Chemical and Metallurgical Engineering*, served as chairman and was joined subsequently by S. D. Kirkpatrick as consulting editor.

After several meetings, this committee submitted its report to the McGraw-Hill Book Company in September 1925. In the report were detailed specifications for a correlated series of more than a dozen texts and reference books which have since become the McGraw-Hill Series in Chemical Engineering and which became the cornerstone of the chemical engineering curriculum.

From this beginning there has evolved a series of texts surpassing by far the scope and longevity envisioned by the founding Editorial Board. The McGraw-Hill Series in Chemical Engineering stands as a unique historical record of the development of chemical engineering education and practice. In the series one finds the milestones of the subject's evolution: industrial chemistry, stoichiometry, unit operations and processes, thermodynamics, kinetics, and transfer operations.

Chemical engineering is a dynamic profession, and its literature continues to evolve. McGraw-Hill and its consulting editors remain committed to a publishing policy that will serve, and indeed lead, the needs of the chemical engineering profession during the years to come.

THE SERIES

CONCEPTUAL DESIGN OF CHEMICAL PROCESSES

James M. Douglas

University of Massachusetts

Boston, Massachusetts   Burr Ridge, Illinois
Dubuque, Iowa   Madison, Wisconsin   New York, New York
San Francisco, California   St. Louis, Missouri

McGraw-Hill

A Division of The McGraw·Hill Companies

This book was set in Times Roman.
The editors were B. J. Clark and James W. Bradley;
the cover was designed by Joseph Gillian;
the production supervisors were Diane Renda and Louise Karam.

CONCEPTUAL DESIGN OF CHEMICAL PROCESSES

13 14 BKMBKM 9 9 8

ISBN 0-07-017762-7

Library of Congress Cataloging-in-Publication Data

Douglas, James M. (James Merrill)
    Conceptual design of chemical processes.

    (McGraw-Hill chemical engineering series)
    Bibliography: p.
    Includes index.
    1. Chemical processes. I. Title. II. Series.
TP155.7.D67 1988   660.2′81   87-21359
ISBN 0-07-017762-7

ABOUT THE AUTHOR

James M. Douglas, Ph.D., is currently a professor of chemical engineering at the University of Massachusetts. Previously he taught at the University of Rochester and at the University of Delaware. Before entering teaching, he spent five years at ARCO, working on reactor design and control problems. He has published extensively in areas of reacting engineering, process control (including two books), and conceptual process design. He won the Post-Doctoral Fellowship Award at ARCO, the Faculty Fellowship Award at the University of Massachusetts, and the Computing and Chemical Engineering Award of AIChE.

CONTENTS

xi

PREFACE

This book describes a systematic procedure for the conceptual design of a limited class of chemical processes. The goal of a conceptual design is to find the best process flowsheet (i.e., to select the process units and the interconnections among these units) and estimate the optimum design conditions. The problem is difficult because very many process alternatives could be considered. In addition, experience indicates that less than 1 % of ideas for new designs ever become commercialized. Thus, there are many possibilities to consider with only a small chance of success.

In many cases the processing costs associated with the various process alternatives differ by an order of magnitude or more, so that we can use shortcut calculations to screen the alternatives. However, we must be certain that we are in the neighborhood of the optimum design conditions for each alternative, to prevent discarding an alternative because of a poor choice of design variables. Hence, we use cost studies as an initial screening to eliminate ideas for designs that are unprofitable. If a process appears to be profitable, then we must consider other factors, including safety, environmental constraints, controllability, etc.

We approach the synthesis and analysis problem by establishing a hierarchy of design decisions. With this approach, we decompose a very large and complex problem into a number of smaller problems that are much simpler to handle. By focusing on the decisions that must be made at each level in the hierarchy (e.g., Do we want to add a solvent recovery system?), we can identify the existing technologies that could be used to solve the problem (e.g., absorption, adsorption, condensation) without precluding the possibility that some new technology (e.g., a membrane process) might provide a better solution. Moreover, by listing the alternative solutions we can propose for each decision, we can systematically generate a list of process alternatives.

In some cases it is possible to use design guidelines (rules of thumb or heuristics) to make some decisions about the structure of the flowsheet and/or to set the values of some of the design variables. We use order-of-magnitude arguments to derive many of these heuristics, and we use a simple analysis of this type to identify the limitations of the heuristics. In many cases, no heuristics are available, and therefore we develop shortcut design methods that can be used as a basis for making decisions.

By following this hierarchical decision procedure, a beginning designer can substitute the evaluation of a number of extra calculations for experience during the development of a conceptual design. Since shortcut calculations are used, however, the penalty paid in the time required to screen more alternatives is not very high. Of course, as a designer gains experience, she or he will be able to recognize what alternatives do not need to be considered for a particular type of process and thereby obtain an increase in efficiency. Note also that experience normally is required for assessing the operability of a design, and therefore a beginner should always get an experienced designer to review the results of the design study.

Organization of the Text

The text is meant to be used in a one-semester, senior-level course in process design for chemical engineering students. We present the material as a lecture course. A single case study is carried throughout the text to illustrate the ideas, and the homework assignments include the evaluation of alternatives for the central case study, as well as several other case studies. The purpose of these other case studies is to help the student understand the similarities and differences between various types of processes (e.g., single reactions versus product distribution problems, cases where gas-recycle costs dominate, cases where liquid separation costs dominate, the choice between recycling or removing by-products formed by secondary reversible reactions, the economic trade-offs encountered when a gas recycle and a purge stream is used, etc.). The focus is on screening calculations, although a computer-aided design program is eventually used to verify the approximations.

Part I discusses a strategy of synthesis and analysis. In Chap. 1 it is noted that only about 1 % of ideas for new designs ever become commercialized, so that we need an efficient procedure for eliminating poor projects. Similarly, since design problems are always underdefined and we can often generate 10^4 to 10^9 alternative processes even for a single-product plant, we need an efficient way of screening process alternatives. These discussions provide the motivation for the use of shortcut calculations. Also, a procedure for decomposing process flowsheets into a hierarchical set of simpler problems is presented.

Chapter 2 presents an introduction to engineering economics, including a discussion of various measures of profitability. In addition, a simple economic model that is useful for conceptual designs is developed.

Chapter 3 presents a very simple design problem (actually a subsystem of what could be a larger design problem). This example illustrates how simple it is to generate process alternatives, the need for design heuristics, the origin of design heuristics, the limitations of design heuristics, the interactions among processing units, the need for a systems viewpoint in place of a unit operations viewpoint, and how shortcut design methods can be developed.

Part II presents the details of the hierarchical decision procedure for the synthesis and analysis of conceptual designs. Chapter 4 describes the information needed to get started, and the decision of designing a batch versus a continuous process is discussed. Chapter 5 presents the important decisions for the input and output structure, the identification of the important design variables at this level of complexity, and shortcut procedures to calculate the stream costs and the costs of a feed compressor (if one is required). Chapter 6 introduces the additional decisions required to fix the overall recycle structure of the flowsheet, i.e., the interaction of the reactor system(s) with the remainder of the process. The reactor cost and any gas-recycle compressor costs are evaluated in terms of the design variables. This discussion is limited to single-product plants.

At present, the systematic preliminary design procedure is also limited to vapor-liquid processes. For this class of processes, the structure of the separation system (i.e., the general structure, vapor recovery system alternatives, and the decisions for the liquid separation system) is described in Chap. 7. Chapter 8 then presents a synthesis procedure for the heat-exchanger network. At this point, a base-case design and an estimate of the optimum design conditions are available.

Our basic design strategy is to develop a base-case design as rapidly as possible, simply listing the process alternatives as we go along, to determine whether there is something about the process that will make all the alternatives unprofitable. Provided that our base-case design appears to be promising, we use the methods in Chap. 9 to screen the process alternatives. Thus, at this point we attempt to identify the best process flowsheet.

Part III presents some other design tools and applications. In the procedure presented in Chaps. 4 through 9, we used case-study calculations to estimate the optimum design conditions because we were continually changing the structure of the flowsheet. Once we have identified the best flowsheet, we can use more sophisticated optimization procedures. However, to assess the degree of sophistication that is desirable, we present an approximate optimization analysis in Chap. 10. This approximate optimization procedure helps to identify the dominant economic trade-offs for each design variable, the dominant design variables, and an indication of how far a design variable is away from the optimum without knowing the exact value of the optimum. This approximate optimization analysis is also very useful for retrofit studies and for optimum steady-state control calculations.

In Chap. 11 we use the same techniques for process retrofits that we used to develop a design for a new plant. A systematic procedure is presented for retrofitting processes, including completely replacing the existing plant with either the same or a better process alternative. The approximate optimization procedure is used to help identify the dominant operating variables and the equipment constraints that prevent the operating costs from being minimized. Then, based on these results, additional equipment capacity is added until the incremental, annualized equipment cost balances the incremental decrease in operating costs.

In Chap. 12 we discuss the use of a computer-aided design program to improve the accuracy of the shortcut calculations. Chapter 13 presents a summary of the design procedure, brief outlines of hierarchical decision procedures for solids and batch processes, and a brief discussion of what remains to be done after a conceptual design has been completed.

The appendixes present some auxiliary information. The shortcut models for equipment design are discussed in Appendix A, and the complete details of a case study are given in Appendix B. Some samples of design data and cost data are given in Appendixes C and E.

Acknowledgments

I am very appreciative of the efforts of A. Eric Anderson (formerly with ARCO), Duncan Woodcock of Imperial Chemical Industries, Edward C. Haun of UOP Inc., Jeff Kantor, University of Notre Dame; Carl F. King from duPont, E. L. Sherk from Exxon, R. Hoch (formerly with Halcon International), John Seinfeld, California Institute of Technology and J. J. Sirola from Tennessee Eastman Co. for their careful review of the text. Similarly, I am grateful to the chemical engineering students at the University of Massachusetts and to the students from Imperial Chemical Industries (United Kingdom), Rohm and Haas, Monsanto, Union Carbide and Celanese, for many valuable comments concerning the course material. In addition, I must acknowledge the numerous contributions that my colleague Mike Malone made to the text, and I want to thank my other colleagues Mike Doherty, Erik Ydstie, and Ka Ng for their feedback when they taught the material. The contributions of my graduate students, particularly Wayne Fisher and Bob Kirkwood, also need to be acknowledged.

Of course, I am especially grateful to my lovely wife, Betsy, to my children, Lynn and Bob, and to my mother, Carolyn K. Douglas, for their support during the preparation of the text. Similarly, Pat Lewis, my administrative assistant, and Pat Barschenski, who did the typing, provided much needed support.

James M. Douglas

CONCEPTUAL DESIGN OF
CHEMICAL PROCESSES

PART
I

THE
STRATEGY
OF PROCESS
SYNTHESIS
AND ANALYSIS

CHAPTER

1

THE
NATURE
OF PROCESS
SYNTHESIS
AND ANALYSIS

1.1 CREATIVE ASPECTS OF PROCESS DESIGN

The purpose of engineering is to create new material wealth. We attempt to accomplish this goal in chemical engineering via the chemical (or biological) transformation and/or separation of materials. Process and plant design is the creative activity whereby we generate ideas and then translate them into equipment and processes for producing new materials or for significantly upgrading the value of existing materials.

In any particular company, we might try to generate new ideas:

To produce a purchased raw material

To convert a waste by-product to a valuable product

To create a completely new material (synthetic fibers, food, bioprocessing)

To find a new way of producing an existing product (a new catalyst, a bioprocessing alternative)

To exploit a new technology (genetic engineering, expert systems)

To exploit a new material of construction (high-temperature- or high-pressure-operation, specialty polymers)

As an indication of the tremendous success of the engineering effort, we note that over 50% of the products sold by most chemical companies were developed during the last decade or two.

Success Rates

Despite this excellent record of success, we should realize that very few new ideas, either for improving existing processes or for developing new processes, lead to new wealth. In fact, the chances of commercialization at the research stage for a new process are only about 1 to 3%, at the development stage they are about 10 to 25%, and at the pilot plant stage they are about 40 to 60%.* Of course, we expect that the success rate for process modifications will be higher than that for completely new processes, but the economic rewards associated with these safer projects will have a significantly lower potential.

It is not surprising that so few ideas in engineering ever prove to be fruitful; the same pattern holds for any type of creative activity. Since experience indicates that only a small number of ideas ever will have a payout, we see that *evaluation* is one of the most significant components of any design methodology. In fact, process synthesis, i.e., the selection of equipment and the interconnections between that equipment which will achieve a certain goal, is really a combination of a synthesis and analysis activity.

Synthesis and Analysis

Perhaps the major feature that distinguishes design problems from other types of engineering problems is that they are underdefined; i.e., only a very small fraction of the information needed to define a design problem is available from the problem statement. For example, a chemist might discover a new reaction to make an existing product or a new catalyst for an existing, commercial reaction, and we want to translate these discoveries to a new process. Thus, we start with only a knowledge of the reaction conditions that we obtain from the chemist, as well as some information about available raw materials and products that we obtain from our marketing organization, and then we need to supply all the other information that we need to define a design problem.

To supply this missing information, we must make assumptions about what types of process units should be used, how those process units will be interconnected, and what temperatures, pressures, and process flow rates will be required. This is the synthesis activity. Synthesis is difficult because there are a very large number (10^4 to 10^9) of ways that we might consider to accomplish the same goal. Hence, design problems are very open-ended.

---

* These values represent the averages of estimates supplied by six friends working in economic evaluation groups of major chemical and petroleum companies.

Normally, we want to find the process alternative (out of the 10^4 to 10^9 possibilities) that has the lowest cost, but we must also ensure that the process is safe, will satisfy environmental constraints, is easy to start up and operate, etc. In some cases, we can use rules of thumb (heuristics) to eliminate certain process alternatives from further consideration, but in many cases it is necessary to design various alternatives and then to compare their costs. Experienced designers can minimize the effort required for this type of evaluation because they can often guess the costs of a particular unit, or group of units, by analogy to another process. However, beginning designers normally must design and evaluate more alternatives in order to find the best alternative.

When experienced designers consider new types of problems, where they lack experience and where they cannot identify analogies, they try to use shortcut (back-of-the-envelope) design procedures as the basis for comparing alternatives. These back-of-the-envelope calculations are used only to screen alternatives. Then if the process appears to be profitable, more rigorous design calculations are used to develop a final design for the best alternative, or the best few alternatives.

Because of the underdefined and open-ended nature of design problems, and because of the low success rates, it is useful to develop a strategy for solving design problems. We expect that the strategy that a beginning designer would use for synthesis and analysis would be different from that of an experienced designer, because a beginner must evaluate many more process alternatives. However, by using shortcut design procedures we can minimize the effort required to undertake these additional calculations.

Engineering Method

If we reflect on the nature of process synthesis and analysis, as discussed above, we recognize that process design actually is an art, i.e., a creative process. Therefore, we might try to approach design problems in much the same way as a painter develops a painting. In other words, our original design procedures should correspond to the development of a pencil sketch, where we want to suppress all but the most significant details of the design; i.e., we want to discover the most expensive parts of a process and the significant economic trade-offs. An artist next evaluates the preliminary painting and makes modifications, using only gross outlines of the subjects. Similarly, we want to evaluate our first guess at a design and generate a number of process alternatives that might lead to improvements. In this way, we hope to generate a "reasonable-looking," rough process design before we start adding much detail.

Then the artist adds color, shading, and the details of various objects in the painting and reevaluates the results. Major modifications may be introduced if they seem to be warranted. In an analogous manner, the engineer uses more rigorous design and costing procedures for the most expensive equipment items, improves the accuracy of the approximate-material and energy-balance calculations, and adds detail in terms of the small, inexpensive equipment items that are necessary for the process operations but do not have a major impact on the total plant cost, e.g., pumps, flash drums, etc.

Thus, we see that both a painting and a process design proceed through a series of successively more detailed synthesis and evaluation stages. Thatcher refers to a solution strategy of this type as successive refinements, and he calls it *the engineering method.** Note that as we make successive refinements, we should always maintain a focus on the *overall problem.*

If we accept this analogy between engineering design and art, then we can recognize some other interesting features of the design process. An artist never really completes a painting; normally the work is terminated whenever the additional effort reaches a point of diminishing returns; i.e., if little added value comes from much additional effort, the effort is not worthwhile. Another feature of art is that there is never a single solution to a problem; i.e., there are a variety of ways of painting a "great" Madonna and Child or a landscape; and in process engineering normally different processing routes can be used to produce the same chemical for essentially the same cost. Still another analogy between engineering design and art is that it requires judgment to decide how much detail should be included in the various stages of painting, just as it does in a process design.

Of course, numerous scientific principles are used in the development of a design, but the overall activity is an art. In fact, it is this combination of science and art in a creative activity that helps to make process design such a fascinating challenge to an engineer.

Levels of Engineering Designs

Now we see that there are a number of levels of engineering designs and cost estimates that we expect to undertake. These vary from very simple and rapid, but not very accurate, estimates to very detailed calculations that are as accurate as we can make. Pikulik and Diaz† classify these design estimates by the categories given in Table 1.1-1.

They also give the relative costs required to obtain these estimates, as shown in Table 1.1-2. From this table we see how rapidly engineering costs increase as we include more detail in the calculations. Obviously, we want to avoid large design costs unless they can be economically justified.

---

* C. M. Thatcher, *The Fundamentals of Chemical Engineering*, Merrill, Columbus, Ohio, 1962, chap. 3.

† A. Pikulik and H. E. Diaz, "Cost Estimating Major Process Equipment," *Chem. Eng.*, **84**(21): 106 (1977). *Note:* These accuracy bounds will vary from one company to another, and the accuracy of the detailed estimates will not be this good during periods of high inflation (the errors might be as much as 8 to 10%, even for a detailed estimate). Also, normally the chance of obtaining positive errors is greater than that for negative errors, so that the order-of-magnitude estimate, i.e., item 1, would be reported as +40 to −25% (design engineers seldom overestimate costs). Similarly, higher contingency fees may be included in the earlier levels (that is, 20 to 25% in item 3 dropping to 10% in item 4) to account for costs not included in the analysis (which is somewhat different from the accuracy of the estimate).

TABLE 1.1-1
Types of design estimates

1. Order-of-magnitude estimate (ratio estimate) based on similar previous cost data; probable accuracy exceeds $\pm 40\%$
2. Study estimate (factored estimate) based on knowledge of major items of equipment; probable accuracy up to $\pm 25\%$
3. Preliminary estimate (budget authorization estimate; scope estimate) based on sufficient data to permit the estimate to be budgeted; probable accuracy within $\pm 12\%$
4. Definitive estimate (project control estimate) based on almost complete data, but before completion of drawings and specifications; probable accuracy within $\pm 6\%$
5. Detailed estimate (contractor's estimate) based on complete engineering drawings, specifications, and site surveys; probably accuracy within $\pm 3\%$

From A. Pikulik and H. E. Diaz, "Cost Estimating Major Process Equipment," *Chem. Eng.*, **84**(21): 106 (1977).

For the case of a new process, where previous cost data are not available, it seems as if it would not be possible to develop an order-of-magnitude estimate. However, an experienced designer can overcome this difficulty by drawing analogies between the new process and other existing processes for which some data are available in the company files. Procedures for developing order-of-magnitude estimates have been described in the literature,* but normally it requires some experience to evaluate the results obtained from this type of calculation.

For a beginning designer, with little or no experience, it would be useful to have a systematic approach for developing order-of-magnitude estimates. We can use order-of-magnitude arguments to simplify many of the design calculations, and we can limit our attention to the major pieces of process equipment as we carry out a preliminary process design. The goal of this text is to develop a systematic

---

* J. H. Taylor, "Process Step-Scoring Method for Making Quick Capital Estimates," *Cost Eng.*, p. 207, July-August 1980. D. H. Allen, and R. C. Page, "Revised Technique for Predesign Cost Estimating," *Chem. Eng.*, **82**(5): 142 (March 3, 1975).

TABLE 1.1-2
Engineering costs to prepare estimates (1977)

Type of estimate	Less than $1 million Plant	$1–$5 million Plant	$5–$50 million Plant
Study ($ thousands)	5–15	12–30	20–40
Preliminary ($ thousands)	15–35	30–60	50–90
Definitive ($ thousands)	25–60	60–120	100–230

From A. Pikulik and H. E. Diaz, "Cost Estimating Major Process Equipment," *Chem. Eng.*, **84**(21): 106 (1977).

procedure of this type and then to show how the results can be extended to a study estimate.

Detailed estimates are considered to be beyond the scope of this text. However, as noted before, the chance that a new idea ever becomes commercialized is only about 1 %, so that we expect to undertake roughly 100 preliminary designs for every detailed design. Hence, the methodology of conceptual process design should be mastered in considerable detail.

Other Applications of the Methodology

Despite the fact that our primary focus is directed to the design and evaluation of new processes, much of the methodology we develop is useful for other engineering tasks, including basic research and technical service. In basic research, we want to spend most of our effort studying those variables that will have the greatest economic impact on the process, and rough process designs will help to identify the high-cost parts of the process and the dominant design variables. Similarly, in technical service activities, we look for ways of improving an existing process. To accomplish this goal, we need to understand the significant economic trade-offs in the process, and it is useful to have procedures available for obtaining quick estimates of the potential payout of new ideas. Thus, the methodology we develop will have numerous applications in the process industries.

1.2 A HIERARCHICAL APPROACH TO CONCEPTUAL DESIGN

The engineering method (or the artist's approach) indicates that we should solve design problems by first developing very simple solutions and then adding successive layers of detail. To see how we can use this approach for process design problems, we consider a typical flowsheet for a petrochemical process, and then we look for ways of stripping away layers of detail until we obtain the simplest problem of interest. By applying this procedure to a number of different types of processes, we might be able to recognize a general pattern that we can use as the basis for synthesizing new processes.

Example: Hydrodealkylation of Toluene (HDA Process)

The example we consider is the hydrodealkylation of toluene to produce benzene.* The reactions of interest are

$$\text{Toluene} + H_2 \rightarrow \text{Benzene} + CH_4 \qquad (1.2\text{-}1)$$

and $$2\text{Benzene} \rightleftharpoons \text{Diphenyl} + H_2 \qquad (1.2\text{-}2)$$

The homogeneous reactions take place in the range from 1150°F (below this

---

* This case study represents a modified version of the 1967 American Institute of Chemical Engineers (AIChE) Student Contest Problem; see J. J. McKetta, *Encyclopedia of Chemical Processing and Design*, vol. 4, Dekker, New York, 1977, p. 182, for the original problem and a solution.

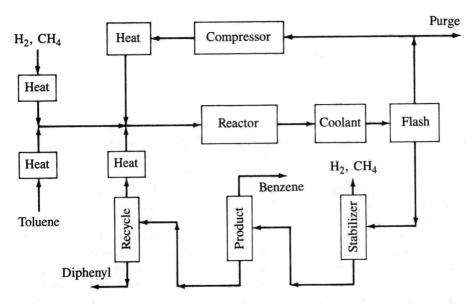

FIGURE 1.2-1
HDA process. [*After J. M. Douglas, AIChE J,* **33**: *353 (1985).*]

temperature the reaction rate is too slow) to 1300°F (above this temperature a significant amount of hydrocracking takes place) and at a pressure of about 500 psia. An excess of hydrogen (a 5/1 ratio) is needed to prevent coking, and the reactor effluent gas must be rapidly quenched to 1150°F in order to prevent coking in the heat exchanger following the reactor.

One possible flowsheet for the process is shown in Fig. 1.2-1. The toluene and hydrogen raw-material streams are heated and combined with recycled toluene and hydrogen streams before they are fed to the reactor. The product stream leaving the reactor contains hydrogen, methane, benzene, toluene, and the unwanted diphenyl. We attempt to separate most of the hydrogen and methane from the aromatics by using a partial condenser to condense the aromatics, and then we flash away the light gases. We use the liquid leaving this flash drum to supply quench cooling of the hot reactor gases (not shown on the flowsheet).

We would like to recycle the hydrogen leaving in the flash vapor, but the methane, which enters as an impurity in the hydrogen feed stream and is also produced by reaction 1.2-1, will accumulate in the gas-recycle loop. Hence, a purge stream is required to remove both the feed and the product methane from the process. Note that no rules of thumb (design guidelines) can be used to estimate the optimum concentration of methane that should be allowed to accumulate in the gas-recycle loop. We discuss this design variable in much greater detail later.

Not all the hydrogen and methane can be separated from the aromatics in the flash drum, and therefore we remove most of the remaining amount in a distillation column (the stabilizer)* to prevent them from contaminating our benzene product.

---

* The term stabilizer has historical roots, but it refers to a column used to remove light components.

The benzene is then recovered in a second distillation column, and finally, the recycle toluene is separated from the unwanted diphenyl. Other, alternative flowsheets can also be drawn, and we discuss some of these as we go through the analysis.

Energy Integration

The process flowsheet shown in Fig. 1.2-1 is not very realistic because it implies that the heating and cooling requirements for every process stream will take place in separate heat exchangers using external utilities (cooling water, steam, fuel, etc.). In the last decade, a new design procedure has been developed that makes it possible to find the minimum heating and cooling loads for a process and the heat-exchanger network that gives the "best" energy integration. This procedure is described in detail in Chap. 8.

To apply this new design procedure, we must know the flow rate and composition of each process stream and the inlet and outlet temperatures of each process stream. One alternative flowsheet that results from this energy integration analysis is shown in Fig. 1.2-2.* Now we see that first the reactor product stream is used to partially preheat the feed entering the reactor. Then the hot reactor gases are used to drive the toluene recycle column reboiler, to preheat some more feed, to drive the stabilizer column reboiler, to supply part of the benzene product column reboiler load, and to preheat some more feed before the gases enter the partial condenser. Also the toluene column is pressurized, so that the condensing temperature for toluene is higher than the boiling point of the bottom stream in the benzene column. With this arrangement, condensing toluene can be used to supply some of the benzene reboiler load, instead of using steam and cooling water from external sources of utilities.

If we compare the energy-integrated flowsheet (Fig. 1.2-2) with the flowsheet indicating only the need for heating and cooling (Fig. 1.2-1), then we see that the energy integration analysis makes the flowsheet more complicated (i.e., there are many more interconnections). Moreover, to apply the energy integration analysis, we must know the flow rate and composition of every process stream, i.e., all the process heat loads including those of the separation system as well as all the stream temperatures. Since we need to fix almost all the flowsheet before we can design the energy integration system and since it adds the greatest complication to the process flowsheet, we consider the energy integration analysis as the last step in our process design procedure.

Distillation Train

Let us now consider the train of distillation columns shown in Fig. 1.2-1. Since the unwanted diphenyl is formed by a reversible reaction (Eq. 1.2-2), we could recycle

---

* This solution was developed by D. W. Townsend at Imperial Chemical Industries, Runcorn, United Kingdom.

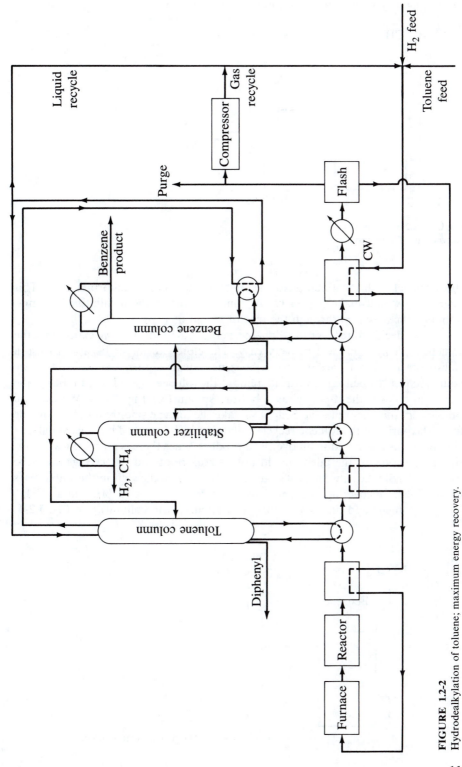

FIGURE 1.2-2
Hydrodealkylation of toluene; maximum energy recovery.

11

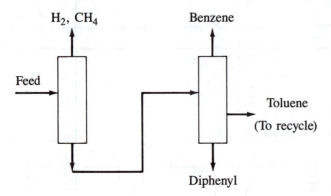

FIGURE 1.2-3
Alternate distillation trains.

the diphenyl with the toluene and let it build up to an equilibrium level. This alternative would make it possible to eliminate one of the distillation columns, although the flow rate through the reactor would increase.

If we decide to recover the diphenyl as Fig. 1.2-1 indicates, we expect that the toluene-diphenyl split will be very easy. Therefore, we might be able to use a sidestream column to accomplish a benzene-toluene-diphenyl split. That is, we could recover the benzene overhead, remove the toluene as a sidestream below the feed, and recover the diphenyl as a bottom stream (see Fig. 1.2-3). We can still obtain very pure benzene overhead if we take the toluene sidestream off below the feed. The purity of the toluene recycle will decrease, however, if it is recovered as a sidestream, as compared to an overhead product. Since there is no specification for the recycle toluene, the purity might not be important and the savings might be worthwhile. Similarly, we expect that the methane-benzene split in the stabilizer is easy. Then, recovering benzene as a sidestream in a H_2 and CH_4-benzene-toluene and diphenyl splitter (a pasteurization column) (see Fig. 1.2-4)

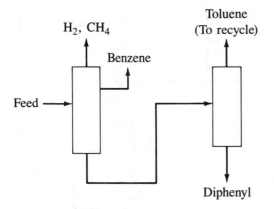

FIGURE 1.2-4
Alternate distillation trains.

might be cheaper than using the configuration shown in the original flowsheet (Fig. 1.2-1).

The heuristics (design guidelines) for separation systems require a knowledge of the feed composition of the stream entering the distillation train. Thus, before we consider the decisions associated with the design of the distillation train, we must specify the remainder of the flowsheet and estimate the process flows. For this reason we consider the design of the distillation train before we consider the design of the heat-exchanger network.

Vapor Recovery System

Referring again to Fig. 1.2-1, we consider the vapor flow leaving the flash drum. We know that we never obtain sharp splits in a flash drum and therefore that some of the aromatics will leave with the flash vapor. Moreover, some of these aromatics will be lost in the purge stream. Of course, we could recover these aromatics by installing a vapor recovery system either on the flash vapor stream or on the purge stream.

As a vapor recovery system we could use one of these:

Condensation (high pressure, or low temperature, or both)
Absorption
Adsorption
A membrane process

To estimate whether a vapor recovery system can be economically justified, we must estimate the flow rates of the aromatics lost in the purge as well as the hydrogen and methane flow in the purge. Hence, before we consider the necessity and/or the design of a vapor recovery system, we must specify the remainder of the flowsheet and we must estimate the process flows. We consider the design of the vapor recovery system before that for the liquid separation system because the exit streams from the options for a vapor recovery system listed above (e.g., a gas absorber) normally include a liquid stream that is sent to the liquid separation system.

Simplified Flowsheet for the Separation Systems

Our goal is to find a way of simplifying flowsheets. It is obvious that Fig. 1.2-1 is much simpler than Fig. 1.2-2, and therefore we decided to do the energy integration last. Similarly, since we have to know the process flow rates to design the vapor and liquid recovery systems, we decided to consider these design problems just before the energy integration. Thus, we can simplify the flowsheet shown in Fig. 1.2-1 by drawing it as shown in Fig. 1.2-5. The connections between the vapor and liquid recovery systems shown in Fig. 1.2-5 are discussed in more detail later.

We now ask ourselves whether all processes can be represented by the simplified flowsheet shown in Fig. 1.2-5. Since this flowsheet contains both gas- and

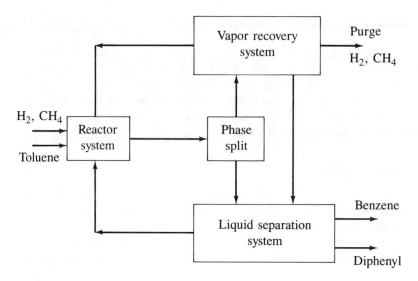

FIGURE 1.2-5
HDA separation system. [*After J. M. Douglas, AIChE J*, **31**: *353 (1985).*]

liquid-recycle loops, but some processes do not contain any gaseous components, we do not expect the results to be general. (See Sec. 7.1 for other alternatives.) However, we can simplify the flowsheet still more by lumping the vapor and liquid separation systems in a single box (see Fig. 1.2-6). Thus, we consider the specification of the general structure of the separation system before we consider the specification of either the vapor or the liquid recovery systems.

Recycle Structure of the Flowsheet

Now we have obtained a very simple flowsheet for the process (Fig. 1.2-6). We can use this simple representation to estimate the recycle flows and their effect on the reactor cost and the cost of a gas-recycle compressor, if any. Moreover, we can try

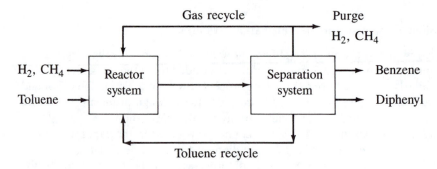

FIGURE 1.2-6
HDA recycle structure. [*After J. M. Douglas, AIChE J*, **31**: *353 (1985).*]

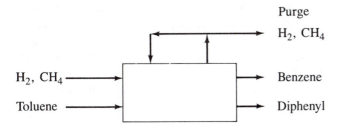

FIGURE 1.2-7
HDA input-output structure. [*After J. M. Douglas, AIChE J,* **31**: *353 (1985).*]

to understand what design questions are important to obtain this simplified representation, without worrying about the additional complexities caused by the separation system or the energy integration network. For example, we can study the factors that determine the number of recycle streams, heat effects in the reactor, equilibrium limitations in the reactor, etc. Thus, continuing to strip away levels of detail, we see that we want to study the recycle structure of the flowsheet before considering the details of the separation system.

Input-Output Structure of the Flowsheet

Figure 1.2-6 provides a very simple flowsheet, but we consider the possibility of obtaining an even simpler representation. Obviously, if we draw a box around the complete process, we will be left with the feed and product streams. At first glance (see Fig. 1.2-7), this representation might seem to be too simple, but it will aid us in understanding the design variables that affect the overall material balances without introducing any other complications. Since raw-material costs normally fall in the range from 33 to 85% of the total product costs,* the overall material balances are a dominant factor in a design. Also, we do not want to spend any time investigating the design variables in the ranges where the products and by-products are worth less than the raw materials. Thus, we consider the input-output structure of the flowsheet and the decisions that affect this structure before we consider any recycle systems.

Possible Limitations

By successively simplifying a flowsheet, we can develop a general procedure for attacking design problems. However, our original flowsheet described a continuous, vapor-liquid process that produced a single product and involved only simple chemicals (no polymers or hydrocarbon cuts). There are a large number of processes that satisfy these limitations, and so we try to develop this systematic

---

* E. L. Grumer, "Selling Price vs. Raw Material Cost," *Chem. Eng.,* **79**(9): 190 (April 14, 1967). Also see H. E. Kyle, *Chem. Eng. Prog.,* **82**(8): 17 (1986), for some data comparing commodity chemical production to speciality chemicals.

procedure in greater detail. However, batch processes may have a somewhat different underlying structure (we often carry out multiple operations in a single vessel), and certainly they are described differently in terms of mathematical models (normally ordinary differential or partial differential equations instead of algebraic equations or ordinary differential equations). Hence, our first decision probably should be to distinguish between batch and continuous processes.

Hierarchy of Decisions

If we collect the results discussed above, we can develop a systematic approach to process design by reducing the design problem to a hierarchy of decisions; see Table 1.2-1. One great advantage of this approach to design is that it allows us to calculate equipment sizes and to estimate costs as we proceed through the levels in the hierarchy. Then if the potential profit becomes negative at some level, we can look for a process alternative or terminate the design project without having to obtain a complete solution to the problem.

Another advantage of the procedure arises from the fact that as we make decisions about the structure of the flowsheet at various levels, we know that if we change these decisions, we will generate process alternatives. Thus, with a systematic design procedure for identifying alternatives we are much less likely to overlook some important choices. The goal of a conceptual design is to find the "best" alternative.

Shortcut Solutions

Experience indicates that it is usually possible to generate a very large number (i.e., often 10^4 to 10^9) of alternative flowsheets for any process if all the possibilities are considered. Hence, it is useful to be able to quickly reduce the number of alternatives that we need to consider. We normally screen these alternatives, using order-of-magnitude arguments to simplify the process material balances, the equipment design equations, and the cost calculations. These shortcut calculations often are sufficiently accurate to eliminate the 90%, or so, of the alternatives that do not correspond to profitable operation. Then if our synthesis and analysis lead

TABLE 1.2-1
Hierarchy of decisions

---

1. Batch versus continuous
2. Input-output structure of the flowsheet
3. Recycle structure of the flowsheet
4. General structure of the separation system
   a. Vapor recovery system
   b. Liquid recovery system
5. Heat-exchanger network

---

to a profitable solution, we repeat all the calculations more rigorously, because then we can justify the additional engineering effort.

The use of shortcut solutions and the hierarchical decision procedure also makes it possible to provide more rapid feedback to the chemist who is attempting to develop a process. That is, alternate chemical routes could be used to make the same product, with a large number of flowsheet alternatives for each route. Hence, quick estimates of the range of conversions, molar ratios of reactants, etc., that are close to the economic optimum for the various routes help the chemist to take data in the range where the most profitable operation might be obtained and to terminate experiments that are outside the range of profitable operation.

Decomposition Procedures for Existing Processes

Of course, we can also use the approach presented above as a decomposition procedure for existing processes, to simplify the understanding of the process, to understand the decisions made to develop the process, or to systematically develop a list of process alternatives. The decomposition procedure we suggest is as follows:

1. Remove all the heat exchangers, drums, and storage vessels.
2. Group all the distillation columns (liquid separation system block).
3. Simplify the general structure of the separation system (similar to Fig. 1.2-5).
4. Lump (group all units in a single box) the complete separation system (similar to Fig. 1.2-6).
5. Lump the complete process.

This decomposition procedure is different from those that break down the flowsheet into discrete subsystems which always retain their identity, i.e., into individual unit operations. To develop process alternatives, we want to modify the subsystems. With our approach we accomplish this task within a framework where we always consider the total plant, although the amount of detail included at various levels changes.

Hierarchical Planning

Our strategy of successive refinements and our hierarchical design procedure are similar to the hierarchical planning strategy discussed in the artificial intelligence (AI) literature. Sacerdoti* states,

> The essence of this approach is to utilize a means for discriminating between important information and details in problem space. By planning in a hierarchy of abstraction spaces in which successive levels of detail are introduced, significant increases in problem-solving power have been achieved.

---

* E. D. Sacerdoti, "Planning in a Hierarchy of Abstraction Spaces," *Artif. Intel.*, **5**: 115 (1974).

The concept can be readily extended to a hierarchy of spaces, each dealing with fewer details than the ground space below it and with more details than the abstraction space above it. By considering details only when a successful plan in a higher level space gives strong evidence of their importance, a heuristic search process will investigate a greatly reduced portion of the search space.

In our hierarchy, the ground state represents the energy-integrated flowsheet, and each level above it contains fewer details. Moreover, if the process appears to be unprofitable as we proceed through the levels in Table 1.2-1, we look for a profitable alternative or we terminate the project before we proceed to the next level. As noted by Sacerdoti, the hierarchy provides an efficient approach for developing a design.

1.3   SUMMARY AND EXERCISES

Summary

Process design problems are underdefined, and only about 1 % of the ideas for new designs ever become commercialized. Hence, an efficient strategy for developing a design is initially to consider only rough, screening-type calculations; i.e., we eliminate poor projects and poor process alternatives with a minimum of effort.

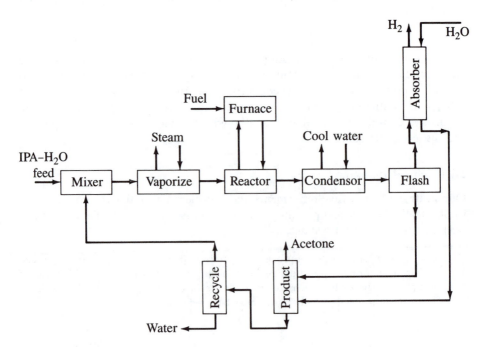

FIGURE 1.3-1
IPA plant. (*After 1947 AIChE Student Contest Problem.*)

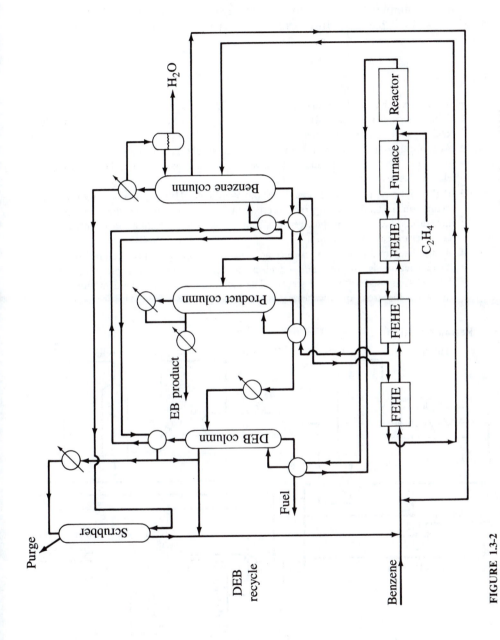

FIGURE 1.3-2
Energy-integrated ethylbenzene process. (*After D. L. Terrill, Ph.D. Thesis, University of Massachusetts, 1985.*)

Then if the results of this preliminary analysis seem promising, we add detail to the calculations and we use more rigorous computational procedures.

We can simplify the design problem by breaking it down into a hierarchy of decisions, as in Table 1.2-1. In this text we discuss this hierarchy of decisions in detail.

Exercises

Recommended exercises are preceded by an asterisk *.

1.3-1. If engineering time costs $100/hr, estimate the worker-hours required to complete each type of design study in Table 1.1-1 for a small plant.

1.3-2. According to the engineering method, what would be the best way to read a textbook that covers a field you have not studied before, (i.e., biotechnology, electrochemistry, etc.)?

*1.3-3. If the diphenyl in the hydrodealkylation of toluene (HDA) process is recycled to extinction, instead of being recovered, show one alternative for the hierarchy of flowsheets, i.e., input-output, recycle, separation system, distillation train (do not consider energy integration).

1.3-4. A flowsheet for a process to produce acetone from isopropanol is given in Fig. 1.3-1. The reaction is isopropanol $\rightarrow$ acetone $+ H_2$, and an azeotropic mixture of IPA–H_2O is used as the feed stream. The reaction takes place at 1 atm and 572°F. Show the hierarchy of flowsheets.

Reaction section

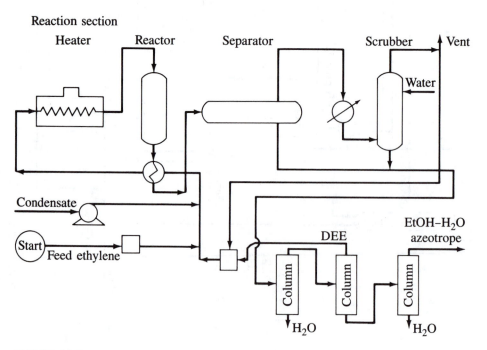

FIGURE 1.3-3
Ethanol synthesis.

1.3-5. An energy-integrated flowsheet for the production of ethylbenzene is given in Fig. 1.3-2. The primary reactions are

$$\text{Ethylene} + \text{Benzene} \rightarrow \text{Ethylbenzene}$$

$$\text{Ethylene} + \text{Ethylbenzene} \rightleftarrows \text{Diethylbenzene}$$

$$\text{Ethylene} + \text{Diethylbenzene} \rightleftarrows \text{Triethylbenzene}$$

The reaction is run with an excess of benzene and almost complete conversion of the ethylene, to try to minimize the formation of di- and triethylbenzene, and it takes place at 300 psig and 820°F over a catalyst. Two reactors are required (one on stream and the other being regenerated because of coke formation). There is 0.94% of ethane in the ethylene feed and 0.28% water in the benzene feed. Develop the hierarchy of flowsheets for this process.

1.3-6. A flowsheet for ethanol synthesis is shown in Fig. 1.3-3. The primary reactions are

$$\text{Ethylene} + H_2O \rightleftarrows \text{Ethanol}$$

$$2\ \text{Ethanol} \rightleftarrows \text{Diethyl Ether} + H_2O$$

The reaction takes place at 560 K and 69 bars, and about 7% conversion of the ethylene is obtained. The equilibrium constant for diethyl ether production at these conditions is about $K = 0.2$. The feed streams are pure water and an ethylene stream containing 90% ethylene, 8% ethane, and 2% methane. Show the hierarchy of flowsheets.

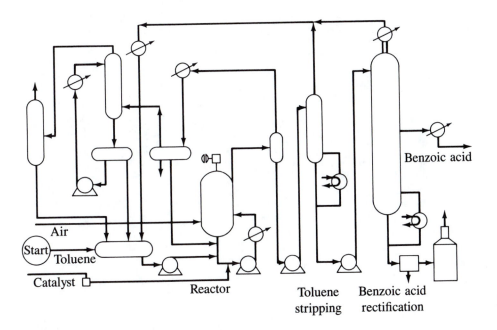

FIGURE 1.3-4
Benzoic acid production. [*After Hydrocarb. Proc.*, **48**(*11*): *156* (*Nov., 1964*).]

1.3-7. A flowsheet for benzoic acid production is shown in Fig. 1.3-4 [from SNIA
VISCOSA Process, *Hydrocarb. Proc.*, **48**(11): 156 (Nov., 1964)]. The primary
reaction is

$$\text{Toluene} + 1.5 O_2 \rightarrow \text{Benzoic Acid} + H_2O$$

However, reversible by-products (benzaldehyde and benzylic alcohol) as well as
heavier ones (assume phenyl benzoate and benzyl benzoate) are also formed at the
reaction conditions of 160°C and 10 atm. Pure toluene and air are used as the raw
materials, and the toluene conversion is kept at 30 to 35%. As shown on the
flowsheet, the toluene is recovered and recycled in one column, and the reversible by-
products are recycled from the overhead of a second. The product is recovered as a
vapor sidestream (with greater than 99% purity), and the heavy components are sent
to fuel. Show the hierarchy of flowsheets.

1.3-8. Select a flowsheet from *Hydrocarbon Processing* (see the November issue of any
year). Develop the hierarchy of flowsheets for the process.

CHAPTER
2

ENGINEERING ECONOMICS

In Chap. 1 we described a systematic approach that can be used to develop a conceptual design. In addition, we listed the types of design estimates that normally are undertaken over the life of a project. The goal of these estimates is to generate cost data, although the accuracy of the calculation procedures and the amount of detail considered are different for each type of estimate.

Since cost estimates are the driving force for any design study, we need to understand the various factors to include. We describe a procedure for generating a cost estimate for a conceptual design in this chapter. We begin by presenting the results from a published case study, in order to gain an overall perspective on the types of cost data required, and then we discuss the details of the cost analysis.

Remember that the cost models that we develop should be used *only* for screening process alternatives. The cost estimates that are reported to management should be prepared by the appropriate economic specialists in the company, because they will include contingency factors based on experience and will include the costs of more items than we consider. Thus, our cost estimates normally will be too optimistic, and they should be kept confidential until they have been verified.

2.1 COST INFORMATION REQUIRED

By considering the results of a published case study, we can get an overview of the kind of information that we need to develop a cost estimate for a conceptual design. Moreover, the framework relating the material and energy balances, equipment sizes and utility flows, capital and operating costs, and process profitability should become more apparent. The particular case study we consider involves the production of cyclohexane by the hydrogenation of benzene*

$$\text{Benzene} + 3\text{H}_2 \rightleftharpoons \text{Cyclohexane} \qquad (2.1\text{-}1)$$

---

* J. R. Fair, *Cyclohexane Manufacture*, Washington University Design Case Study No. 4, edited by B. D. Smith, Washington University, St. Louis, Mo., Aug. 1, 1967.

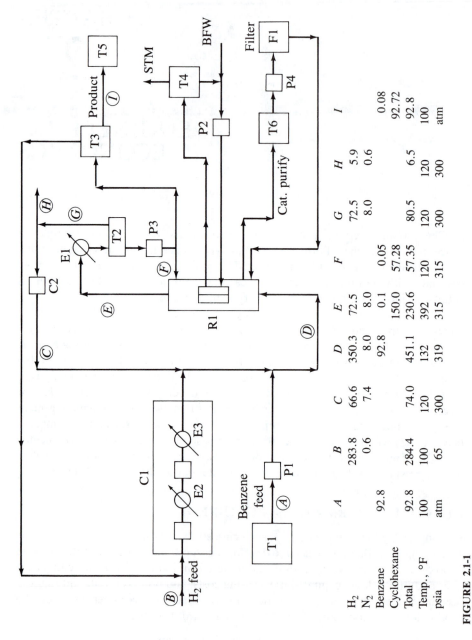

FIGURE 2.1-1

Cyclohexane manufacture. (*From J. R. Fair, Washington University Design Case Study No. 4, edited by B. D. Smith, Washington University, St. Louis, Mo., 1967.*)

	A	B	C	D	E	F	G	H	I
H_2	92.8	283.8	66.6	350.3	72.5		72.5	5.9	
N_2		0.6	7.4	8.0	8.0		8.0	0.6	
Benzene					0.1	0.05			0.08
Cyclohexane				92.8	150.0	57.28			92.72
Total	92.8	284.4	74.0	451.1	230.6	57.35	80.5	6.5	92.8
Temp., °F	100	100	120	132	392	120	120	120	100
psia	atm	65	300	319	315	315	300	300	atm

Our purpose here is not to discuss the details of the design, but merely to see what type of results are generated.

Flowsheet and Stream Table

One of the most important items that we develop during a design is a process flowsheet (see Fig. 2.1-1). The flowsheet shows the major pieces of equipment, and usually each piece of equipment is given a special number or name, as in Fig. 2.1-1. Normally each stream on the flowsheet is also lettered or numbered, and a stream table that contains these letters or numbers often appears at the bottom of the flowsheet. The stream table contains the flows of each component in every stream as well as the stream temperatures and pressures. In some cases, enthalpies, densities, and other information for each stream are included in the stream table.

Operating Costs

Once we know the stream flow rates and the stream temperatures, we can calculate the utility flows for the various units shown on the flowsheet; see Table 2.1-1. Then if we know the unit costs of the utilities, we can calculate the total utility costs. We combine these utilities costs with the raw-materials costs and other operating expenses to obtain a summary of the operating costs; see Table 2.1-2.

TABLE 2.1-1
Utilities summary: Base case

Utility	Item no.	Equipment name	Usage	
			Rate	Annual
Boiler feedwater			gpm	Mgal
	R-1	Reactor (coolant)	10	5,000
Steam, 50 lb.			lb/hr	Mlb
credit	R-1	Waste-heat boiler	5470	45,500
Electricity			kw	kwhr
	C-1	Feed compressor	316	2,620,000
	C-2	Recycle compressor	3.1	26,000
	P-1	Benzene feed pump	5.2	
	P-2	Boiler feed pump	0.5	48,000
	P-3	Reactor reflux pump	0.4	3,000
	P-4	Filter pump	—	
	Lighting	12 hr/day	5	22,000
	Total		330	2,719,000
Cooling water			gpm	Mgal
	E-1	Cooler-condenser	256	128,000
	E-2	Compressor intercooler	19	9,500
	E-3	Compressor aftercooler	19	9,500
	Total			147,000

From J. R. Fair, Washington Univesity Design Case Study No. 4, edited by B. D. Smith, Washington University, St. Louis, Mo., 1967.

TABLE 2.1-2
Operating cost summary: Cyclohexane—base case

ESTIMATED PRODUCTION COST AT ARNOLD, CONSOLIDATED CHEMICAL CO.

C_6H_{12} OUTPUT = 10,000,000 GAL (65,000,000 LB)   PER YEAR (8322 HOURS)

PRODUCT DELIVERED AS LIQUID, 99.9 + %

TOTAL MFG CAPITAL   = $510,000

TOTAL FIXED & WORKING CAPITAL = 693,000

	UNIT	QUANTITY PER YEAR	UNIT PRICE	COST PER YEAR	COST PER 100 LB
RAW MATERIALS					
BENZENE	gal	8,230,000	$0.23	$1,893,000	
HYDROGEN	MCF	900,000	0.23	207,000	
CATALYST	lb	10,800	2.00	21,600	
R. M. HANDLING					
TOTAL R. M.					
CREDITS SPENT CATALYST	lb	10,800	0.50	−5,400	

NET RAW MATERIALS				2,116,200	$3.26
DIRECT EXPENSE					
Labor				26,300	
Supervision				9,600	
Payroll Charges				5,400	
Steam (50 PSIG—CREDIT)	Mlb	45,500	0.50	−22,800	
Electricity	kwh	2,719,000	0.01	27,200	
Comp. Air					
Repairs @ 4% MFG. CAP.				20,400	
Water—Cooling	Mgal	147,000	0.015	2,200	
Water—Process					
Water—BOILER FEEDWATER	Mgal	5,000	0.30	1,500	
Fuel—Gas—Oil					
Fuel—Coal					
Factory Supplies / Laboratory } 2% MFG. CAP.				10,200	
TOTAL D.E.				80,000	0.12

(Continued)

TABLE 2.1-2 (*Continued*)

	UNIT	QUANTITY PER YEAR	UNIT PRICE	COST PER YEAR	COST PER 100 LB
INDIRECT EXPENSE					
Depreciation—M. & E. ⎱ 8% MFG. CAP.				40,800	
Depreciation—Bldg. ⎰					
Taxes & Ins. on Property ⎱ 4% MFG. CAP.				20,400	
Other Indirect ⎰					
TOTAL I.D.E.				61,200	0.09
TOTAL PROD. COST IN BULK ETC.					

From J. R. Fair, Washington University Design Case Study No. 4, edited by B. D. Smith, Washington University, St Louis, Mo., 1967.

TABLE 2.1-3
Equipment schedule

Item no.	No. registered	Name	Size (each)
R-1	1	Reactor*	4.5-in. diam. × 28 ft
C-1	1	Feed compressor	400 bhp, two-stage
C-2	1	Recycle compressor	5 bhp
E-1	1	Cooler-condenser	525 ft²
E-2	1	Intercooler	155 ft²
E-3	1	Aftercooler	155 ft²
P-1	2	Benzene feed pump	17 gpm, 860 ft
P-2	2	Boiler feed pump	11 gpm, 116 ft
P-3	2	Reflux pump	13 gpm, 93 ft
P-4	1	Filter pump	25 gpm, 62 ft
T-1	1	Benzene surge	57,000 gal
T-2	1	Reflux drum	930 gal
T-3	1	Line separator	12-in. diam. × 3 ft
T-4	1	Steam drum	150 gal
T-5	2	Product storage	158,000 gal
T-6	1	Filter charge tank	300 gal
F-1	1	Catalyst filter	35 ft²

*(E-5)	1	Reactor cooling coil	470 ft²

From J. R. Fair, Washington University Design Case Study No. 4, edited by
B. D. Smith, Washington University, St. Louis, Mo., 1967.

Capital Costs

After we have determined the stream flows and stream temperatures, we can calculate the equipment sizes; see Table 2.1-3. Then we can use cost correlations (which are discussed in Sec. 2.2) to estimate the delivered equipment costs. Next we use installation factors to estimate the installed equipment costs (see Table 2.1-4). We must also estimate the working capital required for the plant (see Table 2.1-5). Combining all these costs, we obtain an estimate of the total capital requirements (see Table 2.1-6). Note the magnitude of the compressor costs, as compared to the other kinds of equipment.

Profitability Estimate

We combine the operating and capital costs, along with some other costs, and we use these results to estimate the profitability of the process (see Table 2.1-7). The return on investment is used as criterion of profitability in the case study, but a number of other criteria can be used. These are discussed in Sec. 2.4.

Engineering Economics

Now that we can see what types of costs are included in an economic analysis, how can we generate these cost data? First we consider some of the methods for

TABLE 2.1-4
Manufacturing capital: Base case

Item no.	Delivered cost	Hand factor (see p. 33)	Total
R-1	$ 9,700	4.6	$ 44,600
C-1	76,000	2.8	212,800
C-2	3,000	2.8	8,400
E-1	5,100	4.0	20,400
E-2	2,500	4.0	10,000
E-3	2,500	4.0	10,000
P-1a	1,900	4.6	8,800
P-1b	1,900	4.6	8,800
P-2a	1,200	4.6	5,500
P-2b	1,200	4.6	5,500
P-3a	800	4.6	3,700
P-3b	800	4.6	3,700
P-4	1,200	4.6	5,500
T-1	6,500	4.6	30,000
T-2	2,700	4.6	12,400
T-3	500	4.6	2,300
T-4	600	4.6	2,800
T-5a	10,800	4.6	50,000
T-5b	10,800	4.6	50,000
T-6	775	4.6	3,600
F-1	2,900	4.0	11,500
	$143,370		$510,300
			Use $510,000

From J. R. Fair, Washington University Design Case Study No. 4, edited by B. D. Smith, Washington University, St. Louis, Mo., 1967.

TABLE 2.1-5
Working capital

1. Raw material (50% full)
   C_6H_6: 24,500 gal @ $0.23   5,600
2. Goods in process
   Est. 1750 gal @ $0.23   400
3. Product inventory (50% full)
   Cyclohexane: 145,000 gal @ $0.23 estimated   33,000
4. Other, at 5% gross sales
   10,000,000(0.24)(0.05)   120,000
                                                      $159,000

From J. R. Fair, Washington University Design Case Study No. 4, edited by B. D. Smith, Washington University, St. Louis, Mo., 1967.

TABLE 2.1-6
Estimate of capital requirements: Base case based on construction in 1967

1. Manufacturing Capital	
Equipment	Total cost
Reactor	$ 9,700
Compressors	79,000
Exchangers	10,000
Pumps	9,000
Tanks	32,670
Filter	2,900
Total process equipment	143,370
Total manufacturing capital based on Hand factors	$510,000
2. Nonmanufacturing Capital	
Proportionate share existing capital estimated at 15% manufacturing capital	$ 76,500
3. Total Fixed Capital	
Sum of 1 and 2	$586,000
4. Working Capital	
Raw-material inventory	$ 5,600
Goods in process	400
Finished product inventory	33,000
Store supplies and all other items at 5% gross sales	120,000
Total working capital	159,000
5. Total Fixed and Working Capital	$693,000

From J. R. Fair, Washington University Design Case Study No. 4, edited by B. D. Smith, Washington University, St. Louis, Mo., 1967.

TABLE 2.1-7
Profitability of cyclohexane manufacture

	Base case, 10^7 gal/yr
Manufacturing capital	$ 510,000
Total F&W capital*	693,000
Gross sales per year	2,400,000
Manufacturing cost	2,257,400
Gross profit	142,600
SARE[†] @ 10%	14,300
	128,300
Income tax	64,200
Net profit	64,100
Return on total F&W	9.3%

From J. R. Fair, Washington University Design Case Study No. 4, edited by B. D. Smith, Washington University, St. Louis, Mo., 1967.

* F & W is an acronym for fixed and working capital.

[†] SARE is an acronym for sales, administration, research, and engineering.

calculating capital and operating costs, then we describe the techniques for putting capital and operating costs on the same basis, next we discuss profitability measures, and finally we present a simple model that is useful for screening process alternatives when we develop a conceptual design.

2.2 ESTIMATING CAPITAL AND OPERATING COSTS

In Table 2.1-1 the utility loads for the various pieces of equipment on the flowsheet were itemized, and in Table 2.1-2 the utility costs were calculated. Similarly, in Table 2.1-3 the equipment sizes for the flowsheet were listed, and the costs were calculated in Table 2.1-4. Thus, the first costs we consider are the operating and capital costs associated with the equipment on the flowsheet.

Operating Costs

Operating costs are normally simple to estimate. Once we know the flows of the raw-materials streams and the utility flows (fuel, steam, cooling water, power), we simply multiply the flow by the dollar value of that stream. In companies that operate their utility systems, i.e., steam and power production, as a separate company, the utilities costs factors are simple to obtain. If this is not the case, however, an analysis of the total site is needed to estimate the cost of steam at various pressure levels. For our preliminary designs, we assume that a value is available.

Care must be taken that the utility values are given on a thermodynamically consistent basis; i.e., fuel and electricity should be more expensive than high-pressure steam, which should be more expensive than low-pressure steam, etc. Aberrations in prices do occur at times, so that it might appear that there is a profit in burning feedstocks to make electricity or in using electricity to produce steam. However, designs based on unusual market situations normally pay heavy economic penalties after a few years. One way to keep utility costs uniform is to relate all utility prices (electricity, various steam levels, and cooling-water costs) to an equivalent fuel value; see Appendix E.1.

The costs of chemicals can be obtained from the marketing department in a company. For academic purposes, current prices for most chemicals can be found in the *Chemical Marketing Reporter* or many of the trade publications. Light gases, for example, O_2, N_2, CO, etc., are not listed in the *Chemical Marketing Reporter* beause most are sold locally on long-term contracts. The current prices available in trade publications are often different from the price obtained from the marketing department because of long-term contract arrangements.

Capital Costs

As we might expect, there are a variety of ways of estimating the capital costs of equipment that range from very quick calculations with limited accuracy to very detailed calculations that are very time-consuming but more accurate. The most accurate estimate is simply to obtain a quote from a vendor; i.e., a heat-exchanger manufacturer agrees to sell you a heat exchanger that has a specified performance and that will be delivered on a certain date for a specified price. It pays to shop around because a vendor's quote will depend on how much work is on hand. These vendor's quotes are used as the costs of a final design.

For conceptual designs we need a faster and simpler approach (i.e., we do not want to try to optimize a process based on vendor's quotes). Thus, we normally use equipment cost correlations. For example, the capital cost of a heat exchanger normally is expressed in terms of the heat-exchanger area, and it is not neessary to specify the number of tubes, the number of baffles, the baffle spacing, or any of the details of the design. Similarly, the cost of a furnace is given in terms of the heat duty required, and the cost of a distillation column is specified in terms of the column height and diameter. The cost correlations are obtained by correlating a large number of vendor's quotes against the appropriate equipment size variable.

PURCHASED EQUIPMENT COST CORRELATIONS. A quite extensive set of cost correlations is available in Peters and Timmerhaus.* Other correlations of this type have been published by Chilton, Happel and Jordan, and Guthrie.† The correlations of Peters and Timmerhaus are among the most recent, although an even more recent update is available in ASPEN. Several correlations for various pieces of equipment that are taken from Guthrie can be found in Appendix E.2.

Of course, we are most interested in estimating the total processing costs. Therefore, we must be able to predict the installed equipment costs, rather than the purchased equipment costs. To accomplish this goal, we need to introduce a set of installation factors.

INSTALLED EQUIPMENT COSTS. One of the earliest approaches for estimating the installed equipment costs from the purchased equipment costs was proposed by Lang.‡ He noted that the total installed equipment costs were approximately equal to 4 times the total purchased costs, although different factors could be used for different kinds of processing plants. Hand¶ found that more accurate estimates could be obtained by using different factors for different kinds of processing equipment. For example, the purchased costs of distillation columns, pressure vessels, pumps, and instruments should be multiplied by 4; heat exchangers should be multiplied by 3.5; compressors by 2.5; fired heaters by 2; and miscellaneous equipment by 2.5. The use of Hand's factors is illustrated in Table 2.1-4.

---

* M. S. Peters and K. D. Timmerhaus, *Plant Design and Economics for Chemical Engineers*, McGraw-Hill, New York, 1968, chaps. 13 to 15.

† C. H. Chilton, "Cost Data Correlated," *Chem. Eng.*, **56**(6): 97 (Jan. 1949); J. Happel and D. G. Jordan, *Chemical Process Economics*, Dekker, New York, 1975, chap. 5; K. M. Guthrie, "Capital Cost Estimating," *Chem. Eng.*, **76**(6): 114 (1969).

‡ H. J. Lang, "Simplified Approach to Preliminary Cost Estimates," *Chem. Eng.*, **55**(6): 112 (1948).

¶ W. E. Hand, "From Flow Sheet to Cost Estimate," *Petrol. Refiner*, **37**(9): 331 (1958).

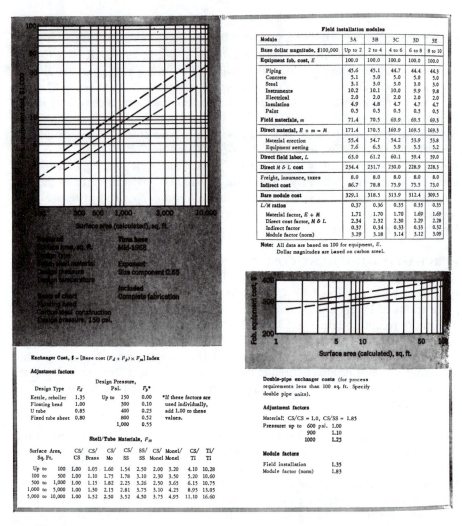

Field installation modules

Module	3A	3B	3C	3D	3E
Base dollar magnitude, $100,000	Up to 2	2 to 4	4 to 6	6 to 8	8 to 10
Equipment fob. cost, E	100.0	100.0	100.0	100.0	100.0
Piping	45.6	45.1	44.7	44.4	44.3
Concrete	5.1	5.0	5.0	5.0	5.0
Steel	3.1	3.0	3.0	3.0	3.0
Instruments	10.2	10.1	10.0	9.9	9.8
Electrical	2.0	2.0	2.0	2.0	2.0
Insulation	4.9	4.8	4.7	4.7	4.7
Paint	0.5	0.5	0.5	0.5	0.5
Field materials, m	71.4	70.5	69.9	69.5	69.3
Direct material, $E + m = M$	171.4	170.5	169.9	169.5	169.3
Material erection	55.4	54.7	54.2	53.9	53.8
Equipment setting	7.6	6.5	5.9	5.5	5.2
Direct field labor, L	63.0	61.2	60.1	59.4	59.0
Direct M & L cost	234.4	231.7	230.0	228.9	228.3
Freight, insurance, taxes	8.0	8.0	8.0	8.0	8.0
Indirect cost	86.7	78.8	75.9	75.5	73.0
Bare module cost	329.1	318.5	313.9	312.4	309.5
L/M ratios	0.37	0.36	0.35	0.35	0.35
Material factor, $E + M$	1.71	1.70	1.70	1.69	1.69
Direct cost factor, M & L	2.34	2.32	2.30	2.29	2.28
Indirect factor	0.37	0.34	0.33	0.33	0.32
Module factor (norm)	3.29	3.18	3.14	3.12	3.09

Note: All data are based on 100 for equipment, E. Dollar magnitudes are based on carbon steel.

Exchanger Cost, $ = [Base cost $(F_d + F_p)$ × F_m] Index

Adjustment factors

Design Type	F_d	Design Pressure, Psi.	F_p*	
Kettle, reboiler	1.35	Up to 150	0.00	*If these factors are used individually, add 1.00 to these values.
Floating head	1.00	300	0.10	
U tube	0.85	400	0.25	
Fixed tube sheet	0.80	800	0.52	
		1,000	0.55	

Shell/Tube Materials, F_m

Surface Area, Sq. Ft.	CS/CS	CS/Brass	CS/Mo	CS/SS	SS/SS	CS/Monel	Monel/Monel	CS/Ti	Ti/Ti
Up to 100	1.00	1.05	1.60	1.54	2.50	2.00	3.20	4.10	10.28
100 to 500	1.00	1.10	1.75	1.78	3.10	2.30	3.50	5.20	10.60
500 to 1,000	1.00	1.15	1.82	2.25	3.26	2.50	3.65	6.15	10.75
1,000 to 5,000	1.00	1.30	2.15	2.81	3.75	3.10	4.25	8.95	13.05
5,000 to 10,000	1.00	1.52	2.50	3.52	4.50	3.75	4.95	11.10	16.60

Double-pipe exchanger costs (for process requirements less than 100 sq. ft. Specify double pipe units).

Adjustment factors

Material: CS/CS = 1.0, CS/SS = 1.85
Pressure: up to 600 psi. 1.00
900 1.10
1000 1.25

Module factors

Field installation 1.35
Module factor (norm) 1.83

FIGURE 2.2-1
Shell-and-tube heat exchangers. (*From K. M. Guthrie, "Capital Cost Estimating," Chem. Eng., p. 114, Mar. 24, 1969.*) Also see p. 572.

GUTHRIE'S CORRELATIONS. An alternate approach was developed by Guthrie,* who published a set of cost correlations which included information both on the purchased cost and on the installed cost of various pieces of process equipment. Guthrie's correlation for shell-and-tube heat exchangers is shown in Fig. 2.2-1. We see that the information for the purchased cost for a carbon-steel

---

*K. M. Guthrie, "Capital Cost Estimating," *Chem. Eng.*, **76**(6): 114 (1969).

exchanger can be read directly from the graph. Then a series of correction factors can be used to account for the type of heat exchanger (fixed tubes, floating head, etc.), the operating pressure of the exchanger, and the materials of construction for both the tubes and the shell.

Moreover, once the purchased cost of the exchanger has been estimated, there is another set of factors available which can be used to find the installed cost. The installation factors provide separate accountings for the piping required, concrete used for the structural supports, conventional instrumentation and controllers, installation of the needed auxiliary electrical equipment, insulation, and paint. Similarly, factors for the labor costs required to install the equipment are listed as well as the indirect costs associated with freight, insurance, taxes, and other overhead costs.

The installation factors listed in the correlations are for carbon-steel exchangers, and we assume that the installation costs are essentially independent of the correction factors for pressure, materials of construction, etc. Hence, we can write the expressions

$$\text{Purchased Cost} = (\text{Base Cost})(F_c)(\text{Index}) \tag{2.2-1}$$

where F_c corresponds to the correction factors for materials, pressure, etc., and

$$
\begin{aligned}
\text{Installed Cost} &= \text{Installed Cost of Carbon-Steel Equipment} \\
&\quad + \text{Incremental Cost for Materials, Pressure, etc.} \\
&= (\text{IF})(\text{Base Cost})(\text{Index}) + (F_c - 1)(\text{Base Cost})(\text{Index}) \tag{2.2-2}
\end{aligned}
$$

where IF is the installation factor and Index is the correction factor for inflation. Hence,

$$\text{Installed Cost} = (\text{Base Cost})(\text{Index})(\text{IF} + F_c - 1) \tag{2.2-3}$$

Guthrie's correlations provide much more information than most other cost correlations, although they are as simple to use as other procedures. Moreover, if we should want a breakdown of the total cost for piping, or instrumentation, for all the process units, we could develop this information on a consistent basis. Some additional examples of Guthrie's correlations are given in Appendix E.2, and several examples illustrating the use of these correlations are given in Appendix B.

THE ASPEN CORRELATIONS. Another new set of cost correlations has been developed by Project ASPEN,* using data supplied by PDQ$, Inc. These correlations are part of a large, computer-aided design program, and therefore the correlations are all in numerical form, rather than the graphs used in most other sources. For example, the expression they use for heat exchangers is

$$C_E = C_B F_D F_{MC} F_P \tag{2.2-4}$$

---

* L. B. Evans, ASPEN Project, Department of Chemical Engineering & Energy Laboratory, MIT, Cambridge, Mass.

where C_E = 1979 exchanger cost; C_B = base cost for a carbon-steel, floating-head exchanger with a 100-psig design pressure and between 150 and 12,000 ft^2 of surface area; F_D = a design-type correction; F_{MC} = materials-of-construction correction factor; and F_P = a pressure correction factor. The expression they use for the base cost is

$$\ln C_B = 8.202 + 0.01506 \ln A + 0.06811(\ln A)^2 \qquad (2.2\text{-}5)$$

Equations for the correction factors are available as well as the cost expressions for a variety of other pieces of equipment. Similarly, the installation factors are given in the form of equations.

Updating Cost Correlations

Chilton's correlations were published in 1949, Guthrie's were published in 1968, and the Peters, Timmerhaus, and ASPEN correlations are more recent. However, it takes about three years to build a chemical plant, and so we must be able to predict future costs. Clearly the cost of almost everything increases with time, and so we must be able to update the cost correlations. Several methods can be used for this purpose, but they are all similar in that they involve multiplying the base cost in a certain year by the ratio of a cost index for some other year to the cost index for the base year.

One of the most popular cost indices of this type is published by Marshall and Swift (M&S) and is updated monthly in *Chemical Engineering*. A plot of the M&S index is shown in Fig. 2.2-2. Similar relationships are the *Engineering News-Record* index, the Nelson refinery index, the *Chemical Engineering* plant construction index, and the materials-and-labor cost index. Some of these indices include

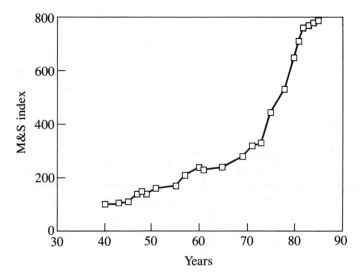

FIGURE 2.2-2
M&S index.

separate factors for labor and materials, which often experience different inflationary forces. Guthrie's correlations have the advantage that it is possible to update the material and labor factors at different rates, or some kind of average factor can be used to account for inflation.

IN-HOUSE COST CORRELATIONS. Many companies have developed their own cost correlations and installation factors. These are frequently updated by using vendor's quotations and recent construction costs. These company cost correlations should *always* be used if they are available. We use Guthrie's correlations because they are available in the published literature.

2.3   TOTAL CAPITAL INVESTMENT AND TOTAL PRODUCT COSTS

There are numerous costs required to build and operate a chemical plant other than the operating costs and the installed equipment costs; see Tables 2.1-2 and 2.1-6. Some of these costs add to the capital investment, whereas others are operating expenses. Fortunately, most of these costs can be related directly to the installed equipment costs through the use of various factors. A very concise summary of these costs was prepared by Peters by Timmerhaus,* and a modified version of their list for the total capital investment is shown in Table 2.3-1. The corresponding breakdown for the total product costs is given in Table 2.3-2.

It is common practice in the development of a design first to calculate the sizes of all the equipment and to estimate the amounts of utilities required. Next, the equipment costs are determined, and the utility costs are calculated. Then the other cost factors are added, and finally a profitability analysis is undertaken. However, for preliminary process design, we prefer to look for processs alternatives as soon as a design appears to be unprofitable. Therefore, we would like to develop simplified cost models for total investment, total processing costs, and process profitability. We develop a simple model of this type as we discuss the individual cost items.

Total Capital Investment

According to Table 2.3-1, the total capital investment (Tot. Inv.) is the sum of the fixed capital investment (Fixed Cap.) and the working capital (Work. Cap.):

$$\text{Tot. Inv.} = \text{Fixed Cap.} + \text{Work. Cap.} \tag{2.3-1}$$

---

* M. S. Peters and K. D. Timmerhaus, *Plant Design and Economics for Chemical Engineers*, 3d ed., McGraw-Hill, New York, 1969, chap. 5.

TABLE 2.3-1

Breakdown of total capital investment and start-up costs

Total capital investment equals the sum of the fixed capital investment plus the working capital.

I. *Fixed capital investment* (FCI) is the costs required to build the process, equal to the sum of the direct costs and the indirect costs.

    A. *Direct costs* equal the sum of the material and labor costs required to build the complete facility; about 70–85% of FCI.

        1. *Onsite costs or ISBL (inside of battery limits)* are the costs of installing the equipment shown on the process flowsheet in a specific geographical location (the battery limits); about 50–60% of FCI.

            a. *Purchased equipment* includes all equipment listed on a complete flowsheet; spare parts and noninstalled equipment spares; surplus equipment, supplies, and equipment allowances; inflation cost allowance; freight charges; taxes, insurance, and duties; allowance for modification during start-up; about 20–40% of FCI.

            b. *Purchased-equipment installation* includes installation of all equipment listed on a complete flowsheet including structural supports, insulation, and paint; about 7.3–26% of FCI or 35–45% of purchased equipment cost.

            c. *Instrumentation and control* includes purchase, installation, and calibration; about 2.5–7.0% of FCI or 6–30% of purchased equipment cost.

            d. *Piping* includes cost of pipe, pipe hangers, fittings, valves, insulation, and equipment; about 3–15% of FCI or 10–80% of purchased equipment cost.

            e. *Electrical equipment and materials* include the purchase and installation of the required electrical equipment including switches, motors, conduit, wire, fittings, feeders, grounding, instrument and control wiring, lighting panels, and associated labor costs; about 2.5–9.0% of FCI or 8–20% of purchased equipment cost.

        2. *Offsite costs or OSBL (outside of battery limits)* include costs directly related to the process but built in separate locations from the main processing equipment.

            a. *Buildings* (including services); about 6–20% of FCI or 10–70% of purchased equipment cost.

                (1) *Process buildings* include substructures, superstructures, stairways, ladders, access ways, cranes, monorails, hoists, elevators. (Some companies include these factors as part of the ISBL costs, and not the OSBL costs.)

                (2) *Auxiliary buildings* include administration and office, medical or dispensary, cafeteria, garage, product warehouse, parts warehouse, guard and safety, fire station, change house, personnel building, shipping office and platform, research laboratory, control laboratory.

                (3) *Maintenance shops* include electrical, piping, sheet metal, machine, welding, carpentry, instruments.

                (4) *Building services* include plumbing, heating, ventilation, dust collection, air conditioning, building lighting, elevators, escalators, telephones, intercommunication system, painting, sprinkler systems, fire alarm.

            b. *Yard improvements* involve site development including site clearing, grading, roads, walkways, railroads, fences, parking areas, wharves and piers, recreational facilities, landscaping; about 1.5–5.0% of FCI.

            c. *Service facilities* (installed); about 8.0–35.0% of FCI.

                (1) *Utilities* include steam, water, power, refrigeration, compressed air, fuel, waste disposal.

                (2) *Facilities* include boiler plant, incinerator, wells, river intake, water treatment, cooling towers, water storage, electric substation, refrigeration plant, air plant, fuel storage, waste disposal plant, fire protection.

      (3) *Nonprocess equipment* composed of office furniture and equipment, safety and medical equipment, shop equipment, automotive equipment, yard material-handling equipment, laboratory equipment, shelves, bins, pallets, hand trucks, fire extinguishers, hoses, fire engines, loading equipment.

      (4) *Distribution and packaging* include raw-material and product storage and handling equipment, product packaging equipment, blending facilities, loading stations.

    d. *Land*; about 1–2% of FCI or 4–8% of purchased equipment costs.

      (1) Surveys and fees.

      (2) Property costs.

  B. *Indirect costs* are expenses not directly involved with material and labor of actual installation; about 15–30% of FCI.

    1. *Engineering and supervision*; about 4–21% of FCI or 5–15% of direct costs.

      a. *Engineering costs* include administrative, process design and general engineering, drafting, cost engineering, processing, expediting, reproduction, communications, scale models, consultant fees, travel.

      b. *Engineering supervision and inspection.*

    2. *Construction expenses*; about 4.8–22.0% of FCI.

      a. *Temporary facilities* composed of construction, operation, and maintenance of temporary facilities; offices, roads, parking lots, railroads, electrical, piping, communications, fencing.

      b. *Construction tools and equipment.*

      c. *Construction supervision* involving accounting, timekeeping, purchasing, expediting.

      d. *Warehouse personnel and guards.*

      e. *Safety, medical, and fringe benefits.*

      f. *Permits, field tests, special licenses.*

      g. *Taxes, insurance, and interest.*

    3. *Contractor's fee*; about 1.5–5.0% of FCI.

    4. *Contingency*—to compensate for unpredictable events such as storms, floods, strikes, price changes, small design changes, errors in estimates, etc.; about 5–20% of FCI.

  C. *Alternate breakdown of FCI.*

    1. *Manufacturing capital investment*—same as onsites.

    2. *Nonmanufacturing capital investment* is offsite plus indirect costs.

II. *Working capital* is the capital required to actually operate the plant; about 10–20% of the total capital investment.

  A. *Raw material* for a one-month supply. (The supply depends on availability, seasonal demands, etc.)

  B. *Finished products* in stock and semifinished products; approximate production costs for one month. (Again, the amount may vary.)

  C. *Accounts receivable*—to give customers 30 days to pay for goods; about the production costs for one month.

  D. *Cash on hand* to meet operating expenses—salaries and wages, raw-material purchases.

  E. *Accounts payable and taxes payable.*

III. *Start-up costs*; about 8–10% of FCI.

  A. *Process modifications* needed to meet design specifications.

  B. *Start-up labor*—more people are needed to start up plant than to keep it running.

  C. *Loss in production* involves loss of revenues during debugging of the process.

Taken from M. S. Peters and K. D. Timmerhaus, *Plant Design and Economics for Chemical Engineers*, McGraw-Hill, New York, 1968.

TABLE 2.3-2

Gross earnings and total product costs

Gross earnings = total income − total production cost.
I. Total income = annual revenues.
II. Total product cost = manufacturing cost + general expenses.
  A. Manufacturing cost = direct production costs + fixed charges + plant overhead.
    1. Direct production costs (about 60% of the total product cost).
      a. Raw materials (about 10–50% of total product cost).
      b. Utilities (about 10–20% of total product cost).
      c. Maintenance and repairs (about 2–10% of FCI).
      d. Operating supplies (about 10–20% of cost for maintenance and repairs or 0.5–1% of FCI).
      e. Operating labor (about 10–20% of total product cost).
      f. Direct supervision and clerical labor (about 10–25% of operating labor).
      g. Laboratory charges (about 10–20% of operating labor).
      h. Patents and royalties (about 0–6% of total product cost).
    2. Fixed charges (about 10–20% of total product cost).
      a. Depreciation (about 10% of FCI).
      b. Local taxes (about 1–4% of FCI).
      c. Insurance (about 0.4–1% of FCI).
      d. Rent (about 10% of value of rented land and buildings).
      e. Interest (about 0–7% of total capital investment).
    3. Plant overhead (about 50–70% of the cost for operating labor, supervision, and maintenance or 5–15% of total product cost); costs include general plant upkeep and overhead, payroll overhead, packaging, medical services, safety and protection, restaurants, recreation, salvage, laboratories, and storage facilities.
  B. General expenses = administrative costs + distribution and selling costs + research and development costs [also called SARE (sales, administration, research, and engineering)].
    1. Administrative costs (about 15% of costs for operating labor, supervision, and maintenance or 2–5% of total product cost); includes costs for executive salaries, clerical wages, legal fees, office supplies, and communications.
    2. Distribution and selling costs (about 2–20% of total product cost); includes costs for sales offices, sales staff, shipping, and advertising.
    3. Research and development costs (about 2–5% of every sales dollar or about 5% of total product cost).

Taken from M. S. Peters and K. D. Timmerhaus, *Plant Design and Economics for Chemical Engineers*, McGraw-Hill, New York, 1968.

Start-up Costs

Many companies also include the start-up costs as part of the capital investment. Other companies consider the fraction of the start-up costs that is allocated to equipment modifications as part of the capital investment, whereas the funds used for additional workforce and materials needed to start up the plant are considered operating expenses. The choice among these various possibilities depends on the tax situation of the company. However, for our purposes we include the start-up costs (Start-up) as part of the investment. Hence, Eq. 2.3-1 becomes

$$\text{Tot. Inv.} = \text{Fixed Cap.} + \text{Work. Cap.} + \text{Start-up} \qquad (2.3\text{-}2)$$

From Table 2.3-1, item IV, we see that

$$\text{Start-up} \sim 0.1(\text{Fixed Cap.}) \qquad (2.3\text{-}3)$$

Working Capital

The working capital represents the funds required to actually operate the plant, i.e., to pay for raw materials, to pay salaries, etc. We attempt to replace the working capital each month out of product revenues. Nevertheless, we must have money available before we commence operations to fill up the tanks and to meet the initial payroll. For this reason the working capital is considered to be part of the total investment.

A breakdown of the working capital is given in Table 2.3-1, and a reasonable first estimate of this cost can be taken as a 3-month supply of raw materials, or products. We can greatly simplify the initial investment analysis, however, if we assume that working capital is related to the investment. For this reason, we let

$$\text{Work. Cap.} \sim 0.15(\text{Tot. Inv.}) \tag{2.3-4}$$

Fixed Capital Investment

From Table 2.3-1 we see that the fixed capital investment is the sum of the direct cost and the indirect costs:

$$\text{Fixed Cap.} = \text{Direct Cost} + \text{Indirect Cost} \tag{2.3-5}$$

The direct costs include the onsite costs (Onsite) or ISBL costs (inside battery limits), and the offsite costs, or OSBL costs (outside battery limits):

$$\text{Direct Cost} = \text{Onsite} + \text{Offsite} \tag{2.3-6}$$

The onsite costs correspond to the installed equipment costs for the items shown on the process flowsheet. All these items are built in a specific geographical area, called the *battery limits*. We can estimate the onsite costs directly from Guthrie's correlations.

The offsite costs, or OSBL costs, refer to the steam plant, cooling towers, and other items listed in Table 2.3-1 that are needed for the operation of the process but are built in a different geographical area. It is common practice to have central areas for cooling towers, steam generation equipment, etc. We note from the table that the variation in the individual offsite costs is much larger than that in the onsite costs. In fact, the offsite costs may vary from as little as 40 to 50% of the onsite costs for an expansion of an existing facility, up to 200 or 400% of the onsite costs for the construction of a grass-roots plant (a brand new facility starting from scratch) or a major plant expansion. This situation is analogous to building an addition to a house versus building a new home. In our studies, we consider only plant expansions, and we assume that

$$\text{Offsite} \sim 0.45 \text{ Onsite} \tag{2.3-7}$$

The indirect costs described in the table often are lumped in two categories: (1) the owner's costs, which include the engineering, supervision, and construction expenses; and (2) contingencies and fees (Conting.) which account both for items overlooked in the preliminary design and funds to pay the contractor. A contingency allowance of at least 5% should be included, even if we have firm quotes on hand from vendors, because something can always go wrong. For our preliminary designs, where we consider only the most expensive pieces of equipment, we include a contingency factor of 20%. Thus, we assume that

$$\text{Indirect Costs} = \text{Owner's Costs} + \text{Conting.} \tag{2.3-8}$$

$$\text{Owner's Cost} \simeq 0.05(\text{Onsite} + \text{Offsite}) \tag{2.3-9}$$

$$\text{Conting.} \sim 0.20(\text{Onsite} + \text{Offsite}) \tag{2.3-10}$$

With these approximations we can write

$$\text{Fixed Cap.} = \text{Onsite} + \text{Offsite} + \text{Owner's Cost} + \text{Conting.}$$
$$= 1.25(\text{Onsite} + \text{Offsite}) \tag{2.3-11}$$

A Simplified Investment Model

The factors we have selected to use in our analysis should give a reasonable estimate of the investment for the type of petrochemical processes that we are considering. However, different assumptions should be made for different processes, and the choice of these factors is an area where design experience is needed. Our goal is to develop a simple method for preliminary process design, so other factors should be used where they are applicable.

When we combine the expressions above, we find that

$$\text{Tot. Inv.} = \text{Fixed Cost} + \text{Work. Cap.} + \text{Start-up}$$
$$= \text{Fixed Cap.} + 0.15(\text{Tot. Inv.}) + 0.1(\text{Fixed Cap.})$$

so that

$$\text{Tot. Inv.} = 1.30(\text{Fixed Cap.}) \tag{2.3-12}$$

Then, from Eq. 2.3-11,

$$\text{Tot. Inv.} = 1.30(1.25)(\text{Onsite} + \text{Offsite})$$

or $$\text{Tot. Inv.} = 1.625(\text{Onsite} + \text{Offsite}) \tag{2.3-13}$$

For the case of a plant expansion, we substitute Eq. 2.3-7 to obtain

$$\text{Tot. Inv.} = 1.625[\text{Onsite} + 0.45(\text{Onsite})] = 2.36(\text{Onsite}) \tag{2.3-14}$$

Hence, once we have estimated the installed equipment costs, it is a simple matter to estimate the total investment, although it is important to remember that the estimate depends on the assumptions made in Eqs. 2.3-3, 2.3-4, 2.3-7, 2.3-9, and 2.3-10.

Total Product Cost

Table 2.3-2 lists a breakdown of the total product cost. Since the total product cost (Tot. Prod. Cost) is the sum of the manufacturing costs (Manu. Cost) and the general expenses (or SARE), we can write

$$\text{Tot. Prod. Cost} = \text{Manu. Cost} + \text{SARE} \qquad (2.3\text{-}15)$$

The SARE costs often are about 2.5% of the sales revenues for chemical intermediates, although they may be higher for finished products sold directly to consumers:

$$\text{SARE} \sim 0.025(\text{Revenue}) \qquad (2.3\text{-}16)$$

The manufacturing cost is the sum of the direct production cost, the fixed charges, and the plant overhead (OVHD):

$$\text{Manu. Cost} = \text{Direct Prod. Cost} + \text{Fixed Charges} + \text{Plant OVHD} \qquad (2.3\text{-}17)$$

The direct production costs include the raw materials, the utilities, maintenance and repairs, operating supplies (Op. Supply), operating labor, direct supervision, laboratory charges, and patents and royalties:

$$\begin{aligned} \text{Direct Prod. Cost} = \text{Raw Matl.} + \text{Util.} + \text{Maint.} + \text{Op. Supply} \\ + \text{Labor} + \text{Supervis.} + \text{Lab.} + \text{Royalty} \qquad (2.3\text{-}18) \end{aligned}$$

We can estimate the raw-materials costs and the utilities based on our preliminary design calculations. From the table we see that the maintenance and repairs and the operating supplies depend on the fixed capital investment, and for our studies we assume that

$$\text{Maint.} = 0.04(\text{Fixed Cap.}) \qquad (2.3\text{-}19)$$

$$\text{Supply} = 0.15(\text{Maint.}) = 0.15(0.04)(\text{Fixed Cap.}) \qquad (2.3\text{-}20)$$

The costs for operating labor, direct supervision, and laboratory charges also can be combined into a single factor. We assume that

$$\text{Labor} + \text{Supervis.} + \text{Lab.} = (1 + 0.2 + 0.15)(\text{Labor}) = 1.35(\text{Labor}) \qquad (2.3\text{-}21)$$

The table indicates that the cost for patents and royalties should be about 3% of the total product cost:

$$\text{Royalty} = 0.03(\text{Tot. Prod. Cost}) \qquad (2.3\text{-}22)$$

When we combine these relationships, we find that

$$\begin{aligned} \text{Direct Prod. Cost} = \text{Raw Matl.} + \text{Util} + 0.046(\text{Fixed Cap.}) \\ + 1.35(\text{Labor}) + 0.03(\text{Tot. Prod. Cost}) \qquad (2.3\text{-}23) \end{aligned}$$

The fixed charges (Fixed Chg.) given in Table 2.3-2 include local taxes, insurance, rent, and interest:

$$\text{Fixed Chg.} = \text{Tax} + \text{Insur.} + \text{Rent} + \text{Interest} \qquad (2.3\text{-}24)$$

Based on the values given in the table we assume that (with $3\%/yr$)

$$\text{Tax} + \text{Insur.} = 0.03(\text{Fixed Cap.}) \tag{2.3-25}$$

The interest charges on borrowed capital depend on the company's financing policy, and for our preliminary deigns we assume that internal funds are used to finance the venture, so

$$\text{Interest} = 0 \tag{2.3-26}$$

Similarly, we assume that we do not rent any facilities

$$\text{Rent} = 0 \tag{2.3-27}$$

The allocation for depreciation* may be calculated in a variety of ways, and so we discuss depreciation allowances in more detail later. With these approximations, we find that

$$\text{Fixed Chg.} = 0.03(\text{Fixed Cap.}) \tag{2.3-28}$$

According to Table 2.3-2, it is reasonable to assume that the plant overhead is roughly 60% of the cost for operating labor, direct supervision, and maintenance. Referring to Eqs. 2.3-19 and 2.3-21, we obtain

$$\begin{aligned}
\text{Plant OVHD} &= 0.6(\text{Labor} + \text{Supervis.} + \text{Maint.}) \\
&= 0.6[\text{Labor} + 0.2(\text{Labor}) + 0.04(\text{Fixed Cap.})] \\
&= 0.72(\text{Labor}) + 0.024(\text{Fixed Cap.}) \tag{2.3-29}
\end{aligned}$$

When we combine all the expressions above, we obtain an expression for the total product cost:

$$\begin{aligned}
\text{Tot. Prod. Cost} &= \text{Manu. Cost} + \text{SARE} \\
&= (\text{Direct Prod.} + \text{Fixed Chg.} + \text{Plant OVHD}) \\
&\quad + 0.025(\text{Revenue}) \\
&= [\text{Raw Matl.} + \text{Util.} + 0.046(\text{Fixed Cap.}) + 1.35(\text{Labor}) \\
&\quad + 0.03(\text{Tot. Prod. Cost})] + 0.03(\text{Fixed Cap.}) + [0.72(\text{Labor}) \\
&\quad + 0.024(\text{Fixed Cap.}] + 0.025(\text{Revenue})
\end{aligned}$$

Thus,

$$\begin{aligned}
\text{Tot. Prod. Cost} &= 1.03(\text{Raw Matl.} + \text{Util.}) + 2.13(\text{Labor}) \\
&\quad + 0.103(\text{Fixed Cap.}) + 0.025(\text{Revenue}) \tag{2.3-30}
\end{aligned}$$

---

* The depreciation allowance is included as a fixed charge in the table, but many companies do not account for depreciation in this way.

Now we would like to eliminate labor and fixed capital from this expression. From our previous analysis we know that

$$\text{Fixed Cap.} = 1.25(\text{Onsite} + \text{Offsite}) = 1.25(1.45)(\text{Onsite})$$
$$= 1.81(\text{Onsite}) \tag{2.3-31}$$

The cost for operating labor primarily depends on the complexity of the process, and it can be "guesstimated" from an inspection of the flowsheet (although some experience is required to make reasonable estimates). An attempt to quantify the reasoning involved was published by Wessel,[*] who correlated operating labor in worker-hours per day per processing step versus plant capacity. The difficulty with this procedure lies in estimating the number of processing steps; i.e., a batch reactor may require a full-time operator, whereas a continuous reactor may require only one-half of an operator's time.

For relatively small processes, such as we consider in this text, between two and four shift positions (operators) would be required. Labor costs per shift position are about \$100,000 (since we operate 24 hr/day for 7 days/wk, we need about 4.5 operators per shift position):

$$\text{Labor} = 100,000 \text{ Operators} \tag{2.3-32}$$

Simplified Cost Model for the Total Product Cost

When we combine Eqs. 2.3-30, 2.3-31, and 2.3-32, we obtain

$$\text{Tot. Prod. Cost} = 1.031(\text{Raw Matl.} + \text{Util.})$$
$$+ 2.13 \times 10^5 \text{ operators} + 0.103(1.81)(\text{Onsite})$$
$$+ 0.0256(\text{Revenue})$$

or $\text{Tot. Prod. Cost} = 1.031(\text{Raw Matl.} + \text{Util.}) + 0.186(\text{Onsite})$
$$+ 2.13 \times 10^5 \text{ Operators} + 0.0256(\text{Revenue}) \tag{2.3-33}$$

Hence, we can use the estimates of the raw-materials cost, the utilities, the revenues, and the installed equipment costs from our preliminary process design, to calculate the total product cost.

Profits

PROFIT BEFORE TAXES. The gross profit before taxes is the revenues minus the total product cost:

$$\text{Profit before Tax} = \text{Revenue} - \text{Tot. Prod. Cost} \tag{2.3-34}$$

---

[*] H. E. Wessel, "New Graph Correlates Operating Labor Data for Chemical Processes," *Chem. Eng.*, **59**(7): 209 (1952).

or, after eliminating the total product cost, we have

$$\text{Profit before Tax} = 0.974(\text{Revenue}) - 1.031(\text{Raw Mat.} + \text{Util.})$$
$$- 0.186(\text{Onsite}) - 2.13 \times 10^5 \text{ Operators} \qquad (2.3\text{-}35)$$

To calculate the profit after taxes, we must consider various depreciation policies.

DEPRECIATION. If we consider buying a car or truck to use for business purposes, it is apparent that the vehicle will wear out over time. Hence, we should set aside part of our revenues in order to accumulate sufficient funds to replace the vehicle when it does wear out, and we should consider these funds to be one of the costs of doing business. We could deposit this replacement allowance in a bank and draw interest, but we hope that we could gain an even higher effective interest rate by investing the funds in another venture of our company.

Fortunately, the government recognizes that it is a legitimate expense to deduct a fraction of the cost of equipment as it wears out, despite the fact that the funds are not actually used for this purpose; i.e., they are invested in other ventures, and a portion of the profits of these other ventures is used to replace the equipment. Thus, to prevent a company from establishing completely arbitrary or unrealistic depreciation schedules, the government specifies the average lifetime that can be expected for various types of processing equipment.

Of course, if pieces of equipment having different lifetimes are combined in a single process, clearly accounting for depreciation can become quite complicated. Since there are often several processes in an integrated plant complex, however, we can consider that there is an average lifetime for the process for preliminary design calculations. In fact, for petroleum processes we often assume a 16-yr life, whereas for chemical plants we often take an 11-yr life.

Once the process lifetime has been fixed, the government still allows us to choose between methods of computing the depreciation: straight-line or ACRS (accelerated cost recovery system). Land does not depreciate in value, and therefore the investment in the land should not be considered in a depreciation calculation. Similarly, if we replace the working capital each month, we will have the same amount of working capital at the end of the project as we started with, so that working capital does not depreciate. Furthermore, the equipment may have some salvage value at the end of the project (often about 10% of the purchased cost, which corresponds to about 3% of the fixed capital investment), so we should account for this salvage value at the end of the project. (If salvage value is not included in the depreciation calculation and the equipment is sold at the end of the project life, then a capital-gains tax must be paid on the value of the equipment sold.)

Straight-line depreciation simply means deducting 33% per year of the value of equipment having a 3-yr life, 20% per year for equipment with a 5-yr life, 10% per year for equipment having a 10-yr life, etc. The ACRS* method is more complex. It only started in 1980, and at that time all equipment had to be grouped in one of four categories—3-yr property, 5-yr property, 10-yr property, and 15-yr property.

---

* The depreciation allowances are still in a state of change.

TABLE 2.3-3
Depreciation allowances

3-yr property		5-yr property		10-yr property	
Year	%	Year	%	Year	%
1	25	1	15	1	8
2	38	2	22	2	14
3	37	3–5	21	3	12
				4–6	10
				7–10	10

The depreciation allowances for the first three categories are given in Table 2.3-3. The 15-yr property deduction allowable depends on the month that the item was placed in service, and is given in Table 2.3-4 for equipment placed in service after 1980 but before March 15, 1984. The U.S. government has been changing both the lifetime (from 15 to 18 and now 19 yr) and the allowance for this type of property every year, so recent tax information must be consulted.

The ACRS method is too complex to use for screening calculations, thus we use the simpler expression

$$\text{Deprec.} = 0.1(\text{Fixed Cap.}) = 0.1(1.81)(\text{Onsite})$$
$$= 0.181(\text{Onsite}) \tag{2.3-36}$$

PROFIT AFTER TAXES. The depreciation allowance is subtracted from the profit before taxes because it represents a cost for replacing equipment. For most large corporations, the income tax rate is 48%*, so that the profit after taxes is

$$\text{Profit after Taxes} = (1 - 0.48)(\text{Profit before Taxes} - \text{Deprec.})$$
$$= (0.52)(\text{Profit before Taxes} - \text{Deprec.}) \tag{2.3-37}$$
$$= 0.507(\text{Revenue}) - [0.536(\text{Raw Matl.} + \text{Util.}) + 0.52(\text{Deprec.})$$
$$+ 0.0967(\text{Onsite}) + 1.108 \times 10^5 \text{ Operators}] \tag{2.3-38}$$

TABLE 2.3-4
Depreciation allowance in % for 15-yr property (start in 1980)

Year	Month placed in service											
	1	2	3	4	5	6	7	8	9	10	11	12
1	12	11	10	9	8	7	6	5	4	3	2	1
2	10	10	11	11	11	11	11	11	11	11	11	12
3	9	9	9	9	10	10	10	10	10	10	10	10
4	8	8	8	8	8	8	9	9	9	9	9	9
5	7	7	7	7	7	7	8	8	8	8	8	8
6	6	6	6	6	6	7	7	7	7	7	7	7

* The tax rate has been changed, and is likely to change again in the near future.

Cash Flow

The actual cash flow (CF) that is retained by the company is the profit after taxes plus the depreciation allowance:

$$CF = \text{Profit after Taxes} + \text{Deprec.}$$
$$= 0.52(\text{Profit before Taxes} - \text{Deprec.}) + \text{Deprec.}$$
$$= 0.52(\text{Revenue} - \text{Tot. Prod. Cost}) + 0.48(\text{Deprec.}) \qquad (2.3\text{-}39)$$

or, after substituting Eqs. 2.3-35 and 2.3-36, we have

$$CF = 0.507(\text{Revenue}) - 0.536(\text{Raw Matl.} + \text{Util.})$$
$$+ 0.0098(\text{Onsite}) + 1.108 \times 10^5 \text{ Operators} \qquad (2.3\text{-}40)$$

Profitability Analysis

Now that we have calculated the cash flow, we have the information required to undertake a profitability analysis. However, since a profitability evaluation involves both capital and operating costs, first we must find some way of putting both types of cost on the same basis. To do this, we need to consider the time value of money.

2.4 TIME VALUE OF MONEY

When we consider process optimization studies, we often encounter trade-offs between capital and operating costs. For example, we can recover more of a solvent entering a gas absorber by increasing the number of trays. Operating costs are measured in $/hr (or more commonly in $/yr), whereas capital costs correspond to a single expenditure of money (i.e., an investment). Then, to trade off capital costs against operating costs, we must be able to place both costs on the same basis. Thus, we can either annualize the capital costs or capitalize the operating costs. In this text we report all costs on an annual basis.

Similar Problems and Strategy

The problem of trading off capital against operating costs is commonly encountered in everyday life. For example, when I bought my last car, I wanted to determine if it was to my advanage to buy a VW Rabbit with a diesel engine for $6400 as compared to a conventional engine for $5200, when diesel fuel cost $0.89/gal as compared to gasoline at $0.94/gal and the diesel engine averages 45 mi/gal as compared to a conventional engine that averages 32 mi/gal. There are different capital and operating costs for the two choices, and we want to find which is cheaper. A similar problem occurs if we want to assess the desirability of installing a solar heater that costs $15,000 in order to save 55 % of an oil bill of $1000/yr. To solve problems of this type, we must consider the time value of money.

We determine the time value of money simply by assuming that we will always borrow the capital that we need from a bank and, of course, that we must pay interest on the money that we borrow. With this approach, we replace the amount of capital investment by the annual payments that we must make to the bank to repay the loan and the interest on the loan. These annual payments have the same units as operating costs, which is what we want to achieve.

Thus, the key to understanding the relationship betwen capital and operating costs is merely to develop a detailed understanding of the repayment of bank loans. There are two parts to this repayment—principal and interest.

Conservation of Money in a Bank Account

Banks lend money at compound interest, and the simplest way of understanding the changing balance in an account is to assume that money in a bank account is a conserved quantity. That is, the money deposited (input) plus the interest paid by the bank to the account (input) minus the money withdrawn (output) must be equal to the amount of money that accumulates. Thus, the conservation of money in a bank account can be treated just as the conservation of mass energy, etc. Of course, this conservation principle is valid only for bank accounts and not for the federal government, because the government can simply print additional money. However, recognizing this restriction, we can write

$$\text{Accumulation} = \text{Input} - \text{Output} \tag{2.4-1}$$

CONTINUOUS INTEREST. Some banks are now offering continuous compounding on money, rather than compounding the interest at discrete intervals. Since the continuous compounding case is similar to other conservation problems that chemical engineers study, we consider it first. We let $S|_t$ be the money we have in the bank at time t. If we make no deposits or withdrawals, the amount we have in the bank will increase to $S|_{t+\Delta t}$ over a time interval Δt because the bank pays us interest. If we let the continuous interest rate be i_c [\$ interest/(\$ in account)(yr)], then the amount the bank pays us in the time interval Δt is $i_c S|_t \, \Delta t$.

Hence, the conservation equation, Eq. 2.4-1, becomes

$$S|_{t+\Delta t} - S|_t = i_c S|_t \, \Delta t \tag{2.4-2}$$

Now if we divide by Δt and take the limit as Δt approaches zero, we obtain

$$\lim_{\Delta t \to 0} \frac{S|_{t+\Delta t} - S|_t}{\Delta t} = \frac{dS}{dt} = i_c S \tag{2.4-3}$$

We can solve this differential equation to obtain

$$S = P_r e^{i_c t} \tag{2.4-4}$$

where P_r is the principal that we put into the bank initially; i.e., at $t = 0$, $S = P_r$. Thus, our money grows exponentially.

DISCRETE COMPOUNDING. It is more common for a bank to compound interest at discrete intervals. If we let $S|_{n+1} - S|_n$ = the accumulation of money in the account over one compounding interval, i = \$ interest/[(\$ in account)(1 period)], the amount of interest in one period = $iS|_n$, and if we make no deposits or withdrawals, then the conservation equation becomes

$$S|_{n+1} - S|_n = iS|_n \tag{2.4-5}$$

The parameter n takes on only integral values, and thus we call Eq. 2.4-5 a first-order, linear finite-difference equation.

Finite differences are not as common in chemical engineering practice as ordinary differential equations. However, the equations describing the compositions in a plate gas absorber or distillation column, where the composition changes from plate to plate instead of continuously, have this same form. If we use finite-difference calculus to solve Eq. 2.4-5, we obtain

$$S = P_r(1 + i)^n \tag{2.4-6}$$

where, again, $S = P_r$ when $n = 0$, so that P_r is the initial amount we deposited with the bank.

Instead of using finite-difference equations to find the compound interest, we can prepare a table showing the amount at the beginning of each compounding period, the amount of interest paid during that period, and the amount at the end of each period; see Table 2.4-1. As this table indicates, it is a simple matter to generalize the results and thus to obtain Eq. 2.4-6.

Comparison between Discrete and Continuous Compounding

As we might expect, the interest rate that a bank would pay using continuous compounding is different from that for quarterly compounding. Similarly, the rate for quarterly compounding is different from the rate for semiannual compounding. Hence, we need to find a way of comparing these various rates.

Suppose that a bank pays 1.5% interest per quarter and compounds the interest 4 times a year. In this case we say that the *nominal interest rate* is 6%/yr,

TABLE 2.4-1
Discrete compound interest

Period	Principal at beginning of period	Interest earned during period	Value of fund at end of period
1	P_r	$P_r i$	$P_r + P_r i = P_r(1 + i)$
2	$P_r(1 + i)$	$P_r(1 + i)i$	$P_r(1 + i) + P_r(1 + i)i = P_r(1 + i)^2$
3	$P_r(1 + i)^2$	$P_r(1 + i)^2 i$	$P_r(1 + i)^2 + P_r(1 + i)^2 i = P_r(1 + i)^3$
n	$P_r(1 + i)^{n-1}$	$P_r(1 + i)^{n-1} i$	$P_r(1 + i)^{n-1} + P_r(1 + i)^{n-1}i = P_r(1 + i)^n$

compounded quarterly. We note that it is essential to include the compounding interval in the description of the interest, because we expect that the *effective interest rate* on an annual basis will be greater than 6%. If we let r be the nominal interest rate and m be the number of corresponding intervals per year, the expression which is analogous to Eq. 2.4-6 for payments for 1 yr is

$$S|_{1\,\text{yr}} = P_r\left(1 + \frac{r}{m}\right)^m \tag{2.4-7}$$

When we compare this result to Eq. 2.4-6 for 1 yr, where we call the interest rate in Eq. 2.4-6 the effective interest rate i_{eff},

$$S|_{1\,\text{yr}} = P_r(1 + i_{\text{eff}}) \tag{2.4-8}$$

we see that

$$1 + i_{\text{eff}} = \left(1 + \frac{r}{m}\right)^m \tag{2.4-9}$$

For very frequent compounding periods, i.e., as m approaches infinity for n yr, Eq. 2.4-7 becomes

$$S = P_r\left(1 + \frac{r}{m}\right)^{mn} = P_r\left(1 + \frac{r}{m}\right)^{(m/r)rn} \tag{2.4-10}$$

However, by definition,

$$\lim_{m \to \infty}\left(1 + \frac{r}{m}\right)^{m/r} = e \tag{2.4-11}$$

so that Eq. 2.4-10 becomes

$$S = P_r e^{rn} \tag{2.4-12}$$

which has the same form as Eq. 2.4-4 if $t = n$ yr.

Thus, if we write Eq. 2.4-8 for n yr and compare it to Eq. 2.4-12, we see that

$$(1 + i_{\text{eff}})^n = e^{rn} \tag{2.4-13}$$

so that

$$i_{\text{eff}} = e^r - 1 \tag{2.4-14}$$

After rearranging this expression and comparing it to Eq. 2.4-4 with $t = n$, we find

$$r = \ln(i_{\text{eff}} + 1) = i_c \tag{2.4-15}$$

Thus, we can find expressions that relate the various interest rates.

Example 2.4-1. If the nominal annual interest rate is 6%, find the value of a $100 deposit after 10 yr with (*a*) continuous compounding, (*b*) daily compounding, (*c*) semiannual compounding, and (*d*) the effective annual interest rate for continuous compounding.

Solution

(a) $S = P_r e^{rn} = 100 e^{0.06(10)} = \$182.21$

(b) $S = P_r\left(1 + \dfrac{r}{m}\right)^{mn} = 100\left(1 + \dfrac{0.06}{365}\right)^{365(10)} = \$182.20$

(c) $S = P_r\left(1 + \dfrac{r}{m}\right)^{mn} = 100\left(1 + \dfrac{0.06}{2}\right)^{2(10)} = \$180.61$

(d) $i_{\text{eff}} = e^r - 1 = e^{0.06} - 1 = 0.0618$

It is interesting to note that continuous compounding is essentially the same as daily compounding.

Annuities

If we return to our example of whether to buy a car with a conventional or a diesel engine, we recognize that we have to make monthly payments on the car and that the bank can reinvest this money every month. Similarly, we usually continue to make deposits in a savings account rather than just make a single deposit. Thus, we need to extend our analysis of interest payments to cover these cases. The method involved is essentially the same as buying an annuity from a life insurance company.

DISCRETE CASE. Suppose we make periodic payments of $\$R$ for a total of n periods and the interest rate for each payment period is i. It is common practice to make the first payment at the end of the first period, so it will accumulate interest for $n - 1$ periods; the second payment will accumulate interest for $n - 2$ periods, etc. Hence, the money accumulated at the end of the n periods will be

$$S = R(1 + i)^{n-1} + R(1 + i)^{n-2} + R(1 + i)^{n-3} + \cdots + R(1 + i) + R \quad (2.4\text{-}16)$$

We can simplify this expression by multiplying by $1 + i$:

$$S(1 + i) = R(1 + i)^n + R(1 + i)^{n-1} + \cdots + R(1 + i) \quad (2.4\text{-}17)$$

and then subtracting Eq. 2.4-16 from Eq. 2.4-17 to obtain

$$iS = R[(1 + i)^n - 1]$$

or
$$S = R\,\frac{(1 + i)^n - 1}{i} \quad (2.4\text{-}18)$$

This same solution can be obtained using finite-difference calculus.

CONTINUOUS COMPOUNDING. For continuous compounding, the input term in Eq. 2.4-12 includes both the payment $R_c\,\Delta t$ and the interest rate on the money accumulated $i_c S|_t\,\Delta t$ during the time interval Δt, so that

$$S|_{t+\Delta t} - S|_t = R_c\,\Delta t + i_c S\,\Delta t$$

In the limit as Δt approaches zero, we find that

$$\frac{dS}{dt} = i_c S + R_c \qquad (2.4\text{-}19)$$

Again, we can separate variables and integrate, and if $S = 0$ at $t = 0$, we find that

$$S = \frac{R_c(e^{i_c t} - 1)}{i_c} \qquad (2.4\text{-}20)$$

which is the continuous analog of Eq. 2.4-18.

> **Example 2.4-2.** A friend buys a VW Rabbit for \$5200, makes a down payment of 10%, and then pays \$151.01/mo for 3 yr. If the nominal interest rate is 10%/yr compounded monthly. What is your friend's total cash outlay for the car?
>
> *Solution.* The total cost is the sum of the down payment and the accumulated value of the monthly payments given by Eq. 2.4-18:
>
> $$\text{Tot. Cost} = 520 + \frac{151.01}{0.10/12}\left[\left(1 + \frac{0.10}{12}\right)^{12(3)} - 1\right] = \$6829.47$$

General Approach to Interest Problems

With a detailed background in conservation equations, chemical engineers might find it simpler to derive interest formulas for other situations in terms of continuous compounding by making money balances based on Eq. 2.4-1. Exactly the same approach can be taken for discrete compounding if we use finite-difference calculus. Numerous tables are available in a variety of books that give the results for the discrete cases, and it is always possible to convert from one procedure to the other by calculating the effective annual interest rate.

Present Value

The interest formulas that we developed earlier describe the amount of money that will be in our bank account after a specified time interval; i.e., they indicate the future value of money. However, we make decisions about investments today, and so we would prefer to know the *present value* of various kinds of investment and payment plans. In other words, we want to ask, How much principal P, would we need to invest today in order to have a certain amount of money S available in the future? Of course, we can answer this question merely by rearranging the equations we derived before. The present value (PV) for discrete compounding is simply

$$\text{PV} = S(1 + i)^{-n} \qquad (2.4\text{-}21)$$

while that for continuous compounding is

$$\text{PV} = Se^{-i_c t} \qquad (2.4\text{-}22)$$

The terms $(1 + i)^{-n}$ and $e^{-i_c t}$ are called the *discount factors*.

Example 2.4-3. On your eighteenth birthday your rich uncle promises to give you $10,000 on the day you are 25. If the nominal interest rate is 8% compounded quarterly, how much money would he need to put into the bank on your eighteenth birthday for him to be able to keep his promise?

Solution. For Eq. 2.4-21 we find that

$$PV = 10,000\left(1 + \frac{0.08}{4}\right)^{-4(7)} = \$5743.75$$

Comparing Capital and Operating Costs

If we want to compare an investment I_1 plus annual payments R_1 to another investment I_2 with different annual payments R_2, to see which is the smaller, we want to compare the present values for each case. The present value is given by

$$PV = I + \frac{R}{i}[1 - (1 + i)^{-n}] \tag{2.4-23}$$

for the discrete compounding or

$$PV = I + \frac{R}{i_c}(1 - e^{-i_c t}) \tag{2.4-24}$$

for the continuous case. The appropriate expressions for other payment periods can be derived in a similar way.

Example 2.4-4. Suppose that we drive a car 15,000 mi/yr, that we keep a car for 7 yr before it rusts away and we junk it, and that we pay our gas bills monthly. If the nominal interest rate is 11 %/yr compounded monthly, is it better to buy a VW Rabbit with a conventional engine or a diesel engine (assume we pay the total purchase price in cash and that we use the cost and mileage conditions given earlier)?

Solution. According to Eq. 2.4-23,

$$\text{Conventional:}\quad PV = 5200 + \frac{(15,000/12)(1/32)(0.94)}{0.11/12}\left[1 - \left(1 + \frac{0.11}{12}\right)^{-12(7)}\right]$$

$$= 5200 + 2144.48 = \$7344.48$$

$$\text{Diesel:}\quad PV = 6400 + \frac{(15,000/12)(1/45)(0.89)}{0.11/12}\left[1 - \left(1 + \frac{0.11}{12}\right)^{-12(7)}\right]$$

$$= 6400 + 1443.85 = \$7843.85$$

Hence, it is better to buy the conventional engine.

Estimating Capital Costs

Now we can use a present-value calculation to compare alternatives that have different capital and operating costs. We can use these results to assess the profitability of a process.

2.5   MEASURES OF PROCESS PROFITABILITY

From the discussion in Sec. 2.3 clearly the cost accounting associated with process economics can become quite complicated. We anticipate the same kind of difficulty in attempts to assess the process profitability. Of course, we prefer to use the simplest possible procedures for our preliminary design calculations, but we would like the results to be as meaningful as possible. Thus, before we select a procedure for estimating profitability, we need to understand the relative advantages and disadvantages of the various techniques.

Return on Investment

In Table 2.1-7, the profitability measure calculated was the return on investment (ROI). We calculate this value by dividing the annual profit by the total investment and multiplying by 100:

$$\% \, \text{ROI} = \frac{\text{Annual Profit}}{\text{Tot. Inv.}} \times 100 \tag{2.5-1}$$

We can base this return on investment on either the profit before taxes or the profit after taxes, so we must be careful to report the basis for the calculation. Also, it is important to remember that the working capital, as well as the portion of the start-up costs considered as an investment for tax purposes, should be included in the total investment.

The return on investment is a very simple measure of the profitability, but it does not consider the time value of money. Moreover, it must be based on some kind of an average year's operation, since variable depreciation allowances (such as the ACRS method), increasing maintenance costs over the project life, changing sales volumes, etc., cannot be accounted for except by averaging. Despite these shortcomings, the return on investment often is used for preliminary design calculations.

Payout Time

Another measure that sometimes is used to assess profitability is the *payout time*, which is the time in years it takes to recover the funds that we invest (after the payout period we are playing poker with someone else's money, which is a desirable situation). We recover the working capital every month, and therefore we neglect the working capital in the calculation. However, the fraction of the start-up costs that is considered to be an investment should be added to the fixed capital investment, to find the amount of money tied up in the project. The funds that we recover from the project are the profit after taxes plus the depreciation allowance, which we call the *cash flow*, so that the payout time is

$$\text{Payout time (yr)} = \frac{\text{Fixed Cap.} + \text{Start-up}}{\text{Profit after Taxes} + \text{Deprec.}} \tag{2.5-2}$$

This criterion also is very simple to calculate, but it suffers from the same limitations as the ROI. Hence, we would like to contrast these simple procedures with a more rigorous analysis that accounts for the time value of money. In this way we gain a better understanding of the additional complexity required to obtain a more accurate estimate.

Discounted-Cash-Flow Rate of Return

Still another way to judge the desirability of investing in a new process is to estimate the maximum amount of interest that we could afford to pay if we borrowed all the investment and the project would just break even. Obviously, if our analysis indicates that we could afford to pay 120% interest, we know that it would be far better to invest in this project rather than in a bank. However, if the interest we could afford to pay was only 2%, we should abandon the project. When we consider interest calculations, we recognize that interest often is compounded at discrete intervals, and therefore we need to consider the time value of money. Hence, we want to evaluate the revenues, costs, depreciation, taxes paid, and the investment on a year-by-year basis.

Normally it takes about 3 yr to build a plant, and for this reason we want our investment costs, raw-material and product prices, utilities costs, etc., to reflect the values at least 3 yr in the future, rather than at the current time. Moreover, the calculation of the process profitability should be based on the income and costs 3 yr after the decision has been made to start construction. In other words, zero time is considered as 3 yr beyond the project approval.

ALLOCATION OF CAPITAL INVESTMENT. Since it requires about 3 yr to build a complex processing plant, the direct costs will be spent over this total period. At the outset, we will have to pay for the land, hire a contractor and construction crew, order the equipment, prepare the site, and start preparing the foundations for the equipment. Then we start installing the equipment as it is delivered. Thus, the direct cost expenditure at time minus 3 yr is about 10 to 15% of the total. During the periods of both -2 and -1 yr, we often spend 40 or 35% of the direct costs each year, and in the last year we normally spend the remaining 10 to 15%. However, the owner's costs, which are for engineering and supervision, and the contingencies and fees may be spent uniformly throughout the construction period.

The working capital and start-up costs are invested at time zero, but remember that the working capital is recovered at the end of the project. Similarly, the salvage value of the equipment can be realized at the end of the project, and this often amounts to 10% of the purchased equipment cost or roughly 3% of the fixed capital investment. Of course, money returned after N yr has a smaller value at time zero, because we could deposit a smaller sum in a bank at time zero and receive the compound interest on these funds for N yr. Again, we see that the time value of money requires us to account for funds in terms of their *present value,* which is just the principal P_r required to accumulate an amount of money S after N yr. The present value of various investment policies can be estimated once the interest rate has been specified by using the relationships we developed earlier.

ALLOCATION OF REVENUES AND COSTS. Most new plants do not reach their full productive capacity in the first few years of operations, often because a market does not exist for all the product. Experience indicates that the production rate increases from about 60 to 90 to 95 % during the first, second, and third years, respectively, of operation. After that time, hopefully, the process operates at full capacity.

Similarly, the depreciation allowance will vary each year, unless a straight-line depreciation schedule is used. Thus, with variable revenues and a variable depreciation allowance, the annual profits, the income taxes, and the net profit will change from year to year. We call the sum of the annual net profit, which is the profit after taxes plus the depreciation allowance, the *cash flow*, because this amount of money is actually retained by the company each year. Of course, the cash flow at the end of the first year, and later years, must be discounted to the present value, again because we could realize the same amount of money at a later date by investing a smaller amount of money at time zero.

DISCOUNTED-CASH-FLOW ANALYSIS. With the background given above we can set up the procedure for calculating the discounted-cash-flow rate of return (DCFROR). This is accomplished by equating the present value of the investment to the present value of the cash flows. If we consider a fairly general case where

1. The allocation of the direct costs can be represented by percentages, such as $a_1 = 0.1$, $a_2 = 0.4$, $a_3 = 0.4$, and $a_4 = 0.1$.
2. The revenues are constant except for the first 3 yr when $b_1 = 0.6$, $b_2 = 0.9$, and $b_3 = 0.95$.
3. The total product costs (or cash operating expenses) are constant.
4. We use straight-line depreciation, so that $d_1 = d_2 = d_3 = \cdots = d_N$.
5. We exclude the depreciation allowance from the total product cost.

then we can develop an expression for the equality of the present values of the expenses and the income.

The direct costs, the owner's costs, and the contingencies are spent over the construction period, but the working capital and the start-up costs are required only at start-up. Thus, the total value at start-up time is

$$\sum_{j=0}^{3} \{[a_j(\text{Direct Cost} + \text{Owner's Cost} + \text{Conting.})](1 + i)^j\}$$

$$+ \text{Work. Cap.} + \text{Start-up} \qquad (2.5\text{-}3)$$

The present value of all the cash flows discounted back to the start-up time plus the discounted value of the working capital and the salvage value at the end of the plant life after N yr is

$$\sum_{j=1}^{N} \left\{ \frac{b_j[0.52(\text{Revenue}_j - \text{Tot. Prod. Cost}_j) + 0.48d_j]}{(1 + i)^j} + \frac{(\text{Work. Cap.} + \text{Salv. Val.})}{(1 + i)^N} \right\} \qquad (2.5\text{-}4)$$

For a DCFROR calculation, we look for the interest rate i that makes these two expressions equal to each other. Unfortunately, there is no simple way to sum the series involved, so that we must use a trial-and-error procedure to find the interest rate i.

APPARENTLY UNPROFITABLE PROCESSES. Of course, there is no sense in attempting to solve the problem by trial and error if the total cash flow over the life of the project plus the salvage value is not adequate to pay for the fixed capital investment plus the start-up costs. In other words, for the interest rate to be positive, we require that

$$(\text{No. Years}) \times \text{Revenue} > \frac{\text{Fixed Cap.} + \text{Start-up} - \text{Sal. Val.} - 0.48 \, m \, \text{Deprec.}}{0.52}$$

$$- 0.52 \, m \, \text{Tot. Prod. Cost} \qquad (2.5\text{-}5)$$

In practice, we expect to encounter this limitation quite frequently; many ideas for new processes simply are not profitable, and the effects of inflation will make it appear that we can never build a plant similar to one that already exists—even if the market expands.

However, we do not want to eliminate projects that may become sound investments when product prices rise because of supply-and-demand considerations. Thus, if Eq. 2.5-5 is not satisfied, we often let i equal 0.15 or 0.2 in Eqs. 2.5-3 and 2.5-4; we substitute our estimates of direct costs, owner's costs, contingencies, working capital, start-up cost, total product cost, and salvage value; and then we solve for the revenue we would need to obtain. Next we estimate the product price that corresponds to these revenues and undertake a supply-and-demand analysis to determine how far in the future we might expect to obtain that price. If the time projection is 20 yr, we might as well put the project in the files for 15 yr or so; but if the time projection is 5 or 6 yr, we might continue to work on the design. Again, judgment is required to make this decision.

SIMPLIFIED MODEL. As we might expect, the analysis becomes much simpler if the investments, cash flows, and depreciation allowances are uniform. With constant cash flows we can use our interest and annuity formulas to sum the series.

The result becomes

$$\left(\frac{\text{Fixed Cap.}}{4}\right)\left[\frac{(1 + i)^4 - 1}{i}\right] + \text{Work. Cap.} + \text{Start-up}$$

$$= [0.521(\text{Revenue} - \text{Tot. Prod. Cost}) + 0.48\text{Deprec.}]\left[\frac{1 - (1 + i)^{-N}}{i}\right]$$

$$+ [\text{Work. Cap.} + \text{Salv. Val.}](1 + i)^{-N}$$

or

$$\left\{\left(\frac{\text{Fixed Cap.}}{4}\right)[(1 + i)^4 - 1] + (\text{Work. Cap.} + \text{Start-up})i\right\}(1 + i)^N$$

$$= [0.52(\text{Revenue} - \text{Tot. Prod. Cost}) + 0.48\text{Deprec.}][(1 + i)^N - 1]$$

$$+ (\text{Work. Cap.} + \text{Salv. Val.})i \tag{2.5-6}$$

Of course, we still need to use a trial-and-error procedure to find i.

Example 2.5-1. Calculate the DCFROR for the allocation of investment and revenue pattern given.

Solution

		$i = 0.15$	$i = 0.2$	$i = 0.168$
	Investment	Discount factor	Discount factor	Discount factor
Year-4	150,000	1.749	2.074	1.861
Year-3	350,000	1.521	1.728	1.593
Year-2	350,000	1.322	1.440	1.364
Year-1	150,000	1.15	1.20	1.168
Working capital	150,000			
Start-up cost	100,000			
PV investment		1,680,032	1,849,840	1,739,540
	Cash flow			
Year 1	240,000	0.870	0.833	0.856
2	360,000	0.756	0.694	0.733
3	380,000	0.658	0.578	0.627
4	400,000	0.572	0.482	0.537
5	400,000	0.497	0.402	0.460
6	400,000	0.432	0.335	0.394
7	400,000	0.376	0.279	0.337
8	400,000	0.327	0.232	0.289
9	400,000	0.284	0.194	0.247
10	400,000	0.247	0.161	0.212
Working capital	150,000	0.247	0.161	0.212
Salvage value	30,000	0.247	0.161	0.212
PV return		1,869,474	1,533,375	1,736,312

Thus, the DCFROR is 16.87%.

Capital Charge Factors

We prefer to avoid trial-and-error calculations in preliminary process designs, and yet we would like to account for the time value of money in some way in our profitability analysis. To accomplish this goal, we define a capital charge factor (CCF) as

$$\text{Revenue} - \text{Tot. Prod. Cost} = \text{CCF}(\text{Tot. Inv.}) \tag{2.5-7}$$

where the total investment includes the working capital and the start-up costs. Now we can use this definition to eliminate Revenue − Tot. Prod. Cost, which involves hourly costs, from Eq. 2.5-6, so that we obtain an expression containing only investments:

$$\left\{\left[\frac{\text{Fixed Cap.}}{4}\right][(1+i)^4 - 1] + (\text{Work. Cap.} + \text{Start-up})i\right\}(1+i)^N$$

$$= [0.52(\text{CCF})(\text{Tot. Inv.}) + 0.48(\text{Deprec.})][(1+i)^N - 1]$$

$$+ (\text{Work. Cap.} + \text{Salv. Val.})i \tag{2.5-8}$$

If we solve for CCF, we obtain

$$\text{CCF} =$$

$$\frac{\left\{\left(\frac{\text{Fixed Cap.}}{4}\right)[(1+i)^4 - 1] + (\text{Work. Cap.} + \text{Start-up})i - 0.48(\text{Deprec.})\right\}}{0.52(\text{Tot. Inv.})[(1+i)^N - 1](1+i)^{-N}}$$

$$+ \frac{-(\text{Work. Cap.} + \text{Salv. Val.})i + 0.48(\text{Deprec.})}{0.52(\text{Tot. Inv.})[(1+i)^N - 1]} \tag{2.5-9}$$

In our discussion of depreciation, we assumed that

$$\text{Salv. Val.} = 0.03(\text{Fixed Cap.}) \tag{2.5-10}$$

Also, we can relate the other quantities appearing in Eq. 2.5-9 to fixed capital, using the expressions we developed earlier:

$$\text{Start-up} = 0.1(\text{Fixed Cap.}) \tag{2.3-3}$$

$$\text{Tot. Inv.} = 1.3(\text{Fixed Cap.}) \tag{2.3-12}$$

$$\text{Work. Cap.} = 0.15(\text{Tot. Inv.}) \tag{2.3-4}$$

$$= 0.15(1.3)(\text{Fixed Cap.}) = 0.195(\text{Fixed Cap.}) \tag{2.5-11}$$

$$\text{Deprec.} = 0.1(\text{Fixed Cap.}) \tag{2.3-36}$$

With these approximations, all the investment terms cancel in Eq. 2.5-9, and we find that

$$\text{CCF} = \frac{[0.25(1+i)^4 + 0.295i - 0.298](1+i)^N - 0.225i + 0.048}{0.676[(1+i)^N - 1]} \tag{2.5-12}$$

Hence, based on these simplifying assumptions, we see that there is a direct relationship between CCF and the discounted-cash-flow rate of return i. If we let $i = 0.15$ and $N = 11$, we find that CCF $= 0.351$; whereas if we let $i = 0.15$ and $N = 16$, we obtain CCF $= 0.306$. In our preliminary design calculations, for the sake of simplicity, we let

$$\text{CCF} = 0.333 \text{ yr}^{-1} \qquad (2.5\text{-}13)$$

A value of $i = 0.15$ is the smallest value we would ever consider for a new project; i.e., a value of 0.2 is more realistic for safe projects. For a project with a high risk, such as in biotechnology, we might let CCF $= 1 \text{ yr}^{-1}$.

A Simplified Profitability Model

Using the definition of the CCF

$$\text{Revenue} - \text{Tot. Prod. Cost} = \text{CCF(Tot. Inv.)} \qquad (2.5\text{-}7)$$

we can substitute Eq. 2.3-14 for total investment and Eq. 2.3-33 for total production cost, to obtain

$$\text{CCF}(2.36 \text{ Onsite}) = \text{Revenue} - [1.031(\text{Raw Matl.} + \text{Util.}) + 0.186(\text{Onsite})$$

$$+ \ 0.0256(\text{Revenue}) + 2.13 \times 10^5 \text{ Operators}] \qquad (2.5\text{-}14)$$

or

$$\text{CCF} = \frac{0.413(\text{Revenue}) - 0.436(\text{Raw Matl.} + \text{Util.})}{\text{Onsite}}$$

$$+ \ \frac{0.0788(\text{Onsite}) + 0.902 \times 10^5 \text{ Operators}}{\text{Onsite}} \qquad (2.5\text{-}15)$$

Hence, from estimates of the revenues, the raw materials, the utilities, and the installed equipment costs (Onsite), we can calculate the CCF corresponding to a design. If this result is greater than 0.333, then the project appears promising and we can justify undertaking a more detailed design.

If the CCF is less than 0.333, we let CCF $= \frac{1}{3}$ in Eq. 2.5-14, and we calculate the revenue:

$$\text{Revenue} = 1.061(\text{Raw Matl.} + \text{Util.})$$

$$+ \ \frac{0.186 + 2.36 \text{ CLF}}{0.975}(\text{Onsite}) + 2.18 \times 10^5 \text{ Operators} \qquad (2.5\text{-}16)$$

From these revenues we calculate the product price required to make the process profitable, and then we undertake a supply-and-demand analysis.

Preliminary Design Profitability Model

For preliminary process designs where we are screening process alternatives, we can approximate the expression above by the equation (where all terms are in \$/yr.)

$$\text{Revenue} = \text{Raw Matl.} + \text{Util.}$$

$$+ \ (0.191 + 2.42 \text{ CCF}) \text{ Onsite} + 2.13 \times 10^5 \text{ Operators} \qquad (2.5\text{-}17)$$

Our first estimate of the economic potential (EP_i), based on current product prices, corresponds to

$$EP_1 = \text{Revenue} - \text{Raw Matl.} \qquad (2.5\text{-}18)$$

After we complete the material and energy balances, we can evaluate the utility costs, and we revise the economic potential:

$$EP_2 = \text{Revenue} - \text{Raw Matl.} - \text{Util.} \qquad (2.5\text{-}19)$$

Then, as we calculate the cost of each piece of equipment, we can subtract them from the revised economic potential:

$$EP_n = \text{Revenue} - \text{Raw Mat.} - \text{Util.} - (0.191 + 2.42\,\text{CCF}) \sum_{j=1}^{n} (\text{Onsite})_j \qquad (2.5\text{-}20)$$

(where all terms are in \$/yr; the onsites have a coefficient of 1/yr.) If this cost becomes significantly less than zero at any point in the analysis, we might want to redirect our efforts and look for less expensive process alternatives, rather than to complete the design calculations. As in any creative activity, judgment is required to make this decision.

In some cases it is easier to make judgments if we write Eq. 2.5-17 in terms of product prices instead of revenues. For a process with a single major product, we divide both sides of Eq. 2.5-17 by the product flow rate (Prod.), so that the left-hand side of the expression just becomes the product price, C_{PR}:

$$C_{PR} = \frac{\text{Raw Matl.}}{\text{Prod.}} + \frac{\text{Util.}}{\text{Prod.}} + \frac{\text{Onsite}}{\text{Prod.}} + 2.13 \times 10^5 \left(\frac{\text{Operators}}{\text{Prod.}} \right) \qquad (2.5\text{-}21)$$

The terms on the right-hand side are the contributions of the various quantities to the total product price; i.e., the units of each term can be ¢/lb product. If any of the terms on the right-hand side are very large compared to the current product prices, we want to consider process alternatives.

For cases where a process produces multiple products, such as a petroleum refinery, the analysis becomes more complex. In these situations, we consider both modifications of the process that lead to different product distributions and processes that can be used to convert one type of product to another. We continue in this way until we have developed as many cost expressions as there are products, and then we look for the optimum process alternative and design conditions.

Optimum Design

In many situations we want to find the values of design variables, such as reactor conversion, that maximize the profitability of the process. To do this, first we look for the values of the design variables that will minimize the product price that guarantees us a 15% DCFROR; i.e., we minimize C_{PR} in Eq. 2.5-21 (or the more exact relationship given by Eq. 2.5-16 divided by Prod.). If the minimum product price that we obtain from this analysis exceeds the current product price, we use a supply-and-demand analysis to decide whether we should terminate the project.

However, if this minimum required product price is less than the current price, it probably is advantageous to build a larger plant and collect more revenues. Since the CCF is directly related to the DCFROR by Eq. 2.5-12, we expect that the maximum CCF would correspond to the maximum DCFROR. To find a design variable x that maximizes CCF, we would write

$$\frac{d\text{CCF}}{dx} = \frac{d}{dx}\left(\frac{\text{Revenue} - \text{Tot. Prod. Cost}}{\text{Tot. Inv.}}\right) = 0$$

$$= \frac{(\text{Tot. Inv.})d(\text{Revenue} - \text{Tot. Prod. Cost})/dx}{(\text{Tot. Inv.})^2}$$

$$- \frac{(\text{Revenue} - \text{Tot. Prod. Cost})d(\text{Tot. Inv.})/dx}{(\text{Tot. Inv.})^2}$$

or $$\frac{d(\text{Revenue} - \text{Tot. Prod. Cost})/dx}{d(\text{Tot. Inv.})/dx} = \frac{\text{Revenue} - \text{Tot. Prod. Cost}}{\text{Tot. Inv.}}$$ (2.5-22)

However, close to the optimum design condition, the incremental return on an incremental investment will become very small. If this is the case, it will be more advantageous to allocate that incremental investment to a project where we would obtain a 15% DCFROR or CCF = 0.333. Hence, from this consideration of incremental return on incremental investment, we require that

$$\frac{d(\text{Revenue} - \text{Tot. Prod. Cost})/dx}{d(\text{Tot. Inv.})/dx} \geq 0.333$$ (2.5-23)

In other words, to find the optimum design conditions for a case where the minimum required product price is less than the current price, first we maximize CCF by solving Eq. 2.5-22. Then we evaluate CCF at the optimum design; and if this value is less than 0.333, we solve the problem by using Eq. 2.5-23. If the optimum CCF does exceed 0.333, we might want to consider the possibility of increasing the plant capacity, since the return on our investment will then be better than for most of our other projects. Of course, marketing considerations may limit this alternative.

Economic Decisions among Process Alternatives

In general, we prefer to select the process alternative that satisfies the production goal and requires the least capital investment, because with a specified CCF this process normally will give the smallest product price. However, if the least expensive process involves a lot of unproven technology, highly corrosive or hazardous materials, an uncertain supply of raw materials, or other similar factors, we must assess the additional costs that we may encounter in overcoming potential problems.

In addition, in some situations we can decrease the losses of either materials or energy from a process by installing additional equipment. For these cases we again require that the incremental return on this additional investment satisfy our investment criterion, i.e., a CCF of 0.333.

Economic Decisions for Process Modifications or Replacements

If our new idea involves the modification or replacement of part of a process by a new technology, we still want to achieve a 15%, or more, return on our investment because this project will be in competition with other projects considered by the company. The investment required is equal to the cost of the new equipment minus the *actual market value* of the equipment we are replacing. Note that we should use the actual market value in the calculation rather than the original cost minus the depreciation we have already recovered, because our original estimates of the equipment life and the depreciation might have been in error. In other words, we always base our economic decisions on present conditions, and we ignore our past mistakes, just as we drop out of a poker game if the cards reveal we have little chance of winning even if we have a large stake in the pot.

The savings we expect to gain from the replacement are the old operating costs plus the depreciation of the old equipment over its expected life as judged from the present (and not the original depreciation calculation) minus the operating costs for the new equipment plus the depreciation for this equipment over its expected life. If these savings provide a 15% return on the net investment, we might want to consider the replacement project using more detailed design and costing procedures.

2.6 SIMPLIFYING THE ECONOMIC ANALYSIS FOR CONCEPTUAL DESIGNS

In Eq. 2.5-17 we presented a very simple economic model that we can use for conceptual designs (i.e., the screening of a large number of flowsheet alternatives by using order-of-magnitude estimates to determine the best flowsheet or the best few alternatives):

$$\text{Revenues} = \text{Raw Matl.} + \text{Util.} + (0.191 + 2.42\,\text{CCF})\,\text{Install. Equip. Cost} + 2.18 \times 10^5\,\text{Operators} \tag{2.6-1}$$

If we choose CCF = 1/3, then $0.191 + 2.42\,\text{CCF} \approx 1.0/\text{yr}$.

Economic Potential

In Chap. 1 we presented a hierarchical decision procedure that would simplify the development of a conceptual design. The approximate cost model presented above fits into the hierarchical framework very nicely. Thus, when we consider the input-output structure of the flowsheet, i.e., level 2 in the hierarchy, we can define an economic potential EP_2 at this level as

$$EP_2 = \text{Revenue} - \text{Raw Matl.}$$

$$- (\text{Power} + (1/\text{yr}) \text{ Cap. Cost of Feed Compress, if any}) \quad (2.6\text{-}2)$$

Similarly, when we consider the recycle structure of the flowsheet, i.e., level 3, and we generate cost estimates for the reactor and a recycle gas compressor (if any), we can write

$$EP_2 = \text{Revenue} - \text{Raw Matl.} - (\text{Feed Compress. Cap.}$$

$$+ \text{Op. Cost}) - \text{Reactor Cost} - (\text{Gas-Recycle}$$

$$\text{Compress. Cap.} + \text{Op. Cost}) \quad (2.6\text{-}3)$$

Thus, as we add more detail to the flowsheet, we merely subtract the new utilities costs and the installed equipment cost of the new equipment that is added. If the economic potential at any level becomes negative, we have three options:

1. Terminate the design study.
2. Look for a better process alternative.
3. Increase the product price so that the economic potential is zero, and continue with the design.

If we follow option 3, we eventually determine a value of the product price that would make the process alternative under consideration profitable. If this new product price were only slightly higher than the current price, we would probably continue with the design. (We need to undertake a supply-and-demand analysis to see how far in the future that we might expect to obtain this higher price.)

However, if the product price required to make the alternative profitable were much greater than the current price at any of the levels in the hierarchy, we would terminate the work on the current alternative and look for one that was cheaper. If none of the alternatives were acceptable, we would terminate the project. This approach is very efficient because it makes it possible to terminate projects with a minimum amount of design effort.

Significant Equipment Items

The case study considered in Sec. 2.1 is somewhat unusual because one piece of equipment (the recycle compressor C-1) comprises almost half of the total purchased (or installed) equipment cost. However, suppose we consider another case study,* for the disproportionation of toluene to produce benzene and xylene.

---

* R. J. Hengstebeck and J. T. Banchero, *Disproportionation of Toluene*, Washington University Design Case Study No. 8, edited by B. D. Smith, Washington University, St. Louis, Mo., June 26, 1969.

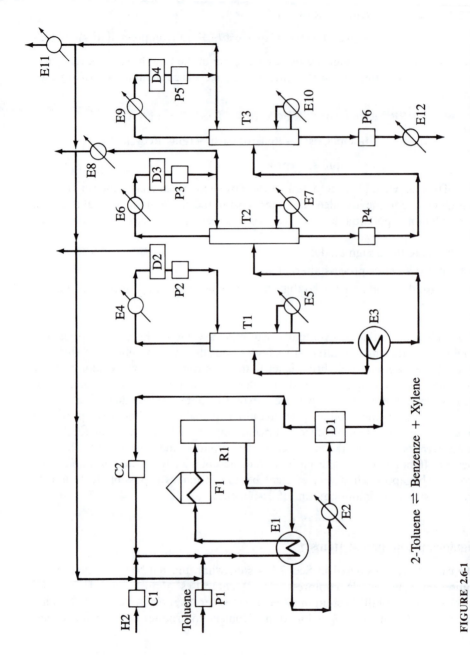

2-Toluene ⇌ Benzenze + Xylene

FIGURE 2.6-1
Disproportionation of toluene. (*From R. J. Hengstebeck and J. T. Banchero, Washington University Design Case Study No. 8, edited by B. D. Smith, Washington University, St. Louis, Mo., 1969.*)

TABLE 2.6-1
Investment summary, $

Pumps (1949)		Furnaces (1969)	209,000
P-1	5,900	Reactor (1969)	29,800
P-2	1,320	Towers (1969)	
P-3	1,950	T-1	25,000
P-4	1,680	T-2	37,600
P-4	2,500	T-3	35,500
P-5	980	Total	98,100
P-6	14,380		
Pumps (1949),		Trays (1969)	
including spares	28,760	T-1	5,800
Pumps (1969)	50,000	T-2	31,200
		T-3	42,000
Exchangers (1968)		Total	79,000
E-1	140,000		
E-2	115,000	Compressors (1969)	
E-3	8,800	C-1	—
E-4	22,000	C-2	313,000
E-5	26,000	Drums (1969)	23,650
E-6	16,000	Installed cost summary	
E-7	9,400	Pumps	185,000
E-8	4,200	Exchangers	1,140,000
E-9	16,000	Reactor	128,000
E-10	17,000	Towers (ex trays)	490,000
E-11	9,300	Trays	395,000
E-12	6,500	Compressors	751,000
	390,200	Drums	130,000
Exchangers (1969)	408,000	Furnace	523,000
			3,742,000

From R. J. Hengstebeck and J. T. Banchero, Washington University Design Case Study No. 8, edited by B. D. Smith, Washington University, St. Louis, Mo., 1969.

The equipment costs for the flowsheet shown in Fig. 2.6-1 are listed in Table 2.6-1, and the operating costs are given in Table 2.6-2. A cost summary for the process is presented in Table 2.6-3.

When we examine Tables 2.1-4 and 2.6-1, we see that the costs of pumps and drums are only a small fraction of the total costs. If we neglect these costs (or simply assume that they are about 10 % of the total), then we can save the effort of designing a large fraction of the total number of pieces of equipment and yet introduce only a small error in our calculations. Similarly, if we assume that the costs of the feed tanks and product storage tanks will be essentially the same for all the process alternatives, then we can omit them from our screening calculations.

Of course, the process will not operate without the pumps, drums, feed tanks, and storage tanks. However, if our screening calculations indicate that the process is not profitable and that the project should be terminated when we do not include these costs, then we never need to design them. Thus, for conceptual designs we

TABLE 2.6-2
Operating cost summary, $1000

Utilities		Taxes, insurance	166
Power	322	Repairs	250
Steam	520	Miscellaneous	83
Fuel	333	Payroll charges	32
Water	30	Total	1850
Total	1205	SARE	150
Labor	95	Catalyst	60
Supervision	19	Total	2060

From R. J. Hengstebeck and J. T. Banchero, Washington University
Design Case Study No. 8, edited by B. D. Smith, Washington University, St. Louis, Mo., 1969.

include only the costs of the significant equipment items. This approach is in
agreement with the engineering method discussed in Chap. 1.

2.7 SUMMARY, EXERCISES, AND NOMENCLATURE

Summary

When we compare process alternatives, normally there are different economic
trade-offs between capital and operating costs. To make valid comparisons

TABLE 2.6-3
Investment and operating summary

Conversion/pass, %	30	Labor and supervision[†]	0.15
Purge gas	No	Taxes and insurance	0.17
		Repairs and miscellaneous	0.33
Investments, $ millions		Catalyst	0.06
ISBL	3.74	SARE	0.15
OSBL	1.12		2.06
	4.86		
Working capital*	1.00	Materials, BCD[‡] (60°F)	
	5.86	Toluene feed	3780
Catalyst inventory	0.06	Products	
	5.92	Benzene	1590
		Xylenes	2000
Operating costs, $1 million/yr		H_2 Feed, 10^6 SCFD[¶]	1.88
Utilities	1.20	Fuel gas, 10^6 Btu/day	1700

From R. J. Hengstebeck and J. T. Banchero, Washington University Design Case Study No. 8, edited by B. D.
Smith, Washington University, St. Louis, Mo., 1969.
* Principally, for a 2-week inventory of feed and products, with the products valued at cost.
† Including payroll charges.
‡ BCD = barrels/calendar day.
¶ SCFD = standard cubic feet/day.

between these two different types of quantities, we must consider the time value of money. Thus, by using interest calculations to determine the present value of two alternatives, we can compare them on the same basis. The present value (PV) of an investment I plus annual payments R with an interest rate i is

$$\text{PV} = I + \frac{R}{i}[1 - (1 + i)^{-n}] \qquad (2.4\text{-}23)$$

Once we find the best alternative, we must evaluate the total cost associated with the process, to see whether additional engineering effort can be justified. That is, we must include the cost of the offsite facilities, maintenance and repairs, working capital, start-up costs, etc. These various factors are discussed in Sec. 2.3, and a profitability model is developed in Sec. 2.5. This model provides an explanation for the simple cost analyses that we use throughout this text.

Note that we still have not considered the control of the process, safety, or environmental factors in adequate detail. Any of these factors might make the process unprofitable. Hence, the profitability calculations for our conceptual design merely provide a basis that we can use to judge whether more detailed design studies can be justified. By including rough estimates of the other processing costs, however, we are better able to make this judgment.

Exercises

2.7-1. Derive an expression for the value of an annuity after n yr if the first payment is made at time zero, rather than at the end of the first year.

2.7-2. A friend of yours joins a Christmas Club at a local bank. She deposits $10/mo starting on January 1 and receives $110 at the beginning of December. If the nominal rate available is 5.75% compounded monthly, how much interest does the bank keep for providing this service? (*Note:* This payment plan is different from the annuity schedule discussed in the text.)

2.7-3. St. Mary's Cemetery in Northampton, Massachussetts, charges $110 for a cemetry plot and $50 for perpetual care of the plot. At a nominal interest rate of 6% compounded monthly, what are the expected annual maintenance charges?

2.7-4. Some universities are fortunate to have endowed chairs for their outstanding faculty, and often these chairs provide $50,000/yr. If the nominal interest rate is 10% compounded continuously, how much money is required to establish a chair?

2.7-5. When we make monthly payments on a car or a house, the amount of the payment is a constant, but different fractions of this payment represent repayment of the principal and the interest. Also, the fraction allocated to each changes over the loan period. Using the terminology below, develop expressions for the principal and interest payments during year y:

M = amount of mortgage p = monthly payment

R = nominal interest rate P = annual payment

$r = R/12$ = monthly interest rate P_y = principal paid during year y

n = no. months for loan I_y = interest paid during year y

x = principal paid at end of 1st month

2.7-6. Estimate the purchased and installed equipment costs of a 3000-ft^2, SS/SS, U-tube heat exchanger operating at 380 psi.

2.7-7. Develop an expression for the total investment if we assume that the working capital is 3 months' worth of the product revenues.

2.7-8. Develop an expression for the profit before taxes if we assume that the labor costs are 15% of the total product costs (see Table 2.3-2).

2.7-9. Develop an expression for the total capital investment for a grass-roots plant (assume offsite costs are 3 times onsite costs).

2.7-10. For an 11-yr plant life and a DCFROR of 25%, what is the capital charge factor?

2.7-11. Calculate the payout time and the DCFROR for the process described in Sec. 2.1. If you need to introduce additional assumptions, clearly state these assumptions.

2.7-12. Calculate the ROI, payout time, and DCFROR for the process described in Sec. 2.6. If you need to introduce additional assumptions, clearly state these assumptions.

Nomenclature

A	Heat-exchanger area
C_B	Base cost for carbon steel
CCF	Capital charge factor
C_E	Exchanger cost
C_{PR}	Cost of product, ¢/lb
Conting.	Contigency costs
Deprec.	Depreciation
d_j	Annual depreciation
EP_i	Economic potential at level i
F_c	Correction factor for pressure, materials of construction, etc.
FCI	Fixed capital investment
F_D	Design-type correction factor
Fixed Cap.	Fixed capital cost
F_{MC}	Material-of-construction correction factor
F_p	Pressure correction factor
i_{eff}	Effective interest rate
i_c	Continuous interest rate
IF	Installation factor
Insur.	Insurance
ISBL	Inside-battery-limits costs = onsite costs
Lab.	Laboratory costs
m	No. payment periods per year
Maint.	Repair and maintenance costs
Manu. Cost	Manufacturing costs
n	No. years
Op. Supply	Operating supplies
OSBL	Outside-battery-limits costs = offsite costs
Plant OVHD	Plant overhead

P_r	Principal
PV	Present value
r	Nominal interest rate
R	Periodic payment
Raw Matl.	Raw-material costs
ROI	Return on investment
R_t	Continuous payment
S	Amount of money in a bank account
Salv. Val.	Salvage value
SARE	Sales, administration, research, and engineering costs
Start-up	Starting costs
Supervis.	Direct supervision
t	Time
TC	Total cost
Tot. Inv.	Total investment
Tot. Prod. Cost	Total product cost
Util.	Utilities costs
Work. Cap.	Working capital
x	Design variables

CHAPTER
3

ECONOMIC DECISION MAKING: DESIGN OF A SOLVENT RECOVERY SYSTEM

To illustrate how process alternatives can be generated and the use of order-of-magnitude calculations to make economic decisions, we consider the very simple example of the design of a solvent recovery system.

3.1 PROBLEM DEFINITION AND GENERAL CONSIDERATIONS

We assume that as part of a process design problem there is a stream containing 10.3 mol/hr of acetone and 687 mol/hr of air that is being fed to a flare system (to avoid air pollution).* The design question of interest is: Should we recover some of the acetone?

---

* This case study is a modified version of the 1935 AIChE Student Contest Problem; see J. J. McKetta, *Encyclopedia of Chemical Processing and Design*, vol. 1, Dekker, New York, 1976, p. 314.

Economic Potential

The first step in the analysis of any design problem is to evaluate the economic significance of the project. Initially we do not know what fraction of the acetone we might attempt to recover, but rather than spend time on this decision we merely base the calculation on complete recovery. Thus, we calculate an economic potential (EP) as

$$EP = \text{Prod. Value} - \text{Raw Matl. Cost} \qquad (3.1\text{-}1)$$

Or, since we are currently burning the acetone,

$$EP = (10.3 \text{ mol/hr})(0.27 \text{ \$/lb})(58 \text{ lb/mol})(8150 \text{ hr/yr})$$
$$= \$1.315 \times 10^6/\text{yr} \qquad (3.1\text{-}2)$$

Operating Time

It is conventional practice to report operating costs or stream costs on an annual basis. Different companies use somewhat different values for the number of operating hours per year, and they may even use different values for different types of projects. We will use 8150 hr/yr for continuous processes and 7500 hr/yr for batch. (This operating time includes scheduled shutdowns for maintenance, unplanned downtime due to mechanical failures, and/or production losses caused by capacity limitations or lack of feed.)

Process Alternatives

The next question we ask is: How can we recover the acetone? From our knowledge of unit operations, we might list the alternatives shown in Table 3.1-1. We might be able to think of other alternatives, and so we need to make some judgment about using conventional technology versus the cost of doing the development work required to design and evaluate unconventional alternatives.

TABLE 3.1-1
Solvent recovery alternatives

1. Condensation
   a. High pressure
   b. Low temperature
   c. A combination of both
2. Absorption
3. Adsorption
4. A membrane separation system
5. A reaction process

Now we come to this question: Which is the cheapest alternative? Fair* suggests that any time the solute concentration in a gas stream is less than 5%, adsorption is the cheapest process. However, many petroleum companies prefer to use condensation or absorption systems because the companies have much more experience designing and operating these types of units. Furthermore, only a few vendors sell adsorption equipment. Thus, we are again required to make a judgment concerning the use of technology where we have a great deal of experience versus using a technology where we have much less experience. We should base this decision on the relative costs, as well as the risks, of the various processes in question.

Of course, we do not know the costs of the various alternatives until we design each of them. We do not necessarily want to develop rigorous designs initially, because (at best) we would build only one of them. Hence, we only want to include sufficient accuracy in our screening calculation to determine which alternative is the cheapest (or to see whether they have about the same costs), and then we will develop a rigorous design if we decide to build the process.

On numerous occasions in design we can develop a number of alternatives that appear as if they will do the job. If we do not know which alternative is the cheapest, we should consider designing them all. By doing quick design calculations, we can simplify the design effort required to make decisions.

3.2   DESIGN OF A GAS ABSORBER: FLOWSHEET, MATERIAL AND ENERGY BALANCES, AND STREAM COSTS

We arbitrarily decide to consider the design of a gas absorber first, although we recognize that we must also consider alternative designs before we decide which process we might build. This very simple design problem illustrates the use of shortcut calculations, rules of thumb, and other process alternatives.

Flowsheets and Alternatives to Gas Absorption

Before we can do any calculations, it is necessary to invent a flowsheet for the process. The simplest possible flowsheet we might imagine is shown in Fig. 3.2-1. We use water as a solvent (because it is cheap) to recover acetone from the air stream in the gas absorber, and then we distill the acetone product from the water and throw away the water (environmental constraints may preclude this possibility, so we include a cost for pollution treatment). Of course, we could recycle the process water to the gas absorber (see Fig. 3.2-2), and in this way we would avoid any environmental problems. If we used anything other than water as a solvent, we would always recover and recycle the solvent.

---

* J. R. Fair, *Mixed Solvent Recovery and Purification*, p. 1, Washington University Design Case Study No. 7, edited by B. D. Smith, Washington University, St Louis, Mo., 1969.

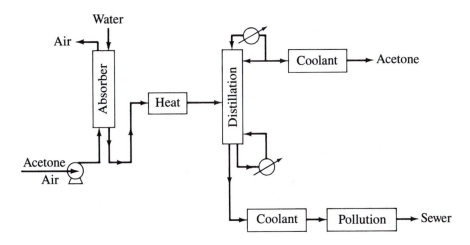

FIGURE 3.2-1
Acetone absorber.

It is reasonable to question whether discarding the process water, as is shown in Fig. 3.2-1, can ever be justified—even when a pollution treatment facility is available. The only justification is based on the temperature of the process water entering the gas absorber. If a recycle process is used, we cool the recycle stream with cooling water. Normally we assume that cooling water is available from the cooling towers at 90°F (on a hot summer day) and that it must be returned to the cooling towers at a temperature less than 120°F (to prevent excessive scale formation on the exchanger surface). Then if we assume a 10°F driving force at the cold end of the exchanger, the recycled process water will enter the gas absorber at 100°F.

However, if we use well water as the solvent, then the temperature of the water fed to the absorber might be 77°F, or possibly less. It is advantageous to operate the absorber at as low a temperature as possible, which we can achieve if

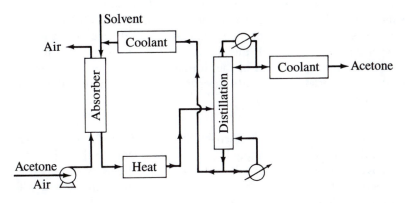

FIGURE 3.2-2
Acetone absorber.

well water is available as a solvent and if we do not recycle the water. This reasoning is the basis for a design heuristic:

> If a raw material component is used as the solvent in a gas absorber, consider feeding the process through the gas absorber. (3.2-1)

Thus, we find that we have two alternative flowsheets. If we do not know which is the cheaper, we should design both. In addition, we must evaluate whether we really want to use water as the solvent. We arbitrarily choose to consider the flowsheet shown in Fig. 3.2-1 first, because it is the simplest.

Material Balances

DISTRIBUTION OF COMPONENTS. Once we have specified a flowsheet, we must try to identify the components that will appear in every stream. The inlet gas flow to the absorber is given in the problem statement as 10.3 mol/hr of acetone and 687 mol/hr of air. If we use well water as a solvent, then the inlet solvent stream is pure water. The gas leaving the absorber will contain air, some acetone (we can never obtain a complete recovery), and some water. Since water is relatively inexpensive, we neglect this solvent loss in our first design calculations. However, if any other solvent is used, it is essential to include the solvent loss in the economic analysis, as we demonstrate later.

The feed stream to the distillation column will contain water, most of the acetone, and some dissolved air. Probably we can neglect this dissolved air in our first set of design calculations, but it is essential to recognize that we must put a vent on the distillation column condenser, to prevent the accumulation of air in this unit. The overhead from the distillation column will then contain acetone and water, while the bottoms will contain water and some acetone.

We now see that acetone leaves the process in three places: the exit gas stream from the absorber, the distillate overhead, and the distillate bottom stream. Normally, we would be given a product specification for the product stream, which would correspond to either the specification that the company established when they used acetone as a raw material or the specification that the marketing group would indicate when they try to sell this acetone. However, even if we fix the acetone composition in the overhead of the distillation column, we still cannot calculate an acetone material balance until we specify some information about the acetone leaving in the other two streams.

RULES OF THUMB. Of course, we can recover 90, or 99, or 99.9%, or whatever, of the acetone in the gas absorber, simply by adding more trays to the top of the absorber. The cost of the gas absorber will continue to increase as we increase the fractional recovery, but the value of the acetone lost to the flare system will continue to decrease. Thus, there is an optimum fractional recovery.

column. As we add more and more plates in the stripping section of this column, the still cost increases, but the value of the acetone lost to the sewer decreases.

We cannot find the optimum fractional recoveries of acetone in either the absorber or the still unless we carry out detailed designs of these columns. Furthermore, it is not clear that we want to find the optimum design conditions until we have decided to build an absorption process. Hence, for our first set of calculations, we base the design a rule of thumb (i.e., a heuristic):

> It is desirable to recover more than 99% of all valuable
> materials (we normally use a 99.5% recovery as a first
> guess). (3.2-2)

We discuss this rule of thumb in greater detail later in this chapter.

We also have to fix the water flow rate to the gas absorber before we can calculate a set of material balances. The greater the solvent flow rate, the fewer trays that are required to achieve a fixed fractional recovery in the absorber, but the greater the load on the distillation column. Thus, there is an optimum solvent flow rate. Again, we prefer not to optimize designs at the screening stage if a heuristic is available, and so we use another rule of thumb to fix the solvent flow:

> For an isothermal, dilute absorber, choose L such that $L = 1.4mG$ (3.2-3)

We also discuss this rule of thumb later.

EXACT MATERIAL BALANCES. With these rules of thumb, it is a straight-forward task to calculate the material balances. For the acetone-water system at 77°F and 1 atm, $\gamma = 6.7$* and $P° = \frac{229}{760}$, so that

$$m = \frac{\gamma P°}{P_T} = \frac{6.7(229)}{760} = 2.02 \qquad (3.2\text{-}4)$$

$$L = 1.4mG = 1.4(2.02)(687) = 1943 \text{ mol/hr} \qquad (3.2\text{-}5)$$

For a 99.5% recovery of acetone in the gas absorber, the acetone lost is

$$0.005(10.3) = 0.0515 \text{ mol/hr} \qquad (3.2\text{-}6)$$

and the acetone flow to the distillation column is

$$0.995(10.3) = 10.25 \text{ mol/hr} \qquad (3.2\text{-}7)$$

If 99.5% of the acetone entering the still is recovered overhead, we obtain

$$0.995(10.25) = 10.20 \text{ mol/hr} \qquad (3.2\text{-}8)$$

Also, if the product composition of acetone is specified to be 99%, the amount of water in the product stream will be

$$\left(\frac{1 - 0.99}{0.99}\right)(10.20) = 0.10 \text{ mol/hr} \qquad (3.2\text{-}9)$$

---

* From Perry's Handbook, pp. 14–15, 5th ed., 1973.

Then the bottom flows of acetone and water are

$$0.005(10.25) = 0.005 \text{ mol/hr} \qquad (3.2\text{-}10)$$

and

$$1943 - 0.1 = 1942.9 \text{ mol/hr} \qquad (3.2\text{-}11)$$

APPROXIMATE MATERIAL BALANCES. Even though these material-balance calculations are very simple and quick, according to the engineering method we do not want to do any calculations unless they provide us with significant information. Thus, we want to explore the possibility of developing a set of material balances that are almost correct (within 10%, or so) with the minimum amount of effort. We need to calculate the solvent feed rate in the same way as we did before, but there is a simpler way of obtaining good estimates of the other flows.

If we assume that 99.5% recovery (or any recovery greater than 99%) is essentially equivalent to a 100% recovery (actually, we are willing to tolerate 10% error), then the acetone flow in the distillation column overhead becomes 10.3 versus 10.2. Also since a 99% acetone composition in the distillate is essentially the same as 100% acetone, the water flow leaving the bottom of the column could be written as 1943, versus 1942.9. Similarly, the feed stream to the still becomes 10.3 mol/hr of acetone and 1943 mol/hr of water, versus 10.25 + 1943. Thus, we see that we essentially make no errors in the stream flows if we merely assume 100% recoveries and pure streams and then just write down the flows.

Of course, we cannot base the design of the gas absorber or the still on 100% recoveries. Moreover, we would like to estimate the acetone losses from the absorber overhead and the still bottoms. However, suppose we use 99.5% recoveries for these calculations. Our material balances will no longer quite balance (although they are certainly within 10% of the exact answers), but our calculation effort will decrease. The savings in time are not significant for this simple problem, but it can be for complex plant designs. Thus, we use approximate material balances for screening calculations throughout this text, although we would make rigorous material balances as we proceeded with final design calculations.

SOLVENT LOSS. For cases where we use a solvent other than water and we recover and recycle this solvent, as is shown in Fig. 3.2-2, it is essential to estimate the loss of this solvent in the absorber exit gas stream very early in the design calculations. For a low-pressure absorber, the fugacity correction factors are negligible, and the vapor–liquid equilibrium relationship for the solvent can be written as

$$P_T y_s = \gamma_s P_s^\circ x_s \qquad (3.2\text{-}12)$$

With greater than 99% recovery of the solute, the solvent composition on the top tray will be essentially

$$x_s \approx 1 \qquad (3.2\text{-}13)$$

If a solvent is used that is in the homologous series with the solute, then

$$\gamma_s = 1 \qquad (3.2\text{-}14)$$

and we can write

$$y_s = P_s^\circ / P_T \tag{3.2-15}$$

Hence, a quick way to estimate the solvent loss is to write

$$m_s = \frac{y_s}{1 - y_s} G \approx y_s G \tag{3.2-16}$$

Stream costs. Once we have estimated the material balances, we calculate the stream costs.

ACETONE-WATER PROCESS. For the acetone-water system with no recycling and 99.5% recoveries, we find from our approximate material balances the following:
Acetone loss in absorber overhead (assume $0.27/lb) is

$$(\$0.27/lb)(58\ lb/mol)(0.0515\ mol/hr)(8150\ hr/yr) = \$6600/yr \tag{3.2-17}$$

Acetone loss in still bottoms is

$$(\$0.27/lb)(58\ lb/mol)(0.0515\ mol/hr)(8150\ hr/yr) = \$6600/yr \tag{3.2-18}$$

Pollution treatment costs (assume $0.25/lb BOD and 1 lb acetone/lb BOD) are

$$(\$0.25/lb\ BOD)(1\ lb\ BOD/lb\ acetone)(58\ lb/mol)(0.0515\ mol/hr)(8150\ hr/yr)$$
$$= \$6100/yr \tag{3.2-19}$$

Sewer charges (assume $0.2/1000 gal) are

$$\left(\frac{\$0.20}{1000\ gal}\right)\left(\frac{1\ gal}{8.34\ lb}\right)(18\ lb/mol)(1943\ mol/hr)(8150\ hr/yr) = \$6800/yr \tag{3.2-20}$$

Solvent water (assume $0.75/1000 gal) is

$$\left(\frac{\$0.75}{1000\ gal}\right)\left(\frac{1\ gal}{8.34\ lb}\right)(18\ lb/mol)(1943\ mol/hr)(8150\ hr/yr) = \$25,600/yr \tag{3.2-21}$$

Since each of these costs is essentially negligible compared to the economic potential of 1.315×10^6 per year, we want to continue developing the design. Also we note that there is little incentive in using more rigorous material balances or in getting very accurate costs for the pollution treatment system. In fact, we can tolerate 100% errors in our calculations, and the conclusion of negligible stream costs will not be affected.

COST OF SOLVENT LOSS. Suppose we also consider using methyl isobutyl ketone (MIBK) as a solvent, and we recycle the MIBK. If $P_s^\circ = 0.0237$ atm at 77°F and $P_T = 1$ atm, from Eq. 3.2-16 we find that

$$\begin{aligned}
\text{MIBK loss (assume } \$0.35/lb &= \$35/mol) \\
&= (\$35/mol)[0.0237(687\ mol/hr)](8150\ hr/yr) \\
&= \$4.464 \times 10^6/yr \tag{3.2-22}
\end{aligned}$$

When we compare this value to the economic potential, we see that we want to drop any idea of using MIBK as a solvent. Moreover, we are glad that we did not take the trouble to design the absorber and still before we calculated the stream costs. Again we find that our decision does not depend to any great extent on the accuracy of the calculation. Thus, we use the economic analysis to help us decide on the accuracy we need for our design calculations.

Energy Balances for the Acetone Absorber

Our original design problem was underdefined, and thus it was necessary to use some rules of thumb (greater than 99% recoveries and $L = 1.4mG$) to be able to calculate a set of material balances. It should not be surprising that the same problem is encountered when we try to write energy balances. Thus, we need to fix each of the stream temperatures in order to estimate the energy flows.

   Since the inlet composition to the gas absorber is quite dilute, we might assume that the absorber will operate isothermally. Hence, if the gas and liquid streams entering the absorber are at 77°F, we assume that the exit streams are at this same temperature (see Fig. 3.2-3).

   We do not want to store our product stream at its boiling point, so we install a product cooler. With cooling water available at 90°F and a 10°F driving force, the temperature of the product stream leaving the product cooler will be 100°F. Our acetone product contains 1% water. But rather than calculate the bubble point of the distillate, we might merely guess that the temperature of the overhead is essentially the same as the boiling point of acetone (135°F); i.e., we expect that the error in calculating the heat load of the product cooler caused by this assumption will be negligible. Similarly, we assume that the bottom stream from the still is at 212°F (because there is only 0.05 mol of acetone as compared to

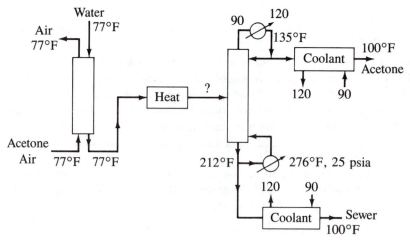

FIGURE 3.2-3
Stream temperatures.

1943 mol of water) and that we must cool this waste stream to 100°F (cooling-water temperature) prior to pollution treatment. (It would be better to assume that the bottom of the column is at 5 to 10 psig, rather than atmospheric pressure, but again the error will be small.)

We still must specify the temperature of the stream entering the distillation column. Saturated-liquid feeds are the most common case, and so we might guess a temperature of 200°F or so (10.3 mol of acetone and 1943 mol of water). Again, we expect that a guessed value will enable us to calculate the load on the still preheater without calculating the bubble point of the feed mixture or correcting for a column operating pressure of 5 to 10 psig.

If we do not preheat the feed stream entering the distillation column to close to the saturated-liquid condition, then we will have a supercooled liquid entering the still and the heat load on the still reboiler will increase. Thus, the total energy demand on the preheater and the still reboiler is essentially constant. However, usually we prefer to preheat the feed because normally we can use a hot process stream that needs to be cooled down as the source of heat, rather than using steam from a utility supply. Hence, we need to consider the energy integration of the process as part of our design activity.

ENERGY BALANCES. Once we have specified the stream temperatures and we have estimated all the stream flows, it is a simple matter to calculate the heat loads of the various streams by using the expression

$$Q_i = F_i C_{P_i}(T_{\text{in}} - T_{\text{out}}) \tag{3.2-23}$$

Then we could decide on a heat-exchanger network and calculate the heat-exchanger areas, the annualized heat-exchanger capital costs, and the utility costs. For a simple process, such as the one we are considering, this would be a reasonable procedure. In general, however, we want to energy-integrate the absorber-stripper heat loads with those of the remainder of the process, and therefore we defer the energy analysis.

PROCESS ALTERNATIVE. In our selection of the stream temperatures, we noted that the still bottom was almost pure water (0.05 mol acetone and 1943 mol of water). For this case, the column reboiler uses 25-psia steam to generate essentially 15-psia steam that is returned to the column. As a process alternative, we could eliminate the reboiler and feed live steam to the column. We pay a penalty with this approach, however, because once the live steam is condensed, it must go to the pollution treatment system and then is lost to the sewer (whereas the steam leaving a reboiler would be vaporized and recycled through the closed steam system). Also, boiler feed water has a higher quality (it is demineralized, etc.) and is therefore more expensive than process water. Hence, we must balance the reboiler savings against the incremental cost of boiler feed water, pollution treatment, and sewer costs to see whether this alternative is worth pursuing.

3.3   EQUIPMENT DESIGN CONSIDERATIONS

In addition to calculating the sizes of the heat exchangers, we must calculate the size and cost of the absorber and the still. Before we begin any calculations, however, we want to understand the cause-and-effect relationships of the design variables and to see whether we can simplify the normal unit-operations models.

Gas Absorber

For *isothermal, dilute* systems, the Kremser equation* can be used to calculate the number of theoretical trays required in the gas absorber;

$$N + 1 = \frac{\ln\left[\left(\dfrac{L}{mG} - 1\right)\left(\dfrac{y_{in} - mx_{in}}{y_{out} - mx_{in}}\right) + 1\right]}{\ln\left(\dfrac{L}{mG}\right)} \tag{3.3-1}$$

If pure water is used as the solvent, then $x_{in} = 0$. From the rules of thumb discussed earlier, we know that

$$\frac{y_{out}}{y_{in}} \leq 1 - 0.99 = 0.01 \tag{3.3-2}$$

and

$$L = 1.4\left(\frac{\gamma P^{\circ}}{P_T}\right) G \tag{3.3-3}$$

We can use the Kremser equation and the rules of thumb to understand the effects of the design variables.

COLUMN PRESSURE. Suppose we double the tower pressure P_T. From Eq. 3.3-3 we see that L decreases by a factor of 2; but since $L/(mG) = 1.4$, the number of plates required in the gas absorber (see Eq. 3.3-1) does not change. Lower values of L mean that the still feed will be more concentrated, the reflux ratio will decrease, the vapor rate in the still will decrease, the column diameter will decrease, the sizes of the condenser and reboiler will decrease, and the steam and cooling-water requirements will decrease. Thus, decreasing the solvent flow to the gas absorber will have a significant effect on the design of the still, but no effect on the number of trays required in the absorber.

The absorber diameter will decrease (because of the density effect and a smaller liquid load), and a feed gas compressor will be required to obtain the increased pressure. Since gas compressors are the most expensive type of processing equipment, normally it does not pay to increase the pressure of the gas absorber with a compressor. (In some cases, a high pressure can be obtained by pumping a liquid stream to a high pressure somewhere upstream of the absorber.)

---

* A. Kremser, *Natl. Petrol. News,* **22**(21): 42 (1930).

EFFECT OF SOLVENT. For the acetone-water system, the value of m was given by

$$m = \frac{\gamma P^{\circ}}{P_T^{\circ}} = \frac{6.7(229)}{760} = 2.02 \qquad (3.3\text{-}4)$$

We see that if we use a solvent such as MIBK that forms an essentially ideal mixture with acetone, so that $\gamma = 1$, then from Eq. 3.3-3 we cut the liquid rate by a factor of 6.7 (and decrease the still cost). However, from Eq. 3.3-1 we see that the required number of plates in the absorber does not change.

EFFECT OF OPERATING TEMPERATURE. If we change the inlet water temperature to 40°C, then $\gamma = 7.8$ and $P^{\circ} = 421$ mmHg. Thus, from Eq. 3.3-3 we see that L will increase (so that the still costs will increase), but the number of absorber trays (see Eq. 3.3-12) will remain the same.

Systems Approach Versus Unit Operations

The simple examples discussed above clearly reveal that the interaction between unit operations is the key feature of process design. Thus, design cannot be accomplished merely by connecting various units and mistakenly thinking that if we properly design each unit, we will obtain a proper design of the whole plant. Instead, we must always look at the behavior of the total system.

Back-of-the-Envelope Design Equation

The Kremser equation, Eq. 3.3-1, is actually a quite simple equation that can be used to design gas absorbers for isothermal operation with dilute feeds. However, in accordance with the engineering method and our basic desire to do calculations only if we gain some significant information from this effort, we would like to review the Kremser equation and to evaluate the significance of each term. We do this by examining the order of magnitude of the various terms in the equation.

First let us consider the left-hand side of Eq. 3.3-1, i.e., the term $N + 1$. We are concerned with obtaining accurate estimates of only the items that are expensive. We expect that relatively expensive absorbers will contain 10 to 20 theoretical trays (the cost of a gas absorber with only 2 or 3 trays will probably be negligible compared to a furnace, a gas compressor, a distillation column with 10 to 20 trays, etc.). If we decide not to undertake a calculation unless it changes the result by more than 10%, then we see that we can simplify the left-hand side of the Kremser equation by writing

$$N + 1 \approx N \qquad (3.3\text{-}5)$$

For pure solvents, $x_{in} = 0$, and the numerator of the right-hand side becomes

$$\ln\left[\left(\frac{L}{mG} - 1\right)\frac{y_{in}}{y_{out}} + 1\right] \qquad (3.3\text{-}6)$$

The rules of thumb indicate that

$$\frac{L}{mG} \approx 1.4 \quad \text{and} \quad \frac{y_{in}}{y_{out}} \approx 100 \tag{3.3-7}$$

Thus,

$$\left(\frac{L}{mG} - 1\right)\left(\frac{y_{in}}{y_{out}}\right) + 1 \approx 40 + 1 \tag{3.3-8}$$

and if we apply the order-of-magnitude criterion ($1 \ll 40$), we can write

$$\ln\left[\left(\frac{L}{mG} - 1\right)\left(\frac{y_{in}}{y_{out}}\right) + 1\right] \approx \ln\left[\left(\frac{L}{mG} - 1\right)\left(\frac{y_{in}}{y_{out}}\right)\right] \tag{3.3-9}$$

The denominator of the right-hand side of the equation, $\ln [L/(mG)]$, can be written as

$$\ln(1 + \varepsilon) \tag{3.3-10}$$

Now, from a Taylor series expansion, we can write

$$\ln\frac{L}{mG} = \ln(1 + \varepsilon) \approx \varepsilon = \frac{L}{mG} - 1 \simeq 0.4 \tag{3.3-11}$$

With these simplifications, and replacing ln by log, we obtain

$$N = \frac{2.3\log\{[L/(mG) - 1](y_{in}/y_{out})\}}{0.4} \tag{3.3-12}$$

Within a 10% error, we note that $2.3/0.4 = 6$ and $(2.3\log 0.4)/0.4 = -2$. Hence, a simplified version of the Kremser equation becomes

$$N + 2 = 6\log\frac{y_{in}}{y_{out}} \tag{3.3-13}$$

Now we have a design equation that we can solve without a calculator. For a 99% recovery. Eq. 3.3-13 predicts 10 trays versus the exact value of 10.1. For a 99.9% recovery, we obtain 16 trays from Eq. 3.3-13 versus 16.6 from the exact equation. In addition to calculating the number of plates in the absorber, we must calculate the height and diameter. These calculations are discussed in Appendix A.3.

Distillation Column

To separate acetone from the solvent water, we use a distillation column. The acetone-water mixture is very nonideal, and we do not know of any shortcut procedures for highly nonideal separations. However, the mixture is only a binary one, and so we can use a McCabe-Thiele diagram to find the number of trays. We also must calculate the still diameter, the size of the condenser and reboiler, and the steam and cooling-water loads. Calculations of this type are discussed in Appendices A.2 and A.3.

Process Alternatives

In addition to completing the gas absorber design, we must design a condensation process and an adsorption process (as well as a membrane process) before we can evaluate whether the acetone is worth recovering and, if so, which process should be selected. Our goal here is merely to indicate the nature of design problems and to illustrate how order-of-magnitude analyses can be used to obtain shortcut design procedures. These shortcut procedures can often be used to simplify the evaluation step in the synthesis and analysis procedure, particularly during the preliminary stages of design when a large number of alternatives is being screened.

3.4   RULES OF THUMB

Originally rules of thumb were developed by experienced designers. A designer might have optimized the design of 10 absorber-stripper systems and found that the optimization always gave values close to $L = 1.4mG$ and optimum recoveries greater than 99%. Then, when the eleventh problem was encountered, the designer simply wrote down the answer. However, today most rules of thumb (or design heuristics) are developed by graduate students who run 500 to 1000 case studies on a computer for a particular problem and then attempt to generalize the results.

Of course, the fact that generalizations of computer optimization studies are possible implies that the optimization calculations are very insensitive to changes in most design and cost parameters. If the design and cost equations are insensitive, then the engineering method indicates that we should be able to simplify the equations. Thus, using the order-of-magnitude arguments to simplify problems, we should be able to derive the rules of thumb. The advantage of a derivation of this type is that the assumptions used in the analysis will clearly indicate the potential limitations of the rule of thumb.

Liquid Flow Rate to Gas Absorbers

For isothermal, dilute gas absorbers, the Kremser equation, Eq. 3.3-1, can be used to calculate the number of trays required for a specified recovery as a function of $L/(mG)$. A plot of the Kremser equation is shown in Fig. 3.4-1. From this graph we see that if we pick L such that $L/(mG) < 1$, we can never get close to complete recovery of the solute even if we use an infinite number of plates (infinite capital cost). If the solute is very valuable or, worse, if it is toxic (such as HCN), certainly we will want to obtain very high recoveries. Thus, we can conclude that we would never choose the liquid flow rate such that $L/(mG) < 1$.

We also see from the graph that if we choose $L/(mG) = 2$, we obtain essentially complete recovery with only five plates. Remember that large solvent flow rates correspond to dilute feeds to the distillation column—and therefore large

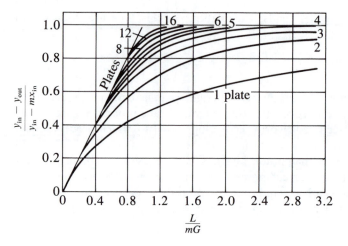

FIGURE 3.4-1
Kremser equation. (*From T. K. Sherwood and R. L. Pigford, Absorption and Extraction, McGraw-Hill, New York, 1952.*)

reflux ratios, high vapor rates, large-diameter columns, large condensers and reboilers, and high steam and cooling-water demands. Hence, if we pick L such that $L/(mG) > 2$, we will obtain tiny, inexpensive absorbers, but very expensive distillation columns.

Based on these simple arguments, we find that we want to choose L such that

$$1 < \frac{L}{mG} < 2 \qquad (3.4\text{-}1)$$

Of course, $L/(mG) = 1.5$ is right in the middle of this range. However, if we inspect the shape of the curves near $L/(mG) = 1.5$ and with high recoveries, we see that we might obtain a better trade-off between a decreasing number of plates required in the absorber (capital cost) and the increasing capital and operating costs of the distillation column by decreasing L. Hence, as a first guess, it seems to be reasonable to choose L such that

$$\frac{L}{mG} = 1.4 \qquad (3.4\text{-}2)$$

which is the common rule of thumb.

Fractional Recovery in Gas Absorbers

For a fixed solvent flow rate, we can always increase the recovery of the solute simply by adding trays in the gas absorber. Hence, there is an economic trade-off between an increasing absorber cost as we add trays versus a decreasing value of the solute lost. One of these is a capital cost (the absorber), and one is an operating cost (the solute loss).

COST MODEL. It is common practice to report operating costs on an annual basis. Thus, to examine the economic trade-off, we must also put the capital cost on an annualized basis. As discussed in Sec. 2.5, we annualize the capital cost by introducing a capital charge factor (CCF) of $\frac{1}{3}$ yr, where the CCF includes all capital-related expenses (depreciation, repairs and maintenance, etc.). A CCF of $\frac{1}{3}$ yr corresponds to about a 15% discounted-cash-flow rate of return (DCFROR); see Eq. 2.5.13.*

Suppose we write a total annual cost (TAC) model as

$$\text{TAC (\$/yr)} = (C_S \text{\$/mol})(Gy_{\text{out}} \text{ mol/hr})(8150 \text{ hr/yr})$$
$$+ [C_N \text{ \$/(plate·yr)}](N \text{ plates}) \qquad (3.4\text{-}3)$$

OPTIMUM DESIGN. Now, if we use our simplified design equation, Eq. 3.3-12 we obtain

$$\text{TAC} = 8150 C_S G y_{\text{in}} \left(\frac{y_{\text{out}}}{y_{\text{in}}}\right) + C_N \left(6 \log \frac{y_{\text{in}}}{y_{\text{out}}} - 2\right) \qquad (3.4\text{-}4)$$

The optimum fractional loss is given by

$$\frac{\partial \text{TAC}}{\partial(y_{\text{out}}/y_{\text{in}})} = 0 = 8150 C_S G y_{\text{in}} - \frac{6 C_N}{y_{\text{out}}/y_{\text{in}}} \qquad (3.4\text{-}5)$$

or

$$\frac{y_{\text{out}}}{y_{\text{in}}} = \frac{6 C_N}{8150 C_S G y_{\text{in}}} \qquad (3.4\text{-}6)$$

If we consider some typical values

$$C_S = \$15.5/\text{mol} \qquad Gy_{\text{in}} = 10 \text{ mol/hr} \qquad C_N = \$850/\text{plate yr}^\dagger \qquad (3.4\text{-}7)$$

we find that

$$\frac{y_{\text{out}}}{y_{\text{in}}} = \frac{6(850)}{8150(15.5)(10)} = 0.004 \qquad (3.4\text{-}8)$$

which corresponds to

$$\text{Fractional recovery} = 99.6\% \qquad (3.4\text{-}9)$$

---

* A 15% DCFROR is a bare minimum for conceptual designs of well-understood processes; i.e., a value of 1/2.5 = 0.4 is more realistic. For high-risk projects, such as in biotechnology, a value of 1/1 = 1 is not unreasonable.

† M. S. Peters and K. D. Timmerhaus, *Plant Design and Economics for Chemical Engineers*, McGraw-Hill, New York, 1968, p. 389, give typical values that range from $1200 to $2700 per plate depending on the column diameter. If we let the cost be $2550 per plate and use a CCF of $\frac{1}{3}$ yr, we obtain $850/yr per plate.

Sensitivity

The significant feature of this elementary analysis is not the relationship for the optimum design, Eq. 3.4-6, which is not exact, but rather the sensitivity of the calculation, Eq. 3.4-8. From Eq. 3.4-8 we see that we can change any of the numbers in the numerator or the denominator by 100%, and the optimum fractional recovery only changes from 99.2% to 99.8%. Thus, the result is very insensitive to any of the design or cost parameters.

   This simple sensitivity analysis clearly demonstrates that there is no incentive for refining the cost data used in the analysis. This same behavior is characteristic of a large number of design problems; i.e., the solutions are often very insensitive to the physical property data, the functional form of the design equation, and design parameters such as heat-transfer coefficients, cost data, etc. Thus, good engineering judgment requires that we obtain some idea of the sensitivity of the solution before we expend a significant amount of time gathering accurate data or attempting rigorous design calculations. That is, we want to spend as little time as possible getting an answer, and we want that answer to have only enough accuracy to make the decision we are faced with.

Limitations of Rules of Thumb

The rules of thumb we developed were both based on the Kremser equation, and we know that the Kremser equation is valid only for isothermal, dilute systems, where both the operating and the equilibrium lines are straight. For a system satisfying these conditions, the minimum solvent flow rate corresponds to the condition at the bottom end of the tower when the exit-liquid composition (for a fixed fractional recovery) is in equilibrium with the entering gas (so that an infinite number of trays is required at the concentrated end of the column): see Fig. 3.4-2a. The economic trade-offs dictate that we want to operate with $L \approx 1.4mG$ and a high fractional recovery. From Fig. 3.4-2c we see that these results correspond to almost a pinch zone at the top (dilute end) of the absorber.

   For concentrated mixtures of solutes, the equilibrium line might be curved, as shown in Fig. 3.4-3. For a fixed fractional recovery, the minimum liquid flow rate is determined by the operating line becoming tangent to the equilibrium curve. As the liquid flow rate is increased, we expect that it will require more trays to get from y_{in} to y_{out} for this case than a corresponding change for the dilute case, Fig. 3.4-2b and c. Hence, we might expect that using a solvent flow such that $L/(mG)$ is somewhat greater than 1.4 and attempting to recover somewhat less solute than for the dilute case will get us closer to the optimum design conditions.

   An even more dramatic difference is encountered for adiabatic absorbers. As the solute leaves the gas stream and is taken up by the solvent, the solute gives up its heat of vaporization. This heat effect causes the temperature of the liquid stream to increase as it approaches the bottom end of the tower. Increasing liquid temperatures increase the vapor pressure of the solute, and from Eq. 3.2-4 we see that the slope of the equilibrium line increases. Thus, the minimum liquid flow rate

Dilute solutions: $Y = y$, $X = x$

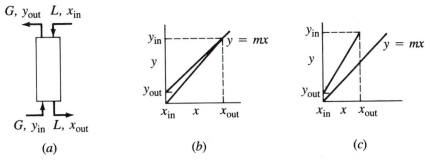

(a)   (b)   (c)

Minimum liquid flow when x_{out} is in equilibrium with y_{in}

FIGURE 3.4-2
Minimum liquid flow rate—isothermal.

occurs when there is a pinch at the bottom of the tower when $y_{in} = m_{out}x_{out}$ (see Fig. 3.4-4).

From Fig. 3.4-4 we see that a very small increase in the solvent flow rate above the minimum will allow us to get from y_{in} to y_{out} with a very few plates. However, the upward-curving nature of the equilibrium line means that the minimum solvent flow rate will be much larger than a corresponding case where we could maintain isothermal operation at the same inlet liquid temperature. In fact,

> The minimum liquid flow rate for an adiabatic absorber
> may be 10 times greater than the rule-of-thumb value
> $L = 1.4mG$ based on the inlet liquid temperature.   (3.4-10)

This example illustrates that the indiscriminate use of rules of thumb may lead to an inoperable design. In general,

> Every rule of thumb has some limitations.   (3.4-11)

From our discussion of the design of an adiabatic absorber, we note that we expect to obtain only a few plates (small absorber cost) and large liquid flows (large still costs), which is not a desirable situation. Thus, it is common practice to put cooling coils or one or more pump-around cooling loops on the bottom two or three trays of a gas absorber, to force it to behave more as an isothermal tower.

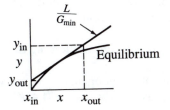

FIGURE 3.4-3
Minimum liquid flow rate—concentrated mixtures.

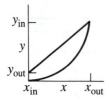

FIGURE 3.4-4
Minimum liquid flow—adiabatic.

Again, it is essential to understand the interaction between process units in order to develop a close to an optimum design. Similarly, structural changes in the flowsheet (cooling coils in the bottom trays of an absorber) normally have a much greater impact on the process economics than exact optimization calculations (the optimum solvent flow rate to an adiabatic absorber). We can therefore propose another heuristic:

> Avoid the use of adiabatic absorbers (unless there is only
> a small temperature rise across the absorber).   (3.4-12)

3.5 SUMMARY, EXERCISES, AND NOMENCLATURE

Summary

A number of important concepts are presented in this chapter:
1. Process alternatives
   a. A large number of alternatives can be generated even for simple processes.
   b. We use shortcut procedures to select the best alternative that we will design rigorously, providing that the process is profitable.
      (1) We want to spend as little time as possible getting an answer.
      (2) We only want to include sufficient accuracy to be able to make a decision.
      (3) We always consider the sensitivity of our calculations.
2. Shortcut design procedures
   a. It is reasonable to base process flows on 100% recoveries in separators and to base equipment designs on 99.5% recoveries, at the screening stage of process design.
   b. Order-of-magnitude arguments can be used to simplify design equations.
3. Systems approach
   a. You should always consider the total problem.
   b. Changes in the design variables in one unit (absorber) might affect the design of some other unit (still), but not the unit under consideration.
4. Rules of thumb—heuristics
   a. If a raw material is used as a solvent in a gas absorber, consider feeding the process through the absorber.
   b. It is desirable to recover more than 99% of valuable components.
   c. Choose the solvent flow for an isothermal, dilute gas absorber as $L = 1.4mG$.

    d. Cooling water is available at 90°F from a cooling tower and must be returned to the tower at 120°F or less.

    e. Assume a 10°F approach temperature for streams cooled with cooling water.
It is important to remember that every rule of thumb has some limitations!

Exercises

3.5-1. If we use the recycle flowsheet shown in Fig. 3.2-2, what are the economic trade-offs that fix the recycle composition of the solvent?

3.5-2. Consider a condensation process for recovering acetone from an air stream.

    (a) Draw a flowsheet for a condensation process for the acetone recovery problem.

    (b) If the condensation process operates at 15 psia, what temperature would be required to recover 99.5 % of the acetone?

    (c) If the condensation process operates at 100°F, what pressure would be required to condense 99.5 % of the acetone?

    (d) Discuss your results.

    (e) Describe the economic trade-offs involved in the design of a condensation process (both low-temperature and high-pressure).

3.5-3. Peters and Timmerhaus* derive an expression for the optimum diameter of a pipe by balancing the cost of the pipe (which increases with the pipe diameter) against the power required to deliver a specified amount of fluid through the pipe (which decreases as the pipe diameter increases). For pipes greater than 1-in. diameter, they give the results

$$D_{i,\text{opt}} = Q_f^{0.448} \rho^{0.132} \mu_c^{0.025} \left[\frac{0.88K(1 + J)H_y}{(1 + F)XEK_f} \right]^{0.158} \qquad (3.5\text{-}1)$$

$$= 3.9 Q_f^{0.45} \rho^{0.13}$$

where D_i = optimum pipe diameter (in.), Q_f = volumetric flow rate (ft³/s), ρ = density (lb/ft³), μ_c = viscosity (cP), $K = \$0.055/\text{kwh}$, $J = 0.35$, $H_y = 8760$ hr/yr, $E = 0.5$, $F = 1.4$, $K_f = 0.2$, and $X = \$0.45/\text{ft}$. Many industrial practitioners use a rule of thumb that the velocity in a pipe is a constant, although they use different values for liquids and gases. Can you use Eq. 3.5-1 to derive a rule of thumb for pipe velocity? What are the limitations of this heuristic? That is, for what cases does it not apply? If we change the annual charge factor K_f from 0.2 to 0.4, how does our estimate change?

3.5-4. A friend of mine in industry tells me that some of the chemists in his company estimate the minimum number of trays in a distillation column for a binary mixture by taking the sum of the boiling points and dividing by 3 times their difference. Can you show that the back-of-the-envelope model is essentially equivalent to Fenske's equation for the minimum number of trays? (*Hint:* Assume ideal, close-boiling mixtures, the Clausius-Clapeyron equation, Trouton's rule, and we want to obtain 97 % purities.)

3.5-5. Several quantitative heuristics have been proposed for deciding on a sequence of distillation columns (i.e., for a ternary mixture we could recover the lightest

---

* M. S. Peters and K. D. Timmerhaus, *Plant Design and Economics for Chemical Engineers*, 3d ed., McGraw-Hill New York, 1980, p. 379.

component in the first column and then split the remaining two, or we could recover the heaviest component first and then split the remaining two). Rod and Marek* use the criterion

$$\frac{\Delta V}{F} = \frac{(\alpha_{AC} + 0.25)\chi_A - 1.25 x_C}{\alpha_{AC} - 1} \tag{3.5-2}$$

whereas Rudd, Powers, and Siirola[†] estimate the relative cost of a distillation separation based on the expression

$$\text{Distillation cost} \sim \frac{\text{Feed Rate}}{\text{Boiling-Point Difference}} \tag{3.5-3}$$

Nadgir and Liu[‡] propose a coefficient of ease of separation (CES), defined as

$$\text{CES} = \frac{D}{F - D}\Delta T = 100\left(\frac{D}{F - D}\right)(\alpha - 1) \tag{3.5-4}$$

while Nath and Motard[¶] and Lu and Motard[∥] present a more complex expression based on the number of trays (proportional to $1/\ln \alpha$) multiplied by combinations of flow rate factors. Can you show that all these expressions have essentially the same dependence on the relative volatility? Derive the simplest expression that you can for the vapor rate in a binary column, assuming that $R/R_m = 1.2$, and compare this result to Rod and Marek's result.

3.5-6. Consider the design of a benzene-toluene distillation column (assume $\alpha = 2.5$) for a case where the feed rate is 100 mol/hr, the benzene feed composition is 0.05, we want to recover 99.5% of the benzene, and we want the benzene purity to be 0.99. Use Smoker's equation to calculate the number of trays required, and assume that $R = 1.2R_m$. Find both the vapor rates and the number of trays required in the rectifying and stripping sections if (a) the feed stream is at 100°F and (b) you heat the feed stream to saturated-liquid conditions (also calculate the load on the heater). Compare the total heat input and the number of trays required for both cases.

3.5-7. A rule of thumb commonly used in design is that the approach temperature in a heat exchanger should be 10°F in the range from ambient to the boiling point of organics (lower values, that is, 3°F, are used for refrigeration conditions and higher values, that is, 50°F, are used for high temperatures). To evaluate this heuristic, consider the simple system shown in Fig. 3.5-1, where we are attempting to recover some heat from a waste stream by producing steam. The total annual cost of this process is the sum of

---

* V. Rod and J. Marek, *Collect. Czech. Chem. Commun.,* **24**, 3240 (1959).

[†] D. F. Rudd, G. J. Powers, and J. J. Siirola, *Process Synthesis*, Prentice-Hall, Englewood Cliffs, N.J., 1973, p. 37.

[‡] V. M. Nadgir and Y. A. Liu, *AIChE J.,* **29**: 926 (1983).

[¶] R. Nath and R. L. Motard, "Evolutionary Synthesis of Separation Processes," AIChE Meeting, Philadelphia, 1978.

[∥] M. D. Lu and R. L. Motard, *Inst. Chem. Eng. Symp. Ser.,* No. 74 (1982).

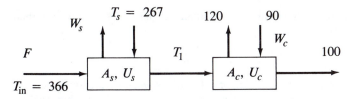

FIGURE 3.5-1
Heat recovery.

the annualized cost of the two exchangers plus the cooling-water cost minus the value of the steam produced:

$$\text{TAC} = C_A A_S + C_A A_C + C_w W_c - C_S S \qquad (3.5\text{-}5)$$

Write the equations for the heat balances for the exchangers, and show that Eq. 3.5-5 can be written as

$$\text{TAC} = C_A \frac{FC_p}{U_c} \left(\frac{T_1 - 100}{T_1 - 130} \right) \ln \frac{T_1 - 120}{100 - 90} + C_A \frac{FC_p}{U_S} \ln \frac{T_{in} - T_s}{T_1 - T_s}$$

$$+ C_w \frac{FC_p}{30 C_{pc}} (T_1 - 100) - C_s \frac{FC_p}{\Delta H_s} (T_{in} - T_1) \qquad (3.5\text{-}6)$$

Since $T_s = 267$ and T_1 must be greater than T_s, we simplify the expression by assuming that $(T_1 - 100)/(T_1 - 130) = 1$. With this approximation show that the optimum value of T_1 is given by

$$0 = - \frac{C_A/U_A}{T_1 - T_s} + \frac{C_A/U_C}{T_1 - 120} + \frac{C_w}{30 C_{pc}} + \frac{C_s}{\Delta H_s} \qquad (3.5\text{-}7)$$

Even though this equation is relatively simple to solve for T_1, suppose that we attempt to bound the answer instead. That is, we know that T_1 must be less than T_{in} and that it must be greater the T_s. Hence, we can write (after we solve for $T_1 - T_s$, which is the most sensitive term in Eq. 3.5-7)

$$\frac{C_A/U_s}{\dfrac{C_A/U_c}{T_s - 120} + \dfrac{C_w}{30 C_{pc}} + \dfrac{C_s}{\Delta H_s}} > T_1 - T_s < \frac{C_A/U_s}{\dfrac{C_A/U_c}{T_{in} - 120} + \dfrac{C_w}{30 C_{pc}} + \dfrac{C_s}{\Delta_s}} \qquad (3.5\text{-}8)$$

Calculate these bounds for a case where $C_A = \$11.38/\text{ft}^2 \cdot \text{yr}$, $F = 51{,}100 \text{ lb/hr}$, $C_p = 1.0 \text{ Btu/(lb} \cdot ^\circ\text{F)}$, $U_c = 30 \text{ Btu/(hr} \cdot \text{ft}^2 \cdot ^\circ\text{F)}$, $U_s = 20 \text{ Btu/(hr} \cdot \text{ft}^2 \cdot ^\circ\text{F)}$, $C_w = \$0.07388/[(\text{lb/hr}) \cdot \text{yr}]$, $\Delta H_s = 933.7 \text{ Btu/lb}$, $T_s = 267^\circ\text{F}$, $T_{in} = 366^\circ\text{F}$, and $C_s = \$21.22/[(\text{lb/hr}) \cdot \text{yr}]$.

3.5-8. Underwood's equation for the minimum reflux ratio for multicomponent mixtures normally requires a trial-and-error solution. That is, for an $AB/(CD)$ split, we first solve the equation below for the value of θ between α_{BC} and 1:

$$\frac{\alpha_{AC} x_{AF}}{\alpha_{AC} - \theta} + \frac{\alpha_{BC} x_{BF}}{\alpha_{BC} - \theta} + \frac{x_{CF}}{1 - \theta} + \frac{\alpha_{DC} x_{DF}}{\alpha_{DC} - \theta} = 1 - q \qquad (3.5\text{-}9)$$

and then we use this result to calculate R_m (assuming a sharp $AB/(CD)$ split and that no D goes overhead):

$$\frac{\alpha_{AC} x_{AD}}{\alpha_{AC} - \theta} + \frac{\alpha_{BC} x_{BD}}{\alpha_{BC} - \theta} + \frac{x_{CD}}{1 - \theta} = R_m + 1 \tag{3.5-10}$$

It is advantageous to avoid trial-and-error calculations when we are attempting to determine the best separation sequence because the number of columns that need to be designed increases rapidly as the number of components increases (the column-sequencing problem is discussed in Chap. 7). Malone* suggested a procedure for bounding the values of θ in Eq. 3.5-9. He first rewrites the equation as

$$\frac{\alpha_{BC} x_{BF}}{\alpha_{BC} - \theta} + \frac{x_{CF}}{1 - \theta} = 1 - q - \frac{\alpha_{AC} x_{AF}}{\alpha_{AC} - \theta} - \frac{\alpha_{DC} x_{DF}}{\alpha_{DC} - \theta} \tag{3.5-11}$$

Since the desired value of θ must be in the range of $\alpha_{BC} < \theta < 1$, he first substitutes α_{BC} for θ on the right-hand side of Eq. 3.5-11 and then solves the remaining quadratic equation. He repeats this procedure with $\theta = 1$ substituted on the right-hand side. The actual value of θ must thus lie between these bounds.

Consider a case where $x_{AF} = x_{BF} = x_{CF} = x_{DF} = 0.25$, $\alpha_{AC} = 4$, $\alpha_{BC} = 2$, $\alpha_{DC} = 0.5$, and $q = 1$, and find the bounds on θ. For a case where $F = 100$, all the A and 99.5% of the B are recovered overhead, and all the D and 99.3% of the C are recovered in the bottoms, find the bounds on R_m.

3.5-9. In the isopropanol-to-acetone process (a single, irreversible reaction) shown in Fig. 1.3-1, there will be a large reactor cost at high conversions, but the recycle flow will be small (and therefore the recycle costs will be small). The opposite situation will hold for low conversions, so that there must be an optimum conversion. Suppose that an estimate of the costs leads to the expression

$$\text{TAC} = \frac{73.3}{x}\left(3082 + 151 \ln \frac{1}{1 - x}\right) \tag{3.5-12}$$

where the term $73.3/x$ corresponds to the recycle flow, 3082 corresponds to the effect of the recycle flow on the costs of the equipment in the recycle loop, and $151 \ln [1/(1 - x)]$ corresponds to the reactor cost. Plot the recycle cost, the reactor cost, and the total annual cost versus conversion. Can you propose a design heuristic for the optimum conversion for first-order, irreversible reactions? For what types of kinetic models would you expect that your heuristic would not be valid?

Nomenclature

A_c	Area of water cooler (ft^2)
A_s	Area of steam exchanger (ft^2)
C_A	Annualized cost of heat exchanger [$/(ft$^2 \cdot$ yr)]
CES	Coefficient of ease of separation
C_N	Annualized cost per plate [$/(plate $\cdot$ yr)]
C_P	Heat capacity [Btu/(mol $\cdot$ °F) or Btu/(lb $\cdot$ °F)]

---

* M. F. Malone, Department of Chemical Engineering, University of Massachusetts, personal communication.

C_s	Cost of steam $\{\$/[(\text{lb/hr}) \cdot \text{yr}]\}$
C_S	Solute value ($/mol)
C_w	Cost of cooling water $\{\$/[(\text{lb/hr}) \cdot \text{yr}]\}$
D	Distillate flow
$D_{i,\text{opt}}$	Optimum pipe diameter
EP	Economic potential
F	Flow rate (mol/hr in Eq. 3.5-2)
G	Gas rate (mol/hr)
L	Solvent rate (mol/hr)
m	Slope of equilibrium line (see Eq. 3.2-4)
N	Number of plates
P_T	Operating pressure; must have same unit as P° (atm, psia, mmHg)
P_i°	Vapor pressure of solute; must have the same units as P_T (atm, psia, mmHg)
q	Feed quality
Q	Heat load (Btu/hr)
Q_f	Volumetric flow rate (ft^3/s)
R_m	Minimum reflux ratio
T	Temperature (°F)
TAC	Total annual cost ($/yr)
T_{in}	Inlet temperature (°F)
T_s	Steam temperature (°F)
T_1	Temperature of intermediate stream (°F)
U_c	Overall heat-transfer coefficient for the water coolant [Btu/(hr $\cdot$ ft^2 $\cdot$ °F)]
U_s	Overall heat-transfer coefficient for the steam exchanger [Btu/(hr $\cdot$ ft^2 $\cdot$ °F)]
W_c	Cooling water flow rate (lb/hr)
W_s	Steam flow rate (lb/hr)
x_i	Mole fraction in liquid phase
x_{iD}	Mole fraction of component i in distillate
x_{iF}	Mole fraction of component i in feed
y_i	Mole fraction in gas phase

Greek symbols

α	Relative volatility
γ	Activity coefficient
ε	Small value
θ	Root of Underwood's equation
ρ	Density = lb/ft^3
μ_c	Viscosity (cP)
ΔH_s	Heat of vaporization of steam (Btu/lb)
ΔT	Boiling-point difference
ΔV	Difference in vapor rate
χ	Conversion

DEVELOPING A CONCEPTUAL DESIGN AND FINDING THE BEST FLOWSHEET

CHAPTER
4

INPUT
INFORMATION
AND BATCH
VERSUS
CONTINUOUS

As we mentioned earlier, the original definition of a design problem is underdefined. Not only must we develop designs based on new reaction schemes invented by our own company, but also we must design all our competitor's new processes, to ensure that our company's technology will remain competitive. Thus, we often must design a process based on a minimum amount of information.

4.1 INPUT INFORMATION

The information that is normally available at the initial stages of a design problem (or that must be gathered) is given in Table 4.1-1. We briefly discuss each.

Reaction Information

The reaction information we need to know is listed in Table 4.1-2. Often it is possible to gather much of this information from the patent literature. In particular, the primary reactions, the temperature and pressure ranges, the catalyst, and the maximum yield often are available. It is essential to work closely with a chemist to gather the remainder of the information or at least to make a best guess of the missing data.

TABLE 4.1-1
Input information

1. The reactions and reaction conditions
2. The desired production rate
3. The desired product purity, or some information about price versus purity
4. The raw materials and/or some information about price versus purity
5. Information about the rate of the reaction and the rate of catalyst deactivation
6. Any processing constraints
7. Other plant and site data
8. Physical properties of all components
9. Information concerning the safety, toxicity, and environmental impact of the materials involved in the process
10. Cost data for by-products, equipment, and utilities

SIDE REACTIONS. It is particularly necessary to write down any side reactions that might take place. Even if only a trace amount of a by-product is produced in a laboratory experiment, this by-product may build up to very large levels in a recycle loop. Hence, all the by-products produced must be known in order to synthesize a separation system. Overlooking side reactions has been a common mistake, and it almost always leads to paying large economic penalties.

MAXIMUM YIELD. Information concerning how the product distribution changes with conversion and/or reactor temperature, molar ratio of reactants, etc., is often difficult to obtain. A chemist's focus is on scouting different catalysts, attempting to define a mechanism, and looking for ways to write a broad patent claim. During these scouting expeditions the chemist normally attempts to find the reaction conditions that maximize the yield. Thus, experience indicates that the existing data base will have most of the points grouped in a small range of conversions close to the maximum yield. (Often the largest amount of data will

TABLE 4.1-2
Reaction information

1. The stoichiometry of all reactions that take place
2. The range of temperatures and pressures for the reactions
3. The phase(s) of the reaction system
4. Some information on the product distribution versus conversion (and possibly reactor temperature, molar ratio of reactants, and/or pressure)
5. Some information about conversion versus space velocity or residence time
6. If a catalyst is used, some information about the state of the catalyst (homogeneous, slurry, packed bed, powder, etc.), some information about the deactivation rate and some idea of the regenerability of the catalyst as well as the method of regeneration (coke burn, solvent wash, etc.)

correspond to a single predefined condition that is used to check the reproducibility of the catalyst, but that is far removed from the optimum economic operating conditions.)

Numerous processes have been designed to operate at the condition of maximum yield, but this operation often does not correspond to the optimum economic conversion. As an example, let us consider a simple, hypothetical reaction system $A \rightarrow B \rightarrow C$, where B is the desired product and C has only fuel value. The concentrations of A, B, and C versus time in a batch reactor or space time in a tubular reactor are shown in Fig. 4.1-1. When B takes on its maximum concentration, point P in Fig. 4.1-1, a considerable amount of undesired C is formed and a large amount of A is converted. However, if we consider operating at point Q on the diagram, only a small amount of C is formed and much less A is converted. Thus, if we operate at point Q, we will require larger recycle flows (and higher recycle costs), but we lose less of our expensive raw material A that is converted to the by-product C which has only fuel value.

We define *selectivity* S as the fraction of the reactant converted that ends up as desired product:

$$S = \frac{\text{Moles of } B \text{ Produced}}{\text{Moles of } A \text{ Converted}} \qquad (4.1\text{-}1)$$

and we refer to the conversion of A to C as a selectivity loss. A diagrammatic sketch of our definition of selectivity for the HDA process (see Sec. 1.2) is shown in Fig. 4.1-2. (Note that a variety of definitions of selectivity and yield are used in the literature, so that it is always necessary to check the definition.)

Normally, raw-materials costs and selectivity losses are the dominant factors in the design of a petrochemical process. Raw-materials costs are usually in the

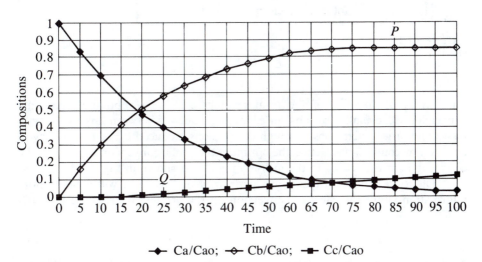

FIGURE 4.1-1
Batch composition profiles.

(a) Reactor

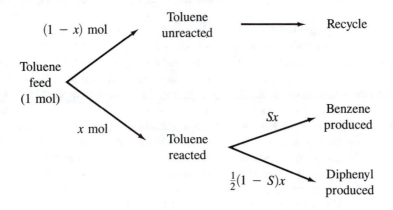

(b) Plant

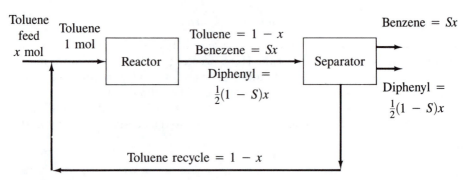

FIGURE 4.1-2
Selectivity.

range from 35 to 85% of the total product cost.* The optimum economic conversion is normally fixed by an economic trade-off between large selectivity losses and large reactor costs at high conversions balanced against large recycle costs at low conversions. Hence, the optimum economic conversion is less than the conversion corresponding to the maximum yield, as shown in Fig. 4.1-3.

Often the scouting experiments performed by a chemist will contain more information about the effect of conversion on the product distribution than the

---

* E. L. Grumer, "Selling Price vs. Raw Material Cost," *Chem. Eng.,* **79** (9): 190 (April 24, 1967).

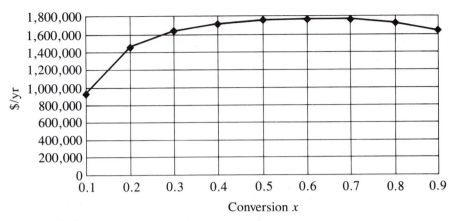

FIGURE 4.1-3
Profit.

bulk of the data which are taken in the neighborhood of the maximum yield. Again, establishing a close relationship with a chemist and providing him or her with feedback about the optimum processing conditions early in the experimental program will lead to more profitable processes.

Unfortunately, most companies are not organized to operate in this manner. Instead, the chemist's apparatus has been completely dismantled, and the chemist has been assigned to another project before a design engineer receives the problem. In this situation, the designer should attempt to estimate the economic incentive for determining the economic optimum conversion, and therefore for doing some additional experiments, rather than just designing a process to operate at maximum yield.

Catalyst Deactivation

Another piece of design data that often is lacking at the early stages of a design is the catalyst deactivation rate. The chemist's efforts are focused on finding a more active or selective catalyst, so that numerous short-time runs with a variety of catalysts are considered. Some catalysts have an operating life of 1 or 2 yr before regeneration or replacement is required, so obviously a time-consuming experiment is required to find the deactivation rate.

In initial designs we expect that there may be large uncertainties in some of the design data. Thus, we examine the sensitivity of the total product cost to these uncertainties, and we use these results to help guide the experimental development program in the direction of the highest potential profitability. We use shortcut techniques for these initial design calculations, because we recognize that it will be necessary to repeat the calculations as more information becomes available.

Production Rate

If we are to design a new plant to meet an expanding market condition, as a first guess of the production rate we consider the largest plant that has ever been built. With this approach we obtain the greatest economy of scale. (Normally, things are cheaper per unit if we buy them in large quantities.)

The maximum size of a plant is usually fixed by the maximum size of one or more pieces of equipment. Often this maximum size is fixed by restrictions on shipping the equipment to the plant site. That is, only a certain size compressor, or whatever, will fit on a railroad flatcar or truck.

We also consider the possibility of exceeding the maximum size of an existing plant. This approach almost always requires the development of new technology and thus has a higher risk. However, by gaining a larger economy of scale, it might be possible to reduce the product price enough to gain a greater share of the market and thereby justify the additional risk. However, if a project gets to be too big, new types of management problems might lead to a significant increase in costs.

Of course, we must also consider how our share of the market might change if we build a new plant. It makes a great deal of difference whether our company has 50 % of the market and some of the largest existing plants versus whether we have 5 % of the market and our existing plant has only one-tenth of the capacity of the largest plant.

The production rate specified for the plant might change (it usually does) during a design. Market conditions are constantly changing, and we must be responsive to these changes. By using shortcut methods for our preliminary designs, we minimize the effort required to change all the calculations.

Product Purity

The product purity normally is also fixed by marketing considerations. In fact, it might be possible to produce a range of product purities at different prices. Again, we must expect that the product price versus purity predictions might change as the design is being developed. It is also essential that a designer inform the marketing department about the very high costs that may be associated with producing some high-purity products early in the development of a new process, so that the marketing department does not raise customer expectations to unrealistic levels.

Raw Materials

A chemist normally uses very pure chemical reagents in laboratory studies, whereas natural or purchased raw materials always contain some impurities. Hence, we need to gather some information from our marketing group about raw-material price versus purity in order to decide whether to include a purification facility as part of the design project. Moreover, we must work with the chemist to see whether the impurities in the raw materials are inert or will affect the reactions. We must also examine their effect on the separation system. In particular, trace amounts of impurities can build up to large values in recycle loops unless the impurities are removed from the process.

Constraints

For the sake of safety we normally want to avoid processing conditions that are within the explosive limits of a mixture. (There is an unwritten law of nature that a spark will always find an explosive mixture and cause an ignition.) There have been a disturbingly large number of serious plant explosions throughout the history of the chemical industry, and we want to avoid another occurrence. However, phthalic anhydride plants, which oxidize zylene with air, do operate within the explosive range, although there are many special safety features.

Similarly, we have to know the processing conditions where some of our materials might polymerize and foul heat-exchange surfaces, or where they might become unstable and rapidly decompose. Numerous materials also form coke and deactivate catalysts, so we need to know how to minimize this coke formation. Very toxic or highly corrosive materials also affect the way in which we approach a design problem. Each of these factors may impose constraints on the design procedure.

Other Plant and Site Data

If we are going to build our new process on an existing site, then we must design the process to be compatible with the facilities that already exist on that site. The battery-limit conditions and costs that we need to know about are given in Table 4.1-3.

Physical Property Data

Conceptual designs often focus on attempts to make new materials, so that in many cases physical property data are not available in the literature. An excellent

TABLE 4.1-3
Plant and site data

1. Utilities
    a. Fuel supply
    b. Levels of steam pressure
    c. Cooling-water inlet and outlet temperatures
    d. Refrigeration levels
    e. Electric power
2. Waste disposal facilities

collection of techniques for estimating physical properties is available.* New estimation procedures based on group contribution methods are an area of active research.

The information we normally need are molecular weights, boiling points, vapor pressures, heat capacities, heats of vaporization, heats of reaction, liquid densities, and fugacity coefficients (or equations of state).

For conceptual designs, we use whatever information is available (which in some cases is a guess), and then we estimate the sensitivity of the total processing costs to these values. This sensitivity evaluation provides a measure of the economic incentive for making the appropriate measurements. In many instances the costs are surprisingly insensitive to the physical property values, but in some cases they are extremely sensitive. The use of shortcut design procedures significantly simplifies the sensitivity analysis.

Cost Data

The capital costs of some pieces of equipment are given in Appendix E.2, and some operating costs are given in Appendix E.1.

Summary

As a general statement about the input information, we can say that

$$\text{You never have the right information!} \qquad (4.1\text{-}2)$$

Some important data will be missing or will be taken in too narrow a range at the wrong temperature and pressure. A chemist's recipe for producing and isolating the product might be available, but not all the side reactions may be known. The chemist might have used a "favorite" solvent for each reaction step without ever considering separation costs.

You should never hesitate to ask the chemist for more information; a close working relationship is essential to developing good designs. And try to use preliminary design calculations to help guide the experimental parts of the development program. In particular, we want to determine how sensitive the design is to physical property data, coking limits or other process constraints, cost factors, purity specifications, etc.

> **Example 4.1-1.** To illustrate our design procedures, we consider a process for producing benzene from toluene. We develop the design from scratch and discuss numerous process alternatives. The input information taken from Sec. 1.2 is given in Table 4.1-4.

---

* R. C. Reid, J. M. Prausnitz, and T. K. Sherwood, *The Properties of Gases and Liquids*, 3d ed., McGraw-Hill, New York, 1977.

TABLE 4.1-4
Input information for the hydrodealkylation of toluene to produce benzene

1. Reaction information
    a. Reactions:

$$\text{Toluene} + H_2 \rightarrow \text{Benzene} + CH_4$$

$$2\text{Benzene} \rightleftharpoons \text{Diphenyl} + H_2 \qquad (4.1\text{-}3)$$

   b. Reaction inlet temperature $> 1150°C$ (to get a reasonable reaction rate); reactor pressure $= 500$ psia.

   c.
$$\text{Selectivity} = \frac{\text{Moles Benzene at Reactor Outlet}}{\text{Moles Toluene Converted}} = S$$

$$\text{Conversion} = \frac{\text{Moles Toluene Converted in Reactor}}{\text{Moles Toluene Fed to Reactor}} = x$$

$$S = 1 - \frac{0.0036}{(1-x)^{1.544}} \qquad x < 0.97 \qquad (4.1\text{-}4)$$

   d. Gas phase
   e. No catalyst
2. Production rate of benzene: 265 mol/hr
3. Product purity of benzene: $x_D = 0.9997$
4. Raw materials: Pure toluene at ambient conditions; H_2 stream containing 95% H_2, 5% CH_4 at 550 psia, 100°F
5. Constraints: H_2/aromatic ≥ 5 at the reactor inlet (to prevent coking); reactor outlet temperature $<$ 1300°F (to prevent hydrocracking); rapidly quench reactor effluent to 1150°F (to prevent coking); $x < 0.97$ for the product distribution correlation
6. Other plant and site data to be given later

4.2 LEVEL-1 DECISION: BATCH VERSUS CONTINUOUS

Continuous processes are designed so that every unit will operate 24 hr/day, 7 days/wk for close to a year at almost constant conditions before the plant is shut down for maintenance. Of course, in some cases equipment failures, or other reasons, cause unexpected shutdowns. In contrast, batch processes normally contain several units (in some cases all the units) that are designed to be started and stopped frequently. During a normal batch operating cycle, the various units are filled with material, perform their desired function for a specified period, are shut down and drained, and are cleaned before the cycle is repeated.

Many batch processes contain one or more units that operate continuously. For example, in numerous cases the products obtained from several batch reactors are temporarily stored, and then the products from this intermediate storage tank are fed to a train of distillation columns that are operated continuously. Similarly, there are cases where a variety of by-products that are produced in small amounts are accumulated continuously, and then when a sufficient amount is available, a batch still is used to separate the products.

The distinction between batch and continuous processes is sometimes somewhat "fuzzy." That is, large, continuous plants that exhibit catalyst deactivation may be shut down every year or so, to regenerate or replace the catalyst. Similarly, a large, continuous plant may include a single adsorption unit, which usually is a batch operation. If there are only one or two batch operations in a plant with large production rates that otherwise operates continuously, we normally refer to the plant as a continuous process.

Guidelines for Selecting Batch Processes

There are some rough guidelines that help to indicate when a batch process may be favored over a continuous process. These are reviewed now.

PRODUCTION RATES. Plants having a capacity of greater than 10×10^6 lb/yr are usually continuous, whereas plants having a capacity of less than 1×10^6 lb/yr are normally batch types. Large-capacity plants can justify a more thorough development program, i.e., a more accurate data base, as well as larger engineering design costs. In contrast, batch plants are usually simpler and more flexible, so that a satisfactory product can be produced with a larger uncertainty in the design. Also, because of their greater flexibility, batch plants are most common when a large number of products are produced in essentially the same processing equipment (e.g. paints).

MARKET FORCES. Many products are seasonal; for example, fertilizer is sold for only about a month in the early spring. If the fertilizer is produced over the complete year, then large inventory costs are incurred in storing it for the single month when sales are generated. However, if the fertilizer could be produced in a month or so in a batch plant and the plant could be used to make other products during the remainder of the year, then the inventory costs could be dramatically reduced. Hence, batch plants often are preferred for products with a seasonal demand.

It requires about 3 yr to build a continuous process, but some products only have an average lifetime of 2 yr (some organic pigments). The greater flexibility of batch plants makes them preferable for products with a short lifetime.

OPERATIONAL PROBLEMS. Some reactions are so slow that batch reactors are the only reasonable alternative. Also, it is very difficult to pump slurries at low flow rates without the solid settling out of the suspension and plugging the equipment. Thus, it is very difficult to build continuous processes when a low capacity of slurries must be handled. Similarly, some materials foul equipment so rapidly that the equipment must be shut down and cleaned at frequent intervals. Batch operation turns out to be ideal for handling materials of this type, because the equipment is periodically started and stopped, and normally it is cleaned after each batch has been processed.

Multiple Operations in a Single Vessel

Another unique feature of batch processes is that it is often possible to accomplish several operations in a single batch vessel, whereas an individual vessel would be needed for each of the operations in a continuous plant. For example, suppose that in a continuous plant we heat a reactant before sending it to a reactor where a catalyst is added, and then we send the product mixture to a distillation column (see Fig. 4.2-1a). In a batch process, we might be able to use the reboiler of a batch still for the heating and reaction steps as well as for the separation; see Fig. 4.2-1b. Thus, a single piece of equipment can be used to replace three pieces of equipment.

When we carry out multiple operations in a single vessel, normally the vessel must be larger than the size required if we had used separate vessels for each operation. For example, if we use one vessel for each operation (heating, reaction, and separation) and it takes 1 hr for each step, we can produce P lb/hr of product by shifting the batches from one unit to another. However, if we use only a single

(a) Continuous

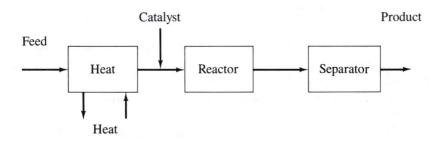

(b) Batch

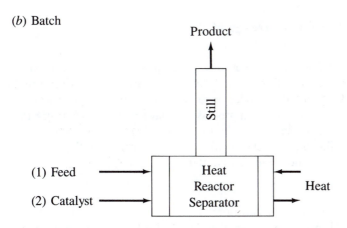

FIGURE 4.2-1
Continuous versus batch.

vessel, then it takes 3 hr to produce any product, and to obtain the same production rate we need to process 3 times as much material. Of course, when we use larger vessels, we often obtain an economy of scale, so that often there is a significant economic incentive to merge processing steps into a single vessel.

Design of Batch versus Continuous Processes

To develop a conceptual design for a continuous process, we must do the following:

1. Select the process units needed.
2. Choose the interconnections among these units.
3. Identify the process alternatives that need to be considered.
4. List the dominant design variables.
5. Estimate the optimum processing conditions.
6. Determine the best process alternative.

For a batch process we must make exactly the same decisions. However, we must also make the following decisions:

7. Which units in the flowsheet should be batch and which should be continuous?
8. What processing steps should be carried out in a single vessel versus using individual vessels for each processing step?
9. When is it advantageous to use parallel batch units to improve the scheduling of the plant?
10. How much intermediate storage is required, and where should it be located?

A Systematic Procedure for the Design of Batch Processes

Clearly it is necessary to make more decisions to design a batch process than to design a continuous process. For this reason, Malone and coworkers* suggest that the best approach to design a batch process is to design a continuous process first. With this approach it is simpler to screen process alternatives and to determine the best process flowsheet. Once the best structure of the flowsheet has been determined, they suggest following the systematic procedure given in Sec. 13.2 to develop the best design for a dedicated batch plant.

---

* O. Irribarren and M. F. Malone, "A Systematic Procedure for Batch Process Synthesis," paper presented at the 1985 Annual AIChE Meeting, Chicago, 1985; C. M. Myriatheos "Flexibility and Targets for Batch Process Designs." M.S. Thesis, University of Massachusetts, Amherst, 1986.

4.3   SUMMARY, EXERCISES, AND NOMENCLATURE

Summary

The input information that we need to undertake a design problem includes

1. The reactions and reaction conditions, including a correlation for the product distribution, a relationship for the conversion and space velocity, and information about catalyst type, deactivation rate, and regeneration
2. The desired production rate, product purity, and value of the product
3. The raw materials available and their costs
4. Any processing constraints
5. Other plant and site data
6. Physical properties and information about the chemicals involved
7. Cost data

Usually the correct data are not available or are uncertain. However, we do the best job that we can, and we evaluate the sensitivity of shortcut designs to changes in the uncertain factors. Estimates of this type can be used to determine the economic incentives for undertaking additional experiments. The data on the product distribution and side reactions are usually critical to a good design.

Level-1 Decision: Batch versus Continuous
   The factors that favor batch operation are
1. Production rate
   a. Sometimes batch if less than 10×10^6 lb/yr
   b. Usually batch if less than 1×10^6 lb/yr
   c. Multiproduct plants
2. Market forces
   a. Seasonal production
   b. Short product lifetime
3. Scale-up problems
   a. Very long reaction times
   b. Handling slurries at low flow rates
   c. Rapidly fouling materials

Exercises

4.3-1. Select a process from *Hydrocarbon Processing, The Encyclopedia of Chemical Processing and Design* by J. J. McKetta, or the *Encyclopedia of Chemical Technology* by Kirk–Othmer, and see how many of the input data that you can find.

4.3-2. The 1967 AIChE Student Contest Problem gives data showing how the selectivity (S = moles of benzene at reactor exit per mole of toluene converted) depends on the

TABLE 4.3-1
Selectivity data for HDA process

S	0.99	0.985	0.977	0.97	0.93
x	0.5	0.6	0.7	0.75	0.85

From the 1967 AIChE Student Contest Problem.

reactor conversion (See Table 4.3-1). Plot the data on arithmetic paper, and make a log–log plot of $1 - S$ versus $1 - x$. Develop correlations for both sets of data. Why is it better to correlate $1 - S$ versus $1 - x$? Also, use the correlation given by Eq. 4.1-4 to calculate the yield of benzene ($Y =$ mol benzene at reactor exit/mol of toluene fed to reactor $= Sx$) as a function of conversion. Estimate the conversion corresponding to the maximum yield.

4.3-3. Selectivity data for a process to produce acetic anhydride from acetone and acetic acid are given in the 1958 AIChE Student Contest Problem.[*] The data are given in Table 4.3-2. The reactions of interest are

$$\text{Acetone} \rightarrow \text{Ketene} + CH_4 \qquad 700°C, 1 \text{ atm}$$

$$\text{Ketene} \rightarrow CO + \tfrac{1}{2}C_2H_4 \qquad 700°C, 1 \text{ atm}$$

$$\text{Ketene} + \text{Acetic Acid} \rightarrow \text{Acetic Anhydride} \qquad 80°C, 1 \text{ atm}$$

and the selectivity is defined as $S =$ mol ketene at reactor exit/mol acetone converted. Develop a correlation for the data. Compare your results to the simple correlation $S = 1 - \tfrac{4}{3}x$. Use your results to estimate the conversion corresponding to the maximum yield.

4.3-4. A simplified version of a process to produce ethylene via ethane cracking has been presented by Bolles.[†] The reactions of interest are

$$C_2H_6 \rightarrow C_2H_4 + H_2$$

$$C_2H_6 \rightarrow \tfrac{1}{2}C_2H_4 + CH_4$$

TABLE 4.3-2
Selectivity data for acetic anhydride process

S	0.88	0.75	0.62	0.49	0.38	0.28	0.19	0.13
x	0.1	0.2	0.3	0.4	0.5	0.6	0.7	0.8

From the 1958 AIChE Student Contest Problem.

---

[*] See J. J. McKetta, *Encyclopedia of Chemical Processing and Design*, vol. 1, Dekker, New York, 1976, p. 271.

[†] W. L. Bolles, *Ethylene Plant Design and Economics*, Washington University Design Case Study No. 6, p. II-32, edited by B. D. Smith, Washington University, St. Louis, Mo., Aug. 1, 1986.

TABLE 4.3-3
Product distribution for ethane cracking

Component	Yield pattern, wt %					
H_2	2.00	2.47	2.98	3.51	4.07	4.64
CH_4	1.28	1.68	2.13	2.66	3.26	3.93
C_2H_4	28.9	35.8	43.2	51.1	59.4	67.8
C_2H_6	67.8	60.1	51.7	42.7	33.3	23.6

From W. L. Bolles, Washington University Design Case Study No. 4.

and some results for the product distribution are given in Table 4.3-3. Convert the data from weight percent to mol percent, and then develop a correlation for the selectivity (moles of C_2H_4 at the reactor exit per mole of C_2H_6 converted).

4.3-5. Wenner and Dybdal* present some product distribution data for styrene production. The reactions they consider are

$$\text{Ethylbenzene} \rightleftharpoons \text{Styrene} + H_2$$

$$\text{Ethylbenzene} \rightarrow \text{Benzene} + C_2H_4$$

$$\text{Ethylbenzene} + H_2 \rightarrow \text{Toluene} + CH_4$$

and points read from their graphs are given in Tables 4.3-4 and 4.3-5. Develop correlations for these data.

4.3-6. Consider two parallel, first-order isothermal reactions in a batch (or tubular) reactor fed with pure reactant

$$A \rightarrow \text{Product} \qquad A \rightarrow \text{Waste}$$

and define selectivity as S = mol product/mol A converted. Use a kinetic analysis to determine how the selectivity depends on the conversion. What are the results if the first reaction is first-order and the second reaction is second-order?

4.3-7. Consider two consecutive, first-order, isothermal reactions in a batch (or tubular) reactor fed with pure reactant:

$$A \rightarrow \text{Product} \qquad \text{Product} \rightarrow \text{Waste}$$

Define the selectivity as S = mol product in reactor effluent/mol A converted, and develop an expression, based on a kinetic analysis, for how the selectivity depends on

---

* R. W. Wenner and E. C. Dybdal, *Chem. Eng. Prog.*, **44**(4): 275 (1948).

TABLE 4.3-4
Moles of benzene per mole of styrene versus conversion

Mol benzene/mol styrene	0	0.005	0.010	0.020	0.030	0.060	0.100	0.140
Conversion x	0	0.10	0.15	0.20	0.25	0.30	0.35	0.40

From R. W. Wenner and E. C. Dybdal, *Chem. Eng. Prog.*, **44**(4): 275 (1948).

TABLE 4.3-5
Moles of toluene per mole of styrene versus conversion

Mol toluene/mol styrene	0	0.006	0.015	0.030	0.045	0.070	0.110	0.160
Conversion x	0	0.10	0.15	0.20	0.25	0.30	0.35	0.40

From R. W. Wenner and E. C. Dybdal, *Chem. Eng. Prog.*, **44** (4): 275 (1948).

conversion. How do the results change if the first reaction is second-order and the second reaction is also second-order?

4.3-8. To better understand the similarities and differences between the designs of a continuous and a batch process, let us consider a very oversimplified design problem where the process consists of only a single reactor. We desire to produce product B by the reaction $A \rightarrow B$. The cost of A is C_f (\$/mol), we operate 8150 hr/yr for a continuous plant, the desired production rate is P mol/hr, the reaction takes place by a first-order isothermal reaction, the separation of the product from unconverted reactants is free, and we cannot recover and recycle any unconverted reactants. We have to pay for the raw materials and reactor, so our cost model becomes

$$\text{TAC} = C_f F_F 8150 + C_v V \tag{4.3-1}$$

The production rate is related to the fresh feed rate F_F and the conversion x by the expression

$$P = F_F x \tag{4.3-2}$$

and the reactor volume is given by

$$V = \frac{F_F}{k\rho_m} \ln \frac{1}{1-x} \tag{4.3-3}$$

Thus, we can write

$$\text{TAC} = \frac{8150 C_f P}{x} + \frac{C_v P}{k\rho_m x} \ln\left(\frac{1}{1-x}\right) \tag{4.3-4}$$

Since the total annual cost becomes unbounded when x approaches either zero or unity, there must be an optimum conversion.

Suppose we do the same process in a batch reactor, where we produce n batches per year for 7500 hr/yr. Derive an expression for the total annual cost in terms of the conversion. Let the time it takes to empty, clean, and refill the reactor be t_d and the reaction time per cycle be t_r. How do the expressions for the batch process compare to the result for the continuous plant?

4.3-9. Swami* considered the problem of making two products in a process that consists only of a reactor, i.e., identical to Exercise 4.3-8 except that we have two reactions $A_1 \rightarrow B_1$ and $A_2 \rightarrow B_2$; the costs of the raw materials are C_{f1} and C_{f2}; the desired

---

* S. Swami, "Preliminary Design and Optimization of Batch Processes," M.S. Thesis, University of Massachusetts, Amherst, 1985.

production rates of B_1 and B_2 are P_1 and P_2 mol/hr, respectively, both reactions are first-order, isothermal, and irreversible with reaction rate constants k_1 and k_2; the densities are ρ_1 and ρ_2; the reactor downtimes are t_{d1} and t_{d2}; the numbers of batches per year are n_1 and n_2; and only one reactor is used to make both products. How do the results for using two separate reactors compare to the results for using a single reactor for a case where the reactor cost is given by the following expression?

$$\text{Reactor Cost} = C_v V^b \qquad b < 1$$

4.3-10. Suppose that two parallel reactors are used in the process described in Exercise 4.3-8. How do the reaction times, numbers of batches per year, raw-materials costs, reactor size, and reactor cost differ from the case of using a single reactor?

4.3-11. Isooctane (gasoline) can be produced by the reactions

$$\text{Butene} + \text{Isobutane} \rightarrow \text{Isooctane}$$
$$\text{Butene} + \text{Isooctane} \rightarrow C_{12}$$

The reactions take place in the liquid phase at 45°F and 90 psia in a continuous stirred tank reactor. Assume that the reaction kinetics agrees with the stoichiometry, and develop an expression for the selectivity (isooctane produced per butene converted).

Nomenclature

A	Reactant
B	Product
C_f	Cost of reactants (\$/mol)
C_v	Annualized cost of reactor volume [\$/(ft$^3 \cdot$ yr)]
F_F	Fresh feed rate (mol/hr)
k	Reaction rate constant (1/hr)
P	Production rate (mol/hr)
S	Selectivity
t_a	Downtime per batch (hr)
t_r	Reaction time per batch (hr)
V	Reactor volume (ft^3)
x	Conversion
x_D	Product purity
ρ_m	Molar density (mol/ft^3)

CHAPTER
5

INPUT-OUTPUT STRUCTURE OF THE FLOWSHEET

In Sec. 1.2 we described a hierarchical decision procedure for inventing process flowsheets and base-case designs. The decision levels are repeated in Table 5.1-1. The batch versus continuous decision of level 1 was discussed in Sec. 4.2, and in the subsequent chapters we discuss the other decision levels in detail.

5.1 DECISIONS FOR THE INPUT-OUTPUT STRUCTURE

To understand the decisions required to fix the input-output structure of a flowsheet, we merely draw a box around the total process. Thus, we focus our attention on what raw materials are fed to the process and what products and by-products are removed. Since the raw materials costs normally fall in the range from 33 to 85% of the total processing costs, we want to calculate these costs before we add any other detail to the design.

Flowsheet Alternatives

Almost every flowsheet has one of the two structures shown in Fig. 5.1-1a and b. There is a rule of thumb in process design (see Sec. 3.4) that it is desirable to recover more than 99% of all valuable materials. For initial design calculations we use the

116

TABLE 5.1-1
Hierarchy of decisions

1. Batch versus continuous
2. Input-output structure of the flowsheet
3. Recycle structure of the flowsheet
4. General structure of the separation system
   a. Vapor recovery system
   b. Liquid separation system
5. Heat-exchanger network

order-of-magnitude argument to say that this rule of thumb is equivalent to requiring that we completely recover and then recycle all valuable reactants. Thus, Fig. 5.1-1*a* indicates that no reactants leave the system.

Of course, if air and water are reactants in a process, they are sufficiently inexpensive, compared to organic materials, that it might be cheaper to lose them in an exit stream rather than to recover and recycle them. Hence, in a few rare cases, Fig. 5.1-1*a* might not be complete.

The other situation in which commonly reactants have been lost from a process occurs when we have a gaseous reactant and either a gaseous feed impurity or a gaseous by-product produced by one of the reactions. We want to recycle the gaseous reactant, but the inert-gas components must be purged from the process so that they do not continue to accumulate in the gas-recycle loop. In the past, it was so expensive to separate gaseous mixtures that some reactant was allowed to leave the process in a gas-recycle and purge stream (see Fig. 5.1-1*b*). However, the new membrane-separation technology, such as Monsanto's prism process, has cut the cost of gaseous separations so that gas recycle and purge might not be necessary.

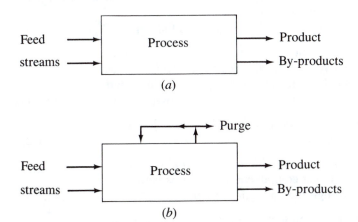

FIGURE 5.1-1
Input-output structure of flowsheet. [*From J. M. Douglas, AIChE J.,* **31**: *353 (1985).*]

TABLE 5.1-2
Level-2 decisions

1. Should we purify the feed streams before they enter the process?
2. Should we remove or recycle a reversible by-product
3. Should we use a gas recycle and purge stream?
4. Should we not bother to recover and recycle some reactants?
5. How many product streams will there be?
6. What are the design variables for the input-output structure, and what economic trade-offs are associated with these variables?

We distinguish between Fig. 5.1-1*a* and *b* because the presence of a gas recycle and purge stream adds a degree of freedom to the design problem; i.e., either the reactant composition of the purge stream or the excess gaseous reactant fed to the process becomes a new design variable. We discuss the design variables later.

Level-2 Decisions

The decisions we must make to fix the input-output structure of the flowsheet are given in Table 5.1-2 and are discussed below.

Purification of Feeds

A decision to purify the feeds before they enter the process is equivalent to a decision to design a preprocess purification system. This is different from a decision to feed the process through a separation system that is required in any event. Since at this stage of our synthesis and analysis procedure we do not know what kind of separation system will be required for the process with no feed impurities, we cannot always make a definite decision.

Some design guidelines to be considered are as follows:

If a feed impurity is not inert and is present in significant quantities, remove it (otherwise, it will lead to raw-material losses, and usually a much more complicated separation system is required to recover the additional by-products). (5.1-1)

If a feed impurity is present in a gas feed, as a first guess process the impurity. (5.1-2)

If a feed impurity in a liquid feed stream is also a by-product or a product component, usually it is better to feed the process through the separation system. (5.1-3)

If a feed impurity is present in large amounts, remove it (there is no quantitative criterion available to indicate how "large" is large). (5.1-4)

If a feed impurity is present as an azeotrope with a reactant,
often it is better to process the impurity.   (5.1-5)

If a feed impurity is inert but is easier to separate from the
product than the feed, it is better to process the impurity.   (5.1-6)

If a feed impurity is a catalyst poison, remove it.   (5.1-7)

PROCESS ALTERNATIVE. Unfortunately, not all these design guidelines are
quantitative, so often we must base our initial decision on our best judgment
(which might be merely a guess).

If we are not *certain* that our decision is correct, we list the
opposite decision as a *process alternative*.   (5.1-8)

With this approach, we have a systematic way of generating a list of process
alternatives.

ECONOMIC TRADE-OFFS FOR FEED PURIFICATION. Our decision of purifying the feed streams before they are processed involves an economic trade-off
between building a preprocess separation system and increasing the cost of the
process because we are handling the increased flow rates of inert materials. Of
course, the amount of inert materials present and where they will enter and leave
the process may have a great impact on the processing costs. Therefore, it is not
surprising that there is no simple design criterion that always indicates the correct
decision.

Recover or Recycle Reversible By-products

The reactions to produce benzene from toluene are

$$Toluene + H_2 \rightarrow Benzene + CH_4$$
$$2\,Benzene \rightleftharpoons Diphenyl + H_2$$

(4.1-3)

Since the second reaction is reversible, we could recycle the diphenyl back to the
reactor and let it build up in the recycle loop until it eventually reached an
equilibrium level. That is, the recycled diphenyl would decompose to form benzene
at the same rate as benzene would be producing diphenyl.

   If we recycle the reversible by-product, we must oversize all the equipment in
that recycle loop, to accommodate the equilibrium flow of the reversible by-
product. However, if we remove it from the process, we pay an economic penalty
because of the increased raw-material cost of reactant (toluene) that was converted
to the reversible by-product (diphenyl), e.g., the raw-material cost of toluene minus
the fuel value of diphenyl. Since our decision involves an economic trade-off
between raw-material losses to less valuable by-products and increased recycle

costs, it should not be surprising that no simple design guideline is available to make this decision. The result will also be sensitive to the equilibrium constant of the by-product reaction. So we generate another process alternative.

Gas Recycle and Purge

If we have a "light" reactant and either a "light" feed impurity or a "light" by-product produced by a reaction, it used to be common practice to use a gas recycle and a purge stream; i.e., we want to recycle the reactant, but we must purge one or more components from the process. We define a *light component* as one which boils lower than propylene ($-55°F$, $-48°C$). We choose propylene as a breakpoint because lower-boiling components normally cannot be condensed at high pressure with cooling water; i.e., both high pressure and refrigeration would be required. Since gaseous reactants normally are less expensive than organic liquids, and since refrigeration is one of the most expensive processing operations, it usually was cheaper to lose some of the gaseous reactants from a gas recycle and purge stream than it was to recover and recycle pure reactant.

However, membrane technology, such as Monsanto's prism process, now makes gas separations less expensive. Unfortunately, there is not a sufficient amount of published information available concerning the costs of membrane processes to be able to develop new design guidelines. Thus, we treat the membrane separator as another process alternative. Because of the lack of published information, we would base our initial design on a process with gas recycle and purge, whenever our design guideline is satisfied:

> Whenever a light reactant and either a light feed impurity
> or a light by-product boil lower than propylene ($-55°F$,
> $-48°C$), use a gas recycle and purge stream. $\qquad$ (5.1-9)

A membrane separation process also should always be considered.

Do Not Recover and Recycle Some Reactants

One of our design guidelines states that we should recover more than 99% of all valuable materials. Since some materials, such as air and water, are much less valuable than organic ones, we normally do not bother to recover and recycle unconverted amounts of these components. Of course, we could try to feed them to the process so that they would be completely converted, but often we feed them as an excess to try to force some more valuable reactant to complete conversion.

For example, in combustion reactions, we usually use an excess amount of air to ensure complete conversion of the fuel. The greater amount of the excess we use, the closer to complete conversion of the other reactant we obtain. However, the capital and operating costs of the blower used to move the air, as well as preheating and cooling costs, increase as we increase the excess amount of air. Thus there is an

TABLE 5.1-3
Destination codes and component classifications

Destination code	Component classification
1. Vent	Gaseous by-products and feed impurities
2. Recycle and purge	Gaseous reactants plus inert gases and/or gaseous by-products
3. Recycle	Reactants Reaction intermediates Azeotropes with reactants (sometimes) Reversible by-products (sometimes)
4. None	Reactants—if complete conversion or unstable reaction intermediates
5. Excess—vent	Gaseous reactant not recovered and recycled
6. Excess—waste	Liquid reactant not recovered or recycled
7. Primary product	Primary product
8. Valuable by-product (*I*)	Separate destination for different by-products
9. Fuel	By-products to fuel
10. Waste	By-products to waste treatment

optimum amount of excess that should be used. Similarly, at some point, the cost of excess water used will become significant. Moreover, pollution treatment costs must be considered.

Number of Product Streams

To determine the number of product streams that will leave the process, first we list all the components that are expected to leave the reactor. This component list usually includes all the components in the feed streams and all reactants and products that appear in every reaction. Then we classify each component in the list and assign a destination code to each. The component classifications and destination codes are given in Table 5.1-3. Finally, we order the components by their normal boiling points, and we group neighboring components with the same destination. The number of groups of all but the recycle streams is then considered to be the number of product steams.

   This procedure for determining the number of product streams is based on the common sense design guideline that

> It is never advantageous to separate two streams and then
> mix them together.   (5.1-10)

Also, it is based on the assumption that the components can be separated by distillation and that no azetropes are formed. Thus, in some cases (i.e., solids) a different set of rules must be used to estimate the number of product streams.

   Example 5.1-1. Suppose we have the 10 components listed in order of their boiling points and with the destination codes indicated. How many product streams will there be?

Component	Destination	Component	Destination
A	Waste	F	Primary product
B	Waste	G	Recycle
C	Recycle	H	Recycle
D	Fuel	I	Valuable by-product 1
E	Fuel	J	Fuel

Solution. The product streams are

1. $A + B$ to waste (do not separate them and then mix them in the sewer)
2. $D + E$ to fuel (do not separate them and then mix them to burn)
3. F—primary product (to storage for sale)
4. I—valuable by-product 1 (to storage for sale)
5. J to fuel (J must be separated from D and E to recover components F, G, H, and I, so we treat J as a separate product stream)

Example 5.1-2 Hydroalkylation of toluene to produce benzene. Find the number of product streams for the HDA process; i.e., see Example 4.1-1.

Solution. In Example 4.1-1 the components in the reactions are toluene, hydrogen, benzene, methane, and diphenyl. No additional components are present in either of the feed streams. If we arrange these components in order of their normal boiling points, we obtain the results shown in Table 5.1-4. Since both hydrogen (a reactant) and methane (both a feed impurity and a by-product) boil at a lower temperature than propylene, we decide to use a gas recycle and purge stream. Benzene is our primary product, and toluene (a reactant) is recycled. Diphenyl is a reversible by-product, and for this example we decided to remove it from the process and to use it as part of our fuel supply. These destination codes also are given in Table 5.1-4.

Thus, there are three product streams: a purge containing H_2 and CH_4, the primary product stream containing benzene, and a fuel by-product stream containing diphenyl. The initial flowsheet is given in Fig. 5.1-2.

TABLE 5.1-4
Toluene to benzene

Component	Boiling point	Destination code
H_2	$-253°C$	Recycle and purge
CH_4	$-161°C$	Recycle and purge
Benzene	$80°C$	Primary product
Toluene	$111°C$	Recycle
Diphenyl	$253°C$	Fuel

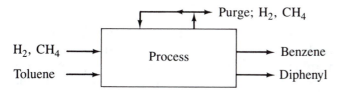

FIGURE 5.1-2
Input-output structure of HDA process. [*From J. M. Douglas, AIChE J.*, **31**: *353 (1985)*.]

Evaluation of the Flowsheet

At each stage of the development of a flowsheet, this is essential:

> Be certain that all products, by-products, and impurities
> leave the process!   (5.1-11)

For example, in our HDA process, there might be some other impurities in the hydrogen feed stream. We would expect them to leave with the purge. We also expect that there will be some impurities in the toluene feed stream. We attempt to identify these impurities by considering the source of our toluene, and then we look for where they will exit the 'process. Furthermore, we need to consider the possibility of other side reactions (such as the formation of terphenyl), and we must ensure that these new by-products leave the process or can be recycled.

If even a trace amount of a component is fed to a recycle loop and is not removed, the process will be inoperable. This is one of the most common mistakes made by inexperienced designers.

5.2 DESIGN VARIABLES, OVERALL MATERIAL BALANCES, AND STREAM COSTS

To calculate the overall process material balances and the stream costs, we must first assess whether the problem definition is complete or whether there are any degrees of freedom that must be specified to calculate the material balances. In general, the problem definition is not complete. Therefore, normally, it is impossible to develop a *unique* set of material balances for a process. Only a process with a single reaction and with no loss of reactants from the process has no degrees of freedom.

Of course, if the material balances are not unique, then the stream costs will also not be unique. Thus, we must develop the material balances and the stream costs in terms of the unknown design variables, and eventually we will look for the economic optimum values of these design variables.

Design Variables

The variables we select to complete the definition of the design problems are what we call the *degrees of freedom*. For complex reactions, it is usually possible to correlate the product distribution measured in the chemist's laboratory as a function of the conversion of the limiting reactant, the molar ratio of reactants, the reactor temperature, and/or the reactor pressure. If the activation energies of all the reactions are equal, then temperature will not appear in this correlation. Similarly, if there are the same number of moles of reactants and products for gas-phase reactions, or if we are considering liquid-phase reactions, then pressure will not appear in this correlation. In addition, we usually attempt to correlate the conversion against the space velocity in order to estimate the reactor size.

For preliminary designs, where kinetic data often are not available, we assume that the reactor configuration used in the process is the same as that used by the chemist (a batch reactor is equivalent to a plug flow reactor), because we cannot evaluate the effect of the reactor configuration on the product distribution without a kinetic model. However, if the preliminary design calculations indicate that additional development effort is justified, part of that development effort should be directed to determining the best reactor configuration.

In addition to the design variables which affect the product distribution, the other design variables that enter into the overall material balances correspond to situations where we do not recover and recycle all the reactants. Thus, if we use an excess amount of air in a combustion reaction, we must specify the amount of excess. Similarly, if we have a gas recycle and purge stream present, we must specify either the excess amount of gaseous reactant fed (i.e., in excess of the reaction requirements) which is lost in the purge stream or the reactant composition in the purge stream.

A list of the level-2 design variables that might affect the overall material balances is given in Table 5.2-1.

Overall Material Balances

Normally it is possible to develop expressions for the overall material balances in terms of the design variables without ever considering any recycle flows. Any time the overall equations are underdefined, it is necessary to look for one or more design variables to complete the problem definition, and these design variables always correspond to *significant process-optimization* problems. Hence, the initial analysis should always focus on the input-output flows only.

TABLE 5.2-1
Possible design variables for level 2

Complex reactions:	Reactor conversion, molar ratio of reactants, reaction temperature and/or pressure
Excess reactants:	Reactants not recovered or gas recycle and purge

TABLE 5.2-2
Procedures for developing overall material balances

1. Start with the specified production rate.
2. From the stoichiometry (and, for complex reactions, the correlation for product distribution) find the by-product flows and the reactant requirements (in terms of the design variables).
3. Calculate the impurity inlet and outlet flows for the feed streams where the reactants are completely recovered and recycled.
4. Calculate the outlet flows of reactants in terms of a specified amount of excess (above the reaction requirements) for streams where the reactants are not recovered and recycled (recycle and purge or air or water).
5. Calculate the inlet and outlet flows for the impurities entering with the reactant streams in step 4.

Material Balance Procedure

To develop the overall material balances for single-product plants, we always start with the given production rate. (If a process in an existing complex is replaced by a new process, then the feed rate may be known instead of the production rate.) From the production rate, the reaction stoichiometry, and (usually) the correlations for the product distribution, we calculate all the by-product flows and the raw-material requirements as functions of the design variables that appear in the product distribution correlations. Then, knowing the impurity compositions of the feed streams, we calculate the inlet and then the outlet flows of the inert materials. For feed streams where an excess of reactant is fed and not recovered and recycled or a feed stream where the reactant exits through a gas recycle and purge stream, we must specify the excess amount of reactant that is lost from the process. Then we can calculate the inlet and outlet flows of any impurities from the composition of these feed streams. A summary of the procedure is given in Table 5.2-2.

Limitations

Note that the preliminary material balances described above are based on the assumption of complete recovery of all valuable materials, instead of the rule-of-thumb value of greater than 99% recovery. There are no rules of thumb available to fix any of the design variables. Moreover, it is always much more important to find the neighborhood of the optimum values of the design variables which fix the inlet and outlet flows than it is to include the losses early in the analysis. However, we will want to revise our initial material balances at some point to include the losses.

Our approach focuses on processes that produce a single product. There are a number of processes that produce multiple products, e.g., chlorination of methane, amine processes, glycol plants, refineries, etc. The overall material balance calculations for these processes are usually more difficult than the procedure described in Table 5.2-1 because the product distribution must match the market demand.

Example 5.2-1   Toluene to benzene. Develop the overall material balances for the HDA process.

Solution. The reactions of interest are

$$Toluene + H_2 \rightarrow Benzene + CH_4$$

$$2Benzene \rightleftharpoons Diphenyl + H_2$$

$$(4.1-3)$$

and we are given that the desired production rate of benzene is $P_B = 265$ mol/hr (Example 4.1-1). If we use a gas recycle and purge stream for the H_2 and CH_4, and if we recover and remove the diphenyl, then there are three product streams; see Example 5.1-2 and the flowsheet given in Fig. 5.1-2.

We note from Fig. 5.1-2 that all the toluene we feed to the process gets converted (i.e., no toluene leaves the system). However, this does not mean that the conversion of toluene in the reactor is unity. A low conversion per pass simply means that there will be a large internal recycle flow of toluene. However, we evaluate these recycle flows at a later stage of our design procedure. Of course, our assumption that all the feed toluene is converted neglects any toluene losses in any of the product streams, but this corresponds to our initial design assumption of complete recovery versus greater than 99% recoveries.

SELECTIVITY AND REACTION STOICHIOMETRY. We define the *selectivity S* as the fraction of toluene converted in the reactor that corresponds to the benzene flow at the reactor outlet. Also, we recover and remove all this benzene. Hence, for a production of P_B mol/hr, the toluene fed to the process F_{FT} must be

$$F_{FT} = \frac{P_B}{S} \tag{5.2-1}$$

Also, from the stoichiometry (Eq. 5.1-3), the amount of methane produced P_{R,CH_4} must be

$$P_{R,CH_4} = \frac{P_B}{S} \tag{5.2-2}$$

If a fraction S of toluene is converted to benzene, a fraction $1 - S$ must be lost to diphenyl. However, from the stoichiometry of Eq. 5.1-3, the amount of diphenyl produced P_D must be

$$P_D = F_{FT} \frac{1 - S}{2} = \frac{P_B}{S} \frac{1 - S}{2} \tag{5.2-3}$$

Since the toluene stream contains no impurities, we find the toluene fresh feed rate and the diphenyl by-product flow rate in terms of selectivity from the reaction stoichiometry and the given production rate. A relationship between selectivity and conversion is given in Example 4.1-1.

RECYCLE AND PURGE. The stoichiometry also indicates the amount of H_2 required for the reaction. If we feed an excess amount of hydrogen, F_E, into the process, this hydrogen will leave with the purge stream. Thus, the total amount of hydrogen fed to the process will be

$$F_E + \frac{P_B}{2S}(1 + S) = y_{FH}F_G \tag{5.2-4}$$

which is equal to the amount of hydrogen in the makeup gas stream $y_{FH}F_G$. Similarly, the methane flow rate leaving the process will be the amount of methane entering the process, $(1 - y_{FH})F_G$, plus the amount of methane produced by the reactions, $P_{R,CH_4} = P_B/S$, or

$$P_{CH_4} = (1 - y_{FH})F_G + \frac{P_B}{S} \tag{5.2-5}$$

The total purge flow rate P_G will then be the excess H_2, F_E, plus the total methane P_{CH_4} or

$$P_G = F_E + (1 - y_{FH})F_G + \frac{P_B}{S} \tag{5.2-6}$$

Rather than using the excess hydrogen feed F_E as a design variable, we normally use the purge composition of the reactant y_{PH}, where

$$y_{PH} = \frac{F_E}{P_G} \tag{5.2-7}$$

This purge composition is always bounded between zero and unity (actually there is a smaller upper bound which depends on the feed composition and sometimes on the conversion). The use of bounded variables makes the preparation of graphs simpler.

We can develop expressions for the makeup gas rate F_G and the purge rate P_G explicitly in terms of the purge composition of reactant y_{PH} either by using Eq. 5.2-7 to eliminate F_E from Eqs. 5.2-4 and 5.2-7 or by writing balances for the hydrogen and methane and then combining them. That is, the amount of hydrogen in the feed must supply the net reaction requirements as well as the purge loss

$$y_{FH}F_G = \frac{P_B}{S} - \frac{P_B}{S}\frac{1 - S}{2} + y_{PH}P_G \tag{5.2-8}$$

and the methane in the feed plus the methane produced must all leave with the purge

$$(1 - y_{FH})F_G + \frac{P_B}{S} = (1 - y_{PH})P_G \tag{5.2-9}$$

Adding these expressions gives

$$P_G = F_G + \frac{P_B}{S}\frac{1 - S}{2} \tag{5.2-10}$$

We can then solve for F_G:

$$F_G = \frac{P_B\left[1 - (1 - y_{PH})\dfrac{1 - S}{2}\right]}{S(y_{FH} - y_{PH})} \tag{5.2-11}$$

If we are given values of P_B, S, and either F_E or y_{PH}, we can use the above equations to calculate the fresh feed rate of toluene, F_{FT} (Eq. 5.2-1), the production rate of diphenyl, P_D (Eq. 5.2-3), the makeup gas rate F_G (Eq. 5.2-4 or 5.2-11), and the purge flowrate, P_G (Eq. 5.2-6 or 5.2-10). Thus, we have determined all the input and output flows in terms of the unknown design variables S and either F_E or y_{PH}.

MATERIAL BALANCES IN TERMS OF EXTENT OF REACTION. It is becoming common practice to describe material balance calculations in terms of the extent of reaction ξ_1 (or fractional extent of reaction). Thus, for the HDA process, we would say that ξ_1 mol (or moles/hr) of toluene react with ξ_1 mol of H_2 to produce ξ_1 mol of benzene plus ξ_1 mol of CH_4. Also, $2\xi_2$ mol of benzene produces ξ_2 mol of diphenyl plus ξ_2 mol of hydrogen. Then, we combine these statements to say that

$$\text{Net benzene produced} = \xi_1 - 2\xi_2 \tag{5.2-12}$$

$$\text{Methane produced} = \xi_1 \tag{5.2-13}$$

$$\text{Diphenyl produced} = \xi_2 \tag{5.2-14}$$

$$\text{Toluene consumed} = \xi_1 \tag{5.2-15}$$

$$\text{Hydrogen consumed} = \xi_1 - \xi_2 \tag{5.2-16}$$

We can generalize these expressions and say that the number of moles (or moles per hour) of any component is given by

$$n_j = n_j^0 + v_{ij}\xi_i \tag{5.2-17}$$

where the v_{ij} are the stoichiometric coefficients, which are positive for products and negative for reactants. Normally (for the purposes of initial design calculations) no products are fed to the process, $n_j^0 = 0$, and no reactants are allowed to leave, $n_j = 0$. Experience indicates that it normally is possible to correlate the extent of each reaction ξ_i against the per-pass conversion of the limiting reactant, although in some cases the molar ratio of reactants, reactor temperature, and/or reactor pressure must be included in the correlation.

EXTENT VERSUS SELECTIVITY. For the case of only two reactions, it often is simpler to describe the product distribution in terms of selectivity. Selectivity can be described in a number of different ways, such as the production of the desirable component divided by the amount of limiting reactant converted or the production of the desired component divided by the production of the undesired component. For the HDA process, in the first case we would have

$$S_a = \frac{\xi_1 - 2\xi_2}{\xi_1} \tag{5.2-18}$$

whereas in the second case we would have

$$S_b = \frac{\xi_1 - 2\xi_2}{\xi_2} \tag{5.2-19}$$

It is essential to ensure that the definition of selectivity that is reported by a chemist (or in the literature) is clearly defined, but it is a simple matter to convert from one definition to another.

Example 5.2-2   Toluene to benzene. Develop the expressions relating the extents of reaction to production rate and selectivity for the HDA process.

Solution. From Eqs. 5.2-15 and 5.2-1 we find that

$$\xi_1 = \frac{P_B}{S} \tag{5.2-20}$$

Also from Eq. 5.2-12 we find that

$$\xi_1 - 2\xi_2 = P_B \tag{5.2-21}$$

Thus we can write

$$\xi_2 = \tfrac{1}{2}(\xi_1 - P_B) = \frac{P_B}{2}\left(\frac{1 - S}{S}\right) \tag{5.2-22}$$

Stream Tables

It is common practice to report material balance calculations in terms of stream tables. That is, the streams are numbered on a flowsheet, and then a table is prepared that gives the component flows in each of these streams which correspond to a particular set of values of the design variables. An example is given in Fig. 5.2-1. The temperatures, pressures, and enthalpy of each stream also are normally listed. Since we do not consider energy balances until the end of the synthesis procedure, we add these values to the stream tables later.

One major difficulty with this conventional practice is that the designer is forced to select values of design variables without knowing the optimum values. Moreover, once a set of values has been selected, it is often difficult to remember that they were selected arbitrarily. For this reason, we recommend that the stream tables list the values of the design variables and that the appropriate material balance equations, such as Eqs. 5.2-1, 5.2-3, 5.2-4 5.2-6, etc., be programmed on a spreadsheet such as LOTUS. With this approach it is very easy to change the production rate and the design variables and then recalculate all the stream flows. We can use this same table as the basis for calculating the stream costs as a function of the design variables.

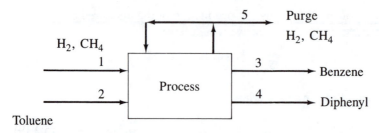

Production rate = 265

Design variables: F_E and x

Component	1	2	3	4	5
H_2	F_{H_2}	0	0	0	F_E
CH_4	F_M	0	0	0	$F_M + P_B/S$
Benzene	0	0	P_B	0	0
Toluene	0	P_B/S	0	0	0
Diphenyl	0	0	0	$P_B(1 - S)/(2S)$	0
Temperature	100	100	100	100	100
Pressure	550	15	15	15	465

where $S = 1 - 0.0036/(1 - x)^{1.544}$ $F_{H_2} = F_E + P_B(1 + S)/2S$
$F_M = (1 - y_{FH})[F_E + P_B(1 + S)/S]/y_{FH}$ $F_G = F_{H_2} + F_M$

FIGURE 5.2-1
Stream table for the HDA process.

Stream Costs: Economic Potential

Since the "best" values of the design variables depend on the process economics, we want to calculate the stream costs, i.e., the cost of all raw materials and product streams in terms of the design variables. Normally, we combine these costs into a single term, which we call the economic potential (EP). We define the economic potential at level 2 as

$$EP_2 = \text{Product Value} + \text{By-product Values}$$
$$- \text{Raw-Material Costs, \$/yr} \quad (5.2\text{-}23)$$

which for an HDA example would be

$$EP = \text{Benzene Value} + \text{Fuel Value of Diphenyl} + \text{Fuel}$$
$$\text{Value of Purge} - \text{Toluene Cost} - \text{Makeup Gas Cost} \quad (5.2\text{-}24)$$

We would also subtract the annualized capital and operating cost of a feed compressor, if one is needed. (The calculations required are discussed in Sec. 6.5.)

TABLE 5.2-3
Cost data for HDA process

Value of benzene	$0.85/gal = $9.04/mol
Value of toluene*	$0.50/gal = $6.40/mol
Value of H_2 feed	$3.00/1000 ft^3 = $1.14/mol
Fuel = $4.0/$10^6$ Btu	
Fuel value: H_2	0.123×10^6 Btu/mol
CH_4	0.383×10^6 Btu/mol
Benzene	1.41×10^6 Btu/mol
Toluene	1.68×10^6 Btu/mol
Diphenyl[†]	2.688×10^6 Btu/mol

* Assume an internal transfer-price value versus the current prices of $1.26/gal.

[†] We also assume that the fuel value of diphenyl is $5.38/mol.

The economic potential is the annual profit we could make if we did not have to pay anything for capital costs or utilities costs. Of course, if the economic potential is negative, i.e., the raw materials are worth more than the products and by-products, then we want to terminate the design project, look for a less expensive source of raw materials, or look for another chemistry route that uses less expensive raw materials.

> **Example 5.2-3   HDA stream costs.** If we use the cost data given in Table 5.2-3, where the values of H_2, CH_4, and diphenyl in the product streams are based on the heats of combustion of the components and a fuel value of $4/$10^6$ Btu, we can calculate the economic potential for the HDA process in terms of the design variables (we use reactor conversion per pass x, instead of S, and y_{PH}). The results are shown in Fig. 5.2-2. The graph indicates that at high conversions the process is unprofitable

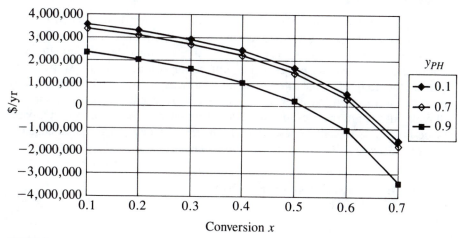

FIGURE 5.2-2
Economic potential—level 2.

TABLE 5.3-1
Alternatives for the HDA process

1. Purify the hydrogen feed stream.
2. Recycle the diphenyl to extinction.
3. Purify the H_2-recycle stream.

(i.e., we convert so much toluene to diphenyl that this selectivity loss exceeds the increased value of the benzene that we produce). Also, at high purge compositions of hydrogen we lose money (we lose so much hydrogen to fuel that we cannot make up for this loss).

According to the graph, the most desirable values (i.e., the greatest profit) of the design variables correspond to $x = 0$ (i.e., no selectivity loss) and a reactant composition of the purge stream y_{PH} equal to zero (i.e., purge pure methane).

As we proceed through the design, however, we find that a zero conversion per pass ($x = 0$) corresponds to an infinitely large recycle flow of toluene and that purging no hydrogen ($y_{PH} = 0$) corresponds to an infinitely large gas-recycle flow. Hence, we develop the optimum values of x and y_{PH} as we proceed through the design.

5.3 PROCESS ALTERNATIVES

In our development of a design for the HDA process (see Fig. 5.1-2), we made the decisions (1) not to purify the hydrogen feed stream, (2) to remove the diphenyl from the process, and (3) to use a gas recycle and purge stream. If we change any of these decisions, we generate process alternatives. It is always good practice to make a list of these alternatives. Such a list is given in Table 5.3-1.

Evaluation of Alternatives

We could attempt to simultaneously develop designs that corresponded to each process alternative. However, if we remember that less than 1 % of ideas for new designs ever become commercialized, our initial goal should be to eliminate, with as little effort as possible, projects that will be unprofitable. Thus, we prefer to complete the design for one alternative as rapidly as possible before we give any consideration to the other alternatives, for we might encounter some factor that will make all the alternatives unprofitable. Then, after we have completed a base-case design, we examine the alternatives. In the terminology of artificial intelligence (AI), we are using a *depth-first*, rather than a *breadth-first*, *strategy*.

5.4 SUMMARY, EXERCISES, AND NOMENCLATURE

Summary

The questions we must answer to fix the input-output structure of the flowsheet include the following:

1. Should we purify the feed stream?
2. Should we remove or recycle a reversible by-product?
3. Should we use a gas recycle and purge stream?
4. Should we use an excess of some reactant that we discard?
5. How many product streams will there be?
6. What are the design variables and the economic trade-offs at this level of analysis?

In some cases heuristics can be used to help make these decisions. When no heuristics are available, we make a guess and then list the opposite decision as a process alternative. We complete a first design based on our original guess before we consider any other alternatives. (Since less than 1 % of ideas for new designs are successful, we might learn something about the process that will make all the alternatives unprofitable.)

Some of the heuristics and design guidelines that were presented include the following:

If a feed impurity is not inert, remove it.

If an impurity is present in a gas feed stream, as a first guess process the impurity.

If an impurity in a liquid feed stream is a product or by-product, usually feed the process through the separation system.

If an impurity is present in large amounts, remove it.

If an impurity is present as an azeotrope with a reactant, process the impurity.

If a feed impurity is an inert, but is easier to separate from the product and by-product than from the feed, it is better to process the impurity.

Whenever there is a light reactant and a light feed impurity or by-product (where light components boil lower than propylene, $-48°C$), use a gas recycle and purge stream for the first design. Also consider a membrane separator later.

If O_2 from air or water is a reactant, consider using an excess amount of this reactant.

For single-product, vapor-liquid processes, we determine the number of product streams by grouping components with neighboring boiling points that have the same exit destinations; i.e., we never separate streams and then remix them.

Be certain that all by-products and impurities leave the process!

The significant design variables are those that affect the product distribution and purge compositions of gas streams.

Raw-material costs are often in the range from 33 to 85 % of the total costs.

Exercises

5.4-1. Draw the input-output flowsheet and plot the economic potential for the HDA process for the case where the diphenyl is recycled.

5.4-2. Draw the flowsheet and plot the economic potential for the HDA process for the case where the H_2 is separated from CH_4 before it is recycled and where diphenyl is removed from the process.

5.4-3. Acetic anhydride* can be produced by the reaction system

$$\left.\begin{array}{c} \text{Acetone} \rightarrow \text{Ketene} + CH_4 \\[2mm] \text{Ketene} \rightarrow CO + \tfrac{1}{2}C_2H_4 \end{array}\right\} \quad 700°\text{C, 1 atm}$$

$$\text{Ketene} + \text{Acetic Acid} \rightarrow \text{Acetic Anhydride} \qquad 80°\text{C, 1 atm}$$

The selectivity (moles of ketene leaving the pyrolysis reactor per mole of acetone converted) is given by $S = 1 - 4x/3$ at low conversions. The desired production rate of anhydride is 16.58 mol/hr at a purity of 99%. The cost data are: acetone = $15.66/mol, acid = $15.00/mol, anhydride = $44.41/mol, and fuel at $4.00/10^6$ Btu. Draw the flowsheet, and plot the economic potential.

5.4-4. A process for producing acetone from isopropanol is discussed in Exercise 1.3-4. The desired production rate is 51.3 mol/hr. The feed azeotrope contains 70 mol % IPA, and the costs are acetone = $15.66/mol, IPA–$H_2O$ azeotrope = $9.53/mol, H_2 as fuel = $0.49/mol, and H_2O as waste = −$0.007/mol. Draw the input-output flowsheet, calculate the overall material balances, and plot the economic potential.

5.4-5. A simplified flowsheet for ethanol synthesis is discussed in Exercise 1.3-6. The desired production rate of the azeotropic product is 783 mol/hr (85.4 mol % EtOH), and the costs are: ethylene feed mixture = $6.15/mol, process water = $0.00194/mol, ethanol as azeotrope = $10.89/mol, and the fuel at $4.00/10^6$ Btu. Draw the input-output structure of the flowsheet, and plot the economic potential.

5.4-6. Styrene can be produced by the reactions

$$\text{Ethylbenzene} \rightleftharpoons \text{Styrene} + H_2$$

$$\text{Ethylbenzene} \rightarrow \text{Benzene} + \text{Ethylene}$$

$$\text{Ethylbenzene} + H_2 \rightarrow \text{Toluene} + CH_4$$

The reactions take place at 1115°F and 25 psia. We want to produce 250 mol/hr of styrene. The costs are: ethylbenzene = $15.75/mol, styrene = $21.88/mol, benzene = $9.04/mol, toluene = $8.96/mol, and fuel at $4.00/10^6$ Btu. Wenner and Dybdal[†] give correlations for the product distribution

$$\frac{\text{Mol Benzene}}{\text{Mol Styrene}} = 0.333x - 0.215x^2 + 2.547x^3$$

$$\frac{\text{Mol Toluene}}{\text{Mol Styrene}} = 0.084x - 0.264x^2 + 2.638x^3$$

where x = styrene conversion. The ethylbenzene feed stream contains 2 mol % benzene. Draw the input-output flowsheet and plot the economic potential.

---

* This problem is a modified version of the 1958 AIChE Student Contest Problem; see J. J. McKetta, *Encyclopedia of Chemical Processing and Design*, vol. 1, Dekker, New York, 1976, p. 271.

† R. R. Wenner and E. C. Dybdal, *Chem. Eng. Prog.*, **44**(4): 275 (1948).

5.4-7. Cyclohexane* can be produced by the reaction

$$\text{Benzene} + 3H_2 \rightleftarrows \text{Cyclohexane}$$

The reaction takes place at 392°F and 370 psia. Pure benzene is used as a feed stream, but the hydrogen stream contains 2% methane. The desired production rate is 100 mol/hr, and the costs are: benzene = $6.50/mol, H_2 = $1.32/mol, cyclohexane = $12.03/mol, and fuel at $4.00/$10^6$ Btu. Draw the input-output flowsheet, and plot the economic potential.

5.4-8. Ethylene can be produced by the thermal cracking of ethane[†]

$$C_2H_6 \rightarrow C_2H_4 + H_2$$
$$C_2H_6 \rightarrow \tfrac{1}{2}C_2H_4 + CH_4$$

The reactions take place at 1500°F and 50 psia. We desire to produce 875 mol/hr of ethylene with 75% purity. Assume that the selectivity is given by

$$S = \frac{\text{mol } C_2H_4 \text{ Formed}}{\text{mol } C_2H_6 \text{ Converted}} = 1 - \frac{0.0381}{(1-x)^{0.241}}$$

The ethane feed contains 5% CH_4 and costs $1.65/mol. Ethylene at 95% composition is worth $6.15/mol. Fuel is worth $4.00/$10^6$ Btu. Draw the input-output flowsheet and plot the economic potential.

5.4-9. Butadiene sulfone* can be produced by the reaction

$$\text{Butadiene} + SO_2 \rightleftarrows \text{Butadiene Sulfone}$$

The reaction takes place in the liquid phase at 90°F and 150 psia. The costs are SO_2 = $0.064/mol, butadiene = $6.76/mol, and butadiene sulfone = $8.50/mol. We want to make 80 mol/hr of product. Draw the input-output flowsheet, and plot the economic potential.

5.4-10. Isooctane (gasoline)[¶] can be produced by the reactions

$$\text{Butene} + \text{Isobutane} \rightarrow \text{Isooctane}$$

$$\text{Butene} + \text{Isooctane} \rightarrow C_{12}$$

The reactions take place in the liquid phase at 45°F and 90 psia; see Exercise 4.3-11. The desired production rate is 918 mol/hr, and the costs are: butene = $14.56/mol, isobutane = $18.59/mol, isooctane = $36.54/mol, and fuel = $4.00/$10^6$ Btu. One feed stream contains 8% C_3, 80% butene, and 12% n-C_4, while the other contains 12% C_3, 73% i-C_4, and 15% n-C_4. Draw the input-output flowsheet, and plot the economic potential.

---

* J. R. Fair, *Cyclohexane Manufacture*, Washington University Design Case Study No. 4, edited by B. D. Smith, Washington University, St. Louis, Mo., Aug. 1, 1967.

† W. L. Bolles, *Ethylene Plant Design and Economics*, Washington University Design Case Study No. 6, edited by B. D. Smith, Washington University, St. Louis, Mo., 1970.

‡ This problem is a modified version of the 1970 AIChE Student Contest Problem; see J. J. McKetta, *Encyclopedia of Chemical Processing and Design*, vol. 5, Dekker, New York, 1977, p. 192.

¶ This problem is a modified version of the 1977 AIChE Student Contest Problem; *AIChE Student Members Bulletin*, AIChE Headquarters, 1977.

Nomenclature

EP	Economic potential
F_E	Excess H_2 fed to the process, mol/hr
F_{FT}	Fresh feed rate of toluene, mol/hr
F_G	Makeup gas rate, mol/hr
F_{H_2}	H_2 consumed by the reactions, mol/hr
n_j	Final moles of component j
n_j^0	Initial moles of component j
P_B	Production rate of benzene, mol/hr
P_{CH_4}	Purge flow of methane, mol/hr
P_D	Diphenyl produced, mol/hr
P_G	Purge flow rate, mol/hr
P_{R,CH_4}	CH_4 produced by the reaction, mol/hr
S	Selectivity, mol benzene produced/mol toluene converted
y_{FH}	Feed composition of H_2
y_{PH}	Mole fraction of H_2 in the purge stream

Greek symbols

ν_{ij}	Stoichiometric coefficients
ξ_i	Extent of reaction i

CHAPTER
6

RECYCLE
STRUCTURE
OF THE
FLOWSHEET

Now that we have decided on the input-output structure of the flowsheet, we want to add the next level of detail. From earlier discussions we know that the product distribution dominates the design, and therefore we add the details of the reactor system. Also, since gas compressors are the most expensive processing equipment, we add the annualized capital and operating costs of any compressors required. However, at this level of the synthesis and analysis procedure, we treat the separation system as just a blackbox, and we consider the details of the separation system later.

6.1 DECISIONS THAT DETERMINE THE RECYCLE STRUCTURE

The decisions that fix the recycle structure of the flowsheet are listed in Table 6.1.-1. Each of these decisions is discussed in detail.

TABLE 6.1-1
Decisions for the recycle structure

1. How many reactor systems are required? Is there any separation between the reactor systems?
2. How many recycle streams are required?
3. Do we want to use an excess of one reactant at the reactor inlet?
4. Is a gas compressor required? What are the costs?
5. Should the reactor be operated adiabatically, with direct heating or cooling, or is a diluent or heat carrier required?
6. Do we want to shift the equilibrium conversion? How?
7. How do the reactor costs affect the economic potential?

Number of Reactor Systems

If sets of reactions take place at different temperatures or pressures, or if they require different catalysts, then we use different reactor systems for these reaction sets. For example, in the HDA process the reactions

$$\left. \begin{array}{l} \text{Toluene} + H_2 \rightarrow \text{Benzene} + CH_4 \\ 2\,\text{Benzene} \rightleftharpoons \text{Diphenyl} + H_2 \end{array} \right\} \quad 1150\text{--}1300°F, 500 \text{ psia} \qquad (6.1\text{-}1)$$

both take place at the same temperature and pressure without a catalyst. Therefore there is only one reactor required. In contrast, in the reaction system

$$\left. \begin{array}{l} \text{Acetone} \rightarrow \text{Ketene} + CH_4 \\ \text{Ketene} \rightarrow CO + \tfrac{1}{2}C_2H_4 \end{array} \right\} \quad 700°C, 1 \text{ atm} \qquad (6.1\text{-}2)$$

$$\text{Ketene} + \text{Acetic Acid} \rightarrow \text{Acetic Anhydride} \qquad 80°C, 1 \text{ atm}$$

the first two reactions take place at a high temperature, whereas the third reaction takes place at a lower temperature. Hence, two reactor systems would be required, and we could call these R1 and R2.

Number of Recycle Streams

From the discussion above, we see that we can associate reaction steps with a reactor number. Then we can associate the feed streams with the reactor number where that feed component reacts; e.g., in the anhydride process, acetone would be fed to the first reactor, whereas acetic acid would be fed to the second reactor. Similarly, we can associate the components in recycle streams with the reactor numbers where each component reacts; e.g., in the anhydride process, acetone would be recycled to the first reactor, whereas acetic acid would be recycled to the second reactor (see Fig. 6.1-1).

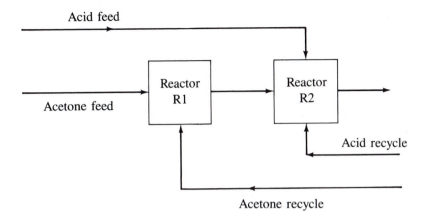

FIGURE 6.1-1
Acetic anhydride.

Now we can take our list of all the components leaving the reactor that has been ordered by the normal boiling points, e.g., Table 5.1-4, and we list the reactor number as the destination code for each recycle stream. Next we group recycle components having neighboring boiling points if they have the same reactor destination. Then the number of recycle streams is merely the number of groups. This simple procedure is based on this common sense heuristic:

Do not separate two components and then remix them at a reactor inlet.   (6.1-3)

We also distinguish between gas- and liquid-recycle streams, because gas-recycle streams require compressors, which are always expensive. We consider a stream to be a gas-recycle stream if it boils at a lower temperature than propylene (i.e., propylene can be condensed with cooling water at high pressure, whereas lower-boiling materials require refrigerated condensers, which require a compressor). Liquid-recycle streams require only pumps. In our initial design calculations we do not include the costs of the pumps because they are usually small compared to compressors, furnaces, distillation columns, etc. (High-head, high-volume pumps can be very expensive, so in some cases we must check this assumption.)

Example 6.1-1 Number of recycle streams. Consider the components and the destinations given below in the order of their normal boiling points:

A. Waste by-product   *F.* Primary product
B. Waste by-product   *G.* Reactant — recycle to R2
C. Reactant — recycle to R1 *H.* Reactant — recycle to R2
D. Fuel by-product   *I.* Reactant — recycle to R1
E. Fuel by-product   *J.* Valuable by-product

There are four product streams ($A + B, D + E, F$, and J) and three recycle streams ($C, G + H$, and I), where the first and last go to R1 and the second goes to R2.

Example 6.1-2   HDA process. The components and their destination for the HDA process are as follows:

Component	NBP, °C	Destination
H_2	-253	Recycle + purge — gas
CH_4	-161	Recycle + purge — gas
Benzene	80	Primary product
Toluene	111	Recycle — liquid
Diphenyl	255	Fuel by-product

Thus, there are three product streams—purge, benzene, and diphenyl—and two recycle streams, $H_2 + CH_4$ and toluene, where the first is a gas and the second is a liquid. A recycle flowsheet is given in Fig. 6.1-2, and it shows the reactor and the recycle gas compressor.

Example 6.1-3   Anhydride process. The component list and the destination codes for the anhydride process are given:

Component	NBP, °F	Destination
CO	-312.6	Fuel by-product
CH_4	-258.6	Fuel by-product
C_2H_4	-154.8	Fuel by-product
Ketene	-42.1	Unstable reactant — completely converted
Acetone	133.2	Reactant — recycle to R1 — liquid
Acetic acid	244.3	Reactant — recycle to R2 — liquid
Acetic anhydride	281.9	Primary product

Thus, there are two product streams, $CH_4 + CO + C_2H_4$ and anhydride, and two liquid-recycle streams are returned to different reactors: acetone is recycled to R1, and acetic acid is recycled to R2. A flowsheet is shown in Fig. 6.1-3.

Excess Reactants

In some cases the use of an excess reactant can shift the product distribution. For example, if we write a very oversimplified model for the production of isooctane via butane alkylation as

$$\text{Butene} + \text{Isobutane} \rightarrow \text{Isooctane}$$
$$\text{Butene} + \text{Isooctane} \rightarrow C_{12}$$

$$(6.1\text{-}4)$$

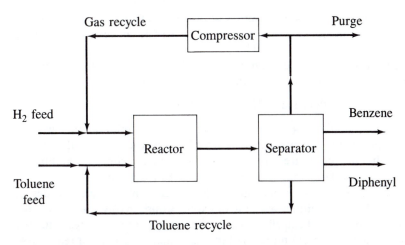

FIGURE 6.1-2
HDA recycle structure. [*From J. M. Douglas, AIChE J.*, **31**: *353 (1985).*]

and if the kinetics match the stoichiometry, then the use of an excess of isobutane leads to an improved selectivity to produce isooctane. The larger the excess, the greater the improvement in the selectivity, but the larger the cost to recover and recycle the isobutane. Thus, an optimum amount of excess must be determined from an economic analysis.

The use of an excess component can also be used to force another component to be close to complete conversion. For example, in the production of phosgene

$$CO + Cl_2 \rightarrow COCl_2 \tag{6.1-5}$$

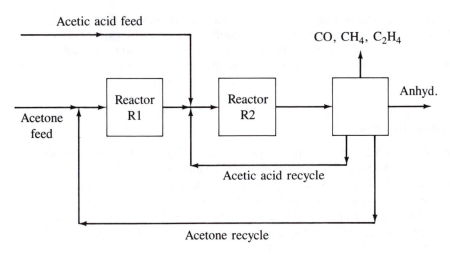

FIGURE 6.1-3
Acetic anhydride recycle.

which is an intermediate in the production of di-isocyanate, the product must be free of Cl_2. Thus, an excess of CO is used to force the Cl_2 conversion to be very high.

Similarly, the use of an excess can be used to shift the equilibrium conversion. For example, in the production of cyclohexane by the reaction

$$\text{Benzene} + 3\,H_2 \rightleftharpoons \text{Cyclohexane} \tag{6.1-6}$$

we want to obtain equilibrium conversions very close to unity so that we can obtain a high conversion of benzene and avoid a benzene-cyclohexane distillation separation (the boiling points are very close together). We can shift the equilibrium conversion to the right by using an excess of H_2 at the reactor inlet.

Thus, the molar ratio of reactants at the reactor inlet is often a design variable. Normally the optimum amount of excess to use involves an economic trade-off between some beneficial effect and the cost of recovering and recycling the excess. Unfortunately, there are no rules of thumb available to make a reasonable guess of the optimum amount of excess, and therefore we often need to carry out our economic analysis in terms of this additional design variable.

Heat Effects and Equilibrium Limitations

In general, the reactor flows need to be available in order to evaluate the reactor heat effects. Also, in many cases, equilibrium calculations are simplified if we have calculated some of the reactor flows. Thus, we defer our discussion of these topics until we have discussed procedures for estimating the process flow rates.

6.2 RECYCLE MATERIAL BALANCES

Our goal is to obtain a quick estimate of the recycle flows, rather than rigorous calculations. We have not specified any details of the separation system as yet, and therefore we assume that greater than 99 % recoveries of reactants are equivalent to 100 % recoveries. This approximation normally introduces only small errors in the stream flows.

Limiting Reactant

First we make a balance on the limiting reactant. For the HDA process (see Fig. 6.2-1), we let the flow of toluene entering the reactor be F_T. Then, for a conversion x the amount of toluene leaving the reactor will be $F_T(1 - x)$. For complete recoveries in the separation system, the flow leaving the reactor will be equal to the recycle flow. Now if we make a balance at the mixing point before the reactor, the sum of the fresh feed toluene F_{FT} plus the recycle toluene will be equal to the flow of toluene into the reactor, or

$$F_{FT} + F_T(1 - x) = F_T \tag{6.2-1}$$

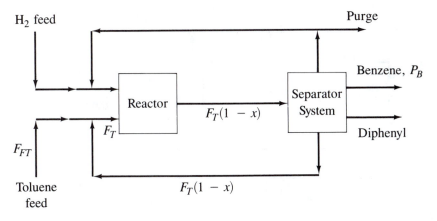

FIGURE 6.2-1
HDA, liquid recycle.

Thus, the feed to the reactor is

$$F_T = \frac{F_{FT}}{x} \tag{6.2-2}$$

This same material balance is always valid for the limiting reactant when there is complete recovery and recycle of the limiting reactant.

In some cases, some of the limiting reactant might leave the process in a gas recycle and purge stream (ammonia synthesis), or it may leave with the product (ethanol synthesis). If we consider a simplified version of the ethanol process, the reactions are

$$CH_2CH_2 + H_2O \rightleftharpoons CH_3CH_2OH$$
$$2CH_3CH_2OH \rightleftharpoons (CH_3CH_2)_2O + H_2O \tag{6.2-3}$$

We suppose that we want to produce 783 mol/hr of an EtOH–H_2O azeotrope that contains 85.4 mol % EtOH, from an ethylene feed stream containing 4% CH_4 and pure water. A recycle flowsheet is shown in Fig. 6.2-2 for the case where we recycle the diethylether and the water.

The overall material balances start with the production rate of the azeotrope

$$P_{azeo} = 783 \text{ mol/hr} \tag{6.2-4}$$

This contains

$$y_{azeo} P_{azeo} = P_{EtOH}$$

or $\qquad P_{EtOH} = 0.854(783) = 669 \text{ mol/hr EtOH} \tag{6.2-5}$

Then the amount of water in the product stream is

$$P_{H_2O} = P_{azeo} - P_{EtOH} = 783 - 669 = 114 \text{ mol/hr } H_2O \tag{6.2-6}$$

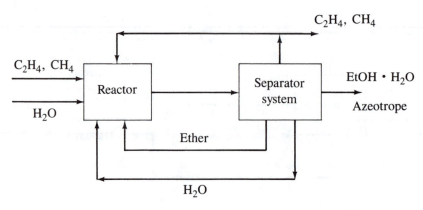

FIGURE 6.2-2
Ethanol synthesis.

Thus, from Eq. 6.2-3 and the results above, the required feed rate of water, which is the limiting reactant, is

$$F_{H_2O} = y_{azeo}P_{azeo} + (1 - y_{azeo})P_{azeo} = 669 + 114 = 783 \text{ mol/hr} \quad (6.2\text{-}7)$$

Suppose that we let the water leaving with the product be $F_{w,P} = 114$ and the fresh feed water required for the reaction be $F_{w,R}$. Now, referring to the schematic in Fig. 6.2-3 for water, we let the amount entering the reactor be F_w, the amount leaving the reactor be $F_w(1 - x)$, the amount leaving with the product be $F_{w,P}$, and the amount recycled be $F_w(1 - x) - F_{w,P}$. Then a balance at the mixing point before the reactor gives

$$(F_{w,P} + F_{w,R}) + [F_w(1 - x) - F_{w,P}] = F_w \quad (6.2\text{-}8)$$

so that
$$F_w = F_{w,R}/x \quad (6.2\text{-}9)$$

This result is identical to Eq. 6.2-2, except that instead of the fresh feed rate we substitute only the amount of material that enters into the reaction. A similar result is obtained for the case where the limiting reactant leaves with a gas purge stream.

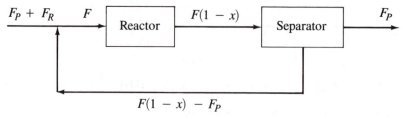

FIGURE 6.2-3
Ethanol synthesis.

Other Reactants

After we have estimated the flow of the limiting reactant, we use a specification of the molar ratio at the reactor inlet to calculate the recycle flows of the other components. For example, in the HDA process (see Fig. 6.2-4), the total amount of hydrogen entering the reactor is the sum of the fresh feed hydrogen $y_{FH}F_G$ and the recycle hydrogen $R_G y_{PH}$. We showed above that the amount of the limiting reactant, toluene, entering the reactor is F_{FT}/x. Thus, if we let the molar ratio of hydrogen to toluene at the reactor inlet be MR, we find that

$$y_{FH}F_G + y_{PH}R_G = MR\left(\frac{F_{FT}}{x}\right) \qquad (6.2\text{-}10)$$

or
$$R_G = \frac{P_B}{Sxy_{PH}}\left(\frac{MR}{x} - \frac{y_{FH}\left[1 - (1 - y_{PH})\dfrac{1-S}{2}\right]}{y_{FH} - y_{PH}}\right) \qquad (6.2\text{-}11)$$

Once we specify the design variables x, y_{PH}, and MR, we can solve this equation for the recycle gas flow R_G.

Design Heuristics

There are no rules of thumb available for selecting x for the case of complex reactions. Similarly, there are no rules of thumb for selecting the purge composition y_{PH} or the molar ratio MR. For the case of single reactions, a reasonable first guess of conversion is $x = 0.96$ or $x = 0.98x_{eq}$:

For single reactions, choose $x = 0.96$ or $x = 0.98x_{eq}$ as a first guess.   (6.2-12)

This rule of thumb is discussed in Exercise 3.5-8.

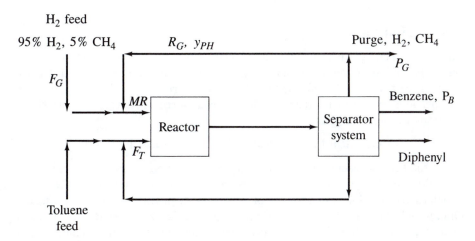

FIGURE 6.2-4
HDA, gas recycle.

By-products formed by Secondary Reversible Reactions

If we recycle a by-product formed by a secondary reversible reaction and let the component build up to its equilibrium level, such as the diphenyl in the HDA process

$$2\,\text{Benzene} \rightleftharpoons \text{Diphenyl} + H_2$$

or the diethylether in ethanol synthesis (Eq. 6.2-3), then we find the recycle flow by using the equilibrium relationship at the reactor exit. That is, at the reactor exit

$$K_{eq} = \frac{(\text{Diphenyl})(H_2)}{\text{Benzene}^2} \qquad (6.2\text{-}13)$$

However, the H_2 and benzene flows can be determined if we neglect the reversible reaction. Hence, we can substitute these values of (H_2) and (Benzene) into Eq. 6.2-13 and solve for the diphenyl flow.

6.3   REACTOR HEAT EFFECTS

We need to make a decision as to whether the reactor can be operated adiabatically, with direct heating or cooling, or whether a diluent or heat carrier is needed. In particular, if we need to introduce an extraneous component as a diluent or heat carrier, then our recycle material balances, and perhaps even our overall material balances, will have to be changed. Moreover, we need to make this decision before we consider the specification of the separation system because the decision to add an extraneous component normally will affect the design of the separation system.

To make the decision concerning the reactor heat effects, first we estimate the reactor head load and the adiabatic temperature change. These calculations provide some guidance as to the difficulty of dealing with the reactor heat effects. Similarly, we consider any temperature constraints imposed on the design problem.

Reactor Heat Load

For single reactions we know that all the fresh feed of the limiting reactant usually gets converted in the process (the per-pass conversion might be small so that there is a large recycle flow, but all the fresh feed is converted except for small losses in product and by-product streams or losses in a purge stream). Thus, for single reactions we can state that

$$\text{Reactor Heat Load} = \text{Heat of Reaction} \times \text{Fresh Feed Rate} \qquad (6.3\text{-}1)$$

where the heat of reaction is calculated at the reactor operating conditions.

For complex reactions, the extent of each reaction will depend on the design variables (conversion, molar ratio of reactants, temperature, and/or pressure). Once we select the design variables, we can calculate the extent of each reaction and then calculate the heat load corresponding to the side reactions. Hence, it is a simple matter to calculate reactor heat loads as a function of the design variables. Some guidelines for reactor heat loads are discussed later in this section.

Example 6.3-1   HDA process. From the overall material balances for the HDA process, we found that only small amounts of diphenyl were produced in the range of conversion where we obtained profitable operation. If we want to estimate the reactor heat load for a case where $x = 0.75$, $P_B = 265$, and $F_{FT} = 273$ mol/hr, we might neglect the second reaction and write

$$Q_R = \Delta H_R F_{FT} = (-21,530)(273) = -5.878 \times 10^6 \text{ Btu/hr} \qquad (6.3\text{-}1)$$

where ΔH_R is the heat of reaction at 1200°F and 500 psia and heat is liberated by the reaction.

Example 6.3-2. Acetone can be produced by the dehydrogenation of isopropanol

$$(CH_3)_2CHOH \rightarrow (CH_3)_2CO + H_2 \qquad (6.3\text{-}2)$$

If we desire to produce 51.3 mol/hr of acetone, then 51.3 mol/hr of IPA is required. The heat of reaction at 570°F and 1 atm is 25,800 Btu/mol, so the reactor heat load is

$$Q_R = 25,800(51.3) = 1.324 \times 10^6 \text{ Btu/hr} \qquad (6.3\text{-}3)$$

and heat is consumed by the endothermic reaction.

Adiabatic Temperature Change

Once we have determined the reactor heat load and the flow rate through the reactor as a function of the design variables, we can estimate the adiabatic temperature change from the expression

$$Q_R = FC_p(T_{R,\text{in}} - T_{R,\text{out}}) \qquad (6.3\text{-}4)$$

Example 6.3-3   HDA process. The flows and heat capacities of the reactor feed stream for a case where $x = 0.75$ and $y_{PH} = 0.4$ are given below.

Stream	Flow, mol/hr	C_p, Btu/(mol·°F)
Makeup gas	496	$0.95(7) + 0.05(10.1) = 7.16$
Recycle gas	3371	$0.4(7) + 0.6(10.1) = 8.86$
Toluene feed	273	48.7
Toluene recycle	91	48.7

Then, from Example 6.3-1 and Eq. 6.3-4 with $T_{R,\text{in}} = 1150$°F,

$$Q_R = -5.878 \times 10^6 = [(273 + 91)48.7 + 496(7.16) + 3371(8.86)](T_{R,\text{in}} - T_{R,\text{out}})$$

$$T_{R,\text{out}} = 1150 + 115 = 1265°F \qquad (6.3\text{-}5)$$

This value is below the constraint on the reactor exit temperature of 1300°F (see Example 4.1-1). Also, the calculation is not very sensitive to the C_p values or to the flows. The relatively small temperature rise is due to the large gas-recycle flow.

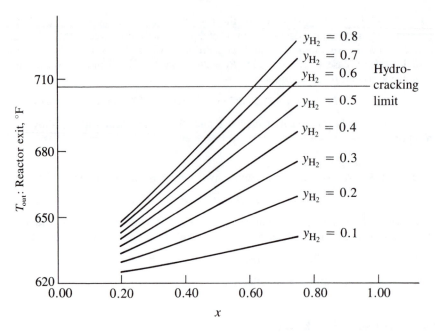

FIGURE 6.3-1

Reactor exit temperature. [*From J. M. Douglas, AIChE J.,* **31**: *353 (1985).*]; Note: $y_{H_2} = y_{PH}$

A plot of the reactor exit temperature as a function of the design variables is given in Fig. 6.3-1, and we see that in certain cases the constraint on the reactor exit temperature is violated.

Example 6.3-4   IPA process. If the feed stream to the acetone process described by Eq. 6.3-2 is an IPA–H_2O azeotrope (70 mol % IPA) and if we recycle an azeotropic mixture, then it is easy to show that 22.0 mol/hr of water enters with the feed. Also, for a conversion of $x = 0.96$, the recycle flow will be 2.1 mol/hr of IPA and 0.9 mol/hr of water. If the reactor inlet temperature is 572°F, then from Eqs. 6.3-1 and 6.3-4 the adiabatic temperature change is

$$Q_R = 1.324 \times 10^6 = [(51.3 + 22.0) + (2.1 + 0.9)](22.0)(572 - T_{R,out})$$

or $T_{R,out} = 572 - 788 = -216°F$ (6.3-6)

Clearly, this is an unreasonable result. Thus, instead of using an adiabatic reactor, we attempt to achieve isothermal operation by supplying the heat of the reaction to the process. In particular, we might attempt to pack the tubes of a heat exchanger with a catalyst.

Heuristic for Heat Loads

If adiabatic operation is not feasible, such as in the isopropanol example, then we attempt to use direct heating or cooling. However, in many cases there is a limit to the amount of heat-transfer surface area that we can fit into a reactor. To get some "feeling" for the magnitude of this area, we consider the case of a high-temperature gas-phase reaction, and we let $U = 20$ Btu/(hr·ft^2·°F) and $\Delta T = 50$°F. Then, for a heat load of 1×10^6 Btu/hr,

$$A = \frac{Q_R}{U \, \Delta T} = \frac{1 \times 10^6}{(20)(50)} = 1000 \text{ ft}^2 \qquad (6.3\text{-}7)$$

The maximum heat transfer area that fits into the shell of a floating-head heat exchanger is in the range of 6000 to 8000 ft^2. Thus, to use a single heat exchanger as a reactor, when we are attempting to remove or supply the heat of reaction by direct heating or cooling, the reactor heat loads are limited to the range of 6.0 to 8.0×10^6 Btu/hr.

Heat Carriers

The reactor heat load is often fixed by the fresh feed rate of the limiting reactant (if only a small amount of by-products is produced so that the secondary reactions are unimportant). The adiabatic temperature change depends primarily on the flow through the reactor. Hence, we can always moderate the temperature change through the reactor by increasing the flow rate.

    If we desire to moderate the temperature change, we prefer to do this by recycling more of a reactant or by recycling a product or by-product. However, where this is not possible, we may add an extraneous component. Of course, the introduction of an extraneous component may make the separation system more complex, and so we normally try to avoid this situation.

    In the HDA process (see Eq. 6.3-5), the methane in the gas-recycle stream (Approximately 60% methane) acts as a heat carrier. Thus, if we purified the hydrogen-recycle stream, the recycle flow would decrease and the reactor exit temperature would increase. If this exit temperature exceeded the constraint of 1300°F, we could no longer use an adiabatic reactor. Instead, we would have to cool the reactor, increase the hydrogen recycle flow, or introduce an extraneous compound as a heat carrier.

    A similar behavior is encountered in many oxidation reactions. If pure oxygen is used as a reactant, the adiabatic temperature rise is normally so large that problems would be encountered. However, if air is used as the reactant stream, the presence of nitrogen moderates the temperature change.

6.4   EQUILIBRIUM LIMITATIONS

In numerous industrial processes equilibrium limitations are important. We discussed a procedure for estimating the flows of reversible by-products when they are recycled and allowed to build up to their equilibrium levels at the reactor outlet. Our focus here is the primary reaction.

Equilibrium Conversion

We can use our previous procedure for calculating the process flows as a function of the design variables (conversion, molar ratio of reactants, etc.). Then we can substitute these flows into the equilibrium relationship to see whether the conversion we selected is above or below the equilibrium value. Of course, if it exceeds the equilibrium conversion, the result has no meaning.

In most cases, however, it is necessary to determine the exact value of the equilibrium conversion (as a function of the design variables), because this value appears in the kinetic model used to determine the reactor size. Our same general approach can be applied, but normally a trial-and-error solution is required. We illustrate this type of problem by considering the cyclohexane process.

Example 6.4-1   Cyclohexane production. Cyclohexane can be produced by the reaction

$$C_6H_6 + 3H_2 \rightleftharpoons C_6H_{12} \tag{6.4-1}$$

We consider a case where we desire to produce 100 mol/hr of C_6H_{12} with a 99.9% purity. A pure benzene feed stream is available, and the hydrogen makeup stream contains 97.5% H_2, 2.0% CH_4, and 0.5% N_2. A flowsheet for the recycle structure is shown in Fig. 6.4-1 for a case where we recycle some of the benzene (which is not necessarily the best flowsheet).

Solution
    Overall balances. Assume no losses. Then

$$\text{Production of } C_6H_{12}\text{: } P_c = 100 \tag{6.4-2}$$

$$\text{Benzene fresh feed: } F_{FB} = P_c = 100 \tag{6.4-3}$$

Assume we use a gas recycle and a purge stream. Let

$$F_E = \text{Excess } H_2 \text{ Fed to Process}$$

$$\text{Total } H_2 \text{ Feed} = 3P_c + F_E = 0.975 F_G \tag{6.4-4}$$

$$\text{Purge composition of } H_2\text{: } y_{PH} = \frac{F_E}{F_E + 0.025 F_G} \tag{6.4-5}$$

$$\text{Makeup gas rate: } F_G = 3P_c \frac{1 - y_{PH}}{0.975 - y_{PH}} \tag{6.4-6}$$

$$\text{Purge rate: } P_G = F_E + 0.025 F_G \tag{6.4-7}$$

    Recycle balances

$$\text{Benzene fed to reactor: } F_B = \frac{P_c}{x} \tag{6.4-8}$$

Let molar ratio of H_2 to benzene be MR. Then

$$\text{Recycle gas flow: } R_G = \frac{1}{y_{PH}} \left(\frac{MRP_c}{x} - 0.975 F_G \right) \tag{6.4-9}$$

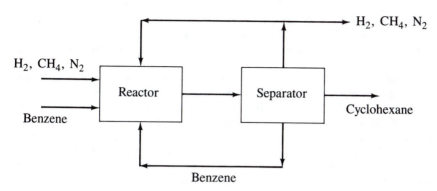

FIGURE 6.4-1
Cyclohexane flowsheet.

Reactor exit flows

$$\text{Cyclohexane} = P_c \tag{6.4-10}$$

$$\text{Benzene} = \frac{P_c(1-x)}{x} \tag{6.4-11}$$

$$H_2 = MR_B - 3P_c = \left(\frac{MR}{x} - 3\right)P_c \tag{6.4-12}$$

$$\text{Inerts} = 0.025F_G + (1-y_{PH})R_G = \frac{1-y_{PH}}{y_{PH}}\left(\frac{MR}{x} - 3\right)(P_c) \tag{6.4-13}$$

$$\text{Total flow} = P_c\left[\frac{1}{x} + \left(\frac{MR}{x} - 3\right)\frac{1}{y_{PH}}\right] \tag{6.4-14}$$

Equilibrium relationship

$$K_e = \frac{\bar{f}_C}{\bar{f}_B \bar{f}_H^3} = \frac{v_C y_C}{p_{\text{tot}}^3 v_B v_H^3 y_B y_H^3} \tag{6.4-15}$$

From the *Washington University Design Case Study No. 4*, p. 4–3, Part II,[*]

$$v_H = 1 \qquad v_C/v_B = 1.13 \tag{6.4-16}$$

Then

$$1.13 P_{\text{tot}}^3 K_e = \frac{x_e}{1-x_e}\left[\frac{1 + MR - 3x_e}{(MR - 3x_e)y_{PH}}\right] \tag{6.4-17}$$

Discussion. Since benzene and cyclohexane are very close boilers, we would like to avoid a benzene-cyclohexane distillation separation. This can be accomplished by operating the reactor at a sufficiently high conversion that we can leave any

---

[*] J. R. Fair, "Cyclohexane Manufacture," Washington University Design Case Study No. 4, edited by B. D. Smith, Washington University, St. Louis, Mo., 1967.

unconverted benzene as an impurity in the product. However, to obtain high benzene conversions, we must force the equilibrium conversion to be very close to unity. Equation 6.4-17 indicates the dependence of the equilibrium conversion on the design variables.

Economic trade-offs. From Eq. 6.4-17 we see that we can increase the equilibrium conversion by increasing the reactor pressure P_{tot}, by increasing the molar ratio of hydrogen to benzene at the reactor inlet MR, or by decreasing the reactor temperature (the reaction is exothermic). However, high reactor pressures correspond to a large feed compressor and more expensive equipment because of an increased wall thickness. Large molar ratios of hydrogen correspond to larger gas-recycle flows (see Eq. 6.4-9) and therefore a more expensive recycle compressor. Lower reactor temperatures correspond to larger reactors, because of the decreased reaction rate. Thus, an optimization analysis is required to determine the values of P_{tot}, x, T_{react}, MR, and y_{PH}.

Approximate model. If we expect that x_e will be close to unity in Eq. 6.4-17, then we can write

$$x_e \approx 1 - \frac{1}{1.13 K_e P_{tot}^3} \left[\frac{MR - 2}{(MR - 3) y_{PH}} \right]^3 \qquad (6.4\text{-}18)$$

which provides a simpler model to use in preliminary optimization studies.

Separator Reactors

If one of the products can be removed while the reaction is taking place, then an apparently equilibrium-limited reaction can be forced to go to complete conversion. Two examples of this type are discussed now.

Example 6.4-2   Acetone production. Acetone can be produced by the dehydrogenation of isopropanol

$$\text{Isopropanol} \rightleftharpoons \text{Acetone} + H_2 \qquad (6.4\text{-}19)$$

in the liquid phase as well as the gas phase. At 300°F the equilibrium conversion for the liquid-phase process is about $x_{eq} = 0.32$. However, by suspending the catalyst in a high-boiling solvent and operating the reactor at a temperature above the boiling point of acetone, both H_2 and acetone can be removed as a vapor from the reactor. Thus, the equilibrium conversion is shifted to the right. A series of three continuous stirred tank reactors, with a pump-around loop containing a heating system that supplies the endothermic heat to reaction, can be used for the process.

Example 6.4-3   Production of ethyl acrylate. Ethyl acrylate can be produced by the reaction

$$\text{Acrylic Acid} + \text{Ethanol} \rightleftharpoons \text{Ethyl Acrylate} + H_2O \qquad (6.4\text{-}20)$$

Both acrylic acid and ethyl acrylate are monomers, which tend to polymerize in the reboilers of distillation columns. We can eliminate a column required to purify and recycle acrylic acid from the process if we can force the equilibrium-limited reaction to completion, say, by removing the water. Hence, we use an excess of ethanol to shift the equilibrium to the right, and we carry out the reaction in the reboiler of a rectifying column. With this approach, the ethanol, water, and ethyl acrylate are taken overhead, and the acrylic acid conversion approaches unity.

Reversible Exothermic Reactions

There are several important industrial reactions that are reversible and exothermic. For example,

$$\text{Sulfuric acid process:} \quad SO_2 + \tfrac{1}{2}O_2 \rightleftharpoons SO_3 \qquad (6.4\text{-}21)$$

In ammonia synthesis,

$$\text{Water-gas shift:} \quad CO + H_2O \rightleftharpoons CO_2 + H_2 \qquad (6.4\text{-}22)$$

$$\text{Amonia synthesis:} \quad N_2 + 3H_2 \rightleftharpoons 2NH_3 \qquad (6.4\text{-}23)$$

High temperatures correspond to small reactor volumes, but for these exothermic reactions the equilibrium conversion decreases as the reactor temperature increases. Hence, these reactions are often carried out in a series of adiabatic beds with either intermediate heat exchangers to cool the gases or a bypass of cold feed to decrease the temperatures between the beds. With these procedures we obtain a compromise between high temperatures (small reactor volumes) and high equilibrium conversions.

Diluents

From the discussion above we have found that temperature, pressure, and molar ratio can all be used to shift the equilibrium conversion. However, in some cases an extraneous component (a diluent) is added which also causes a shift in the equilibrium conversion. For example, styrene can be produced by the reactions

$$\text{Ethylbenzene} \rightleftharpoons \text{Styrene} + H_2 \qquad (6.4\text{-}24)$$

$$\text{Ethylbenzene} \rightarrow \text{Benzene} + \text{Ethylene} \qquad (6.4\text{-}25)$$

$$\text{Ethylbenzene} \rightarrow \text{Toluene} + \text{Methane} \qquad (6.4\text{-}26)$$

where the reactions take place at about $1100°F$ and 20 psia. The addition of steam (or methane) at the reactor inlet lowers the partial pressure of styrene and H_2 and so decreases the reverse reaction rate in Eq. 6.4-24. The steam serves in part as a heat carrier to supply endothermic heat of reaction.

Steam is often used as a diluent because water-hydrocarbon mixtures are usually immiscible after condensation. Hence, the separation of water can be accomplished with a decanter (and usually a stripper to recover the hydrocarbons dissolved in the water, if the water is not recycled).

6.5   COMPRESSOR DESIGN AND COSTS

Whenever a gas-recycle stream is present, we will need a gas-recycle compressor. The design equation for the theoretical horsepower (hp) for a centrifugal gas compressor is

$$hp = \left(\frac{3.03 \times 10^{-5}}{\gamma}\right) P_{in} Q_{in} \left[\left(\frac{P_{out}}{P_{in}}\right)^{\gamma} - 1\right] \qquad (6.5\text{-}1)$$

TABLE 6.5-1

Values of γ

Monotonic gases	0.40
Diatomic gases	0.29
More complex gases (CO_2, CH_4)	0.23
Other gases	$\dfrac{R}{C_p}$

Source: Taken from J. Happel and D. G. Jordan, *Chemical Process Economics*, 2d ed., Dekker, New York, 1975, p. 454.

where $P_{in} = lbf/ft^2$, $Q_{in} = ft^3/min$, and $\gamma = (C_p/C_v - 1)/(C_p/C_v)$. The exit temperature from a compression stage is

$$\frac{T_{out}}{T_{in}} = \left(\frac{P_{out}}{P_{in}}\right)^{\gamma} \tag{6.5-2}$$

(where the temperatures and pressures must be in absolute units). Values of γ that can be used for first estimates of designs are given in Table 6.5-1.

Efficiency

For first designs, we assume a compressor efficiency of 90% to account for fluid friction in suction and discharge valves, ports, friction of moving metal surfaces, fluid turbulence, etc. Also we assume a driver efficiency of 90% to account for the conversion of the input energy to shaft work.

Spares

Compressors are so expensive that spares are seldom provided for centrifugal units (although reciprocating compressors may have spares because of a lower service factor). In some instances two compressors may be installed, with each providing 60% of the load, so that partial operation of the plant can be maintained in case one compressor fails and additional flexibility is available to respond to changes in process flows.

Multistage Compressors

It is common practice to use multistage compressors. The gas is cooled to cooling-water temperatures (100°F) between stages. Also knockout drums are installed between stages to remove any condensate. It is essential to ensure that no condensation takes place inside the compressor, since liquid droplets condensing on the vanes which are rotating at very high speeds might cause an imbalance and wild vibrations.

For a three-stage compressor with intercooling, the work required is

$$\text{Work} = MRT_{\text{in}}\left[\left(\frac{P_2}{P_1}\right)^{\gamma} + \left(\frac{P_3}{P_2}\right)^{\gamma} + \left(\frac{P_4}{P_3}\right)^{\gamma} - 1\right] \tag{6.5-3}$$

The intermediate pressures that minimize the work are determined from

$$\frac{\partial \text{Work}}{\partial P_2} = \frac{\partial \text{Work}}{\partial P_3} = 0 \tag{6.5-4}$$

which leads to the results

$$\frac{P_2}{P_1} = \frac{P_3}{P_2} = \frac{P_4}{P_3} \tag{6.5-5}$$

Thus, we obtain another design heuristic:

The compression ratios for each stage in a gas compressor should be equal.
$$\tag{6.5-6}$$

Installed Cost

The brake horsepower bph is obtained by introducing the compressor efficiency into Eq. 6.5-1:

$$\text{bhp} = \frac{\text{hp}}{0.8} \tag{6.5-7}$$

Then, Guthrie's correlation (see item 4 in Appendix E.2) or some equivalent correlation can be used to calculate the installed cost for various types of compressors:

$$\text{Installed Cost} = \left(\frac{\text{M \& S}}{280}\right)(517.5)(\text{bhp})^{0.82}(2.11 + F_c) \tag{6.5-8}$$

where F_c is given in Appendix E.2 and M & S = Marshall and Swift inflation index (which is published each month in *Chemical Engineering*).

Note: Guthrie's correlations and capital charge factors are discussed in detail in Chap. 2, and if we let CCF $= \frac{1}{3}$ (yr^{-1}) our cost model Eq. 2.5-20, will include a factor of 1 yr^{-1} before the installed costs of the onsite equipment.

Operating Cost

By dividing the brake horsepower by the driver efficiency, we can calculate the utility requirement. Then from the utility cost and using 8150 hr/yr, we can calculate the operating cost.

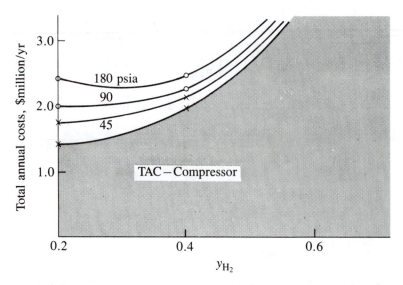

FIGURE 6.5-1
Sensitivity of recycle compressor ΔP.

Sensitivity

At the preliminary stages of a design, we do not have a complete flowsheet. Thus, we cannot obtain a good estimate for the compression ratio P_{out}/P_{in} for recycle streams. Our approach is simply to make a guess and then to evaluate the sensitivity of that guess. In most cases, the results are fairly insensitive. An example for the HDA process with $x = 0.75$ and $y_{PH} = 0.4$ is given in Fig. 6.5-1.

6.6   REACTOR DESIGN

At the very early stages in a new design, a kinetic model normally is not available. Thus, we base our material balance calculations on a correlation of the product distribution. Also, we assume that we will use the same type of reactor in the plant that the chemist used in the laboratory, and we often base a first estimate of the reactor size on the reaction half-life measured by the chemist. For adiabatic reactors we might base the design on an isothermal temperature which is the average of the inlet and outlet temperatures or an average of the inlet and outlet rate constants.

This type of a kinetic analysis is very crude, but in most cases the reactor cost is not nearly as important as the product distribution costs. Thus, again, we merely look at sensitivities until we can justify additional work and a kinetic model becomes available. We estimate the costs of plug flow reactors in the same way as we do pressure vessels (see Appendix E.2); See p. 526 for an example.

TABLE 6.6-1
Design guidelines for reactors

I. Single irreversible reactions (not autocatalytic)
    A. Isothermal—always use a plug flow reactor.
    B. Adiabatic.
        1. Plug flow if the reaction rate monotonically decreases with conversion
        2. CSTR operating at the maximum reaction rate followed by a plug flow section
II. Single reversible reactions—adiabatic
    A. Maximum temperature if endothermic
    B. A series of adiabatic beds with a decreasing temperature profile if exothermic
III. Parallel reactions—composition effects
    A. For $A \rightarrow R$ (desired) and $A \rightarrow S$ (waste), where the ratio of the reaction rates is $r_R/r_S = (k_1/k_2)C_A^{a_1 - a_2}$
        1. If $a_1 > a_2$, keep C_A high.
            a. Use batch or plug flow.
            b. High pressure, eliminate inerts.
            c. Avoid recycle of products.
            d. Can use a small reactor.
        2. If $a_1 < a_2$, keep C_A low.
            a. Use a CSTR with a high conversion.
            b. Large recycle of products.
            c. Low pressure, add inerts.
            d. Need a large reactor.
    B. For $A + B \rightarrow R$ (desired) and $A + B \rightarrow S$ (waste), where the ratio of the rates is $r_R/r_S = (k_1/k_2)C_A^{a_1 - a_2}C_B^{b_1 - b_2}$
        1. If $a_1 > a_2$ and $b_1 > b_2$, both C_A and C_B high.
        2. If $a_1 < a_2$ and $b_1 > b_2$, then C_A low, C_B high.
        3. If $a_1 > a_2$ and $b_1 < b_2$, then C_A high, C_B low.
        4. If $a_1 < a_2$ and $b_1 < b_2$, both C_A and C_B low.
        5. See Fig. 6.6-1 for various reactor configurations.
IV. Consecutive reactions—composition effects
    A. $A \rightarrow R$ (desired); $R \rightarrow S$ (waste)—minimize the mixing of streams with different compositions.
V. Parallel reactions—temperature effects $r_R/r_S = (k_1/k_2)f(C_A, C_B)$
    A. If $E_1 > E_2$, use a high temperature.
    B. If $E_1 < E_2$, use an increasing temperature profile.
VI. Consecutive reactions—temperature effects $A \overset{k_1}{\rightarrow} R \overset{k_2}{\rightarrow} S$
    A. If $E_1 > E_2$, use a decreasing temperature profile—not very sensitive.
    B. If $E_1 < E_2$, use a low temperature.

Reactor Configuration

Since the product distribution can depend on the reactor configuration, we need to determine the best configuration. A set of design guidelines has been published by Levenspiel.* For the sake of completeness some of these guidelines are reviewed in Table 6.6-1. As this table indicates in some cases we obtain complex reactor configurations; see Fig. 6.6-1.

---

* O. Levenspiel, *Chemical Reaction Engineering*, 2d ed., Wiley, New York, 1972, chaps. 7 and 8.

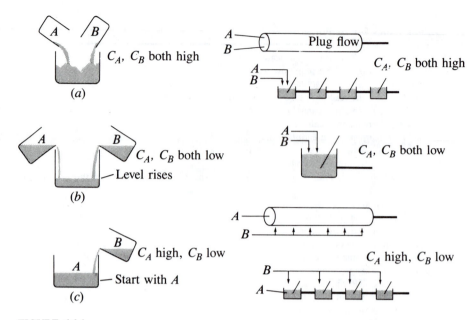

FIGURE 6.6-1
Parallel reactions. (*From O. Levenspiel, Chemical Reaction Engineering, 2d ed., Wiley, New York, 1972, chaps. 7 and 8.*)

6.7 RECYCLE ECONOMIC EVALUATION

Our economic analysis for the input-output structure considered only the stream costs, i.e., products plus by-products minus raw-material costs. The results for the HDA process indicated that the most profitable operation was obtained when the conversion was zero (we made no diphenyl by-product) and when the purge composition of hydrogen was zero (we both purged and recycled pure methane).

However, when we consider the recycle, Eq. 6.2-2 shows that we need an infinite recycle flow of toluene when the conversion is equal to zero, and Eq. 6.2-11 shows that an infinite recycle flow of gas is required when the hydrogen purge composition is equal to zero. Thus, if we subtract the annualized reactor cost and the annualized compressor costs, both capital and power, from the economic potential, then we expect to find both an optimum conversion and an optimum purge composition. Figure 6.7-1 shows this result for the HDA process. (We would also subtract the annualized capital and operating cost of a feed compressor, if one was required and we did not include it in the level-2 calculation.)

The values for the optimum shown in Fig. 6.7-1 are not the true optimum values because we have not included any separations or heating and cooling costs. Hence, we expect that the true optimum economic potential will be smaller and will be shifted to lower recycle flows. However, we can see that our simple analysis is

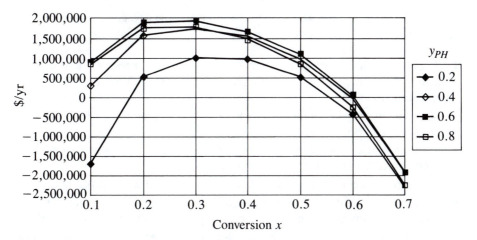

FIGURE 6.7-1
Economic potential—level 3; MR = 5.0 to satisfy the coking constraint

rapidly restricting the range of the design variables where we obtain a positive profit. Thus, our calculations for the separation system and the heat-exchanger network are simplified.

6.8 SUMMARY, EXERCISES, AND NOMENCLATURE

Summary

The decisions that need to be made to fix the recycle structure of the flowsheet are as follows:

1. How many reactors are required? Should some components be separated between the reactors?
2. How many recycle streams are required?
3. Do we want to use an excess of one reactant at the reactor inlet?
4. Is a gas-recycle compressor required? How does it affect the costs?
5. Should the reactor be operated adiabatically, with direct heating or cooling, or is a diluent or heat carrier needed?
6. Do we want to shift the equilibrium conversion? How?
7. How do the reactor costs affect the economic potential?

The design guidelines we use to make some of these decisions for *first* designs are as follows:

1. If reactions take place at different temperatures and pressures and/or they require different catalysts, then a separate reactor system is required for each operating condition.
2. Components recycled to the same reactor that have neighboring boiling points should be recycled in the same stream.
3. A gas-recycle compressor is required if the recycled components boil at a temperature lower than that of propylene.
4. If an excess reactant is desirable, there is an optimum amount of the excess.
5. If the reactor temperature, pressure, and/or molar ratio are changed to shift the equilibrium conversion, there must be an optimum value of these variables.
6. For endothermic processes with a heat load of less than 6 to 8×10^6 Btu/hr, we use an isothermal reactor with direct heating. For larger heat loads we may add a diluent or heat carrier.
7. For exothermic reactions we use an adiabatic reactor if the adiabatic temperature rise is less than 10 to 15% of the inlet temperature. If the adiabatic temperature rise exceeds this value, we use direct cooling if the reactor heat load is less than 6 to 8×10^6 Btu/hr. Otherwise, we introduce a diluent or a heat carrier.
8. For single reactions we choose a conversion of 0.96 or 0.98 of the equilibrium conversion.
9. The most expensive reactant (or the heaviest reactant) is usually the limiting reactant.
10. If the equilibrium constant of a reversible by-product is small, recycle the reversible by-product.
11. Several design guidelines for reactors are given in Table 6.6-1.
12. The recycle flow of the limiting reactant is given by $F = F_R(1 - x)/x$, where F_R is the amount of the limiting reactant needed for the reaction and x is the conversion.
13. The recycle flow of other components can be determined by specifying the molar ratio(s) at the reactor inlet.

Exercises

6.8-1. Develop the recycle structure for the HDA process with diphenyl recycled (see Eq. 6.2-13). Plot the economic potential versus the design variables, assuming that $K_{eq} = 0.2396$. (Also see Appendix B.)

6.8-2. Develop the recycle structure for an acetic anhydride process (see Eq. 6.1-2 and Exercise 5.4-3). If the acetone pyrolysis reactor costs are calculated as a furnace cost, and if the anhydride reactor cost is negligible, plot the economic potential versus the design variables. (Assume $\Delta H_{R,\,acet} = 34,700$ Btu/mol, $\Delta H_{R,\,ket} = -27,000$ Btu/mol, and $\Delta H_{R,\,anhyd} = 20,700$ Btu/mol.)

6.8-3. Develop the recycle structure for the gas-phase process that produces acetone from isopropanol (Exercise 5.4-4). Assume that $\Delta H_R = 25,800$ Btu/mol and that the reactor cost can be estimated as a heat exchanger with $U = 10$ Btu(hr·ft²·°F). The heat required for the reaction is supplied from a Dowtherm furnace with Dowtherm at 600°F. Plot the economic potential versus the significant design variables.

6.8-4. Develop the recycle structure for the ethanol synthesis problem (see Exercise 5.4-5). Assume that $\Delta H_{R,\text{EtOH}} = -19,440$ Btu/mol and $\Delta H_{\text{DEE}} = -5108$ Btu/mol; that the forward reaction rate constant is given by

$$k_1 = 1.4 \times 10^9 \exp\{-53,700/[RT(°R)]\} \text{ hr}^{-1}$$

and is first-order in water; and that

$$K_{\text{eq}} = (1.679 \times 10^{-7}) \exp[10, 119/T(°R)].$$

Plot the economic potential versus the significant design variables.

6.8-5. Develop the recycle structure for the styrene process (see Exercise 5.4-6). Assume that $\Delta H_{\text{styr}} = 50,530$ Btu/mol, $\Delta H_{R,\text{benz}} = 45,370$ Btu/mol, and $\Delta H_{R,\text{tol}} = -23,380$ Btu/mol; that the reaction rate constant for the primary reaction is given by the expression $k_1 = (383) \exp\{-20,440/[RT(°R)]\} \text{ hr}^{-1}$; and that $K_{\text{eq}} = 7.734 \exp[-27,170/T(°R)]$. Plot the economic potential versus the significant design variables.

6.8-6. Develop the recycle structure for the cyclohexane process (see Exercise 5.4-7). Assume that $\Delta H_R = 93,200$ Btu/mol, that $K_{\text{eq}} = (2.67 \times 10^{-21}) \exp[4.72 \times 10^4/T(°R)]$, and that the forward reaction rate constant is given by $(7.88 \times 10^5) \exp\{-14,400/[RT(°R)]\}$, where the forward reaction is first-order in benzene. Plot the economic potential versus the significant design variables.

6.8-7. Develop the recycle structure for the ethylene process (see Exercise 5.4-8). Assume that $\Delta H_1 = 58,650$ Btu/mol, that $\Delta H_2 = 15,320$ Btu/mol, and that the reactor cost can be estimated as a pyrolysis furnace. Plot the economic potential versus the design variables.

6.8-8. Develop the recycle structure for the butadiene sulfone process (see Exercise 5.4-9). Assume that $\Delta H_R = -48,000$ Btu/mol, $K_{\text{eq}} = (6.846 \times 10^{-11}) \exp[+36,940/RT(°R)]$, $k_{-1} = (8.172 \times 10^{15}) \exp[-52,200/RT(°R)]$, and $k_1 = k_{-1}K_{\text{eq}}[\text{mol}/(\text{ft}^3 \cdot \text{hr})]$; that the reaction rate corresponds to the stoichiometry; and that we use a CSTR for the reactor. Consider variable density effects and assume that the annualized, installed reactor cost is given by $3150V_R^{0.558}$ [$/(ft³·yr)]. Plot the economic potential versus the significant design variables.

6.8-9. Develop the recycle structure for the butane alkylation process (see Exercise 5.4-10). Assume that $\Delta H_1 = -27,440$ Btu/mol, $\Delta H_2 = -25,180$ Btu/mol, $k_1 = (9.56 \times 10^{13}) \exp\{-28,000/[RT(°R)]\} \text{ hr}^{-1}$, and $k_2 = (2.439 \times 10^{17}) \exp\{-35,000/[RT(°R)]\} \text{ hr}^{-1}$. Use a CSTR with the cost correlation given in Exercise 6.8-8.

Nomenclature

$A, B, R, S,$	Reactive and product components
a_i, b_i	Order of reaction for component i
A	Heat-exchanger area (ft²)
bhp	Brake horsepower
C_i	Concentration of component i

C_p	Heat capacity [Btu/(mol·°F)]
E_i	Activation energy for component i
$f(C_A, C_B)$	Function of composition
f_i	Fugacity of component i
F	Reactor feed rate (mol/hr)
F_c	Correction factor for a gas compressor
F_E	Feed rate in excess of reaction requirements (mol/hr)
F_{FT}	Fresh feed rate of toluene (mol/hr)
F_G	Makeup gas rate of H_2 (mol/hr)
F_i	Flow rate of component i (mol/hr)
F_T	Reactor feed rate of toluene (mol/hr)
hp	Horsepower
k_i	Reaction rate constant
K_{eq}	Equilibrium constant
M	Molecular weight
MR	Molar ratio
M & S	Marshall and Swift index for inflation
P_i	Production rate of component i (mol/hr)
P_{in}, P_{out}	Inlet and outlet pressures for a gas compressor
P_j	Pressure at stages of a gas compressor
P_{tot}	Total pressure of reactor
Q_{in}	Volumetric flow rate (ft^3/min)
Q_R	Reactor heat load (Btu/hr)
r_i	Reaction rate
R	Gas constant
R_G	Recycle gas flow (mol/hr)
T_{in}, T_{out}	Inlet and outlet temperatures from a gas compressor (°R)
$T_{R,i}$	Reactor temperature (°F)
U	Overall heat-transfer coefficient [Btu/(hr·ft^2·°F)]
x	Conversion
y_{FH}	Feed mole fraction of H_2
y_i	Mole fraction of component i
y_{PH}	Purge composition of H_2
ΔH_R	Heat of reaction at reaction temperature and pressure (Btu/mol)

Greek symbols

v_i	Fugacity coefficient of component i
γ	$\dfrac{C_p/C_v - 1}{C_p/C_v}$

SEPARATION SYSTEM

Here we consider only the synthesis of a separation system to recover gaseous and liquid components. Our discussion is broken down into three separate parts: general structure, vapor recovery system, and liquid separation system. Also keep in mind that we need to determine the best separation system as a function of the design variables, i.e., the range of the design variables in Fig. 6.7-1 where we obtain profitable operation. Thus, our previous economic studies help to simplify the computational effort.

7.1 GENERAL STRUCTURE OF THE SEPARATION SYSTEM

To determine the general structure of the separation system, we first determine the phase of the reactor effluent stream (see Fig. 7.1-1). For vapor-liquid processes, there are only three possibilities:

1. If the reactor effluent is a liquid, we assume that we only need a liquid separation system (see Fig. 7.1-2). This system might include distillation columns, extraction units, azeotropic distillation, etc., but normally there will not be any gas absorber, gas adsorption units, etc.
2. If the reactor effluent is a two-phase mixture, we can use the reactor as a phase splitter (or put a flash drum after the reactor). We send the liquids to a liquid separation system. If the reactor is operating above cooling-water temperature, we usually cool the reactor vapor stream to $100°F$ and phase-split this stream

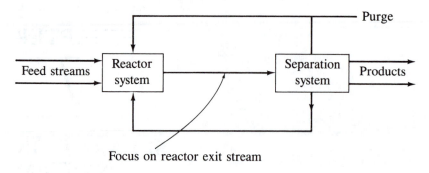

Focus on reactor exit stream

FIGURE 7.1-1
Phase of the reactor effluent stream.

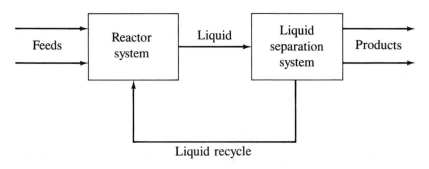

FIGURE 7.1-2
Reactor exit is liquid. [*From J. M. Douglas, AIChE J.,* **31**: *353 (1985).*]

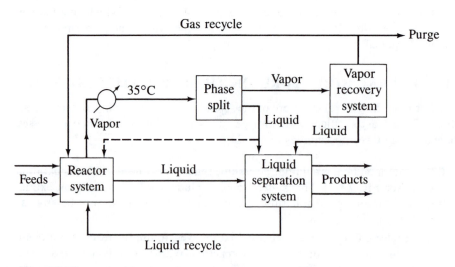

FIGURE 7.1-3
Reactor exit is vapor and liquid. [*From J. M. Douglas, AIChE J.,* **31**: *353 (1985).*]

(see Fig. 7.1-3). If the low-temperature flash liquid we obtain contains mostly reactants (and no product components that are formed as intermediates in a consecutive reaction scheme), then we recycle them to the reactor (we have the equivalent of a reflux condenser). However, if the low-temperature flash liquid contains mostly products, we send this stream to the liquid recovery system. The low-temperature flash vapor is usually sent to a vapor recovery system. But, if the reactor effluent stream contains only a small amount of vapor, we often send the reactor effluent directly to a liquid separation system (i.e., distillation train).

3. If the reactor effluent is all vapor, we cool the stream to 100°F (cooling-water temperature) and we attempt to achieve a phase split (see Fig. 7.1-4) or to completely condense this stream. The condensed liquid is sent to a liquid recovery system, and the vapor is sent to a vapor recovery system.

   If a phase split is not obtained, we see whether we can pressurize the reactor system so that a phase split will be obtained. (We see whether a high pressure can be obtained by using only pumps on liquid feed streams, and we check to see that the pressure does not affect the product distribution.) If a phase split is still not obtained, then we consider the possibility of using both high pressure and a refrigerated partial condenser. In case no phase split can be obtained without refrigeration, we also consider the possibility of sending the reactor effluent stream directly to a vapor recovery system.

   We need to ensure that the same structure is obtained for the complete range of design variables under consideration. These rules are based on the heuristic that phase splits are the cheapest method of separation and the assumption that some type of distillation separation is possible.

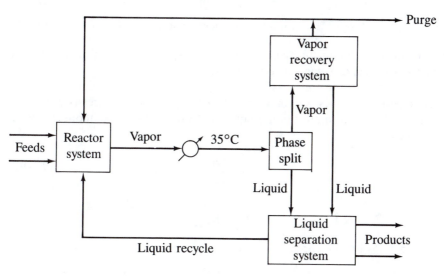

FIGURE 7.1-4
Reactor exit is vapor. [*From J. M. Douglas, AIChE J.,* **31**: *353 (1985).*]

Approximate Flash Calculations

To determine the phase of the reactor effluent, in some cases, we can use a sharp-split approximation procedure to avoid the trial-and-error solutions associated with flash calculations. The flash equations can be written as

Overall balance: $$F = V + L \tag{7.1-1}$$

Component balance: $$Fz_i = Vy_i + Lx_i \tag{7.1-2}$$

Equilibrium: $$y_i = K_i x_i \tag{7.1-3}$$

When we combine these expressions, we obtain

$$y_i = \frac{z_i}{V/F + (1 - V/F)(1/K_i)} \tag{7.1-4}$$

or

$$x_i = \frac{z_i}{(K_i - 1)V/F + 1} \tag{7.1-5}$$

If $K_i \gg 1$ in Eq. 7.1-4, we see that

$$Vy_i \approx Fz_i \tag{7.1-6}$$

and if $K_i \ll 1$ in Eq. 7.1-5, we find that

$$Lx_i \approx Fz_i \tag{7.1-7}$$

Thus, as a first estimate of the vapor and liquid flow rates, we can write

$$V = \sum f_i \quad \text{for all components where } K_i > 10 \tag{7.1-8}$$

$$L = \sum f_j \quad \text{for all components where } K_j < 0.1 \tag{7.1-9}$$

These expressions are equivalent to a perfect split, providing there are no components with a K_i value between 0.1 and 10.

The perfect split expressions ignore the vapor-liquid equilibrium. However, we can superimpose these equilibrium relationships on our expressions for the flows. Thus, the liquid composition in equilibrium with a vapor component having a mole fraction

$$y_i = \frac{f_i}{V} = \frac{f_i}{\sum f_i} \tag{7.1-10}$$

is

$$x_i = \frac{y_i}{K_i} = \frac{f_i}{K_i \sum f_i} \tag{7.1-11}$$

The liquid flow of this component is then

$$l_i = Lx_i = \frac{f_i \sum f_j}{K_i \sum f_i} \tag{7.1-12}$$

TABLE 7.1-1
HDA flash

Component	f_i	K_i	Approximate v_i	Approximate l_i	Exact v_i	Exact l_i
H_2	1549	99.07	1547	2	1548	1
CH_4	2323	20.00	2312	11	2313	10

Component	f_j	K_j	Approximate v_j	Approximate l_j	Exact v_j	Exact l_j
Benzene	265	0.01040	29.6	235.4	28.2	236.8
Toluene	91	0.00363	3.6	87.4	3.6	87.4
Diphenyl	4	0.00008	0	4	0	4.0

Now, we can go back and adjust the vapor flow for this loss:

$$v_i = f_i - l_i = f_i \left(1 - \frac{\sum f_j}{K_i \sum f_i} \right) \qquad (7.1\text{-}13)$$

The corresponding expressions for components that are predominantly in the liquid phase are

$$v_j = \frac{K_j f_j \sum f_i}{\sum f_j} \qquad (7.1\text{-}14)$$

and

$$l_j = f_j \left(1 - \frac{K_j \sum f_i}{\sum f_j} \right) \qquad (7.1\text{-}15)$$

Table 7.1-1 compares the approximate and exact solutions for the HDA process. We see that the approximate solution is satisfactory for preliminary designs. However, the results are valid only if there are no components having K values in the range from 0.1 to 10.

AN ALTERNATE APPROXIMATE PROCEDURE FOR FLASH CALCULATIONS.
Another shortcut procedure for flash calculations was published by King.* If we again consider Eqs. 7.1-1 through 7.1-3, we can write

$$f_i = v_i + \frac{L}{K_i V} v_i \qquad (7.1\text{-}16)$$

where

$$f_i = F z_i \qquad v_i = V y_i \qquad (7.1\text{-}17)$$

---

* C. J. King, *Separation Processes*, 2d ed., McGraw-Hill, New York, 1980.

By rearranging Eq. 7.1-16 we obtain

$$\frac{f_i}{v_i} - 1 = \frac{L}{K_i V} \qquad (7.1\text{-}18)$$

Now, if we divide Eq. 7.1-18 by a similar expression for component j, we obtain

$$\frac{f_i/v_i - 1}{f_j/v_j - 1} = \frac{K_j}{K_i} = \frac{1}{\alpha_{ij}} \qquad (7.1\text{-}19)$$

If we specify the fractional recovery v_j/f_j for one component, we can use Eq. 7.1-14 to calculate the fractional recovery for every other component.

Even though this analysis is rigorous for constant-α systems, the results for a specified fractional recovery of one component normally will not correspond to a flash temperature of 100°F and the flash pressure. Thus, some iteration might be required.

Nonideal Mixtures

There are no shortcut calculation procedures for nonideal mixtures. However, all CAD packages, i.e., FLOWTRAN, PROCESS, DESIGN 2000, ASPEN, etc., will handle these problems.

7.2 VAPOR RECOVERY SYSTEM

When we attempt to synthesize a vapor recovery system, we need to make two decisions:

1. What is the best location?
2. What type of vapor recovery system is cheapest?

Location of Vapor Recovery System

There are four choices for the location of the vapor recovery system:

1. The purge stream
2. The gas-recycle stream
3. The flash vapor stream
4. None

The rules we use to make this decision are (see Fig. 7.2-1) as follows:

1. Place the vapor recovery system on the purge stream if significant amounts of valuable materials are being lost in the purge. The reason for this heuristic is that the purge stream normally has the smallest flow rate.

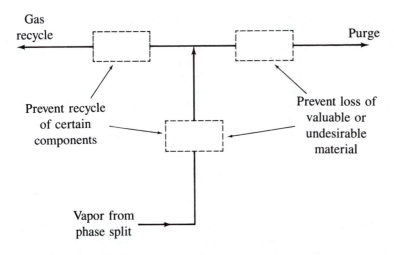

FIGURE 7.2-1
Vapor recovery system location.

2. Place the vapor recovery system on the gas-recycle stream if materials that are deleterious to the reactor operation (catalyst poisoning, etc.) are present in this stream or if recycling of some components degrades the product distribution. The gas-recycle stream normally has the second smallest flow rate.
3. Place the vapor recovery system on the flash vapor stream if both items 1 and 2 are valid, i.e., the flow rate is higher, but we accomplish two objectives.
4. Do not use a vapor recovery system if neither item 1 nor item 2 are important.

Adjust the Material Balances?

Note that unless item 3 is chosen, our simple material balance equations will not be valid; i.e., some materials that we assumed were recovered as liquids will be lost in the purge stream or recycled with the gas stream (which will change the compressor size). However, in many cases the errors introduced are small, so that our previous approximations still provide good estimates. We expect to develop rigorous material balances if we proceed with a final design, and therefore we use our engineering judgment to see whether corrections need to be made at this point.

Example 7.2-1   HDA process. Do we need a vapor recovery system for the HDA process?

Solution. For a conversion $x = 0.75$ and a purge composition $y_{PH} = 0.4$, the vapor flows from the phase splitter are given in Table 7.1-1. The purge and recycle flows for this case were 496 and 3371 mol/hr, respectively. Hence, we can estimate that the benzene and toluene flows lost in the purge are 3.79 and 0.46 mol/hr, respectively. On an annual basis and by neglecting the fuel values of these components, this loss represents 0.304×10^6/yr. This value is small compared to our economic potential, and so we decide not to include a vapor recovery system at this point in the design development. We might well reconsider this decision after we have determined whether the HDA process is profitable; i.e., if we decide to abandon the project, we want to minimize the engineering effort that we invest.

Since the reaction we are considering is homogeneous, no components in the gas-recycle stream can cause catalyst deactivation. However, there is a significant amount of benzene in the flash vapor stream (12% of the benzene flow); see Table 7.1-1. And most of this benzene (29.6 mol/hr − 4.3 mol/hr lost in the purge) will be recycled to the reactor. The benzene is formed as the intermediate in a consecutive reaction scheme

$$\text{Toluene} + H_2 \rightarrow \text{Benzene} + CH_4$$
$$2\text{Benzene} \rightleftharpoons \text{Diphenyl} + H_2$$

(7.2-1)

Therefore, we would expect that some (or most) of any benzene that is recycled to the reactor will be converted to diphenyl. Unfortunately, the selectivity data we are using (see Example 4.1-1) do not include the effect of any benzene feed to the reactor, so we cannot estimate the amount of benzene lost to diphenyl. This difficulty could be overcome if we had a kinetic model available.

With the available data, however, we would need to put a vapor recovery system either on the gas-recycle stream (to prevent the loss of some of the benzene by reaction to diphenyl) or on the flash vapor stream (to prevent loss of benzene both in the purge stream and by reaction). Another alternative could be to recycle the diphenyl to extinction, rather than recovering and removing the diphenyl. With this alternative, we would avoid the selectivity loss of toluene to diphenyl altogether, and we can tolerate the presence of benzene in the gas-recycle stream.

Some of the benzene in the gas-recycle stream can be recovered in the compressor knockout drums before the gas-recycle stream enters the reactor. The flash vapor stream is a saturated vapor, so that as we raise the pressure of this stream in each of the three stages of the gas-recycle compressor, some of the benzene will condense. Normally, we cool the exit from each compressor stage to 100°F with cooling water, and then we include a knockout drum (i.e., a flash drum) to collect the condensible materials. We can send this condensed benzene to the liquid separation system.

Rather than attempt to evaluate all these various alternatives at this time, we merely make some decision and continue to develop a base case. We list all the other alternatives as items that need to be considered after we have estimated the profitability of the process and have a better understanding of the allocation of the costs. Of course, we minimize our effort by guessing that most of the benzene in the gas-recycle stream will be recovered in the knockout drums associated with the compressor or will not be converted to diphenyl if it is recycled to the reactor. However, it is essential to check this assumption later.

PROCESS FLOWS. If we do not recover the benzene and toluene from the flash vapor streams, our assumptions concerning the overall and recycle material balances are no longer valid. In particular, the amount of benzene leaving in the flash liquid stream is not adequate to meet the plant production rate, although the benzene recovered in the compressor knockout drums will decrease the magnitude of the error.

We could go back and revise all our material balance calculations. However, it will be necessary to revise them again after we have specified a liquid separation system. Since the changes we introduced in the process flows are not too large, we decide to continue with the analysis. Of course, we are starting to accumulate errors, and we know that if we decide not to abandon the project, we must revise our calculations. We describe a procedure for correcting the material balances in Sec. 7.5.

Type of Vapor Recovery System

The most common choices (with current technology) are

1. Condensation—high pressure or low temperature, or both
2. Absorption
3. Adsorption
4. Membrane separation process
5. Reaction systems

Shortcut design procedures for gas absorbers were discussed in Chap. 3. The economic trade-offs for the design of a condensation process are considered in Exercise 3.5-2. A design procedure for adsorption processes has been presented by Fair.* Neither a design procedure† nor a cost correlation for membrane recovery processes seem to be available in the open literature, although vendors of membranes will provide this service. Reactions are sometimes used to remove CO_2 from gas streams, and H_2S is recovered with amines.

Strategy

We design the vapor recovery system before we consider the liquid separation system because each of the vapor recovery processes usually generates a liquid stream that must be further purified. For the case of a gas absorber, where we need to supply a solvent to the absorber, we also introduce a new recycle loop between the separation systems (see Fig. 7.2-2). Normally we need to estimate the size and costs of each unit to determine which is the cheapest.

---

* J. R. Fair, "Mixed Solvent Recovery and Purification," p. 1, Washington University Design Case Study No. 7, edited by B. D. Smith, Washington University, St. Louis, Mo., 1969.

† A simple model that can be used to estimate the area of a membrane process has been published by J. E. Hogsett and W. H. Mazur, *Hydrocarb. Proc.* August 1983, p. 52.

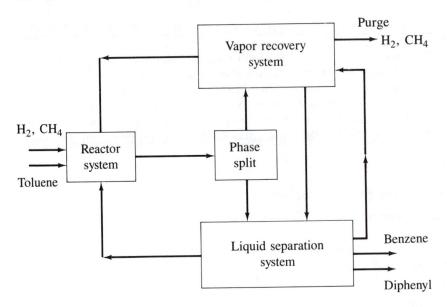

FIGURE 7.2-2
Separation system recycle loop.

Combining the Vapor Recovery System with the Liquid Separation System

If we use a partial condenser and a flash drum to phase-split the reactor effluent, some of the lightest liquid components will leave with the flash vapor (i.e., a flash drum never yields perfect splits) and therefore will not be recovered in the liquid recovery system. However, if there is only a small amount of vapor in the stream leaving the partial condenser and if the first split in the liquid separation system is chosen to be distillation, we could eliminate the phase splitter and feed the reactor effluent stream directly into the distillation column.

The diameter of a distillation column with the two-phase feed will need to be larger (to handle the increased vapor traffic) than a column that follows a flash drum. However, this increased cost may be less than the costs associated with using a vapor recovery system to remove the liquid components from the flash vapor stream. There does not seem to be a heuristic available for making this decision, and so we need to add another process alternative to our list.

7.3 LIQUID SEPARATION SYSTEM

The decisions that we need to make to synthesize the liquid separation system include the following:

1. How should light ends be removed if they might contaminate the product?
2. What should be the destination of the light ends?

3. Do we recycle components that form azeotropes with the reactants, or do we split the azeotropes?
4. What separations can be made by distillation?
5. What sequence of columns do we use?
6. How should we accomplish separations if distillation is not feasible?

Each of these decisions is discussed below. Remember that we want to make these decisions as a function of the design variables over the range of potentially profitable operation.

Light Ends

Some light ends will be dissolved in the liquid leaving the phase splitters shown in Figs. 7.1-3 and 7.1-4, and normally some will be dissolved in the liquid streams leaving the vapor recovery systems. If these light ends might contaminate the product, they must be removed.

ALTERNATIVES FOR LIGHT-ENDS REMOVAL. The choices we have for removing light ends are these:

1. Drop the pressure or increase the temperature of a stream, and remove the light ends in a phase splitter, i.e., another flash drum.
2. Use a partial condenser on the product column.
3. Use a pasteurization section on the product column.
4. Use a stabilizer column before the product column.

The last three alternatives are shown in Fig.7.3-1.
    The options are listed in the order of expected increasing cost, and therefore we prefer to use the earlier entries. However, to make a decision for light-ends removal, it is necessary to know the flow rates of the light ends and to make some shortcut calculations or some CAD runs to estimate the amount recovered:

1. Flash calculations. These are discussed in Sec. 7.1.
2. Partial condensers. CAD programs handle these problems, or in some cases the approximate flash calculations given in Sec. 7.1 can be used.
3. Pasteurization columns. A shortcut design procedure has been published by Glinos and Malone* (see Appendix A.5).
4. Stabilizer columns. This is a normal distillation column that removes light ends.

---

* K. Glinos and M. F. Malone, *Ind. Eng. Chem. Proc. Des. Dev.*, **24**: 1087 (1985).

If product quality is unacceptable, options are:

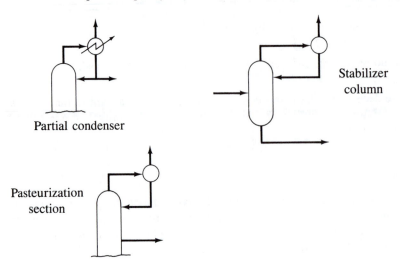

Note: Recycle vapor stream to vapor recovery system if possible.

FIGURE 7.3-1
Alternatives for removing light ends.

DESTINATION OF LIGHT ENDS. For the destination of the light ends, we can vent them (possibly to a flare system), send the light ends to fuel, or recycle the light ends to the vapor recovery system or the flash drum. If the light ends have very little value, we want to remove them from the process through a vent. If this venting causes air pollution problems, we try to vent them through a flare system to burn the offending component. If most of the light ends are flammable, we try to recover the fuel value. However, if the light ends are valuable, we want to retain them in the process. If we recycle them to the vapor recovery system, we introduce another recycle stream into the process.

SUMMARY FOR LIGHT ENDS. If light ends will not contaminate the product, we merely recycle them to the reactor with a reactant-recycle stream or remove them from the process with a by-product stream that is sent to the fuel supply. If light ends will contaminate the product, they must be removed from the process. The method of removal and the destination of the light ends depend on the amount of light ends. Hence, we must determine the amount of light ends as a function of the design variables before we can make a decision.

Azeotropes with Reactants

If a component forms an azeotrope with a reactant, we have the choice of recycling the azeotrope or splitting the azeotrope and just recycling the reactant. Splitting the azeotrope normally requires two columns and therefore is expensive. However, if we recycle the azeotrope, we must oversize all the equipment in the recycle loop to handle the incremental flow of the extra components. A general design heuristic does not seem to be available for making this decision, and so we usually need to evaluate both alternatives. Azeotropic systems are discussed in more detail in the next section.

Applicability of Distillation

In general, distillation is the least expensive means of separating mixtures of liquids. However, if the relative volatilities of two components with neighboring boiling points is less than 1.1 or so, distillation becomes very expensive; i.e., a large reflux ratio is required which corresponds to a large vapor rate, a large column diameter, large condensers and reboilers, and large steam and cooling water costs. Whenever we encounter two neighboring components having a relative volatility of less than 1.1 in a mixture, we group these components together and we treat this group as a single component in the mixture. In other words, we develop the best distillation sequence for the group and the other components, and then we separate the lumped components by using other procedures (see Fig. 7.3-2).

Column Sequencing—Simple Columns

For sharp splits of a three-component mixture (with no azeotropes) we can either recover the lightest component first or the heaviest component first, and then we split the remaining two components (see Fig. 7.3-3). When the number of components increases, the number of alternatives increases very rapidly (see Table 7.3-1). The splits that can be made in the 14 alternatives for a five-component mixture are listed in Table 7.3-2.

   It appears as if it will be a major task to decide which distillation column sequence to select for a particular process, particularly since the best sequence

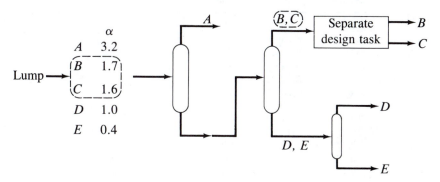

FIGURE 7.3-2
Distillation separations.

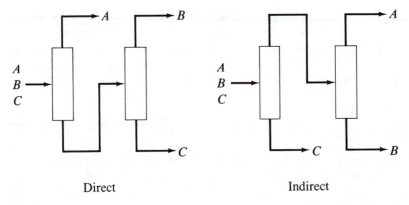

Direct   Indirect

FIGURE 7.3-3
Distillation alternatives for a ternary mixture.

TABLE 7.3-1
Number of alternatives

Number of components	2	3	4	5	6
Number of sequences	1	2	5	14	42

TABLE 7.3-2
Column sequences for five product streams

	Column 1	Column 2	Column 3	Column 4
1	A/BCDE	B/CDE	C/DE	D/E
2	A/BCDE	B/CDE	CD/E	C/D
3	A/BCDE	BC/DE	B/C	D/E
4	A/BCDE	BCD/E	B/CD	C/D
5	A/BCDE	BCD/E	BC/D	B/C
6	AB/CDE	A/B	C/DE	D/E
7	AB/CDE	A/B	CD/E	C/D
8	ABC/DE	D/E	A/BC	B/C
9	ABC/DE	D/E	AB/C	A/B
10	ABCD/E	A/BCD	B/CD	C/D
11	ABCD/E	A/BCD	BC/D	B/C
12	ABCD/E	AB/CD	A/B	C/D
13	ABCD/E	ABC/D	A/BC	B/C
14	ABCD/E	ABC/D	AB/C	A/B

TABLE 7.3-3
General heuristics for column sequencing

1. Remove corrosive components as soon as possible.
2. Remove reactive components or monomers as soon as possible.
3. Remove products as distillates.
4. Remove recycle streams as distillates, particularly if they are recycled
   to a packed bed reactor.

might change as we alter the design variables. To simplify this effort, we might want to look for heuristics for column sequencing. There has been a considerable research effort in this area over the past decade or so, and some of the results are given below.

GENERAL HEURISTICS. There are some general heuristics that can be used to simplify the selection procedure for column sequences (see Table 7.3-3). The first heuristic in this list is based on the fact that the material of construction of the column is much more expensive than carbon steel if corrosive components are present. Thus, the more columns that a corrosive component passes through, the more expensive will be the distillation train.

Reactive components will change the separation problem and thus should be removed as soon as possible. Monomers foul reboilers, so it is necessary to run the columns at vacuum conditions in order to decrease the column overhead and bottom temperatures, so that the rate of polymerization is decreased. Vacuum columns are more costly than pressure columns, and we also prefer to avoid the increased cleaning costs.

We prefer to remove products and recycle streams to packed bed reactors as a distillate to avoid contamination of the product or recycle stream with heavy materials, rust, etc., which always accumulate in a process. If it is necessary to remove a product or recycle stream as a bottom stream, it is often taken as a vapor from a reboiler and then condensed again. At the same time a small, liquid purge stream may be taken from the reboiler to prevent the buildup of contaminants.

COLUMN SEQUENCING HEURISTICS FOR SIMPLE COLUMNS. A number of other heuristics for selecting sequences of simple columns (i.e., columns with one top and one bottom stream) have been published; a short list is given in Table 7.3-4.

TABLE 7.3-4
Heuristics for column sequencing

1. Most plentiful first.
2. Lightest first.
3. High-recovery separations last.
4. Difficult separations last.
5. Favor equimolar splits.
6. Next separation should be cheapest.

However, the first and fifth heuristics in this list depend on feed compositions, whereas the second and fourth depend on relative volatilities. Hence, we expect that these heuristics will lead to contradictions; i.e., if the most plentiful component is the heaviest, there is a conflict between the first and second heuristics.

A longer list of heuristics has been published by Tedder and Rudd,[*] and some investigators have tried to order the importance of the heuristics, to resolve the conflicts.[†] A survey of the literature has been presented by Nishida, Stephanopoulos, and Westerberg,[‡] and a detailed discussion of the limitations of these heuristics has been published by Malone et al.[§] Some additional discussion of the heuristics is given below.

We might also note that as we change the conversion in a process, we expect that the unconverted reactant will go from being the most plentiful component at very low conversions to the least plentiful at very high conversions. Hence, the heuristics in Table 7.3-4 imply that the best column sequences will change as we alter the design variables. Similarly, note that the studies used to develop the heuristics were limited to sequences of simple columns having a single feed stream that were isolated from the remainder of the process, so that different results may be obtained when we consider the interactions between a distillation train and the remainder of the plant.

INTERACTIONS BETWEEN THE SEPARATION SYSTEM AND THE PROCESS. For example, suppose we consider the two flowsheet alternatives shown in Fig. 7.3-4a and b. We might consider these two configurations to be two of the alternatives in a sequencing problem. However, there is a different number of columns in the liquid-recycle loop for the two systems, and therefore the recycle costs will be different. Hence, the optimum conversion, which usually corresponds to a trade-off between selectivity losses and recycle costs, will be different for the two cases. Of course, we should compare alternatives at the optimum processing conditions of each alternative, rather than on an identical feed-stream condition for the two alternatives.

From this simple argument we see that the problem of selecting the best separation sequence cannot always be isolated from the design of the remainder of the process: i.e., the least expensive sequence for a fixed feed-stream condition might not be the least expensive sequence (becauce the feed-stream condition should be changed to correspond to the optimum flow). In fact, there might be another heuristic:

$$\text{Select the sequence that minimizes the number of columns in a recycle loop.} \qquad (7.3\text{-}1)$$

[*] D. W. Tedder and D. F. Rudd, *AIChE J.*, **24**: 303 (1978).

[†] J. D. Seader and A. W. Westerberg, *AIChE J.*, **23**: 951 (1977).

[‡] N. Nishida, G. Stephanopoulos, and A. W. Westerberg, *AIChE J.*, **29**: 326 (1981).

[§] M. F. Malone, K, Glinos, F. E. Marquez, and J. M. Douglas, *AIChE J.*, **31**: 683 (1985).

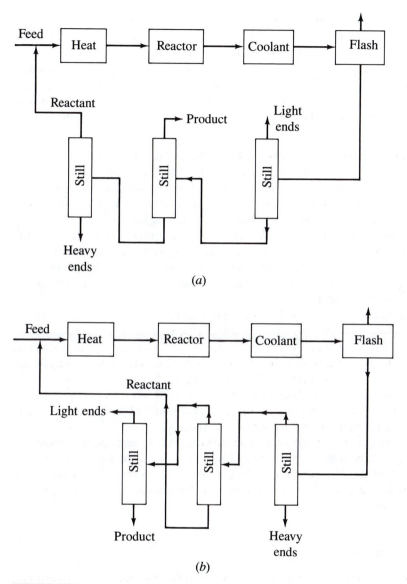

FIGURE 7.3-4
Sequence selection changes recycle costs.

MULTIPLE SEPARATION SEQUENCES. Suppose we consider separation systems that correspond to the general flowsheets given in Fig. 7.1-2 or 7.1-3; i.e., we need both a vapor and a liquid recovery system. A flash drum never gives sharp splits, so that some of the most volatile "liquid" components will leave with the flash vapor, and often they need to be recovered and sent to a liquid separation system. However, the flash liquid might contain a large number of much heavier components as well as those that are returned from the vapor recovery system. In situations such as this, it might be better to split the sequencing problem into two parts. That is, we would split the flash liquid into one portion containing the components returned from the vapor recovery system and one portion containing the heavier components. Then we would design one separation system having a single feed stream for the heavy components and one separation system having multiple feed streams for the components returned from the vapor recovery system.

AN ALTERNATE APPROACH TO SELECTING COLUMN SEQUENCES. The reason for attempting to develop heuristics is that the number of alternative sequences increases very rapidly as the number of components increases (see Table 7.3-1). However, there are a large number of plants where four or less distillation columns are needed to accomplish the separation. Four simple columns (one top and one bottom stream) are needed to separate a five-component mixture into five pure streams, but only four columns are needed to separate six components, if two components with neighboring boiling points leave in the same stream. Thus, for five exit streams and using only simple columns, we need to consider the 14 sequences shown in Table 7.3-5.

An examination of this table indicates that 20 column designs are required to evaluate all the possibilities. Twenty column designs requires a considerable amount of effort if each of the designs is rigorous. However, by using shortcut procedures (see Appendix A.4), it is possible to significantly simplify the calculations. Using shortcut techniques, Glinos* demonstrated that the evaluation of the 14 sequences was almost instantaneous on a VAX 11-780; i.e., the results appeared as soon as the program was run. Kirkwood[†] has shown that the 14 sequences can be evaluated in only a few seconds on an IBM-PC XT.

The results of Glinos and Kirkwood indicate that for modest-size sequencing problems it is better to develop computer codes that evaluate the costs of sequence alternatives than it is to use heuristics. Moreover, the running times for these codes are sufficiently small that the best sequence can be determined as a function of the design variables.

Complex Columns

Rather than consider only sequences of simple columns (one overhead and one bottom stream), we can consider the use of sidestream columns, sidestream strippers and reboilers, prefractionators, etc. One set of heuristics for columns of

---

* K. Glinos, "A Global Approach to the Preliminary Design and Synthesis of Distillation Trains," Ph.D. thesis, University of Massachusetts, 1984.

[†] R. L. Kirkwood, "PIP—Process Invention Procedure," Ph.D. Thesis, University of Massachusetts, Amherst, 1987.

TABLE 7.3-5
Heuristics for complex columns—Tedder and Rudd

Criteria

Ease-of separation index (ESI) $= \dfrac{K_A K_C}{K_B K_B} = \dfrac{\alpha_{AB}}{\alpha_{BC}}$

If ESI <1, the A/B split is harder than the B/C split. If ESI >1, the A/B split is easier than the B/C split.

Heuristics for ESI <1.6
1. If 40 to 80% is middle product and nearly equal amounts of overhead and bottoms are present, then favor design 5.
2. If more than 50% is middle product and less than 5% is bottoms, then favor design 6.
3. If more than 50% is middle product and less than 5% is overheads, then favor design 7.
4. If less than 15% is middle product and nearly equal amounts of overheads and bottoms are present, then favor design 3.
5. Otherwise favor design 1 or 2, whichever removes the most plentiful component first.

Heuristics for ESI >1.6
1. If more than 50% is bottoms product, then favor design 2.
2. If more than 50% is middle product and from 5 to 20% is bottoms, then favor design 5.
3. If more than 50% is middle product and less than 5% is bottoms, then favor design 6.
4. If more than 50% is middle product and less than 5% is overheads, then favor design 7.
5. Otherwise, favor design 3.

Other Heuristics
1. Thermally coupled designs 3 and 4 should be considered as alternatives to designs 1 and 2, respectively, if less than half the feed is middle product.
2. Designs 3, 4, 6, and 7 should be considered for separating all mixtures where a low middle-product purity is acceptable.

Strategy
1. Reduce N-component separations to sequences of pseudoternary separations, and perform the most difficult ternary separation last.
2. This heuristic does not guarantee structural optimality or explicitly consider all complex column alternatives.

From D. W. Tedder and D. F. Rudd, *AIChE J.*, **24**: 303 (1978).

this type has been published by Tedder and Rudd* (see Table 7.3-5 and Figs. 7.3-5 and 7.3-6). Another set has been presented by Glinos and Malone[†] (see Table 7.3-6). Some shortcut design procedures that are useful for complex columns are given in Appendix A.5.

COMPLEX COLUMNS IN SEQUENCES. Our goal is to complete a base-case design as rapidly as possible in order to make a preliminary evaluation of whether

* D. W. Tedder and D. F. Rudd, *AIChE J.*, **24**: 303 (1978).

[†] K. Glinos and M. F. Malone, "Complex Column Alternatives in Distillation Systems," Paper submitted to *Chem. Eng. Res. Des.*, 1985.

TABLE 7.3-6
Heuristics for complex columns—Glinos and Malone

1. Simple sequences
   a. Use the direct sequence if $x_{AF}/(x_{AF} + x_{CF}) > (\alpha_{AB} - 1)/(\alpha_{AC} - 1)$.
   b. Use the indirect sequence if $x_{AF}/(x_{AF} + \chi_{CF}) < 1/(\alpha_{AC} + 1)$.
   c. Calculate the vapor rates if $(\alpha_{AB} - 1)/(\alpha_{AC} - 1) > x_{AF}/(x_{AF} + \chi_{CF}) > 1/(\alpha_{AC} + 1)$.

2. Sidestream columns
   a. Always consider using a sidestream column when χ_{AF} and/or $\chi_{CF} < 0.1$.
   b. Consider using a sidestream column when the intermediate is recycled and a high purity is not required.
   c. Consider using a sidestream column when the volatilities are not evenly distributed.
   d. Consider a sidestream above the feed when the intermediate is more difficult to separate from the heavy than from the light. Otherwise, consider a sidestream below the feed.

3. Sidestream strippers and rectifiers
   a. Consider using a sidestream column when less than 30% of the feed is the intermediate.
   b. As χ_{AF} approaches $(\alpha_{AB} - 1)/(\alpha_{AC} - 1)$, the savings are expected to increase.
   c. The maximum savings are 50%, independent of the relative volatilities.
   d. Consider using a sidestream column when the volatilities are not evenly distributed.

4. Petlyuk columns (see p. 472).
   a. The maximum vapor savings for a Petlyuk column are 50% and are approached when $\chi_A = (\alpha_{AB} - 1)/(\alpha_{AC} - 1)$ and $\chi_B \to 0$.
   b. The vapor rate savings are higher for a Petlyuk column than for any other kind of complex column.
   c. For large or moderate χ_B, a Petlyuk column is favored when
      (1) The volatilities are balanced and both splits are difficult; that is, $\alpha_{AB} \approx \alpha_{BC} \leq 2$.
      (2) The split A/B is difficult, and the B/C split is easy; that is, $\alpha_{AB} < \alpha_{BC}$.
   d. For low χ_B consider using a Petlyuk column when χ_A is close to $(\alpha_{AB} - 1)/(\alpha_{AC} - 1)$, although a side-section column may be better in this case (it results in about the same vapor savings but has fewer trays).
   e. Petlyuk columns may be advantageous for moderate or high χ_B, especially when the A/B split is not much easier than the B/C split or when $\chi_A > 0.5$.
   f. Heuristics recommending the use of Petlyuk columns for large values of χ_B are not always correct, because the performance depends on the volatilities. If the volatilities are evenly distributed, the Petlyuk column and prefractionator should be considered.

5. Prefractionators
   a. Do not consider a prefractionator if a Petlyuk column can be used.
   b. The maximum savings depend on the volatilities and which feed is controlling.
      (1) If the upper feed controls, the maximum savings are $(\alpha_{AC} - \alpha_{BC})/(\alpha_{AC} - 1)$, which occurs as $\chi_B \to 1$.
      (2) If the lower feed controls, the maximum savings are $(\alpha_{BC} - 1)/(\alpha_{AC} - 1)$, which occurs when $\chi_C \to 0$.

From K. Glinos and M. F. Malone, *Chem. Eng. Res. Des.*, submitted paper.

the process is profitable. Thus, normally we include only sequences of simple columns in our first designs, However, a complex column is often cheaper than two simple columns, and therefore we need to consider these possibilities at some point in our design procedure. Since we can replace any two neighboring columns in a sequence by a complex column, we can generate a large number of process alternatives. To avoid getting bogged down in a large number of alternative

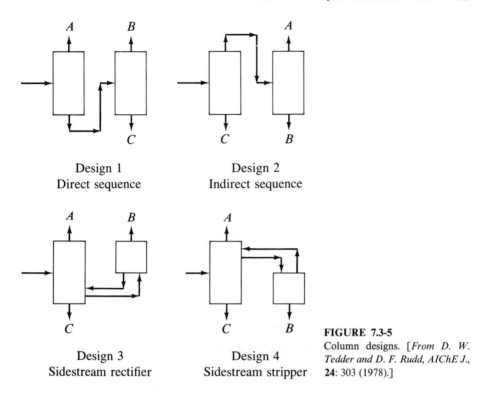

Design 1
Direct sequence

Design 2
Indirect sequence

Design 3
Sidestream rectifier

Design 4
Sidestream stripper

FIGURE 7.3-5
Column designs. [*From D. W. Tedder and D. F. Rudd, AIChE J.,* **24**: 303 (1978).]

evaluations, we defer a consideration of complex columns until we consider other process alternatives, which we discuss in Chap. 9.

Other Types of Separations

If distillation is too expensive to use to separate liquid mixtures, that is, $\alpha < 1.1$, the other choices that are normally the next least expensive are listed in Table 7.3-7. In most cases, these separation procedures require multiple distillation columns to replace a conventional distillation, and so they are normally more expensive. A brief description of each type of separation is given below.

EXTRACTION. To separate a mixture of B and C having a feed composition corresponding to point 1 on Fig. 7.3-7, we countercurrently contact the feed with a solvent S, corresponding to point 2 on Fig. 7.3-7, in an extraction column. Normally, we attempt to recover 99% or more of component C, from the original feed, which corresponds to point 3 on the figure. We remove the solvent from this stream by using a distillation column to obtain the product stream for component B, shown as point 5. The other stream leaving the extraction unit corresponds to point 4, and when we use distillation to remove the solvent from this mixture, we obtain the conditions at point 6.

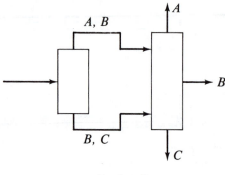

Design 5

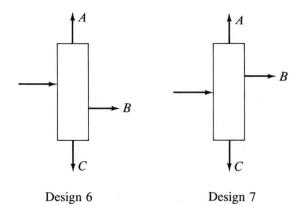

Design 6   Design 7

FIGURE 7.3-6
Column configurations. [*From D. W. Tedder and D. F. Rudd, AIChE J.,* **24**: 303 (1978).]

We note that point 6 corresponds to a binary mixture of B and C, which was the original separation that we were trying to make, except that it is more concentrated than the original feed, point 1. Also, when we separate B and C by normal distillation, we can set the bottom specification to give us almost pure C, point 7, and the overhead composition as the original feed mixture, point 1. Thus, with extraction, we must carry out the same B-C distillation as we would with just distillation, although the degree of separation required is reduced. Of course, this

TABLE 7.3-7
Alternatives to distillation

1. Extraction
2. Extractive distillation
3. Azeotropic distillation
4. Reactive distillation
5. Crystallization
6. Adsorption
7. Reaction

reduction in the degree of separation must decrease the cost sufficiently to pay for the extraction column and the other two distillation columns. In some cases this is possible.

EXTRACTIVE DISTILLATION. If we attempt to separate HNO_3 and H_2O by extractive distillation, we add a heavy component, H_2SO_4, near the top of the tower. The presence of the heavy component changes the vapor-liquid equilibrium (for this example the activity coefficients will be changed), which in some cases will simplify the separation. We obtain a pure component, HNO_3, overhead in the first column (see Fig. 7.3-8). Then we recover the other component overhead in a second column, and we recycle the extractive entrainer, H_2SO_4, back to the first column. We see that two distillation columns are required.

AZEOTROPIC DISTILLATION. In azeotropic distillation we add a relatively light component that again changes the vapor-liquid equilibrium of the original liquid mixture, often by forming a new azeotrope with one of the feed components. Thus, to split the ethanol-water azeotrope, we can add benzene, which forms a ternary azeotrope. With this modification, we can remove pure ethanol from the bottom of the first column and recover the ternary azeotrope overhead (see Fig. 7.3-9).

Since the ternary azeotrope is a heterogeneous mixture when it is condensed, we use the benzene-rich layer as reflux to the first column, and we use the other layer as the feed to a second column. In the second column, we again take the ternary azeotrope overhead, and we recover an ethanol-water mixture as the

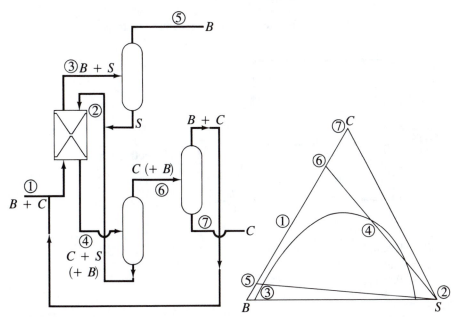

FIGURE 7.3-7
Extraction.

Add nonvolatile component to modify γ's

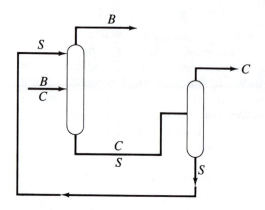

e.g., B = HNO$_3$
  C = H$_2$O
  S = H$_2$SO$_4$

FIGURE 7.3-8
Extractive distillation.

Add volatile component that forms an azeotrope
with one or more of feed components

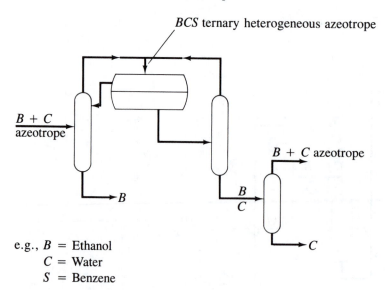

e.g., B = Ethanol
  C = Water
  S = Benzene

FIGURE 7.3-9
Azeotropic distillation.

Add reactive component to modify γ's

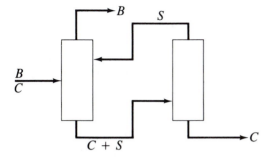

e.g., B, C = xylenes: $\alpha = 1.03$

S = organometallic: B, CS: $\alpha \simeq 30$

FIGURE 7.3-10
Reactive distillation.

bottom stream. Now, in a third column, we recover pure water, our second product, as the bottom stream, along with the original binary azeotrope overhead. This binary azeotrope is recycled to the first column, and we obtain pure products from the system of three columns.

REACTIVE DISTILLATION. In some cases it is possible to add an entrainer that reacts with one component in a mixture that is difficult to separate. For example, the relative volatility between *meta-* and *para-*xylene is only 1.03. However, if sodium cumene is added to a mixture of the xylene isomers, it reacts with the para isomer, and then the relative volatility between the meta-xylene and the organome-tallic complex that is produced becomes 30. The reaction can be reversed in a second column, and the entrainer is recycled (see Fig. 7.3-10). Thus, the original separation is greatly simplified, but at the expense of handling sodium cumene. If entrainers that are simpler to handle can be found, the reactive distillation will become a more important separation alternative.

CRYSTALLIZATION. The separation of xylene isomers is difficult by distillation, so often it is cheaper to use the difference in freezing points to separate the mixture. Thus, by freezing, separation of the liquid-solid mixture, and often using some recycle, the desired separation can be achieved (see Fig. 7.3-11).

DISCUSSION. Extraction, extractive distillation, and azeotropic distillation all involve the separation of nonideal liquid mixtures. Until recently there has been no simple design procedure that could be used for the quick screening of these alternatives. A procedure of this type has recently been developed by Doherty and coworkers, and some of the basic ideas of this procedure are discussed next.

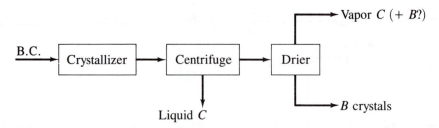

FIGURE 7.3-11
Crystallization.

Example 7.3-1   HDA process. The flows of the flash liquid stream that are fed to the distillation train are given in Table 7.1-1 (100°F and 465 psia). If we let the light ends leave with the product and recover all the product, the composition of the product stream will be

$$x_D = \frac{235.4}{2 + 11 + 235.4} = 0.947 \tag{7.3-1}$$

which is less than the required product purity of 0.997. Hence, we must remove the light ends.

   We could attempt to recover the light ends in a partial condenser at the top of the product column, but since the required product purity is so high, we expect that we will need to use a stabilizer column to remove the light ends. The design of this column is discussed in Appendix B. Since the stabilizer must operate at an elevated

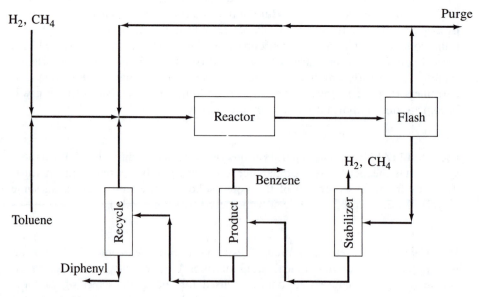

FIGURE 7.3-12
HDA process.

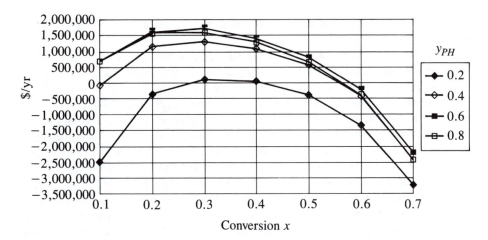

FIGURE 7.3-13
Economic potential—level 4.

pressure, we remove the hydrogen and methane first. We send the light ends to the fuel supply.

The flows of the components remaining after we recover the light ends are: benzene = 235.4, toluene = 87.4, and diphenyl = 4. The heuristics of lightest first, most plentiful first, and favor equimolar splits all favor the direct sequence (recover benzene first), and shortcut design calculations verify this result.

When we add the purge losses and the cost of the distillation columns (see Fig. 7.3-12) to our economic potential calculations for level 3, we obtain the revised economic potential for level 4 shown in Fig. 7.3-13. The range of the design variables where we observe profitable operation has been further decreased, which simplifies the problem of adding a heat-exchanger network (see Chap. 8).

7.4   AZEOTROPIC SYSTEMS

In the previous section where we discussed the sequencing of trains of distillation columns, we assumed that it was possible to split the feed mixture between any two components (see Table 7.3-2). However, if azeotropes are present, it is often impossible to achieve certain splits. Thus, for azeotropic mixtures it is essential to be able to identify when distillation boundaries are present that make certain splits impossible. Most of the discussion below concerning the behavior of these systems has been taken from the papers of Doherty and coworkers.

Distillation Boundaries

For ideal, ternary mixtures we can use either the direct or the indirect sequence (see Fig. 7.3-3) to obtain three pure products. However, for azeotropic mixtures, the feasible separations often depend on the feed composition. Hence, it is necessary to understand the behavior of these processes in much greater detail than the ideal case.

For example, suppose we consider the ternary mixture of acetone, chloroform, and benzene. We can plot the compositions on a triangular diagram, and we note the fact that there is a maximum boiling azeotrope for acetone-chloroform binary mixtures. We also plot the boiling temperature; see Fig. 7.4-1.

Now if we suppose that we put a binary mixture having a composition corresponding to point A in a simple still and continue to increase the temperature in the still, the composition of the material remaining in the still will move in the direction of the arrow shown on Fig. 7.4-1 (toward the binary azeotrope). Also mixtures rich in acetone would be recovered from the top of the still.

In contrast, starting with a binary mixture corresponding to point B on Fig. 7.4-1, as the still temperature is increased, the material left in the still will again be the binary azeotrope, but the overhead will be rich in chloroform. Binary mixtures of acetone and benzene at point C or chloroform and benzene at point D will both lead to final still mixtures of pure benzene.

Suppose now that we consider ternary mixtures corresponding to points A and B on Fig. 7.4-2. As we increase the temperature in a simple still, the still

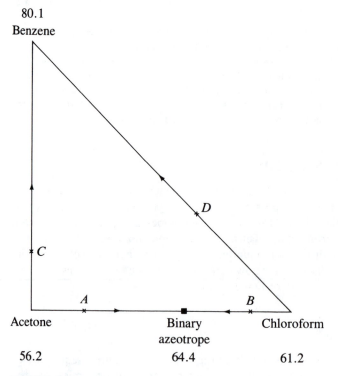

80.1
Benzene

Acetone

Binary
azeotrope

Chloroform

56.2

64.4

61.2

FIGURE 7.4-1
Acetone-chloroform-benzene system—binary mixtures.

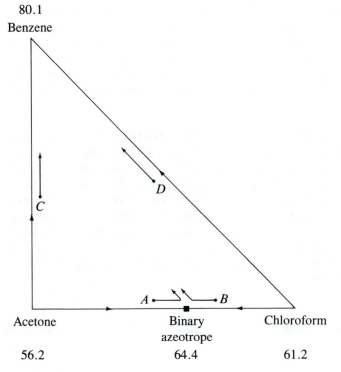

FIGURE 7.4-2
Ternary mixtures.

compositions for each mixture will approach that of the binary azeotrope, until at some point they will tend to collide. Since benzene has a higher boiling point than the binary azeotrope, as we continue to increase the still temperature, the trajectories (residue curves) will both turn toward benzene. Thus, the final composition in the still pot for both cases will be benzene. Ternary mixtures corresponding to points C and D on Fig. 7.4-2 will also yield benzene as the final still composition.

There are rigorous proofs for this type of behavior; see Levy, van Dongen, and Doherty.* However, if we merely say that for every source there must be a sink, then we can develop reasonable pictures of the behavior; i.e., each trajectory (residue curve) must have a stopping point that is either a pure component or an azeotrope (these points correspond to the singular points of the set of differential equations describing a simple still).

---

* S. G. Levy, D. B. van Dongen, and M. F. Doherty, "Design and Synthesis of Azeotropic Distillation. II. Minimum Reflux Calculations," *I&EC Fundamentals*, **24**: 463 (1985).

When we consider more starting conditions, we obtain the results shown in Fig. 7.4-3. Now we see that there is a distillation boundary going from the binary azeotrope to benzene that divides the composition triangle into two distinct regions. Feed mixtures to the left of this boundary produce acetone-rich mixtures overhead and lead to pure benzene in the bottom, whereas feed mixtures to the right of the boundary lead to chloroform-rich mixtures overhead and pure benzene in the bottom.

SEPARATION DEPENDS ON THE FEED COMPOSITION. Suppose that we now consider the separation of a feed mixture having a composition of x_{F1} in Fig. 7.4-4 in two continuous columns using the indirect sequence. It can be shown that the distillate, feed, and bottoms compositions for a single column must fall on a straight line (this is the material balance expression). Hence, if we remove essentially pure benzene from the bottom of the first column and recover essentially all the benzene in the bottoms, the overhead composition will correspond to point A in Fig. 7.4-4. Now, if we split the binary mixture corresponding to point A in a second column, we will obtain essentially pure acetone overhead and the binary azeotrope as a bottoms stream.

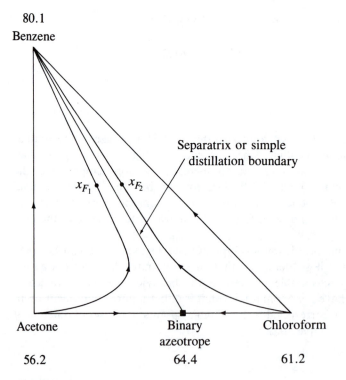

FIGURE 7.4-3

Residue curve map. [*From M. F. Doherty and G. A. Caldarola, I&EC Fundamentals,* **24**: *474 (1985), with permission of the American Chemical Society.*]

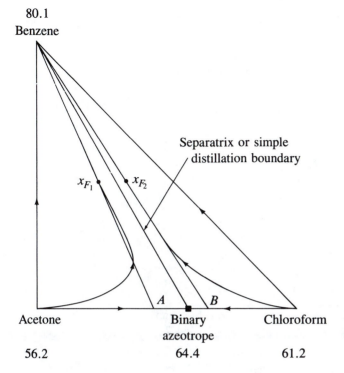

80.1
Benzene

Separatrix or simple
distillation boundary

x_{F_1} x_{F_2}

Acetone   Binary   Chloroform
azeotrope

56.2   64.4   61.2

A   B

FIGURE 7.4-4
Sequence of two continuous columns.

However, if we start with a composition corresponding to x_{F2} in Fig. 7.4-4, then we will obtain pure benzene as a bottoms stream from the first column and a binary mixture corresponding to point B overhead. When we split this binary mixture in a second column, we obtain pure chloroform overhead and the binary azeotrope as a bottoms stream. Hence, the products we obtain depend on the feed composition wherever a distillation boundary is present.

Another way of stating this result is that pure chloroform cannot normally be obtained in a sequence of two columns if the feed composition lies to the left of the distillation boundary, whereas pure acetone cannot normally be obtained if the feed composition lies to the right of the boundary. The relative volatility of components in a mixture close to the boundary changes from a value greater than unity to a value of less than unity if we just move across the boundary. Of course, if we split the binary azeotrope in an additional column system, we could recover all three components in pure form.

MORE COMPLEX SYSTEMS. As a more complex system, we can consider ternary mixtures of methyl acetate, methanol, and hexane. Now each binary pair exhibits an azeotrope, and all three are minimum-boiling azeotropes. If we put binary mixtures corresponding to any points on the edges of the triangle in a simple

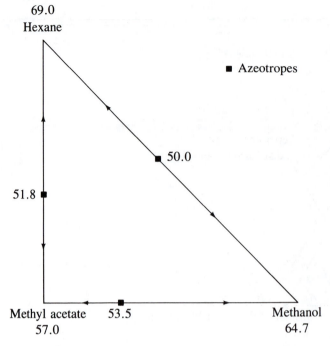

FIGURE 7.4-5
Methyl acetate–methanol–hexane binary mixtures.

still, the composition of the material remaining in the still will move in the same direction as an increased still temperature. Thus, the arrows in Fig. 7.4-5 correspond to the direction of increasing liquid compositions remaining in the still pots.

When we consider a set of ternary mixtures that are close to the sides of the triangles and recognize that the still composition must move in the direction of increasing temperatures, we obtain the results shown in Fig. 7.4-6. Since there must be a source in the interior of the triangle for the trajectories to behave in this way, there must be a ternary azeotrope which has a lower boiling point than any of the binary azeotropes.

A residue curve map showing more trajectories (residue curves) is shown in Fig. 7.4-7. Now we see that there are distillation boundaries that divide the triangle into three distinct regions: *ADGE*, *BDGF*, and *CEGF*. Depending on where our feed composition falls in these regions, we will obtain different products.

Minimum Reflux Ratio

For ideal systems, we can use Underwood's equation to calculate the minimum reflux ratio. Then, after we select a reflux ratio of 20% or so, larger than the

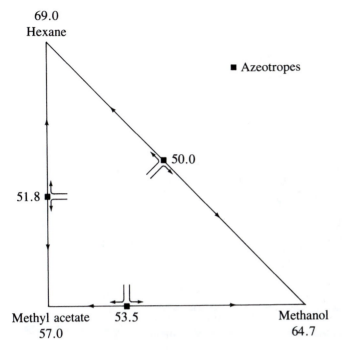

69.0
Hexane

■ Azeotropes

■ 50.0

51.8 ■

Methyl acetate   **53.5**   **Methanol**
    57.0   **64.7**

FIGURE 7.4-6
Ternary mixtures.

minimum value, we can calculate the vapor and liquid flows throughout the column. We use this information to design the column.

A procedure for calculating the minimum reflux ratio for nonideal mixtures, including azeotropic systems, has been developed by Doherty and coworkers.* We discuss this procedure now.

SYSTEM EQUATIONS. The material balance equation for the stripping section can be written as

$$x_{i,m+1} = \left(\frac{s}{s+1}\right) y_{i,m} + \left(\frac{1}{s+1}\right) x_{i,B} \qquad (7.4\text{-}1)$$

where s is the reboil ratio. If we subtract $x_{i,m}$ from both sides of the equation, we obtain

$$x_{i,m+1} - x_{i,m} = \left(\frac{s}{s+1}\right) y_{i,m} - x_{i,m} + \left(\frac{1}{s+1}\right) x_{i,B} \qquad (7.4\text{-}2)$$

---

* S. G. Levy, D. B. van Dongen, and M. F. Doherty, "Design and Synthesis of Azeotropic Distillation. II. Minimum Reflux Calculations," *I&EC Fundamentals*, **24**: 463 (1985).

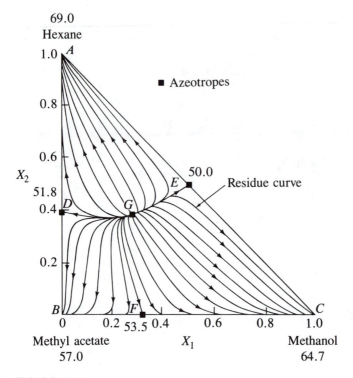

FIGURE 7.4-7
Residue curve map for the system methanol–hexane–methyl acetate at 1 = atm total pressure. Arrows point in the direction of increasing time (or temperature). [*From D. B. van Dongen, and M. F. Doherty, I&EC Fundamentals,* **24**: *454 (1985), with permission from the American Chemical Society.*]

Now we approximate the left-hand side of the equation by a derivative with respect to column height:

$$\frac{dx_i}{dh} = \left(\frac{s}{s+1}\right) y_i - x_i + \left(\frac{1}{s+1}\right) x_{i,B} \tag{7.4-3}$$

Similarly, for the rectifying section we obtain

$$\frac{dx_i}{dh} = \left(\frac{r+1}{r}\right) y_i - x_i - \left(\frac{1}{r}\right) y_{i,D} \tag{7.4-4}$$

where r is the reflux ratio.

At infinite reflux, Eq. 7.4-4 reduces to

$$\frac{dx_i}{dh} = y_i - x_i \tag{7.4-5}$$

which is just the equation for a simple still (which appears in most texts on distillation, but is not discussed here). Also, pinch points exist when $dx_i/dh = 0$, or $x_{i,m+1} = x_{i,m}$, and for this condition Eq. 7.4-3 becomes identical to Eq. 7.4-1. Thus, the differential equation will provide a rigorous calculation of the pinch compositions.

An overall material balance for the column gives

$$y_{i,D} = \left(\frac{r + s + 1}{s}\right)(z_{i,F} - x_{i,B}) + x_{i,B} \tag{7.4-6}$$

We can use these results to develop a procedure for calculating the minimum reflux ratio.

IDEAL MIXTURES. To understand Doherty's procedure, we first consider the case of a hexane-heptane-nonane ideal mixture. If we are attempting to make the sharp split A/BC for the feed composition shown in Fig. 7.4-8, then the distillate, feed, and bottoms compositions all must fall on a straight line (the column material balance must be satisfied). Every component will distribute to some extent, and the component specifications are given on the figure.

If we fix the column splits and the reflux ratio, we can use Eq. 7.4-6 to calculate the reboil ratio. Now, we integrate Eqs. 7.4-3 and 7.4-4, starting from the ends of the column. If the reflux ratio chosen is below the minimum, then the two trajectories do not intersect: see Fig. 7.4-8 for the hexane-heptane-nonane example.

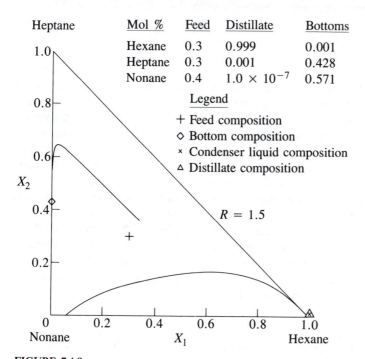

Mol %	Feed	Distillate	Bottoms
Hexane	0.3	0.999	0.001
Heptane	0.3	0.001	0.428
Nonane	0.4	1.0×10^{-7}	0.571

Legend
+ Feed composition
◇ Bottom composition
× Condenser liquid composition
△ Distillate composition

$R = 1.5$

FIGURE 7.4-8
Less than minimum reflux. [*From S. Levy, D. B. van Dongen, and M. F. Doherty, I&EC Fundamentals,* **24**: 463 (1985), with permission from the American Chemical Society.]

If the selected reflux ratio exceeds the minimum value, then the profiles for the stripping section and rectifying sections cross, and we can obtain the desired separation (see Fig. 7.4-9). If the reflux ratio corresponds to the minimum, then the pinch zone for the stripping section just ends on the profile for the rectifying section (see Fig. 7.4-10).

Now if we change the mole fraction of the heaviest component in the distillate from $x_N = 0.001$ to $x_N = 1 \times 10^{-11}$ and repeat the calculations, we obtain the results shown in Fig. 7.4-11. We note that the minimum reflux ratio for this case is $R_m = 1.54$ instead of $R_m = 2.15$. We also note that the rectifying and stripping profiles exhibit very sharp corners, which correspond to pinch zones.

Another feature in Fig. 7.4-11 is that the pinch composition in the rectifying section (point P on Fig. 7.4-11), the pinch point for the stripping section (point Q), and the feed composition F are collinear. This result can be rigorously proved for ideal systems, and the result can also be shown to be equivalent to Underwood's equations. However, the same result is approximately valid for nonideal systems.

COMPLEXITY OF THE PROBLEM. There are 113 types of residue curve maps that can be drawn for ternary mixtures.* The maps that are useful for selecting entrainers for binary mixtures with minimum boiling azeotropes, a total of 35 possibilities, are shown in Fig. 7.4-12.[†] The number of possibilities grows very rapidly as the number of components present in the mixture increases. Hence, azeotropic systems present a formidable challenge as a separation problem. However, the geometric ideas presented above can be used to develop an expression for the minimum reflux ratio for azeotropic systems.

NONIDEAL SYSTEMS. Now suppose we consider the system of acetone, chloroform, and benzene.[‡] We pick a reflux ratio below the minimum value, we use the terminal compositions shown in Fig. 7.4-13, and we apply the procedure described above. Then the profiles for the rectifying and stripping sections do not intersect (see Fig. 7.4-13). However, if we are above the minimum reflux, the curves cross (see Fig. 7.4-14); and if we are at minimum reflux, the pinch region for the stripping section just ends on the profile for the rectifying section (see Fig. 7.4-15).

When we change the end compositions, we obtain the results shown in Fig. 7.4-16, and the minimum reflux ratio decreases. Moreover, we note from Fig. 7.4-16 that the pinch zone for the rectifying section at minimum reflux (point P on Fig. 7.4-16), the pinch point for the stripping section Q, and the feed composition F are essentially collinear. (This example shows the largest deviation that has been observed for numerous case studies.) This collinearity condition provides a criterion for calculating the minimum reflux ratio.

---

* H. Matsuyama and H. J. Nishimura, *J. Chem. Eng. Jpn.*, **10**: 181 (1977).

[†] M. F. Doherty and G. A. Caldarola, *I&EC Fundamentals*, **24**: 474 (1985).

[‡] S. G. Levy, D. B. van Dongen, and M. F. Doherty, "Design and Synthesis of Azeotropic Distillation. II. Minimum Reflux Calculations," *I&EC Fundamentals*, **24**: 463 (1985).

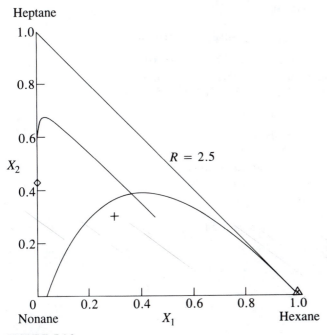

FIGURE 7.4-9
Greater than minimum reflux. [*From S. Levy, D. B. van Dongen, and M. F. Doherty, I&EC Fundamentals*, **24**: *463 (1985), with permission from the American Chemical Society.*]

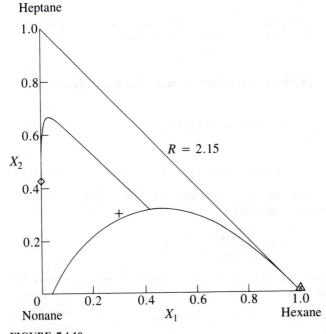

FIGURE 7.4-10
Minimum reflux. [*From S. Levy, D. B. van Dongen, and M. F. Doherty, I&EC Fundamentals*, **24**: *463 (1985), with permission from the American Chemical Society.*]

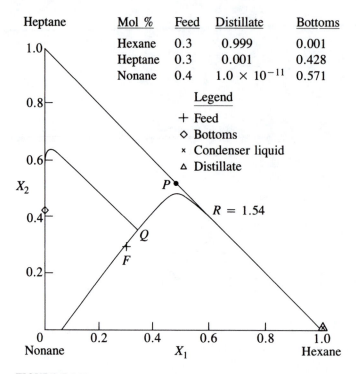

Heptane

Mol %	Feed	Distillate	Bottoms
Hexane	0.3	0.999	0.001
Heptane	0.3	0.001	0.428
Nonane	0.4	1.0×10^{-11}	0.571

Legend

+ Feed
◇ Bottoms
× Condenser liquid
△ Distillate

$R = 1.54$

X_2

FIGURE 7.4-11
Minimum reflux—higher-purity split. [*From S. Levy, D. B. van Dongen, and M. F. Doherty, I&EC Fundamentals,* **24**: *463 (1985), with permission from the American Chemical Society.*]

EQUATIONS FOR MINIMUM REFLUX. We can write the equations describing the rectifying pinch

$$rx_s - (r + 1)y_s + y_D = 0 \qquad (7.4\text{-}7)$$

and the feed pinch

$$sy_e - (s + 1)x_e + x_D = 0 \qquad (7.4\text{-}8)$$

where each of these equations contains expressions for the two key components. The reflux and reboil ratios are related by

$$s = (r + 1)\left(\frac{x_{B,1} - x_{F,1}}{x_{F,1} - y_{D,1}}\right) \qquad (7.4\text{-}9)$$

and we use a vapor-liquid equilibrium model to relate x_i and y_i. Then, the collinearity result requires that

$$(x_{e,1} - x_{F,1})(x_{s,2} - x_{F,2}) - (x_{e,2} - x_{F,2})(x_{s,1} - x_{F,1}) = 0 \qquad (7.4\text{-}10)$$

The value of r that satisfies this set of equations corresponds to the minimum reflux ratio. Underwood's equations are a special case of this new general formalism.

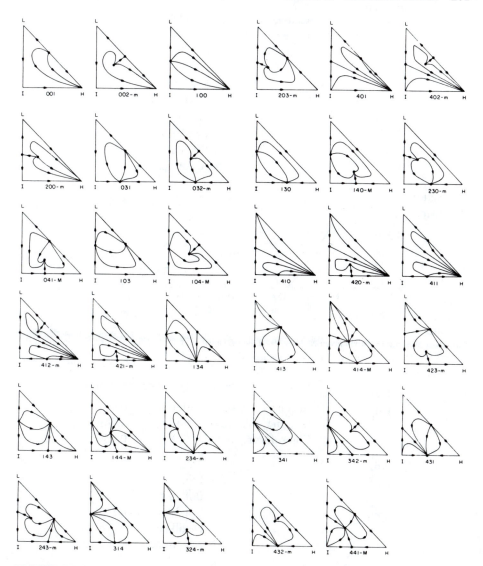

FIGURE 7.4-12
Residue curve maps. [*From M. F. Doherty and G. A. Caldarola, I&EC Fundamentals,* **24**: *474 (1985), with permission from the American Chemical Society.*]

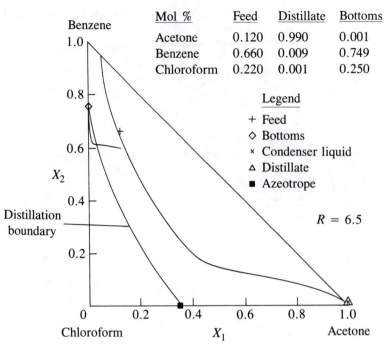

Mol %	Feed	Distillate	Bottoms
Acetone	0.120	0.990	0.001
Benzene	0.660	0.009	0.749
Chloroform	0.220	0.001	0.250

Legend

+ Feed
◇ Bottoms
× Condenser liquid
△ Distillate
■ Azeotrope

$R = 6.5$

FIGURE 7.4-13
Less than minimum reflux. [*From S. Levy, D. B. van Dongen, and M. F. Doherty, I&EC Fundamentals,* **24**: *463 (1985), with permission from the American Chemical Society.*]

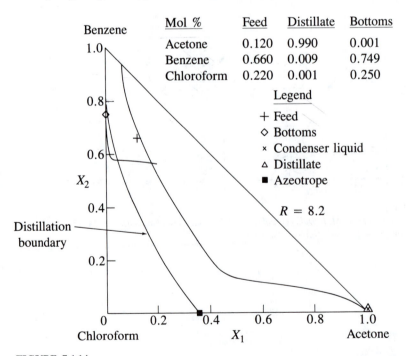

Mol %	Feed	Distillate	Bottoms
Acetone	0.120	0.990	0.001
Benzene	0.660	0.009	0.749
Chloroform	0.220	0.001	0.250

Legend

+ Feed
◇ Bottoms
× Condenser liquid
△ Distillate
■ Azeotrope

$R = 8.2$

FIGURE 7.4-14
Greater than minimum reflux. [*From S. Levy, D. B. van Dongen, and M. F. Doherty, I&EC Fundamentals,* **24**: *463 (1985), with permission from the American Chemical Society.*]

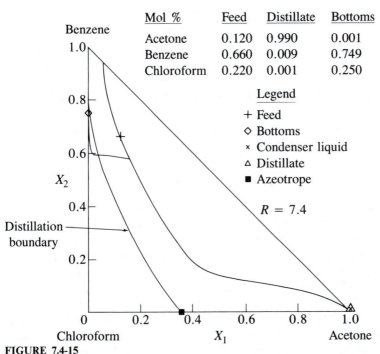

Mol %	Feed	Distillate	Bottoms
Acetone	0.120	0.990	0.001
Benzene	0.660	0.009	0.749
Chloroform	0.220	0.001	0.250

Legend
+ Feed
◇ Bottoms
× Condenser liquid
△ Distillate
■ Azeotrope

$R = 7.4$

FIGURE 7.4-15
Minimum reflux. [*From S. Levy, D. B. van Dongen, and M. F. Doherty, I&EC Fundamentals,* **24**: *463 (1985), with permission from the American Chemical Society.*]

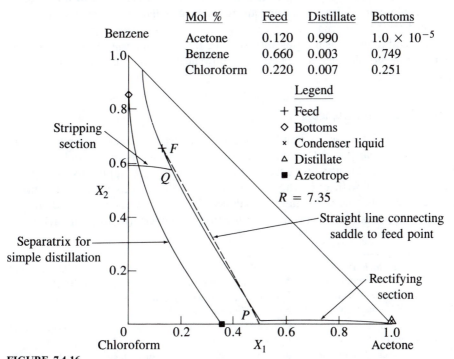

Mol %	Feed	Distillate	Bottoms
Acetone	0.120	0.990	1.0×10^{-5}
Benzene	0.660	0.003	0.749
Chloroform	0.220	0.007	0.251

Legend
+ Feed
◇ Bottoms
× Condenser liquid
△ Distillate
■ Azeotrope

$R = 7.35$

Straight line connecting saddle to feed point

Rectifying section

FIGURE 7.4-16
Minimum reflux with decreased heavy component in the overhead. [*From S. Levy, D. B. van Dongen, and M. F. Doherty, I&EC Fundamentals,* **24**: *463 (1985), with permission from the American Chemical Society.*]

203

EXTENSIONS OF THE METHOD. This approach for nonideal systems has been extended to multiple-feed streams,[*] columns with nonnegligible heat effects,[†] heterogeneous azeotropic systems,[‡] procedures for selecting entrainers, and optimum design and sequencing. Once the minimum reflux ratio has been calculated, we can let $R = 1.2R_m$ and then design the column, following the same procedure as we used for ideal mixtures. A procedure for calculating the minimum reflux for systems with 4, or more, components has recently become available.

7.5   RIGOROUS MATERIAL BALANCES

After we have selected a liquid separation system, we have completely fixed all the units in the flowsheet where the component flows change. These units include mixers (for fresh feed and recycle streams), splitters (for purge streams), reactors, flash drums (phase splitters), gas absorbers (and/or other vapor recovery units), and distillation columns (and/or other liquid separation systems). Thus, we can now develop a set of rigorous material balances.

Of course, if our rigorous balances differ significantly from our earlier, approximate results, then we will need to review the decisions that we made. We could have revised the material balance calculations at any stage of our development of the design, and clearly there is a trade-off between the time required to perform all the calculations and the accuracy of the answer. Our goal is to complete the design as rapidly as possible, providing that major errors are not introduced, and to explore the alternatives using approximate calculations. Then after we have identified the best alternative, we will use rigorous calculation procedures. However, remember that it is not possible to make rigorous material balances until we have completely defined the parts of a flowsheet where the component flows change.

Linear Material Balancing

The procedure we use to develop rigorous material balances is called *linear material balancing* (the set of equations generated is always linear and therefore easy to solve), and it was first described by Westerberg.[§] To apply this procedure, first we draw a flowsheet so that it contains only those units where component flows change. Then we write material balances for each component individually in terms of the molar flow rates and the fractional recovery (or loss) in each unit.

---

[*] S. G. Levy, and M. F. Doherty, "Design and Synthesis of Homogeneous, Azeotropic Distillations. IV. Minimum Reflux Calculations for Multiple Feed Columns," *I&EC Fundamentals*, **25**: 269 (1985).

[†] J. R. Knight and M. F. Doherty, "Design and Synthesis of Homogeneous Azeotropic Distillations. V. Columns with Nonnegligible heat Effects," *I&EC Fundamentals*, **25**: 279 (1985).

[‡] H. N. Pham and M. F. Doherty, "Design and Synthesis of Heterogeneous Azeotropic Distillation. I. Heterogeneous Phase Diagrams," *Chem. Eng. Sci.* (1985).

[§] A. W. Westerberg, "Notes for a Course on Chemical Process Design," taught at the Institute de Desanolo Tecnologico para la Industria Quimica (INTEC), Santa Fe, Argentina, August 1978.

These equations are always linear, and therefore they are simple to solve by either matrix methods or simple substitution. Normally, we start with a balance for the limiting reactant, and then we consider in turn the primary product, other reactants, by-product components, and inert materials.

Not all the fractional recoveries (or losses) of the components in various units can be chosen independently. For example, the simple flash calculation procedure described by King (Eq. 7.1-19) shows that if the fractional recovery of one component is fixed, then all the other fractional recoveries can be calculated. Similarly, the fractional recoveries for a product column must be fixed so that the product purity specification is satisfied, and in some cases the fractional recoveries for purge streams must be chosen so that constraints on molar ratios at the reactor inlet can be satisfied. Hence, in some cases some iteration might be required.

Example 7.5-1   HDA process. The procedure is best illustrated in terms of an example, and for this purpose we choose the HDA process. The flowsheet is shown in Fig. 7.5-1. Now, we write balances for the component flows of each stream, starting with the limiting reactant.

Toluene balances. The toluene entering the reactor $TOL_{R,in}$ is the sum of the fresh feed toluene TOL_{FF}, the toluene in the gas-recycle stream TOL_{GR}, and the toluene in the liquid-recycle stream TOL_{LR}:

$$TOL_{R,in} = TOL_{FF} + TOL_{GR} + TOL_{LR} \qquad (7.5\text{-}1)$$

The toluene leaving the reactor $TOL_{R,out}$ is the toluene that was not converted in the reactor:

$$TOL_{R,out} = TOL_{R,in} (1 - x) \qquad (7.5\text{-}2)$$

If we let $f_{TOL,FV}$ be the fraction of the toluene leaving with the flash vapor TOL_{FV}, then a fraction $1 - f_{TOL,FV}$ leaves with the flash liquid TOL_{FL}:

$$TOL_{FV} = f_{TOL,FV}\, TOL_{R,out} \qquad (7.5\text{-}3)$$

$$TOL_{FL} = (1 - f_{TOL,FV})\, TOL_{R,out} \qquad (7.5\text{-}4)$$

If we let f_{PG} be the fraction of toluene lost in the purge TOL_{PG}, then a fraction $1 - f_{PG}$ of the toluene will be in the gas-recycle stream TOL_{GR}:

$$TOL_{PG} = f_{PG}\, TOL_{FV} \qquad (7.5\text{-}5)$$

$$TOL_{GR} = (1 - f_{PG})\, TOL_{FV} \qquad (7.5\text{-}6)$$

If we let $f_{TOL,ST}$ be the fraction of toluene that leaves with the stabilizer distillate $TOL_{ST,D}$, then a fraction $1 - f_{TOL,ST}$ will leave with the stabilizer bottoms $TOL_{ST,B}$:

$$TOL_{ST,D} = f_{TOL,ST}\, TOL_{FL} \qquad (7.5\text{-}7)$$

$$TOL_{ST,B} = (1 - f_{TOL,ST})\, TOL_{FL} \qquad (7.5\text{-}8)$$

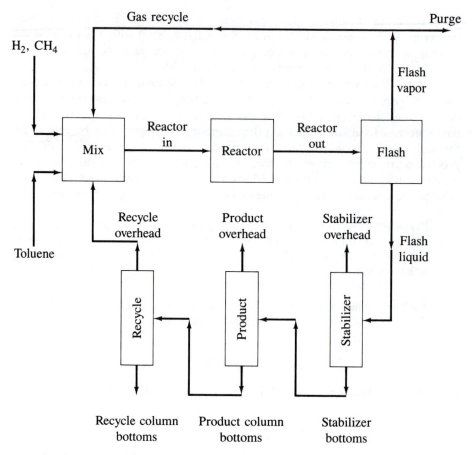

FIGURE 7.5-1
HDA process.

If a fraction $f_{TOL,PR}$ leaves with the benzene product $TOL_{PR,D}$, then a fraction $1 - f_{TOL,PR}$ will leave the product column in the bottoms $TOL_{PR,B}$:

$$TOL_{PR,D} = f_{TOL,PR}\, TOL_{ST,B} \tag{7.5-9}$$

$$TOL_{PR,B} = (1 - f_{TOL,PR})\, TOL_{ST,B} \tag{7.5-10}$$

Finally, if a fraction $f_{TOL,RC}$ is lost with the diphenyl by-product stream from the recycle column TOL_D, then a fraction $1 - f_{TOL,RC}$ is recycled to the reactor TOL_{LR}:

$$TOL_D = f_{TOL,RC}\, TOL_{PR,B} \tag{7.5-11}$$

$$TOL_{LR} = (1 - f_{TOL,RC})\, TOL_{PR,B} \tag{7.5-12}$$

We try to select the fractional recoveries in these equations such that f_i will be a small number. However, the purge split f_{PG} is the same for all components, and the splits of the components in the flash drum are related to one another.

Now if we combine Eqs. 7.5-6, 7.5-3, and 7.5-2 to solve for the gas-recycle flow, we obtain

$$TOL_{GR} = TOL_{R,\text{in}}(1 - f_{PG})(f_{TOL,FV})(1 - x) \qquad (7.5\text{-}13)$$

Also, if we combine Eqs. 7.5-12, 7.5-10, 7.5-8, 7.5-4, and 7.5-2 to calculate the liquid-recycle flow, we obtain

$$TOL_{LR} = TOL_{R,\text{in}}(1 - f_{TOL,RC})(1 - f_{TOL,PR})(1 - f_{TOL,ST})(1 - f_{TOL,FV})(1 - x) \qquad (7.5\text{-}14)$$

Next we substitute Eqs. 7.5-13 and 7.5-14 into Eq. 7.5-1, to obtain

$$TOL_{R,\text{in}}\{1 - [(1 - f_{PG})(f_{TOL,FV})$$
$$+ (1 - f_{TOL,RC})(1 - f_{TOL,PR})(1 - f_{TOL,ST})(1 - f_{TOL,FV})]$$
$$(1 - x)\} = TOL_{FF} \qquad (7.5\text{-}15)$$

or

$$TOL_{R,\text{in}} = \frac{TOL_{FF}/(1 - x)}{1 - [(1 - f_{PG})f_{TOL,FV} + (1 - f_{TOL,RC})(1 - f_{TOL,PR})(1 - f_{TOL,FV})]} \qquad (7.5\text{-}16)$$

We can use this result to solve for all the other toluene flows.

Note that if there is no loss of toluene from the process, i.e.,

$$f_{PG} = 0 \qquad f_{TOL,RC} = 0 \qquad f_{TOL,PR} = 0 \qquad f_{TOL,ST} = 0$$

then Eq. 7.5-16 reduces to

$$TOL_{R,\text{in}} = \frac{TOL_{FF}}{x} \qquad (7.5\text{-}17)$$

which is the simplified approximation that we used previously.

BENZENE BALANCES. The balances for benzene are essentially the same, except for the reactor equation. That is, at the reactor inlet we obtain

$$BZ_{R,\text{in}} = BZ_{FF} + BZ_{GR} + BZ_{LR} = BZ_{GR} + BZ_{LR} \qquad (7.5\text{-}18)$$

where the fresh feed flow of benzene BZ_{FF} is equal to zero. According to our selectivity correlation, a fraction S of the toluene converted appears as benzene, although it is important to remember that this correlation was based on a pure toluene feed stream. Thus, we expect that some of the benzene recycled to the reactor will be converted to diphenyl, and if the benzene recycle flow is significant, we should revise our correlation. Neglecting this discrepancy until we estimate the benzene-recycle flow, we can write that the toluene converted in the reactor is simply

$$\text{Toluene Converted} = TOL_{R,\text{in}}x \qquad (7.5\text{-}19)$$

where we can substitute Eq. 7.5-16 for $\text{TOL}_{R,\text{in}}$. Hence the benzene leaving the reactor is the benzene produced $[xS(\text{TOL}_{R,\text{in}})]$ plus the benzene fed to the reactor:

$$BZ_{R,\text{out}} = BZ_{R,\text{in}} + xS(\text{TOL}_{R,\text{in}}) \tag{7.5-20}$$

Letting $f_{BZ,FV}$ be the fraction of benzene going overhead in the flash drum (which is related to $f_{TOL,FV}$ by Eq. 7.1-19) and f_{PG} be the fraction of benzene lost in the purge (which is the same for all components), we can show that the gas-recycle flow of benzene is

$$BZ_{GR} = (1 - f_{PG})f_{BZ,FV}[BZ_{R,\text{in}} + xS(\text{TOL}_{R,\text{in}})] \tag{7.5-21}$$

Similarly, if we let $f_{BZ,ST}$ be the fraction of benzene lost overhead in the stabilizer and $f_{BZ,PR}$ be the fraction of benzene lost in the bottoms of the product column, and if we assume that all the benzene goes overhead in the recycle column, then the liquid-recycle flow of benzene is

$$BZ_{LR} = f_{BZ,PR}(1 - f_{BZ,ST})(1 - f_{BZ,FV})[BZ_{R,\text{in}} + xS(\text{TOL}_{R,\text{in}})] \tag{7.5-22}$$

Substituting Eqs. 7.5-21 and 7.5-22 into Eq. 7.5-18 gives

$$BZ_{R,\text{in}}[1 - f_{BZ,FV}(1 - f_{PG}) - f_{BZ,PR}(1 - f_{BZ,ST})(1 - f_{BZ,FV})]$$
$$= xS(\text{TOL}_{R,\text{in}})[f_{BZ,FV}(1 - f_{PG}) + f_{BZ,PR}(1 - f_{BZ,ST})(1 - f_{BZ,FV})] \tag{7.5-23}$$

or

$$BZ_{R,\text{in}} = \frac{xS(\text{TOL}_{R,\text{in}})[f_{BZ,FV}(1 - f_{PG}) + f_{BZ,PR}(1 - f_{BZ,ST})(1 - f_{BZ,FV})]}{1 - f_{BZ,FV}(1 - f_{PG}) - f_{BZ,PR}(1 - f_{BZ,ST})(1 - f_{BZ,FV})} \tag{7.5-24}$$

We can now use this result to calculate all the other benzene flows.

Other Component Flows

The material balances for the other components are developed in the same way, with a few exceptions. That is, we assume that there is a negligible amount of diphenyl in the flash vapor stream (see Table 7.1-1). Also, we assume that all the hydrogen and methane in the flash liquid that is not recovered in the stabilizer leaves with the benzene product, i.e., there is no hydrogen or methane in the liquid-recycle stream.

Linear Material Balances

From the discussion above we see that by writing balances for the molar flow of each component in terms of the fractional recoveries obtained in each process unit, we obtain a set of linear equations in terms of the conversion of the limiting reactant and the selectivity (which is related to the conversion). These equations are simple (although somewhat tedious) to solve for the recycle flows of each component. Once we have calculated the recycle flows of each component, we can calculate all the other flows of that component.

An inspection of the resulting equations indicates that we must specify the fresh feed rate of toluene, the fresh feed rate of hydrogen, the fresh feed rate of methane, the reactor conversion, the split fractions of the components in the flash drum (which are related to each other by Eq. 7.1-19 and depend on the temperature and pressure of the flash drum), the split fraction of the purge stream, and the fractional recoveries of the components in the distillation train.

Optimization Variables

In our previous approximate material balances, we specified the production rate of benzene, the product purity of benzene, the purge composition of hydrogen (which we showed was equivalent to specifying the fresh feed rates of hydrogen and methane), the conversion, and the molar ratio of hydrogen to aromatics at the reactor inlet. For our linear material balance problem we can assume that the conversion and makeup gas flows are optimization variables (since the feed composition of the makeup gas stream is fixed, specifying the makeup gas flow fixes the fresh feed rates of both hydrogen and methane). As we discussed earlier, these are the dominant optimization variables.

The fractional loss of benzene overhead in the stabilizer also corresponds to an optimization problem (loss of benzene to fuel versus the number of trays in the rectifying section and the column pressure). The fractional losses of toluene and diphenyl overhead in the stabilizer are then fixed by the column design. Specifying the fractional loss of methane in the stabilizer bottoms will fix the design of the stabilizer (small losses correspond to a large number of trays in the stripping section), and once the column design is fixed, the hydrogen loss in the bottoms is fixed. However, we expect that all the hydrogen and methane leaving in the stabilizer bottoms stream will also leave with the benzene product. Then the fraction of toluene that goes overhead in the product column and leaves with the benzene product stream plus the hydrogen and methane leaving with this stream is fixed by the specified production rate and product purity.

To obtain small amounts of toluene overhead in the product column, we must include a large number of trays in the rectifying section of this column. Thus, there is a trade-off between using a large number of trays in the stripping section of the stabilizer (to keep the hydrogen and methane flows in the product stream small) balanced against using a large number of trays in the rectifying section of the product column [to keep the toluene (and diphenyl) flows in the product stream small], for a case where the sum of the hydrogen, methane, toluene, and diphenyl flows is fixed.

The fractional loss of benzene in the bottoms of the product column is also an optimization variable (trays in the stripping section balanced against the cost of recycling benzene through the reactor system), as are the fractional loss of toluene in the bottom of the recycle column (toluene lost to fuel versus trays in the stripping section) and the fractional loss of diphenyl overhead in the recycle column (recycle costs of diphenyl back through the reactor versus trays in the rectifying section).

To avoid all these separation system optimizations, we fix the fractional recoveries of the keys to correspond to the rule-of-thumb value of greater than 99 % and we fix the fractional losses of the nonkeys arbitrarily as 0.15 to 0.3 times the fractional losses of the keys. Alternatively, we could use Fenske's equation to estimate the fractional loss of the nonkeys. Thus, our material balances are not rigorous, but since we expect that these loss terms to be small, we do not introduce much error.

Constraints

Most of the flows can be written in terms of the fresh feed rate of toluene TOL_{FF} (see Eq. 7.5-16). However, we want to solve the design problem in terms of the production rate PROD of benzene. Hence, we need to sum the flows of benzene, hydrogen, methane, toluene, and diphenyl leaving the top of the product column and then eliminate TOL_{FF} from these expressions and replace it with PROD. This procedure will remove the production rate constraint.

In addition, we must write the expression for the hydrogen-to-aromatics ratio at the reactor inlet, set this value equal to 5/1, and then solve for the fractional split of the purge stream f_{PG} that satisfies this expression. This procedure removes the other process constraint.

Unfortunately, the algebra required to remove these constraints is tedious. Thus, it might be easier to solve for the recycle flows of each component, solve for all the other component flows, and then adjust the solutions, i.e., iterate, until the constraints are satisfied. Alternatively, one of the computer-aided design programs, such as FLOWTRAN, PROCESS, DESIGN 2000, ASPEN, etc., can be used to revise the material balance calculations. We discuss the use of the CAD programs to revise the material balances later in the text.

> **Example 7.5-2   HDA process.** The expression for toluene feed rate to the reactor is given by Eq. 7.5-16. To evaluate this flow, we must specify the terms in the equation.
>
> TOL_{FF} Our original design problem specifies the desired production rate of benzene, rather than the fresh feed rate of toluene. However, from our shortcut balances (with no losses) we found that $F_{FT} = P_B/S$ (see Eq. 5.2-1). For a case where $P_B = 265$, $x = 0.75$, and $S = 0.9694$ (see Appendix B) $F_{FT} = TOL_{FF} = 273.4$. We can use this estimate in the first solution and then use iteration to correct the value.
>
> f_{PG} Using our shortcut calculations, we found that the purge flow rate was 496 mol/hr and that the gas-recycle flow was 3371 mol/hr for a case where $x = 0.75$ and $y_{PH} = 0.4$. Hence, the fraction of the flash vapor that is purged from the process is $496/(496 + 3371) = 0.128$. We use this as a first guess, and then we iterate to match the problem specifications.
>
> $f_{TOL,FV}$ The results of the shortcut flash calculations are given in Table 7.1-1, and we see that $f_{TOL,FV} = 3.6/91 = 0.0396$. Again, we need to iterate to match the flash drum operating conditions.
>
> $f_{TOL,RC}$ The fraction of toluene taken overhead in the recycle column is an optimization variable. For our first design we choose $f_{TOL,RC} = 0.995$.
>
> $f_{TOL,PR}$ The fraction of toluene taken overhead in the product column is also an optimization variable (the amount of toluene plus methane taken overhead is fixed by the product specifications, but either composition can be adjusted). For our first

design we might assume that the impurities in the product are a 50/50 mixture of methane and toluene.

$f_{TOL,ST}$ The fraction of toluene leaving overhead in the stabilizer depends on the sharpness of the split between methane and benzene. Since we must take some benzene overhead in this column to ensure an adequate supply of reflux, we do not expect to obtain a sharp split. For a first design we fix the effluent cooling-water temperature in the partial condenser used in this column as $110°F$; we choose the condensing temperature as 115 or $120°F$; and we fix the column pressure so that the K value of benzene is $K_B = 0.05$. If these results are reasonable, then we find the K value of toluene in the reflux drum, and we can estimate the toluene loss.

For this example, the amount of effort required to solve the rigorous material balances by using linear material balances probably exceeds the effort required to use a CAD program.

7.6 SUMMARY, EXERCISES, AND NOMENCLATURE

Summary

The decisions we must make to synthesize a separation system fall into three categories: the general structure, the vapor recovery system, and the liquid separation system. These decisions are listed here.

1. General structure
   a. Do we need both liquid and vapor recovery units, or just liquid?
2. Vapor recovery systems
   a. Should the vapor recovery system be placed on the purge stream, the gas-recycle stream, or the flash vapor stream? Or, is it better not to include one?
   b. Should we use a condensation process, absorption, adsorption, a membrane process, or a reactor system as the vapor recovery system?
3. Liquid separation system
   a. How should the light ends be separated if they might contaminate the product?
   b. What should be the destination of the light ends?
   c. Do we recycle components that form azeotropes with a reactant, or do we split the azeotrope?
   d. What separations can be made by distillation?
   e. What sequence of columns should we use?
   f. How should we accomplish separations if distillation is not feasible?

Some design guidelines that are helpful in making the decisions above are listed here:

1. The general structure we choose for the separation system depends on whether the phase of the reactor effluent is a liquid, a two-phase mixture, or a vapor.

The three types of flowsheet are shown in Figs. 7.1-2 through 7.1-4. In cases where the reactor effluent is a vapor and we do not obtain a phase split when we cool the effluent to 100°F, either we pressurize the reactor (if the feed and recycle streams are liquid) or we install a compressor and/or a refrigeration system to accomplish a phase split. If a phase split results in only small amounts of either vapor or liquid, we might delete the phase splitter and send the reactor effluent to either a vapor recovery or a liquid recovery system.

2. We install a vapor recovery system on the purge stream if we lose valuable materials with the purge.

3. We install a vapor recovery system on the gas-recycle stream if some recycle components would be deleterious to the reactor operation or degrade the product distribution.

4. We install a vapor recovery system on the flash vapor stream if both items 2 and 3 above are important.

5. We do not use a vapor recovery system if neither item 2 nor item 3 above is important.

6. Our choices for a vapor recovery system are condensation (high-pressure or low-temperature or both), absorption, adsorption, or a membrane recovery process. (A reactor system could also be considered.)

7. If the light ends contaminate the product, they must be removed. Our options are to drop the pressure of the feed stream and flash off the light ends, to remove the light ends by using a partial condenser on the product column, to remove the light ends in a pasteurization section in the product column, or to use a stabilizer column to remove the light ends.

8. We recycle components that form azeotropes with the reactants if the azeotropic composition is not too high, but there is no heuristic available to set the exact level.

9. We normally do not use distillation to split adjacent components when $\alpha < 1.1$.

10. Instead of using heuristics to select column sequences, we usually calculate the costs of all the sequences.

11. If distillation is too expensive, we consider azeotropic distillation, extractive distillation, reactive distillation, extraction, or crystallization.

Exercises

7.6-1. For one of the design problems that you have considered, determine the following:
   (a) The general structure of the separation system.
   (b) Whether a vapor recovery system is required and, if so, where it should be located. If necessary, determine the design of one of the alternatives.
   (c) Several alternative distillation trains. Design one of these. (*Caution*: The IPA and ethanol processes are not ideal and require activity coefficient models and the use of a CAD program.)

7.6-2. Sketch your best guess of a separation system for one of the processes below (i.e., guess the general structure of the separation system); guess whether a vapor recovery system might be needed, where it should be placed, and what type might be the best; and guess the distillation sequencing alternatives that might be the best. Describe in as much detail as you can the reasons for your guesses, and indicate in detail what calculations you would need to do to verify your guesses.

(*a*) The cyclohexane process (see Exercises 5.4-7 and 6.8-6)

(*b*) The butane alkylation process (see Exercises 5.4-10 and 6.8-9)

(*c*) The styrene process (see Exercises 5.4-6 and 6.8-5)

(*d*) The acetic anhydride process (see Exercises 5.4-3 and 6.8-2)

(*e*) The benzoic acid process (see Exercise 1.3-4)

7.6-3. If the HDA process with diphenyl recovered were run at very high conversions, we might obtain a 50/50 mixture of toluene and diphenyl that would be fed to the recycle column. If we select an overhead composition of toluene as $x_D = 0.9$ and we recover 99% of the toluene overhead, how many trays are required in the distillation column (assume $\alpha = 25$)?

7.6-4. Suppose that the flow rate to the distillation train in a butane alkylation process (see Exercises 5.4-10 and 6.8-9) when $x = 0.9$, $T = 40°F$, mol $i\text{-}C_4$/mol $O\text{-}C_4$ at reactor inlet $= 9$ is given by $C_3 = 310$ mol/hr, $i\text{-}C_4 = 11892$ mol/hr, $O\text{-}C_4 = 143$ mol/hr, $n\text{-}C_4 = 419$ mol/hr, $i\text{-}C_8 = 918$ mol/hr, and $C_{12} = 184$ mol/hr (where we do not split $i\text{-}C_4$ from $O\text{-}C_4$). Use heuristics to suggest alternative sequences of distillation to consider. Should sidestream columns be considered?

7.6-5. Suppose that in Exercise 7.6-4 there is no $n\text{-}C_4$ in the feed. Calculate the vapor rates required in each column in a sequence where we remove the lightest component first. Compare this result to a case where we recover the C_3 first, flash the bottoms stream from the C_3 splitter, and send the flash liquid to a distillation train where we recover the lightest component first. Assume that the pressure of the depropanizer is 230 psia and that the pressure of the debutanizer is 96 psia.

7.6-6. Consider a process that produces 100 mol/hr of xylene and 100 mol/hr of benzene by toluene disproportionation

$$2\,\text{Toluene} \rightleftharpoons \text{Benzene} + \text{Xylene} \qquad (7.6\text{-}1)$$

The reaction is acutally equilibrium-limited. But, neglecting this equilibrium limitation, find the amount of toluene in the feed to a distillation train where the direct and the indirect sequences would have the same cost. Would you expect that a complex column would ever be less expensive?

7.6-7. A residue curve map for mixtures of acetone, isopropanol, and water is given in Fig. 7.6-1. For the conditions given in Example 6.3-4, estimate the composition at the bottom of the first tower if the direct sequence is used and at the top of the first tower if the indirect sequence is used.

7.6-8. A model for a simple plant is given in detail in Sec. 10.3 for the case where a direct column sequence is used. If we neglect the optimization of the reflux ratio and the fractional recovery in the second tower and if we use the indirect rather than the direct column sequence, what are the optimum design conditions? How do the costs for the two alternatives compare? Compare the reactor exit compositions at the optimum conditions of each alternative. At these values of the reactor exit compositions, how do the stand-alone direct and indirect sequences compare?

7.6-9. The reaction (see Exercises 5.4-9 and 6.8-8)

$$\text{Butadiene} + SO_2 \rightleftharpoons \text{Butadiensulfone}$$

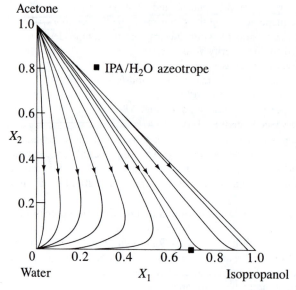

FIGURE 7.6-1
Acetone–IPA–H_2O residue curves.
[*From S. Levy, D. B. van Dongen, and M. F. Doherty, I&EC Fundamentals*, **24**: *463 (1985), with permission from the American Chemical Society.*]

has a significant reverse reaction rate at the boiling point of the product; so we do not want to use distillation to recover and recycle the reactants. Suggest another separation system (not included in our general set of rules) for this process. Plot the economic potential in terms of the significant design variables.

Nomenclature

BZ	Benzene molar flow
ESI	Ease-of-seperation index
F	Feed rate (mol/hr)
f_i	Fractional recovery of component i
f_i, f_j	Component flows of light and heavy materials (mol/hr)
h	Column height
K_i	Distribution coefficient
L	Liquid flow rate (mol/hr)
l_i, l_j	Component flows of liquid (mol/hr)
r	Reflux ratio
s	Reboil ratio
TOL	Toluene molar flow (mol/hr)
V	Vapor rate (mol/hr)
v_i, v_j	Component flows of vapor (mol/hr)
x	Conversion
x_i	Liquid mole fraction
x_{iD}	Mole fraction of component i in distillate
x_{iF}	Mole fraction of component i in feed

y_i Vapor mole fraction
z_i Feed mole fraction

Greek symbols

α_{ij} Relative volatility of component i with respect to component j, $\alpha_{ij} = K_i/K_j$
θ Root of Underwood's equation
γ Activity coefficient

Subscripts

B	Bottoms
D	Distillate
e	Feed pinch
F	Feed
FF	Fresh feed
FL	Flash liquid
FV	Flash vapor
GR	Gas recycle
LR	Liquid recycle
m	Plate number
PG	Purge
PR	Product column
PR,B	Product column bottoms
PR,D	Product column distillate
R,in	Reactor inlet
R,out	Reactor exit
ST,B	Stabilizer bottoms
ST,D	Stabilizer distillate
s	Stripping pinch
TOL,PR	Toluene leaving the product column
TOL,RL	Toluene leaving the recycle column
TOL,ST	Toluene leaving the stabilizer

CHAPTER
8

HEAT-EXCHANGER
NETWORKS

Energy conservation has always been important in process design. Thus, it was common practice to install feed-effluent exchangers around reactors and distillation columns. However, a dramatically different approach that takes into consideration energy integration of the total process has been developed over the past two decades. The basic ideas of this new approach are presented now.

8.1 MINIMUM HEATING AND COOLING REQUIREMENTS

The starting point for an energy integration analysis is the calculation of the minimum heating and cooling requirements for a heat-exchanger network. These calculations can be performed without having to specify any heat-exchanger network. Similarly, we can calculate the minimum number of exchangers required to obtain the minimum energy requirements without having to specify a network. Then the minimum energy requirements and the minimum number of exchangers provide targets for the subsequent design of a heat-exchanger network.

In any process flowsheet, a number of streams must be heated, and other streams must be cooled. For example, in the HDA process in Fig. 8.1-1, we must heat the toluene fresh feed, the makeup hydrogen, the recycle toluene, and the recycle gas stream up to the reaction temperature of 1150°F. Also, we must cool the reactor effluent stream to the cooling-water temperature to accomplish a phase split, and we must cool the product stream from its boiling point to cooling-water

216

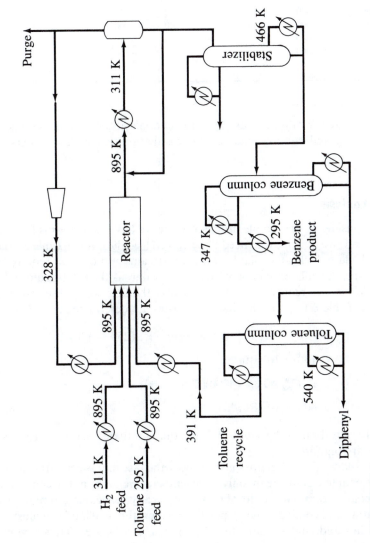

FIGURE 8.1-1
Hydrodealkylation of toluene.

217

TABLE 8.1-1
First-law calculation

Stream No.	Condition	FC_P, Btu/(hr·°F)	T_{in}	T_{out}	Q available, 10^3 Btu/hr
1	Hot	1000	250	120	130
2	Hot	4000	200	100	400
3	Cold	3000	90	150	−180
4	Cold	6000	130	190	−360
					−10

temperatures because we do not want to store materials at their boiling points. We also have heating and cooling loads on the distillation-column condensers and reboilers.

First-Law Analysis

Suppose we consider a very simple problem where we have two streams that need to be heated and two streams that need to be cooled (see the data in Table 8.1-1). If we simply calculate the heat available in the hot streams and the heat required for the cold streams, the difference between these two values is the net amount of heat that we would have to remove or supply to satisfy the first law. These results are also shown in Table 8.1-1, and the first two entries are determined as follows:

$$Q_1 = F_1 C_{p1} \, \Delta T_1 = [1000 \text{ Btu/(hr·°F)}](250 - 120)$$
$$= 130 \times 10^3 \text{ Btu/hr} \tag{8.1-1}$$

$$Q_2 = F_2 C_{p2} \, \Delta T_2 = (4000)(200 - 100)$$
$$= 400 \times 10^3 \text{ Btu/hr} \tag{8.1-2}$$

Thus, 10×10^3 Btu/hr must be supplied from utilities if there are no restrictions on temperature-driving forces.

This first-law calculation does not consider the fact that we can transfer heat from a hot stream to a cold stream only if the temperature of the hot stream exceeds that of the cold stream. Hence, to obtain a physically realizable estimate of the required heating and cooling duties, a positive temperature driving force must exist between the hot and cold streams. In other words, any heat-exchanger network that we develop must satisfy the second law as well as the first law.

Temperature Intervals

A very simple way of incorporating second-law considerations into the energy integration analysis was presented by Hohmann, Umeda et al., and Linnhoff and

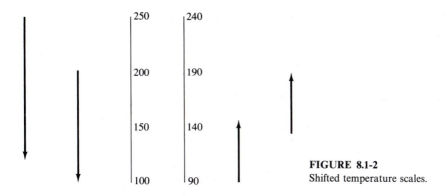

FIGURE 8.1-2
Shifted temperature scales.

Flower,* and we describe their analyses. If we choose a minimum driving force of 10°F between the hot and cold streams, we can establish two temperature scales on a graph, one for the hot streams and the other for the cold streams, which are shifted by 10°F. Then we plot the stream data on this graph (Fig. 8.1-2). Next we establish a series of temperature intervals that correspond to the heads and the tails of the arrows on this graph, i.e., the inlet and outlet temperatures of the hot and cold streams given in Table 8.1-1 (see Fig. 8.1-3).

In each temperature interval we can transfer heat from the hot streams to the cold streams because we are guaranteed that the temperature driving force is adequate. Of course, we can also transfer heat from any of the hot streams in the high-temperature intervals to any of the cold streams at lower-temperature

---

* E. C. Hohmann, "Optimum Networks for Heat Exchange," Ph.D. Thesis, University of Southern California, (1971); T. Umeda, J. Itoh, and K. Shiroko, *Chem. Eng. Prog.*, **74** (9): 70 (1978); B. Linnhoff and J. R. Flower, *AIChE J.*, **24**: 633, 642 (1978).

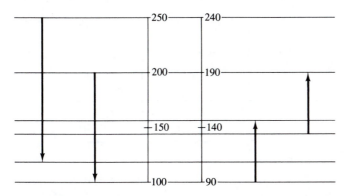

FIGURE 8.1-3
Temperature intervals.

intervals. However, as a starting point we consider the heat transfer in each interval separately. The expression we use is

$$Q_i = \left[\sum (FC_p)_{\text{hot},i} - \sum (FC_p)_{\text{cold},i} \right] \Delta T_i \qquad (8.1\text{-}3)$$

for each interval. Thus, for the first three intervals we obtain

$$Q_1 = (1000)(250 - 200) = 50 \times 10^3 \qquad (8.1\text{-}4)$$

$$Q_2 = (1000 + 4000 - 6000)(200 - 160) = -40 \times 10^3 \qquad (8.1\text{-}5)$$

$$Q_3 = (1000 + 4000 - 3000 - 6000)(160 - 140)$$

$$= -80 \times 10^3 \qquad (8.1\text{-}6)$$

The other values are shown in Fig. 8.1-4. We also note that the summation of the heat available in all the intervals $(50 - 40 - 80 + 40 + 20 = -10)$ is -10×10^3 Btu/hr, which is identical to the result obtained for the first law calculation, i.e., the net difference between the heat available in the hot streams and that in the cold streams.

Cascade Diagrams

One way we could satisfy the net heating and cooling requirements in each temperature interval is simply to transfer any excess heat to a cold utility and to supply any heat required from a hot utility (see Fig. 8.1-5). From this figure, we see that we would need to supply 120×10^3 Btu/hr $(40 + 80)$ and that we would have to reject 110×10^3 Btu/hr $(50 + 40 + 20)$. Again, the difference is the first-law value.

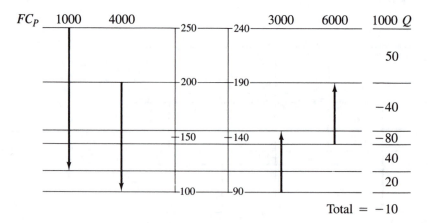

FIGURE 8.1-4
Net energy required at each interval.

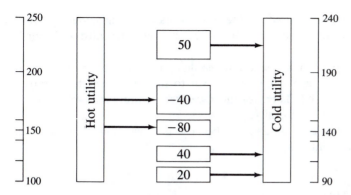

FIGURE 8.1-5
Heat transfer to and from utilities for each temperature interval.

Of course, the arrangement shown in Fig. 8.1-5 would correspond to very poor engineering practice because we are transferring heat from the highest possible temperature interval directly to a cold utility, rather than using this available heat to supply some of the energy requirements at lower-temperature intervals. Thus, instead of using the arrangement shown in Fig. 8.1-5, we take all the heat available at the highest temperature interval (200 to 250°F) and we transfer it to the next lower interval (160 to 200°F) (see Fig. 8.1-6). Since we are transferring this heat to lower-temperature intervals, we always satisfy the second-law constraint.

From Fig. 8.1-6 we see that there is sufficient heat available in the highest temperature interval to completely satisfy the deficiency in the second interval (40 × 10³ Btu/hr) and to also supply 10 × 10³ of the 80 × 10³ requirement for the third interval. However, in this third interval we must supply 70 × 10³ Btu/hr from

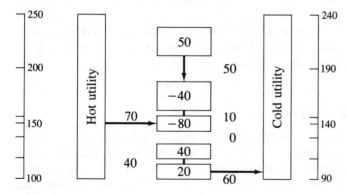

FIGURE 8.1-6
Cascade diagram.

a hot utility because we have used all the heat available at higher-temperature intervals. Then there would be no transfer of heat between the third and fourth temperature intervals.

For the fourth temperature interval, we could either reject the excess heat to cold utility, as shown in Fig. 8.1-5, or transfer it to the next lower temperature interval, as shown in Fig. 8.1-6. Then, for the lowest temperature interval we reject all the remaining heat to a cold utility.

We call Fig. 8.1-6 a *cascade diagram* because it shows how heat cascades through the temperature intervals.

Minimum Utility Loads

From Fig. 8.1-6 we see that the minimum heating requirement is 70×10^3 Btu/hr and the minimum cooling requirement is 60×10^3 Btu/hr. The difference between these values still corresponds to the first-law requirement, but now our minimum heating and cooling loads have been fixed also to satisfy the second law.

Pinch Temperature

We also note from Fig. 8.1-6 that there is no energy transfer between the third and fourth temperature intervals. We call this the *pinch temperature* ($140°F$ for the hot streams and $130°F$ for the cold streams, or sometimes we use the average value of $135°F$). Thus, the pinch temperature provides a decomposition of the design problem. That is, above the pinch temperature we only supply heat, whereas below the pinch temperature we only reject heat to a cold utility.

Dependence on the Minimum Approach Temperature

If we change the minimum approach temperature of $10°F$ that we used as our second-law criterion, then we shift the temperature scales in Fig. 8.1-2. The heat loads in each of the intervals shown in Fig. 8.1-4 will also change, and the minimum heating and cooling loads will alter. It is easy to visualize these changes if we construct a temperature-enthalpy diagram.

Temperature-Enthalpy Diagrams

To construct a temperature-enthalpy diagram, first we calculate the minimum heating and cooling loads, using the procedure described above. Then we define the enthalpy corresponding to the coldest temperature of any hot stream as our base condition; i.e., at $T = 100°F$ (see Fig. 8.1-4), $H = 0$. Next we calculate the cumulative heat available in the sum of all the hot streams as we move to higher-temperature intervals. Thus, from Fig. 8.1-4 we obtain the following:

Hot streams, °F		Cumulative H
$T = 100$	$H_0 = 0$	0
$T = 120$	$H_1 = 4000(120 - 100) = 80,000$	80,000
$T = 140$	$H_2 = (1000 + 4000)(140 - 120) = 100,000$	180,000
$T = 160$	$H_3 = (1000 + 4000)(160 - 140) = 100,000$	280,000
$T = 200$	$H_4 = (1000 + 4000)(200 - 160) = 200,000$	480,000
$T = 250$	$H_5 = 1000(250 - 200) = 50,000$	530,000

Now we plot the cumulative H versus T (see Fig. 8.1-7). We call this a *composite curve* for the hot streams because it includes the effect of all the hot streams.

Of course, since the FC_p values are constant, we could have replaced the calculations for H_2, H_3, and H_4 by a single expression

$$H_{2,3,4} = (1000 + 4000)(200 - 120) = 400,000$$

Thus, we only need to calculate values at the temperature levels when the number of hot streams changes.

At the lowest temperature of any of the cold streams (90°F on Fig. 8.1-4), we choose the enthalpy as the minimum cooling requirement $Q_{c,\min}(60 \times 10^3$ Btu/hr

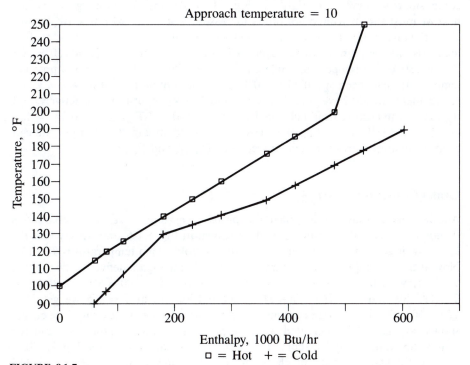

FIGURE 8.1-7
Temperature-enthalpy diagram.

on Fig. 8.1-6). Then we calculate the cumulative enthalpy in each temperature interval:

Cold streams, °F		Cumulative H
$T = 90$	$H_0 = 60,000$	60,000
$T = 130$	$H_1 = 3000(130 - 90) = 120,000$	180,000
$T = 150$	$H_2 = (3000 + 6000)(150 - 130) = 180,000$	360,000
$T = 190$	$H_3 = 6000(190 - 150) = 240,000$	600,000

These results also are plotted on Fig. 8.1-7.

From Fig. 8.1-7 we note that the enthalpy of the hot streams that must be rejected to a cold utility is $Q_C = 60 \times 10^3$ Btu/hr, and the amount of heat that must be supplied from a hot utility is $Q_H = 70 \times 10^3$ Btu/hr. Moreover, when $T_H = 140°$ and $T_C = 130°$, we see that the minimum approach temperature exists, i.e., the heating and cooling curves are closest together. Thus, this temperature-enthalpy diagram gives us exactly the same information as we generated previously.

Suppose now that we set the base enthalpy of the cold curve equal to 110,000, instead of 60,000, and we repeat the calculations for the cold curve. This shifts the composite cold curve to the right (see Fig. 8.1-8). We note from the figure that the heat we must supply from a hot utility increases by 50×10^3 to 120×10^3 Btu/hr. Thus, the increase in the heating load is exactly equal to the increase in the cooling load. Also at the point of closest approach between the curves ($T_{hot} = 150°F$ and $T_{cold} = 130°F$) the temperature difference is $20°F$. Thus, if the minimum approach temperature had been specified as $20°F$, then the minimum heating and cooling requirements would have been 120×10^3 and 110×10^3 Btu/hr, respectively, and the pinch temperature would change from $T_{hot} = 140$ and $T_{cold} = 130$ to $T_{hot} = 150$ and $T_{cold} = 130°F$. By sliding the curve for the cold streams to the right, we can change the minimum approach temperature, $Q_{H,min}$ and $Q_{C,min}$.

Grand Composite Curve

Another useful diagram is called the *grand composite curve*. To prepare this diagram, we start at the pinch condition shown in Fig. 8.1-6, and we say that the heat flow is zero at the average of the hot and cold pinch temperatures $T = 135$. Now at the next higher temperature interval, which we again define by the average $T = 155$, we calculate that the net heat flow is $180 - 100 = 80$. Similarly, at $T = 195$ we find that $H = 80 + 240 - 200 = 120$, and at $T = 245$ we get $H = 120 - 50 = 70$. These points are just the differences between the composite curves shown on Fig. 8.1-7, calculated with the pinch as a starting point. We call the results the grand composite curve above the pinch temperature (see Fig. 8.1-9).

Again, starting at the pinch and moving to colder temperatures, at $T = 115$ we let $H = 40$, and at $T = 95$ we let $H = 20 + 40 = 60$. These points define the

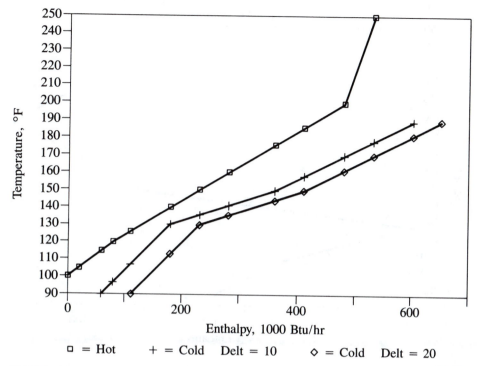

FIGURE 8.1-8
Temperature-enthalpy diagram.

curve below the pinch (see Fig. 8.1-9). This grand composite curve clearly shows that minimum heating requirements are $Q_H = 70 \times 10^3$ Btu/hr and that the minimum cooling load is $Q_C = 60 \times 10^3$ Btu/hr. The grand composite curve is particularly useful for profile matching during heat and power integration studies.

Relationship of Minimum Heating and Cooling to the First-Law Requirement

The first-law analysis indicates that the difference between the heat available in the hot streams and that required by the cold streams is 10×10^3 Btu/hr, which must be removed to a cold utility. The second-law analysis with a 10°F approach temperature indicates that we must supply a minimum of 70×10^3 Btu/hr and remove 60×10^3 Btu/hr. Hence, we see that any incremental heat that we put in from a hot utility must also be removed by a cold utility. Moreover, we recognize that if we put in more than the minimum amount of energy (see Fig. 8.1-10), then we will have to pay more than necessary for both a hot utility and a cold utility (because we will have to remove this excess heat).

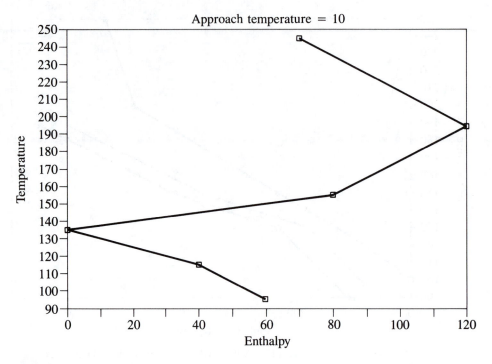

FIGURE 8.1-9
Grand composite curve.

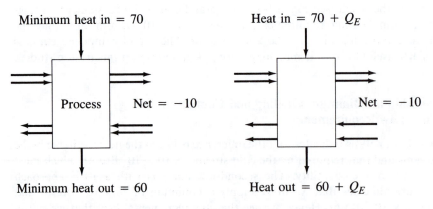

If we put excess heat into the process,
we must remove this excess heat.

FIGURE 8.1-10
Relationship to first law.

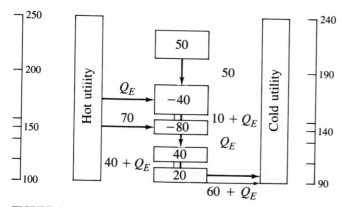

FIGURE 8.1-11
Excess heating and cooling.

From Fig. 8.1-11 we see that if we transfer an amount of heat Q_E across the pinch, we must put this additional heat into the process from a hot utility somewhere in the network. Furthermore, we must also reject this amount of heat to a cold utility. Hence, we obtain a rule of thumb:

$$\text{Do not transfer heat across the pinch!} \qquad (8.1\text{-}7)$$

Other rules of thumb that we have developed are these:

$$\text{Add heat only above the pinch.} \qquad (8.1\text{-}8)$$

$$\text{Cool only below the pinch.} \qquad (8.1\text{-}9)$$

Industrial Experience

The calculation of the minimum heating and cooling requirements is a very simple task, and yet it indicates that significant energy savings are possible compared to past practice. In particular, Imperial Chemical Industries in the United Kingdom and Union Carbide in the United States have both reported the results of numerous case studies that indicate that 30 to 50% energy savings, compared to conventional practice, are possible even in retrofit situations.* Hence, this energy integration design procedure is a very valuable tool.

Multiple Utilities

In the previous analysis, we considered the case of a single hot utility and a single cold utility. However, the analysis is also valid for multiple utilities. If we shift the

---

* D. Boland and E. Hindmarsh, "Heat Exchanger Network Improvements," *Chem. Eng. Prog.*, **80**(7): 47 (1984); B. Linnhoff and D. R. Vredeveld, "Pinch Technology Comes of Age," *Chem. Eng. Prog.*, **80**(7): 33 (1984).

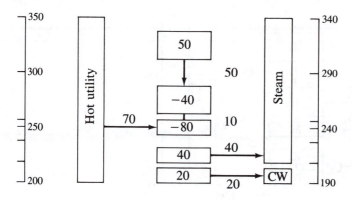

FIGURE 8.1-12
Multiple utilities.

temperature range in our previous example upward by 110°F, we obain the cascade diagram shown in Fig. 8.1-12. Now we see that as a hot utility we need to use steam having a temperature in excess of 260°F. Also, we could use steam at 210°F as one cold utility and cooling water as a second cold utility. With this procedure we would reject 40×10^3 Btu/hr to the steam and 20×10^3 Btu/hr to cooling water.

Note that there is no heat transfer between the bottom two temperature intervals when we use multiple utilities. Thus, we introduce another pinch, which we call a *utility pinch*, into the network. An additional utility pinch is added for each new utility considered. The effect of multiple utilities on a *T-H* diagram is shown on Fig. 8.1-13.

Also recognize that there are some obvious heuristics associated with the use of multiple utilities:

> Always add heat at the lowest possible temperature level relative to the process pinch. (8.1-10)

> Always remove heat at the highest possible temperature level relative to the process pinch. (8.1-11)

Phase Changes

The procedure requires that the FC_p values of the streams be constants. We can incorporate phase changes that take place at constant temperature into this formalism simply by assuming a 1°F temperature change at the temperature of the phase change and then calculating a fictitious FC_p value that gives the same heat duty as the phase change; i.e., if the heat corresponding to the phase change is $F \Delta H_v$, we write

$$F_f C_{pf}(1) = F \Delta H_v \qquad (8.1-12)$$

where F_f and C_{pf} are the fictitious values.

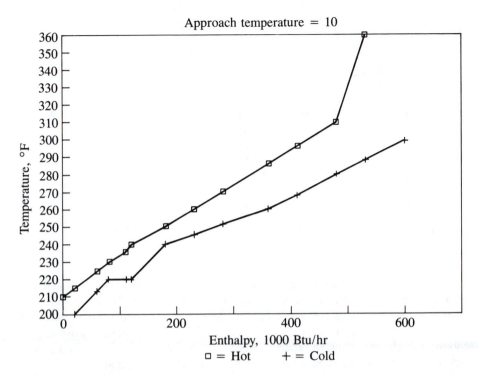

FIGURE 8.1-13
T-H diagram—multiple utilities.

For the case of mixtures, where a plot of enthalpy versus temperature is curved, we merely linearize the graph and select fictitious FC_p values that have the same heat duty (see Fig. 8.1-14). Thus, phase changes simply increase the number of temperature intervals considered.

Limitations of the Procedure

The calculation of the minimum heating and cooling loads requires the following:

$$\text{The } FC_p \text{ values of all streams are known.} \qquad (8.1\text{-}13)$$

$$\text{Inlet and outlet temperatures of all streams are known.} \qquad (8.1\text{-}14)$$

However, the design variables that fix the process flows (i.e., conversion, purge composition, molar ratio of reactants, etc.) must be determined from an optimization analysis. For each variable, the optimization involves recycle costs which depend on the heat-exchanger network. Thus, the optimum process flows depend on the heat-exchanger network, but we must know the flows to determine the network. We resolve this dilemma by calculating networks as a function of the flows to estimate the optimum design conditions.

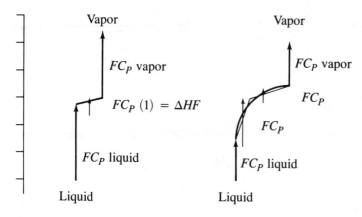

FIGURE 8.1-14
Phase changes.

Similarly, if a heat exchanger is used to preheat a stream leaving a flash drum or a gas absorber and entering a distillation column, then we must include this stream in our analysis. Moreover, if a process stream is used to drive the reboiler of a distillation column, rather than steam, then the optimum reflux ratio in that column will change. Thus, the energy integration analysis is coupled with the total design problem, and often some iterative case studies are required.

8.2   MINIMUM NUMBER OF EXCHANGERS

Our previous analysis allowed us to determine the minimum heating and cooling requirements for a heat-exchanger network. We use these results as a starting point to determine the minimum number of heat exchangers required. The analysis follows the procedures described in earlier references in this chapter.

First-Law Analysis

Suppose we consider the heating and cooling loads for each of the process streams as well as the minimum utility requirements that correspond to the second-law analysis (see Fig. 8.2-1). Now we ignore the minimum approach temperature and just consider how many paths (heat exchangers) are required to transfer the heat from the sources to the sinks. If we transfer 70×10^3 Btu/hr from the hot utility into stream 3, we still have a deficiency of 110×10^3 Btu/hr in stream 3. If we supply this deficiency from stream 1, we still have 20×10^3 Btu/hr of heat available in stream 1. If we transfer this excess to stream 4, we are left with a deficiency of 340×10^3 Btu/hr in stream 4.

The other calculations are shown in Table 8.1-1, and we find that there are five paths, or that five heat exchangers are required. We note that the heat loads just balance, which must always be the case because our minimum heating and

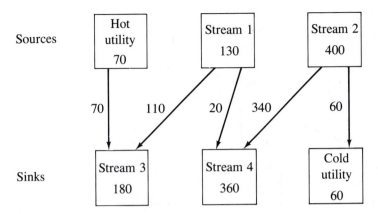

(a) Heat loads balance exactly: results of first-law analysis

(b) $\left(\begin{array}{c}\text{Number of}\\\text{Exchangers}\end{array}\right)=\left(\begin{array}{c}\text{Number of}\\\text{Streams}\end{array}\right)+\left(\begin{array}{c}\text{Number of}\\\text{Utilities}\end{array}\right)-1$

FIGURE 8.2-1
First law, minimum number of exchangers.

cooling loads satisfy the first-law requirement. We can generalize the result and state that normally

$$\left(\begin{array}{c}\text{Number of}\\\text{Exchangers}\end{array}\right)=\left(\begin{array}{c}\text{Number of}\\\text{Streams}\end{array}\right)+\left(\begin{array}{c}\text{Number of}\\\text{Utilities}\end{array}\right)-1 \qquad (8.2\text{-}1)$$

Independent Problems (Exact matches between subproblems)

Equation 8.2-1 is not always correct, as we can see by examining Fig. 8.2-2. In this example, we have merely increased the utility requirements, but the first-law analysis is still satisfied. If we transfer the heat between the sources and the sinks as shown in Fig. 8.2-2, then we require only four exchangers instead of five. However, we could also redraw the figure so that there are two, completely independent problems. This is also a general result, and a more rigorous statement of Eq. 8.2-1 is

$$\left(\begin{array}{c}\text{Number of}\\\text{Exchangers}\end{array}\right)=\left(\begin{array}{c}\text{Number of}\\\text{Streams}\end{array}\right)+\left(\begin{array}{c}\text{Number of}\\\text{Utilities}\end{array}\right)-\left(\begin{array}{c}\text{Number of}\\\text{Independent}\\\text{Problems}\end{array}\right) \qquad (8.2\text{-}2)$$

Loops

If we return to our original example and consider the arrangement shown in Fig. 8.2-3, we see that we can still satisfy the heat-transfer requirements between the sources and the sinks for any value of Q_E. However, for this configuration we need six exchangers. Also there is a loop in the network (i.e., we can trace a path through the network that starts at the hot utility, goes to stream 3, goes to stream 1, goes to

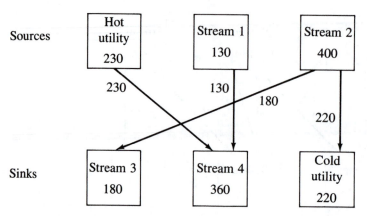

(a) Heat loads balance exactly: results of first-law analysis

(b) $\left(\begin{array}{c}\text{Number of}\\\text{Exchangers}\end{array}\right)=\left(\begin{array}{c}\text{Number of}\\\text{Streams}\end{array}\right)+\left(\begin{array}{c}\text{Number of}\\\text{Utilities}\end{array}\right)-\left(\begin{array}{c}\text{Number of}\\\text{Problems}\end{array}\right)$

FIGURE 8.2-2
Independent problems.

stream 4, and then goes back to the hot utility). Any time we can trace a path that starts at one point and returns to that same point, we say that we have a *loop* in the network. Each loop introduces an extra exchanger into the network.

Effect of Pinch: Second-Law Analysis

As part of our calculation of the minimum heating and cooling requirements, we found that there was a pinch temperature that decomposed the problem into two

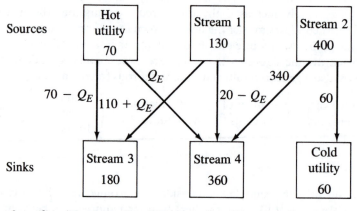

$\left(\begin{array}{c}\text{Number of}\\\text{Exchangers}\end{array}\right)=\left(\begin{array}{c}\text{Number of}\\\text{Streams}\end{array}\right)+\left(\begin{array}{c}\text{Number of}\\\text{Utilities}\end{array}\right)+\left(\begin{array}{c}\text{Number of}\\\text{Loops}\end{array}\right)-\left(\begin{array}{c}\text{Number of}\\\text{Problems}\end{array}\right)$

FIGURE 8.2-3
Loops.

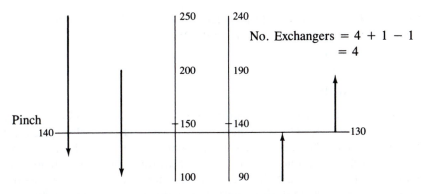

$$\text{No. Exchangers} = 4 + 1 - 1$$
$$= 4$$

(Number of Exchangers) = (Number of Streams) + (Number of Utilities) − 1
$$= 3 + 1 - 1 = 3$$

FIGURE 8.2-4
Effect of pinch.

distinct parts. That is, above the pinch we only supply heat from a utility, whereas below the pinch we only remove heat to a utility. Thus, to include the second-law analysis in our calculation of the minimum number of exchangers, we must apply Eq. 8.2-1 (or 8.2-2) to the streams above and below the pinch. From Fig. 8.2-4 we see that there are four streams above the pinch for our example, so that Eq. 8.2-1 (assuming no loops and no independent problems) gives

$$\text{Above pinch: } N_E = N_S + N_U - 1$$

$$= 4 + 1 - 1 = 4 \qquad (8.2\text{-}3)$$

Below the pinch temperature there are only three streams, and so

$$\text{Below pinch: } N_E = N_S + N_U - 1$$

$$= 3 + 1 - 1 = 3 \qquad (8.2\text{-}4)$$

Thus, to satisfy the minimum heating and cooling requirements requires a total of seven exchangers. However, to satisfy the first law requires only five. Then, we expect that the network for the minimum energy requirements will have two loops that cross the pinch (we introduce an additional exchanger for each loop). If we are willing to sacrifice some energy by transferring heat across the pinch, we can eliminate up to two exchangers from the network. Hence, there is a capital-operating cost trade-off that must be evaluated.

8.3   AREA ESTIMATES

We have been able to estimate the minimum heating and cooling requirements for a process without even specifying a heat-exhanger network (see Sec. 8.1). We can use these results to estimate the utility costs for a plant. It would be very desirable to estimate the capital costs associated with a heat-exchanger network without having to design a network. Fortunately, a technique for making this estimate has been presented by Townsend and Linnhoff* (which is an extension of a previous result by Hohmann).[†]

Estimating Areas

In Sec. 8.1 we developed a temperature-enthalpy plot (see Fig. 8.1-7). Suppose now that we include vertical lines whenever there is a change in the slope (see Fig. 8.3-1), and we consider that each interval represents one or more heat exchangers in parallel. From the graph we can read the heat duty for each exchanger and the values of the temperature driving forces at each end. Then if the heating and cooling curves correspond to a single stream, we can estimate the individual heat-transfer coefficient for each stream as well as the overall coefficient:

$$\frac{1}{U} = \frac{1}{h_i} + \frac{1}{h_0} \tag{8.3-1}$$

where the individual film coefficients include the fouling factors. The area of the heat exchanger is given by

$$A = \frac{Q}{U \, \Delta T_{LM}} \tag{8.3-2}$$

However, if there are multiple streams in any interval, then we must develop an appropriate expression for the overall heat-transfer coefficient. Suppose we consider the interval where two hot streams are matched against two cold streams. If we matched streams 1 with 3 and 2 with 4 (see Fig. 8.1-3), our results would be as shown in Fig. 8.3-2a. However, if we matched streams 1 with 4 and 2 with 3, then we would obtain the results shown in Fig. 8.3-2b. The heat loads and the log-mean temperature driving forces for each of the exchangers will be the same. For the case given in Fig. 8.3-2a, we find that

$$\frac{1}{U_{a_1}} = \frac{1}{h_1} + \frac{1}{h_3} \qquad \frac{1}{U_{a_2}} = \frac{1}{h_2} + \frac{1}{h_4} \tag{8.3-3}$$

so that the total area becomes

$$A_{Ta} = A_{a_1} + A_{a_2} = \frac{Q}{\Delta T_{LM} \, U_{a_1}} + \frac{Q}{\Delta T_{LM} \, U_{a_2}}$$

$$= \frac{Q}{\Delta T_{LM}} \left(\frac{1}{h_1} + \frac{1}{h_2} + \frac{1}{h_3} + \frac{1}{h_4} \right) \tag{8.3-4}$$

---

* D. W. Townsend and B. Linnhoff, "Surface Area Targets for Heat Exchanger Networks," Annual meeting of the Institution of Chemical Engineers, Bath, United Kingdom, April 1984.
[†] E. C. Hohmann, "Optimum Networks for Heat Exchange," Ph.D. Thesis, University of Southern California, 1971.

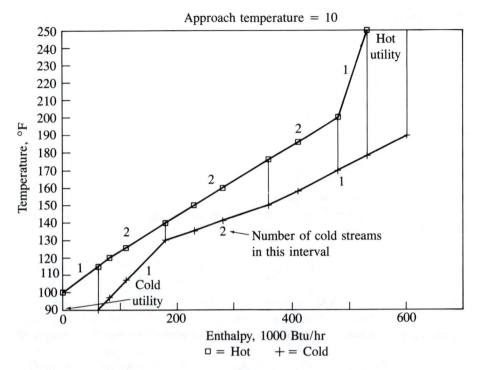

FIGURE 8.3-1
Temperature-enthalpy diagram.

(*a*) Match 1 and 3, 2 and 4   (*b*) Match 1 and 4, 2 and 3

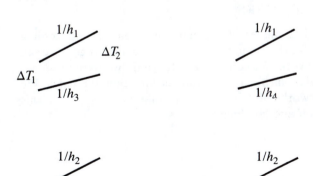

FIGURE 8.3-2
Overall heat-transfer coefficients.

For the configuration shown in Fig. 8.3-2*b* we find that

$$\frac{1}{U_{b_1}} = \frac{1}{h_2} + \frac{1}{h_3} \qquad \frac{1}{U_{b_2}} = \frac{1}{h_1} + \frac{1}{h_4} \tag{8.3-5}$$

Then the total area is

$$A_{\text{tot}} = A_{b_1} + A_{b_2} = \frac{Q}{\Delta T_{LM}}\left(\frac{1}{U_{b_1}} + \frac{1}{U_{b_2}}\right)$$

$$= \frac{Q}{\Delta T_{LM}}\left(\frac{1}{h_2} + \frac{1}{h_3} + \frac{1}{h_1} + \frac{1}{h_4}\right) \tag{8.3-6}$$

which is identical to our previous result.

This result is general, so that we can write an expression for the area in any interval as

$$A = \frac{Q}{\Delta T_{LM}}\left(\sum_i^{\text{hot}} \frac{1}{h_i} + \sum_j^{\text{cold}} \frac{1}{h_j}\right) \tag{8.3-7}$$

and we can estimate the total area simply by adding the results for all the intervals.

Of course, this approximate procedure does not give the same results as those obtained by designing a specific network (normally there are too many exchangers). Nevertheless, Eq. 8.3-7 does provide a reasonable estimate of the area required. This shortcut procedure is particularly useful when we are attempting to find the effect of the process flows on the capital cost of the heat-exchanger network. Once we have estimated the optimum flows, however, we need to undertake a detailed design of a heat-exchanger network.

8.4 DESIGN OF MINIMUM-ENERGY HEAT-EXCHANGER NETWORKS

Now that we have obtained estimates of the minimum heating and cooling requirements and an estimate for the minimum number of heat exchangers, we can design the heat-exchanger network. We consider the design in two parts: First we design a network for above the pinch and then another for below the pinch. We expect that the combined network will have two loops that cross the pinch. This analysis is taken from Linnhoff and Hindmarsh.[*]

Design above the Pinch

As the first step in the design procedure, we calculate the heat loads between either the inlet or the outlet temperature and the pinch temperature for each stream.

---

[*] B. Linnhoff and E. Hindmarsh, *Chem. Eng. Sci.*, **78**: 745 (1983).

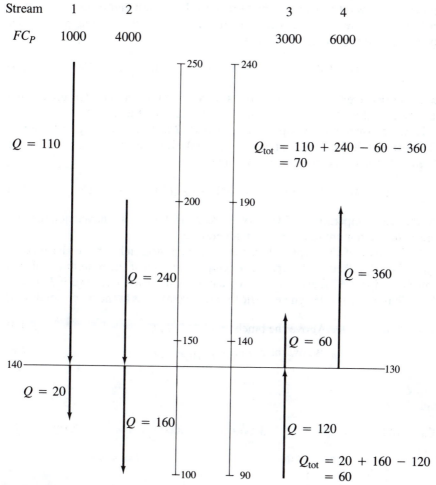

FIGURE 8.4-1
Heat load for streams.

Thus, for the first stream (see Fig. 8.4-1) we obtain

$$\text{Above pinch: } Q = FC_p\,\Delta T = 1000(250 - 140) = 110 \times 10^3 \qquad (8.4\text{-}1)$$

$$\text{Below pinch: } Q = 1000(140 - 120) = 20 \times 10^3 \qquad (8.4\text{-}2)$$

The results for the other streams are shown in Fig. 8.4-1.

Feasible Matches

If we attempt to match stream 1 above the pinch ($Q_H = 110$) with stream 3 ($Q_C = 60$), it is apparent that the maximum amount of heat transfer that is possible is the smaller of the two values ($Q = 60$). The approach temperature is just 10°F at the pinch, so we want to transfer the heat from the coldest end of the hot stream. Then if we calculate the temperature of the hot stream that would be the inlet temperature to the exchanger, we obtain

$$Q = 60 \times 10^3 = FC_p \, \Delta T = 1000(T_H - 140) \qquad T_H = 200 \qquad (8.4\text{-}3)$$

Since the outlet temperature of the cold stream is 150°F, the temperature driving force is 50°F and we have a feasible heat exchanger (see Fig. 8.4-2).

However, we might attempt to match stream 2 with stream 3. Again from Fig. 8.4-2 we see that the limiting heat load is $Q_C = 60$. However, when we calculate the inlet temperature of the hot stream, we obtain

$$Q = 60 \times 10^3 = FC_p \, \Delta T = 4000(T_H - 140) \qquad T_H = 155 \qquad (8.4\text{-}4)$$

Since the exit temperature of the cold stream is 150°F, we have violated our criterion for the minimum approach temperature.

A violation of this type will always occur immediately above the pinch if $(FC_p)_C > (FC_p)_H$. That is, the approach temperature is just the minimum value at the pinch, and the ΔT between the two curves will always decrease if $F_C C_{pC} > F_H C_{pH}$. Thus, there is a design heuristic for feasible matches at the pinch condition:

$$\text{Above the pinch: } F_H C_{pH} \leq F_C C_{pC} \qquad (8.4\text{-}5)$$

$$\text{Below the pinch: } F_H C_{pH} \geq F_C C_{pC} \qquad (8.4\text{-}6)$$

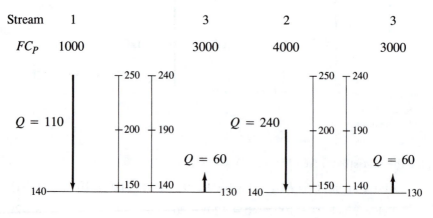

Stream	1	3	2	3
FC_P	1000	3000	4000	3000

$Q = 60{,}000 = 1000(T_H - 140)$
$T_H = 200$
Outlet $T_C = 150$
Match is feasible

$Q = 60{,}000 = 4000(T_H - 140)$
$T_H = 155$
Outlet $T_C = 150$
Match is not feasible
Violates minimum ΔT

FIGURE 8.4-2
Matches above the pinch.

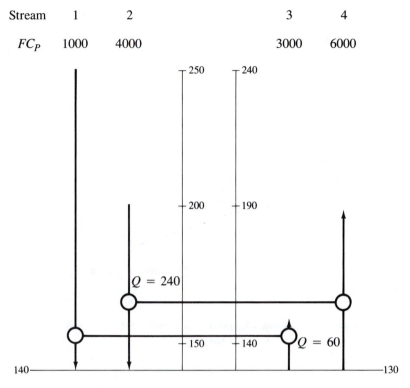

Stream	1	2		3	4
FC_P	1000	4000		3000	6000

(a) Put in the matches at the pinch.
(b) Maximize the heat loads to eliminate streams.
(c) See what is left.

FIGURE 8.4-3
Pinch matches.

Pinch Matches

From our feasibility criterion and Fig. 8.4-1, we see that above the pinch we can match stream 1 with either stream 3 or 4, and we can only match stream 2 with stream 4. Hence, we match stream 1 with stream 3 and stream 2 with stream 4. Also, we transfer the maximum amount of heat possible for each match in an attempt to eliminate streams from the problem. These pinch matches are shown in Fig. 8.4-3.

Next we consider the heat loads remaining. The criteria given by Eqs. 8.4-5 and 8.4-6 are not applicable away from the pinch, and we know that above the pinch we are allowed to add only heat. Hence, we must transfer all the heat remaining in stream 1 to stream 4, which is the only cold stream still available. The heat remaining in stream 1 is $(110 - 60) \times 10^3 = 50 \times 10^3$, and the remaining heating requirement of stream 4 is $(360 - 240) \times 10^3 = 120 \times 10^3$, so that we can install this heat exchanger (see Fig. 8.4-4). The remaining heating requirement of

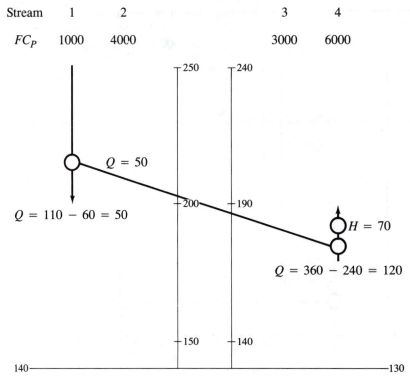

Stream	1	2		3	4
FC_P	1000	4000		3000	6000

$Q = 50$

$Q = 110 - 60 = 50$

$H = 70$

$Q = 360 - 240 = 120$

FIGURE 8.4-4
Matches away from the pinch.

70×10^3, which is just the minimum heating requirement, is supplied from a hot utility.

The complete design above the pinch is shown in Fig. 8.4-5. There are four exchangers, which is the minimum required value, and we have satisfied the minimum heating requirement. Thus, we have satisfied the design targets. The stream temperatures are also shown on Fig. 8.4-5, and the temperature driving force at the ends of every heat exchanger is 10°F or greater.

Alternatives

For the example under consideration, the pinch matches are unique. However, for the other matches away from the pinch the utility heater can be placed either before or after the heat exchanger connecting streams 1 and 4. Figure 8.4-6a shows the location of the heater after the other heat exchanger has been inserted, but Fig. 8.4-6b shows the result with the last two heat exchangers interchanged. Calculations show that both alternatives are feasible. However, the driving forces for the heat

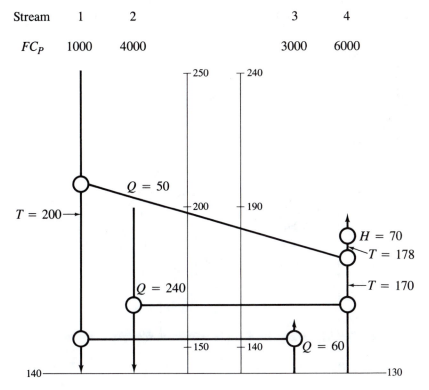

FIGURE 8.4-5
Design above the pinch.

exchanger are largest (a lower exchanger area) when the utility heater is at the highest temperature, although heat must be supplied at a higher temperature level. Thus, in some cases one alternative may have a lower cost than another.

Design below the Pinch

We use exactly the same design procedure below the pinch. For a feasible match we require that $F_H C_{pH} \geq F_C C_{pC}$ (Eq. 8.4-6). Therefore, for our example, we can only match stream 2 with stream 3 (see Fig. 8.4-7). We put in this exchanger and maximize the load to eliminate a stream from the problem, $Q = 120 \times 10^3$ (see Fig. 8.4-8). When we examine what is left, we have only hot streams that need to be cooled. We are only allowed to reject heat to a cold utility below the pinch, so we install two coolers (see Fig. 8.4-9).

The complete design for below the pinch is shown in Fig. 8.4-10. We see that the total amount of heat rejected to cold utility is $Q_c = (20 + 40) \times 10^3 = 60 \times 10^3$, which is identical to the minimum cooling requirement. The number of

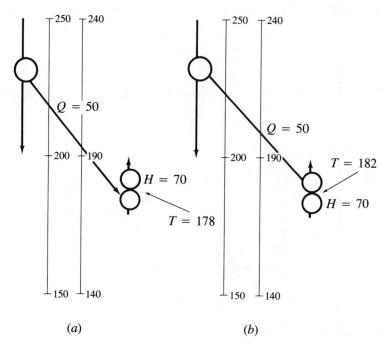

FC_P 1000   6000   1000   6000

FIGURE 8.4-6
Design alternatives.

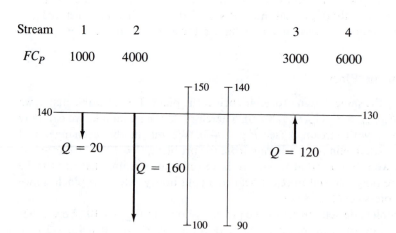

Stream   1   2   3   4

FC_P 1000   4000   3000   6000

FIGURE 8.4-7
Design below the pinch.

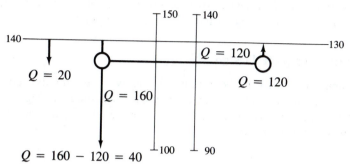

FIGURE 8.4-8
Pinch matches.

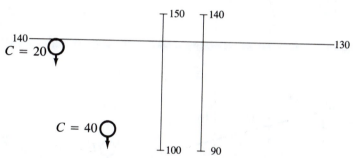

FIGURE 8.4-9
Add coolers.

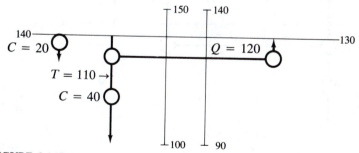

FIGURE 8.4-10
Complete design below the pinch.

exchangers used is 3, which is the minimum number. Also, the temperature driving force at each end of every exchanger is 10°F or greater, so the design is feasible. Thus, we have established one design alternative below the pinch.

Minimum Energy: Complete Design

A complete design that satisfies the minimum energy requirements and the minimum number of exchangers above and below the pinch is shown in Fig. 8.4-11. The total heating load is 70×10^3 Btu/hr, while the total cooling load is 60×10^3 Btu/hr. There are seven exchangers.

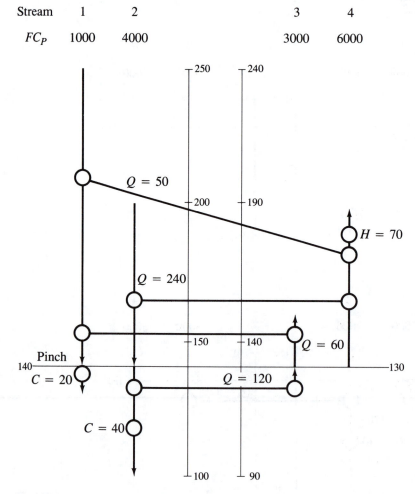

FIGURE 8.4-11
Complete minimum energy design.

As we mentioned earlier, if we apply Eq. 8.2-1 to our example, we predict that we need only five exchangers (although the minimum energy requirement cannot be satisfied with less than seven exchangers). Therefore we anticipated that there would be two loops that crossed the pinch. (A loop is a path that we can trace through the network which starts from some exchanger and eventually returns to that same exchanger.) A loop may pass through a utility (see Fig. 8.2-3). After examining Fig. 8.4-11, we find that there are three loops (see Fig. 8.4-12). Two of these loops pass through the cold utility, and we show later that if we break one of these loops, the other will also be broken. Hence, there are two independent loops, and it should be possible to remove two exchangers from the network shown in

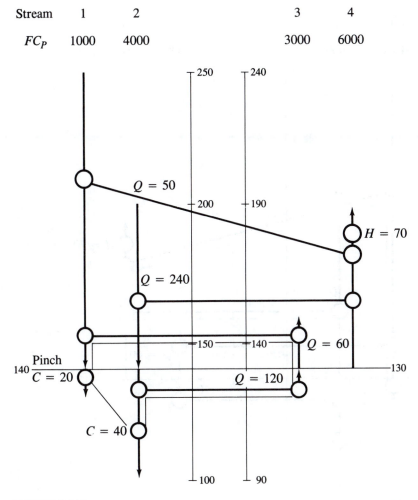

FIGURE 8.4-12a
Loops.

Fig. 8.4-11 by supplying more energy to the process (and removing more). We discuss this procedure in the next two sections.

Optimum Value of the Minimum Approach Temperature

As we noted earlier, the minimum heating and cooling loads change as we change the minimum approach temperature. However, since the heat-exchanger area in the neighborhood of the pinch will change in the opposite direction, there will be an

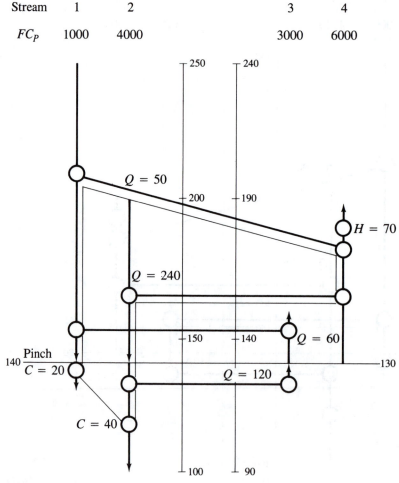

FIGURE 8.4-12b
Loops.

optimum value of the minimum approach temperature; i.e., as we decrease ΔT_{min}, we increase the area. Moreover, this optimum value will change with the process flows. We discuss this optimization problem in more detail later in this chapter.

Additional Complexities in the Design Procedure

The design problem is not always as simple as the example considered. Thus, in some cases it is necessary to split streams. These additional complexities are discussed in Sec. 8.7.

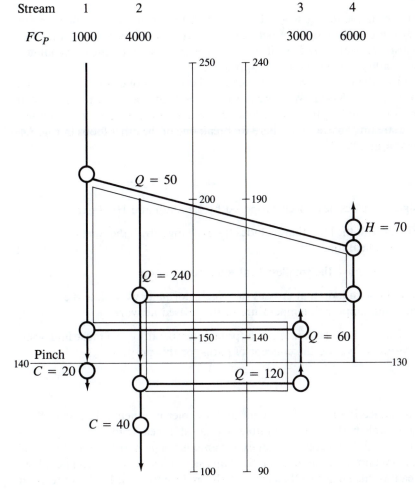

FIGURE 8.4-12c
Loops.

8.5 LOOPS AND PATHS

Loops and paths provide ways of shifting heat loads through a network.

Loops

A loop is a set of connections that can be traced through a network that starts from one exchanger and returns to the same exchanger (see Fig. 8.5-1). A loop may also pass through a utility (see Fig. 8.5-2). The existence of a loop implies that there is an extra exchanger in the network. That is, if we break the loop, we can remove an exchanger.

Breaking Loops

Consider the example in Fig. 8.5-3. The energy requirements are satisfied for any value of Q_E (an excess amount of heat that we put into a network, and therefore that we must remove). However, if we set $Q_E = 20$, one of the heat exchangers (connecting paths) in the network disappears.

Of course, the loop shown in Fig. 8.5-3 is the same as one of the loops in our design problem, Fig. 8.4-12. We always satisfy the heat loads of each stream by subtracting an amount Q_E from one exchanger, but adding it to another exchanger on the same stream. An example where we break one of the other loops in Fig. 8.4-12 is shown in Fig. 8.5-3.

Heuristics

Three design heuristics have been proposed by Linnhoff and Hindmarsh:

> First, break the loop that includes the exchanger with the smallest possible heat load. (8.5-1)
>
> Always remove the smallest heat load from a loop. (8.5-2)
>
> If we break a loop that crosses the pinch, normally we violate the minimum approach temperature in the revised network. (8.5-3)

Of course, if we violate the minimum approach temperature, we must find some way of restoring it. We use the concept of paths for this purpose.

Paths

A path is a connection between a heater and a cooler in a network. Figure 8.5-4 shows two possible paths for our example. We can shift heat loads along a path, as shown in Fig. 8.5-5. We merely add an excess amount of heat to the hot utility and subtract it from another exchanger on the same stream (so that the total heat load for the stream is unchanged). Of course, we also reduce the heat load on the other stream that passes through this exchanger. Thus, we must add heat to this stream in either another exchanger or a cooler.

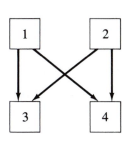
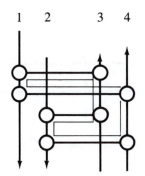

FIGURE 8.5-1
Loops.

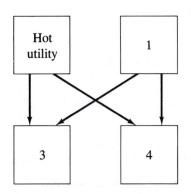
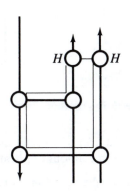

FIGURE 8.5-2
Loop through a utility.

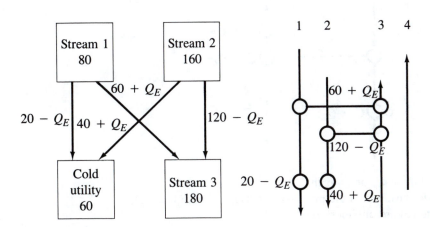

FIGURE 8.5-3
Breaking loops.

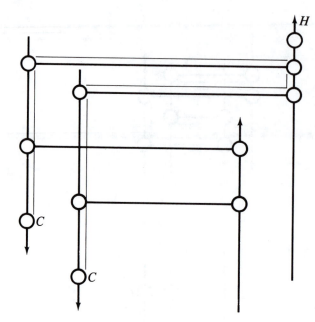

FIGURE 8.5-4
Paths.

Note that

> When we add heat to a heater and shift it along a path, we must
> remove the same amount of heat in a cooler. (8.5-4)

We often shift heat along a path to restore a minimum approach temperature; this
procedure always increases the energy consumption of the process.

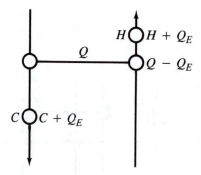

1. Can shift heat along a path.
2. Transfer heat across the pinch—more heat in, more heat out.
3. Use to restore minimum ΔT.

FIGURE 8.5-5
Shift heat along a path.

8.6 REDUCING THE NUMBER OF EXCHANGERS

We can summarize some general rules concerning the design procedure:

> The number of exchangers required for the overall process is always less than or equal to that for the minimum energy network.   (8.6-1)

> If the design procedure for the minimum energy network is used, there will normally be loops across the pinch.   (8.6-2)

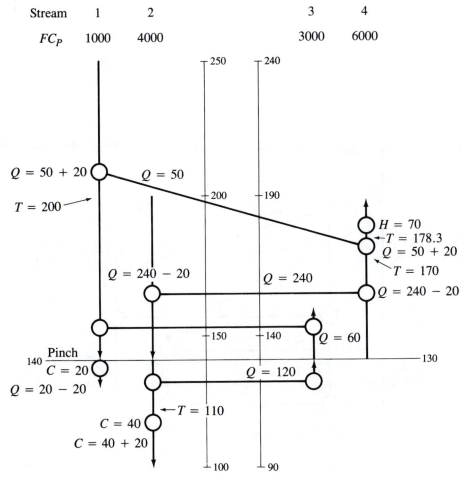

Break loop containing smallest heat load $Q = 20$

FIGURE 8.6-1
Break a loop in minimum energy design.

We can break these loops by transferring heat across the pinch, but we will introduce at least one violation of the specified $\Delta T_{min} = 10°F$. (8.6-3)

We can restore ΔT_{min} by shifting heat along a path, which increases the energy consumption of the process. (8.6-4)

Hence, we have a procedure for reducing the number of heat exchangers (which we expect will reduce the capital cost) at the expense of consuming more energy (which will increase the operating costs). Obviously, we want to find the heat-exchanger network (as a function of the process flows) which has the smallest total annual cost.

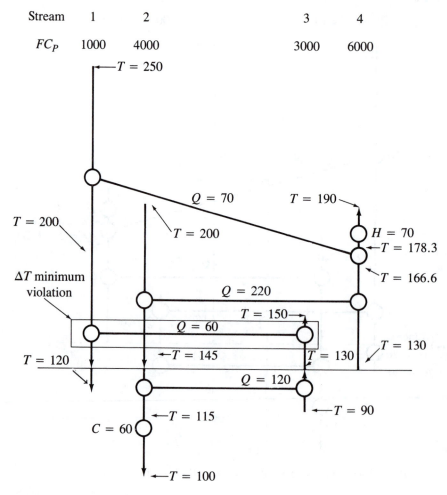

FIGURE 8.6-2
Cooler removed.

Breaking the Loop with the Smallest Heat Load

From Fig. 8.4-12 we see that the smallest heat load in any of the loops is that in the cooler, where $Q_C = 20 \times 10^3$. We arbitrarily decide to break the loop shown in Fig. 8.4-12b. Thus, we start at the cooler and subtract and add $Q_E = 20$ as we proceed around the loop (see Fig. 8.6-1). Now we calculate the new heat loads and the new values of the intermediate temperatures; see Fig. 8.6-2. From Fig. 8.6-2 we see that the exit temperature of stream 1 is actually 10°F lower than the inlet stream for the exchanger with $Q = 60$; i.e., the approach temperature is -10°F, which is impossible.

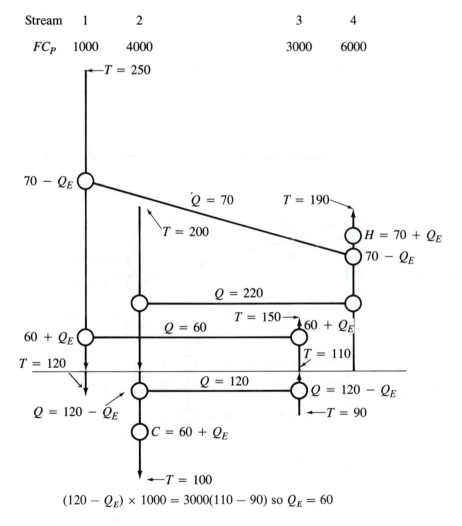

FIGURE 8.6-3
Use path to restore ΔT_{min}.

Restoring $\Delta T_{\min}$

We restore the minimum approach temperature at this point in the network by shifting heat along a path; see Fig. 8.6-3. Since the outlet temperature of stream 1 is 120°F, we want the new inlet temperature to be 110°F. Then we calculate the amount of heat that we must shift along the path to obtain this intermediate temperature. From Fig. 8.6-3,

$$(120 - Q_E) \times 10^3 = 3000(110 - 90)$$
$$Q_E = 60 \times 10^3 \tag{8.6-5}$$

The revised network is shown in Fig. 8.6-4.

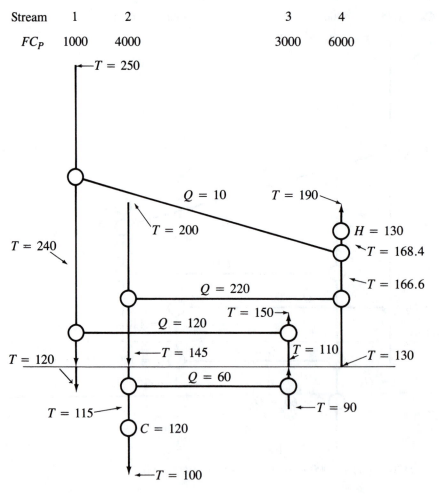

Stream	1	2		3	4
FC_P	1000	4000		3000	6000

FIGURE 8.6-4
Revised network.

Breaking the Second Loop

Breaking one loop through a cooler (see Fig. 8.4-12) has broken both loops through the cooler. However, there is still a second loop remaining in Fig. 8.6-4. The smallest heat load in this loop is 10×10^3 Btu/hr, and so we subtract and add $Q_E = 10$ as we proceed around the loop. The result is shown in Fig. 8.6-5, and the new stream temperatures are included on this figure. For this case we do not encounter another violation of ΔT_{min} (although we often do). In fact, the minimum approach temperature for this design exceeds 10°F ($120 - 107 = 13$°F).

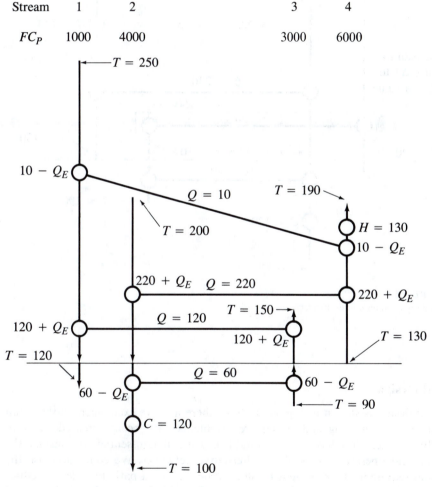

Stream	1	2		3	4
FC_P	1000	4000		3000	6000

FIGURE 8.6-5
Break second loop.

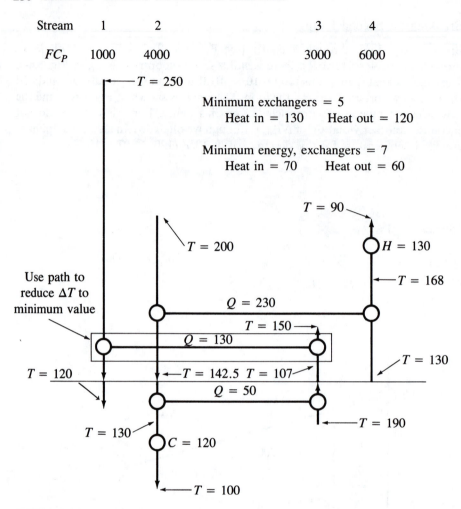

FIGURE 8.6-6
Final design—minimum exchangers.

Final Design

A final design is shown in Fig. 8.6-6. Now there are five exchangers, rather than seven, but the heating and cooling requirements have been increased by 60×10^3 Btu/hr, i.e., they have almost doubled. Thus, it is essential to consider the capital and operating costs of the alternative networks. We could decrease the utilities requirements somewhat by shifting heat along a path, in order to reduce the minimum approach temperature to 10°F from 13°F after we break the second loop. Clearly, we need to optimize the design.

8.7 A MORE COMPLETE DESIGN ALGORITHM—STREAM SPLITTING

There are some situations where our design procedure does not seem to work. However, these additional complications can always be accommodated by splitting a stream. Several examples of this type are presented below, and then a more general design algorithm is discussed.

Number of Hot and Cold Streams

Consider the example shown in Fig. 8.7-1. The values of $F_H C_{pH}$ for both hot streams are less than the $F_C C_{pC}$ value for the cold stream. However, if we install any heat exchanger, such as one between streams 2 and 3, then the increase in

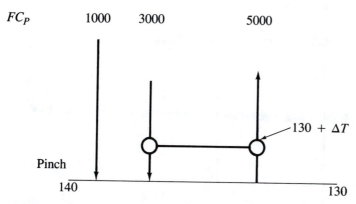

FIGURE 8.7-1
A counterexample.

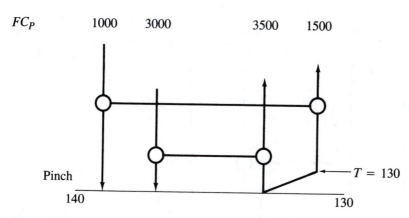

FIGURE 8.7-2
Split the cold stream.

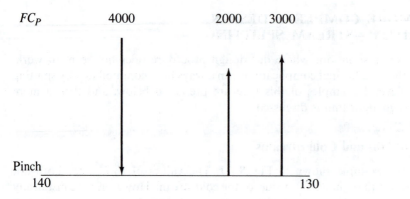

FIGURE 8.7-3
Another counterexample.

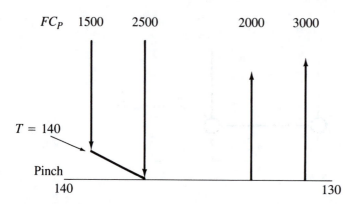

FIGURE 8.7-4
Split the hot stream.

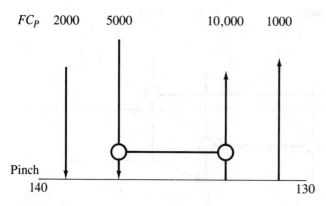

FIGURE 8.7-5
Another counterexample.

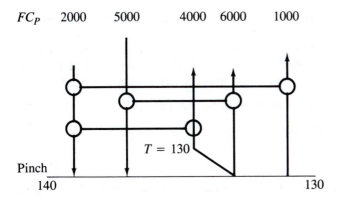

FIGURE 8.7-6
Split a cold stream.

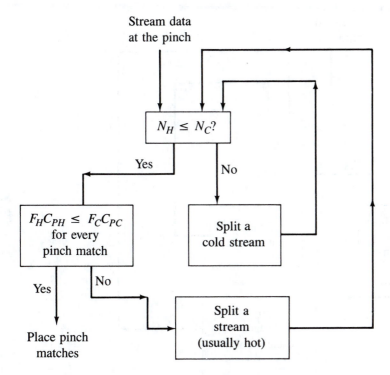

FIGURE 8.7-7
Design procedure above the pinch. (*From B. Linnhoff et al., 1982.*)

temperature of stream 3 will prevent any match for stream 1 because the minimum approach temperature will not be satisfied.

We can resolve this difficulty very simply just by splitting stream 3 (see Fig. 8.7-2), although we must be careful that $F_H C_{pH} \leq F_C C_{pC}$ remains satisfied after the split. From this example we recognize that there are new design heuristics:

$$\text{Above the pinch, we must have } N_H \leq N_C. \qquad (8.7\text{-}1)$$

$$\text{Below the pinch, we must have } N_H \geq N_C. \qquad (8.7\text{-}2)$$

$F_H C_{pH}$ versus $F_C C_{pC}$

As another example consider the problem in Fig. 8.7-3. The criterion $N_H < N_C$ is satisfied, but now $F_H C_{pH} > F_C C_{pC}$, which prevents a feasible match. However, if we split the hot stream (see Fig. 8.7-4) and adjust the flows such that $F_H C_{pH} \leq F_C C_{pC}$, we obtain a satisfactory solution.

Splitting Hot versus Cold Streams

We usually split a hot stream as in Fig. 8.7-4, but not always. Consider the problem shown in Fig. 8.7-5. If we split the hot stream, then $N_H > N_C$, which is not

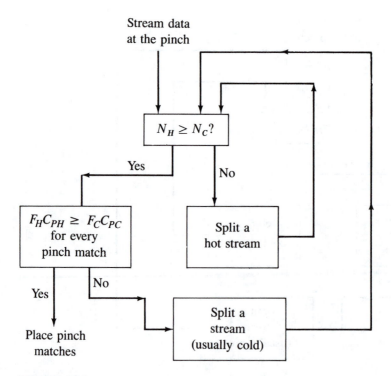

FIGURE 8.7-8
Design procedure below the pinch. (*From B. Linnhoff et al., 1982.*)

allowable. However, if we split a cold stream, we can make a match between streams 1 and 4 away from the pinch and thereby avoid the $F_H C_{pH} \le F_C C_{pC}$ constraint (see Fig. 8.7-6).

General Design Procedure

A general design algorithm for conditions above and below the pinch is shown in Figs. 8.7-7 and 8.7-8. After we put in the pinch matches correctly, it usually is a simple task to complete the design.

8.8   HEAT AND POWER INTEGRATION

According to the first law, heat and power are related. Thus, it should not be surprising that the energy integration procedure can be extended to give results for heat and power integration. A detailed analysis of heat and power integration can be found in Towsend and Linnhoff.* Here, we only outline the basic ideas.

Heat Engines

Every text on thermodynamics discusses the performance and efficiency of heat engines as individual entities. However, if we take a total systems viewpoint, we obtain a totally different perspective. This systems approach was previously discussed for gas absorber/distillation processes in Chap. 3.

Suppose we consider a cascade diagram for the process shown in Fig. 8.8-1. We put heat into the network, no heat is allowed to cross the pinch, and we remove heat below the pinch. Now suppose that we install a heat engine above the heat input to the cascade diagram (Fig. 8.8-1 shows this arrangement at the highest possible temperature, although from Fig. 8.1-5 we see that it could be a lower temperature). If we add an incremental amount of heat W to this engine, recover an amount of work W, and reject the remaining heat Q_{in} to the network (which is the amount of heat that we were required to add in any case), then the efficiency of the heat engine based on the incremental amount of energy input is 100%.

Thermodynamics texts imply that the efficiency of a heat engine is always less than 100% because some of the heat output must be wasted. However, with the arrangement shown in Fig. 8.8-1, this waste heat is just what is required for another task. Thus, again, a systems viewpoint leads to different conclusions from the consideration of a particular unit in isolation.

From Fig. 8.8-2 we see that if we install a heat engine below the pinch, we can also obtain an incremental efficiency of 100%. We convert some of the heat that would be discarded to a cold utility in any event into useful power, and then we

---

* D. W. Townsend and B. Linnhoff, "Heat and Power Networks in Process Design; Parts I and II," *AIChE J.*, **29**: 742, 748 (1983).

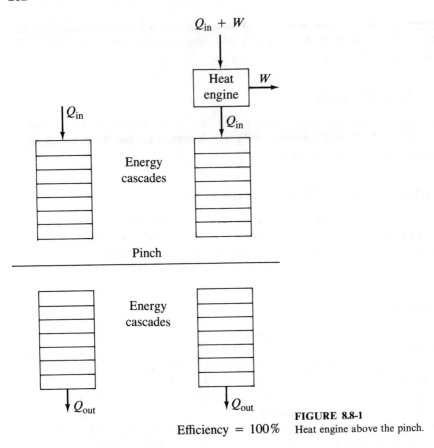

Efficiency $= 100\%$

FIGURE 8.8-1
Heat engine above the pinch.

discard a smaller amount of heat to the utility. However, if a heat engine takes in energy above the pinch and discards it below the pinch (see Fig. 8.8-2), then we gain nothing from heat and power integration; i.e., the efficiency of the heat engine is exactly the same as it would be if the heat engine were isolated from the remainder of the process. Hence, we obtain this heuristic:

> Place heat engines either above or below the pinch, but not across the pinch.
>
> (8.8-1)

Design Procedures for Heat and Power Integration

A design procedure for heat and power integration has been presented by Townsend and Linnhoff. The procedure is basically an attempt to match the enthalpy-temperature profile of various types of heat engines with the profile for the process (the grand composite curve discussed in Sec. 8.1-1 is used for the matching). The details of the procedure can be found in Townsend and Linnhoff's paper.

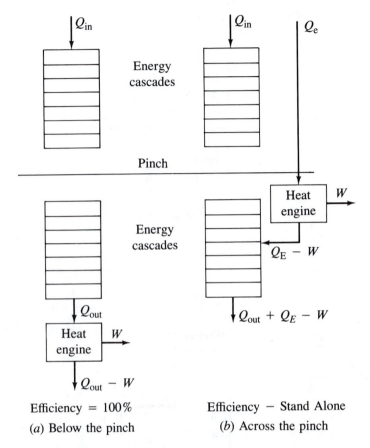

FIGURE 8.8-2
Heat engines versus the pinch.

The heat and power integration procedure is also very useful for the design of utilities systems. Thus, if we energy integrate a whole petrochemical complex, including the utility system, we can often obtain large energy savings.

Heat Pumps

Heat pumps are the opposite of heat engines. We put work into a heat pump to raise the temperature level of the available heat. From Fig. 8.8-3b we see that if we place a heat pump across the pinch, we reduce the heating and cooling requirements of the process. However, as shown in Fig. 8.8-3c, placing a heat pump above the pinch does not provide any benefit. Moreover, as shown in Fig. 8.8-3d, placing a heat pump below the pinch increases the energy requirement of the process and the amount of energy rejected to the cold utility. Thus, we obtain another heuristic:

$$\text{Place heat pumps across the pinch.} \qquad (8.8\text{-}2)$$

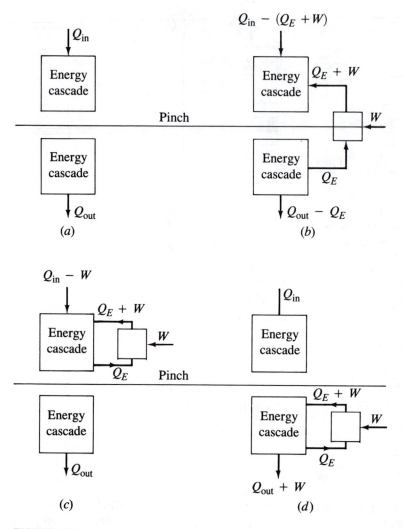

FIGURE 8.8-3
Heat pumps.

8.9 HEAT AND DISTILLATION

Distillation columns are one of the major energy consumers in a chemical plant. Referring to Fig. 8.9-1, we put heat into the reboiler and remove heat in the condenser. In all the previous work on energy integration we plotted temperature scales with the high temperature at the top of the graphs, and so we will turn our distillation column upside down and consider a heat input–heat output diagram, as shown in Fig. 8.9-1b.

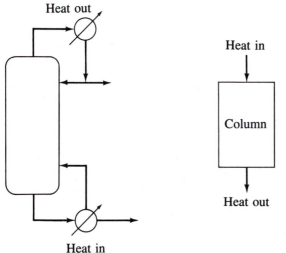

FIGURE 8.9-1
Distillation column.

Distillation Columns above or below the Pinch

Suppose we consider the cascade diagram shown in Fig. 8.9-2. We borrow a certain amount of energy Q_E from the cascade, use it in our distillation column, and then return it to the cascade above the pinch.* With this approach we run our column on borrowed energy, with no external utilities. Hence, the operating costs for the column are reduced significantly. Exactly the same behavior is obained if we can borrow the energy to run the column from below the pinch and then return it below the pinch (see Fig. 8.9-2). However, if the column appears to be energy-integrated but it crosses the pinch, then the result is no better than if we had installed the column as a stand-alone unit; i.e., we must supply the extra energy from a hot utility and remove it to a cold utility (see Fig. 8.9-3).

Of course, if a base-case design indicates that a column falls across the pinch, we might be able to shift it above the pinch by raising the column pressure or to shift it below the pinch by dropping its pressure. Industrial case studies at Imperial Chemical Industries and Union Carbide indicate that substantial energy savings can be obtained by this pressure-shifting idea.

The results above give us a new design heuristic:

Place distillatiᴜn columns either above or below the pinch.   (8.9-1)

---

* B. Linnhoff, H. Dunford, and R. Smith, "Heat Integration of Distillation Columns into Overall Processes," *Chem. Eng. Sci.*, **38**: 1175 (1983).

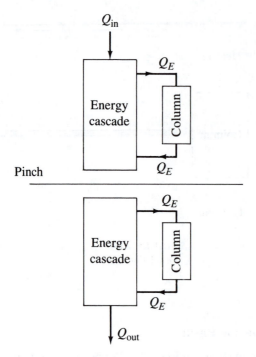

FIGURE 8.9-2
Cascade diagram and distillation. [*From B. Linnhoff, H. Dunford, and R. Smith, Chem. Eng. Sci.* **33**: *1175 (1985).*]

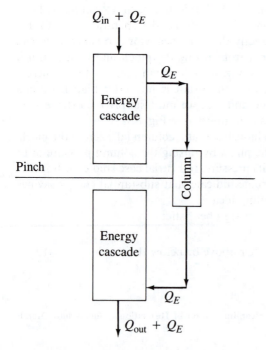

Same as a stand-alone column

FIGURE 8.9-3
Distillation across the pinch. [*From B. Linnhoff, H. Dunford, and R. Smith, Chem. Eng. Sci.* **33**: *1175 (1985).*]

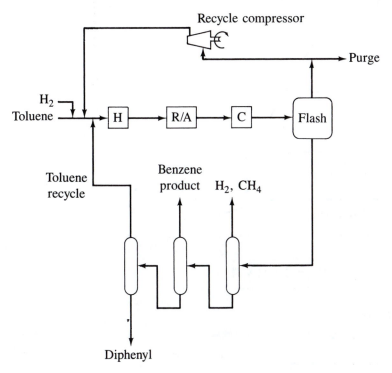

FIGURE 8.9-4
HDA process.

Design Procedure

A design procedure for the energy integration of a train of distillation columns with a process has been presented by Hindmarsh and Townsend.* If we consider the HDA process[†] (see Fig. 8.9-4) and develop the composite enthalpy-temperature diagram for a particular set of flows, we obtain the results shown in Fig. 8.9-5. This diagram shows that the column condensers and reboilers fall across the pinch, which we said was not a desirable situation.

To see how to pressure-shift the columns and to best integrate the columns with the remainder of the process, it is simpler first to remove the columns from the process (see Fig. 8.9-6) and to consider just the energy integration of the process with no columns (see Fig. 8.9-7). The corresponding *T-H* curves for the columns are shown in Fig. 8.9-8, and we can move these curves around by pressure-shifting

---

* E. Hindmarsh and D. W. Townsend, "Heat Integration of Distillation Systems into Total Flowsheets —A Complete Approach," Paper presented at the Annual AIChE Meeting, San Franciso, Calif., 1984.

[†] This example was developed by the ICI Process Synthesis Team, D. W. Townsend, E. Hindmarsh, H. Dunford, A. Patel, D. C. Woodcock, and A. P. Rossiter.

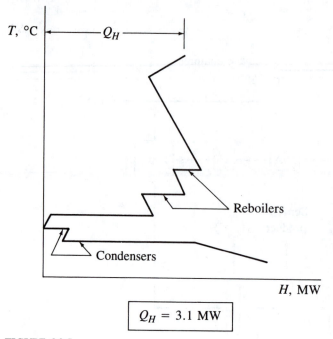

FIGURE 8.9-5
Total flowsheet T-H plot.

$$\boxed{Q_H = 3.1 \text{ MW}}$$

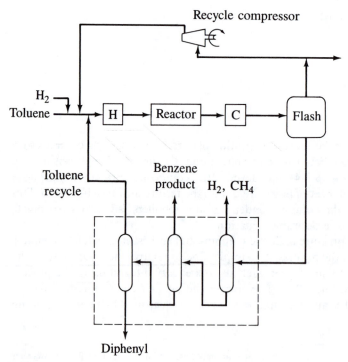

FIGURE 8.9-6
Simplified flowsheet.

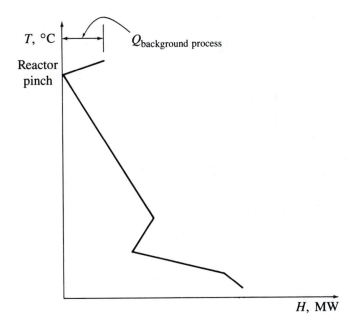

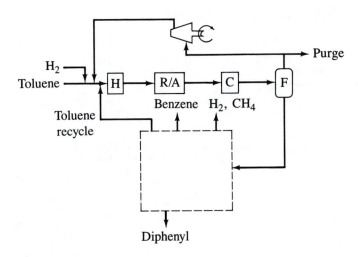

FIGURE 8.9-7
Background process.

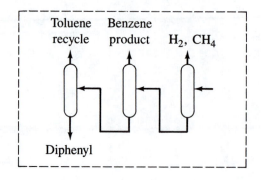

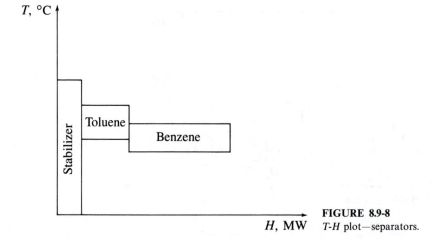

FIGURE 8.9-8
T-H plot—separators.

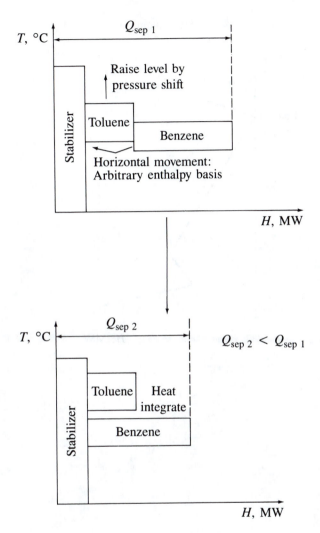

$Q_{\text{sep 2}} < Q_{\text{sep 1}}$

FIGURE 8.9-9
Integration of "stand-alone" separation system.

Profile Matching   HDA plant

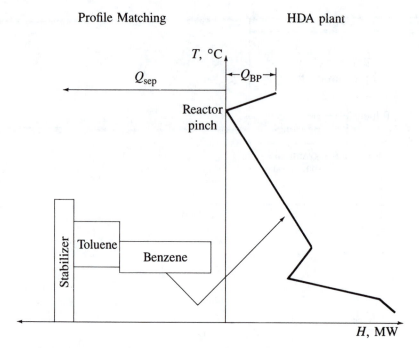

FIGURE 8.9-10
Intergrating the columns with the process.

the columns (see Fig. 8.9-9). Next we try to use as much heat as possible from the process to satisfy the energy requirements of the columns (see Fig. 8.9-10), and we pressure-shift one or more columns to minimize the external energy requirements (see Fig. 8.9-11). This pressure-shifting procedure is discussed in detail for the energy integration of stand-alone columns, or for the columns that cannot be easily integrated with a process (see Fig. 8.9-11), by Andrecovich and Westerberg* (see Appendix A.6). The general approach is to split columns and pressure-shift the sections to use the full range of utilities that are available.

---

* M. J. Andrecovich and A. W. Westerberg, "A Single Synthesis Method on Utility Bonding for Heat-Integrated Distillation Sequences," *AIChE J.*, **31**: 363 (1985).

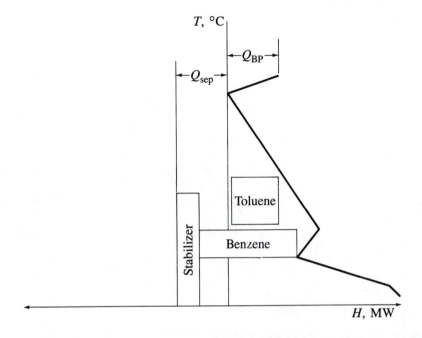

$$Q_{sep} + Q_{BP} = 1.6 + 0.9 \text{ MW}$$
$$= 2.5 \text{ MW}$$
$$\text{Cf. base case of} \quad 3.1 \text{ MW}$$

FIGURE 8.9-11
Final design.

8.10   HDA PROCESS

A study of the sensitivity of the total processing costs to heat-exchanger network alternatives was undertaken by Terrill and Douglas.* They developed a heat-exchanger network for a base-case design ($x = 0.75$, $y_{pH} = 0.4$) for the HDA process. The T-H diagram is shown in Fig. 8.10-1. They also developed six alternative heat-exchanger networks, all of which had close to the maximum energy recovery (see Figs. 8.10-2 through 8.10-7). (Note that the quench stream after the reactor is not shown on these graphs.) Most of the alternatives include a pressure shifting of the recycle column, and the other distinguishing feature is the number of column reboilers that are driven by the hot reactor products.

---

* D. L. Terrill and J. M. Douglas, *I&EC Research*, **26**: 685 (1987).

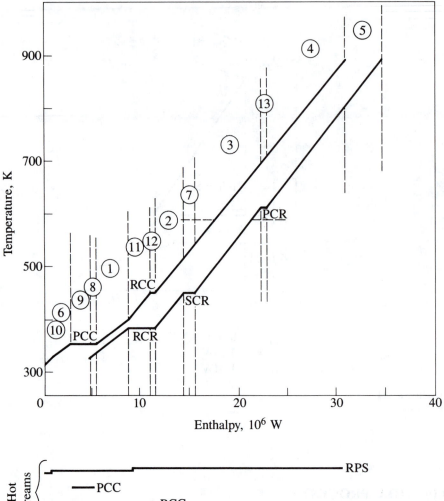

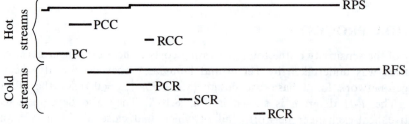

FIGURE 8.10-1

T-H diagram for the HDA process. PCC = product column condenser; RCC = recycle column condenser; RCR, SCR, and PCR are the reboilers for the recycle, stabilizer, and product columns, respectively; RFS and RPS are the reactor feed and product streams. [*From D. L. Terrill and J. M. Douglas, I&EC Research,* **26**: *685 (1987), with permission of the American Chemical Society.*]

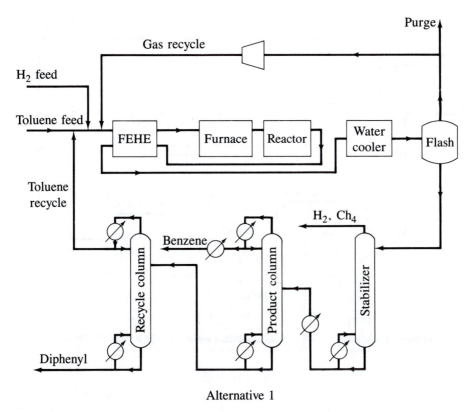

Alternative 1

FIGURE 8.10-2
HDA process with a feed-effluent heat exchanger. (*Note*: The quench stream following the reactor is not shown on these graphs.) [*From D. L. Terrill and J. M. Douglas, I&EC Research,* **26**: *685 (1987), with permission of the American Chemical Society.*]

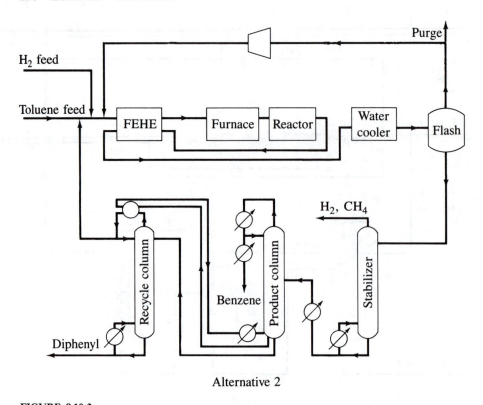

Alternative 2

FIGURE 8.10-3
Energy integration alternative 2. [*From D. L. Terrill and J. M. Douglas, I&EC Research*, **26**: 685 (*1987*), *with permission of the American Chemical Society.*]

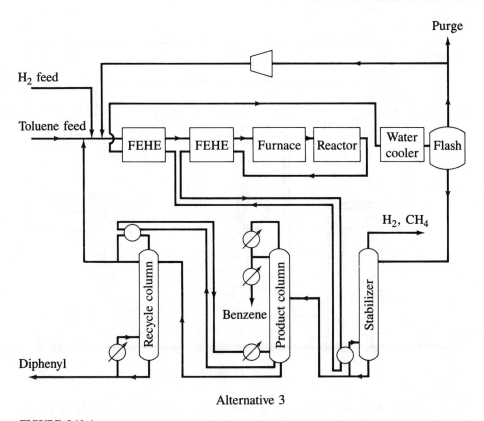

Alternative 3

FIGURE 8.10-4
Energy integration alternative 3. [*From D. L. Terrill and J. M. Douglas, I&EC Research,* **26**: *685* (*1987*), *with permission of the American Chemical Society.*]

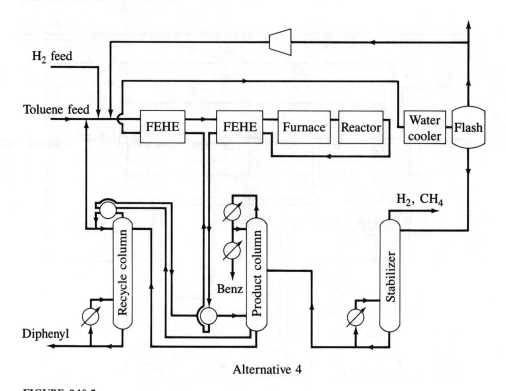

Alternative 4

FIGURE 8.10-5
Energy integration alternative 4. [*From D. L. Terrill and J. M. Douglas, I&EC Research,* **26**: *685* (*1987*), *with permission of the American Chemical Society.*]

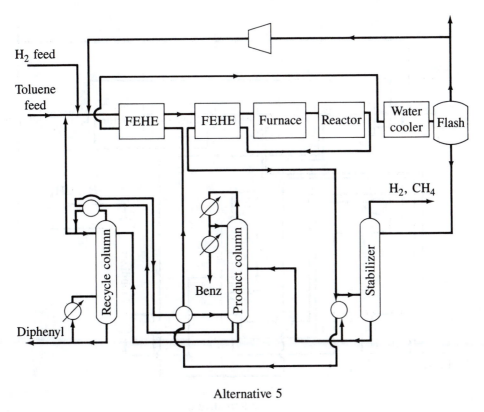

Alternative 5

FIGURE 8.10-6
Energy integration alternative 5. [*From D. L. Terrill and J. M. Douglas, I&EC Research,* **26**: *685 (1987),
with permission of the American Chemical Society.*]

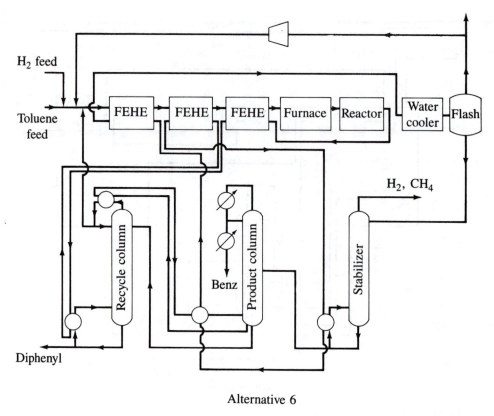

Alternative 6

FIGURE 8.10-7
Energy integration alternative 6. [*From D. L. Terrill and J. M. Douglas, I&EC Research,* **26**: *685 (1987), with permission of the American Chemical Society.*]

Effect of Energy Integration versus Optimization

The benefit obtained from energy integration with the base-case flow rates for the six alternatives is given in Table 8.10-1. The energy savings from the energy integration fall between 29 and 43%, which is in the range obtained by ICI and Union Carbide, but the cost savings are in the range from -1 to 5%. The cost savings are not as dramatic because the raw-material costs dominate the process economics.

If we optimize each of the process alternatives, we obtain the results shown in Table 8.10-2. We obtain savings in the range from 21 to 26%, which is quite dramatic. The improved energy integration has allowed us to increase the recycle flows, which decreases the raw-material costs. The increased recycle flows actually increase the utilities consumption, but the raw-material savings more than compensate for the increased energy costs. Similar results were obtained by Duran

TABLE 8.10-1
Energy integration for the HDA process

	Base case	Alternative					
		1	2	3	4	5	6
1. TAC ($10^6/yr) base-case flows	6.38	6.40	6.45	6.38	6.11	6.04	6.03
2. Utilities usage (MW), base-case flows	12.7	9.06	7.68	7.34	7.30	7.30	7.30
3. Energy savings, %		29	40	42	43	43	43
4. Cost savings, %		−0.3	−1	0	4	5	5

From D. L. Terrill and J. M. Douglas, *I&EC Research*, **26**: 685 (1987), with permission from the American Chemical Society.

TABLE 8.10-2
Energy integration and optimization of HDA process

	Base case	Alternative					
		1	2	3	4	5	6
1. TAC ($10^6/yr) base-case flows	6.38	6.40	6.45	6.38	6.11	6.04	6.03
2. TAC ($10^6/yr) optimized		5.03	5.06	4.91	4.76	4.73	4.74
3. Cost savings, %		21	21	23	25	26	26
4. Utilities (MW) optimized		14.0	13.2	11.5	10.7	10.4	10.3
5. Utilities (MW), base-case	12.7	9.06	7.68	7.39	7.30	7.30	7.30

From D. L. Terrill and J. M. Douglas, *I&EC Research*, **26**: 685 (1987), with permission from the American Chemical Society.

and Grossman.* Thus, we see that we can trade savings from improved energy integration for savings in raw materials.

The relative importance of raw materials and energy is simple to assess by looking at a cost diagram (see Fig. 8.10-8), where the raw-materials costs are expressed in terms of exceeding the stoichiometric requirements. From this diagram it is apparent that raw-materials savings are much more important than energy savings for the HDA process.

Sensitivity of the Optimum Savings

Some of the values for the optimum design variables for the six alternatives, as well as the approximation procedure discussed in Sec. 8.3, are shown in Table 8.10-3. The results, for this case study, indicate that the optimum flows are quite insensitive

---

* M. A. Duran and I. E. Grossman, *AIChE J.*, **32**: 592 (1986).

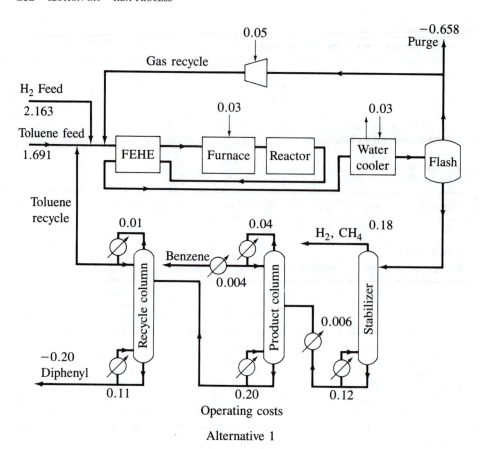

Alternative 1

FIGURE 8.10-8
Operating costs for the HDA process. [*From D. L. Terrill and J. M. Douglas, I&EC Research,* **26**: *685 (1987), with permission of the American Chemical Society.*]

TABLE 8.10-3
Optimization results for HDA process with diphenyl by-product

	Base case	Alternative						Approximate model
		1	2	3	4	5	6	
TAC (10^6/yr)	6.38	5.03	5.06	4.91	4.76	4.73	4.74	4.83
Annualized capital cost (10^6/yr)	1.10	1.50	1.54	1.52	1.48	1.47	1.49	1.60
Annualized operating cost (10^6/yr)	5.28	3.53	3.52	3.39	3.28	3.26	3.25	3.23
Conversion (%)	75.0	67.3	67.6	66.9	67.4	67.5	67.2	67.7
H_2 composition in gas recycle (%)	53.0	31.9	31.7	32.4	32.0	32.3	32.5	32.9
FEHE energy recovery (%), cold stream basis	85.4	94.0	93.6	93.6	91.9	87.5	85.8	85.7
DT_{min} (K)	10	32	38	25	14	9	13	16
Number of units	9	10	10	10	10	10	11	13
Stabilizer column								
Fractional loss of benzene (%)	0.5	0.3	0.3	0.5	0.3	0.4	0.4	0.4
Product column								
Fractional recovery (%)	99.0	98.7	98.7	98.6	98.9	98.9	98.9	98.9
Reflux ratio	1.20	1.45	1.44	1.58	1.91	1.85	1.85	1.70
Feed cooler (MW)	0	1.2	1.19	0	0	0	0	0
Recycle column								
Fractional recovery, overhead (%)	98.6	99.9	99.8	99.8	99.8	99.8	99.8	99.8
Fractional recovery, bottoms (%)	80.7	96	89	89	89	89	89	89
Pressure (kPa)	101	101	507	507	507	507	507	101

From D. L. Terrill and J. M. Douglas, *I&EC Research*, **26**: 685 (1987), with permission from the American Chemical Society.

to changes in the heat-exchanger network, provided that close to the maximum energy recovery is obtained. Moreover, as expected, the significant design variables correspond to the conversion and the purge composition.

8.11 SUMMARY, EXERCISES, AND NOMENCLATURE

Summary

A very simple procedure exists that makes it possible to calculate the minimum heating and minimum cooling requirements for a process. Also, simple procedures exist for calculating the minimum number of exchangers required and for estimating the heat-exchanger area required. These calculations are possible without even specifying a heat-exchanger network and therefore are ideal for screening purposes.

The results indicate that normally a process has a pinch temperature, and above this temperature heat should only be added to the process, whereas below this temperature heat should only be removed. Since this problem decomposition was not widely known a decade ago, it is often possible to significantly reduce the energy requirements of existing processes (that is, 30 to 50% energy savings have been obtained in industry).

A design procedure for heat-exchanger networks is discussed as well as the extension of the basic ideas to heat and power integration and heat and distillation integration. Several new heuristics were also presented:

1. Only add heat to a process above the pinch temperature.
2. Only remove heat from a process below the pinch temperature.
3. A feasible exchanger just above the pinch requires that $F_H C_{pH} \leq F_C C_{pC}$, while the opposite is true below the pinch.
4. To eliminate a heat exchanger from a network, we prefer to break a loop that includes the smallest heat load.
5. When we break loops that cross the pinch in order to eliminate heat exchangers from a network, we often violate the ΔT_{min} condition.
6. If we add extra heat to a process, we must remove this same amount of heat to a cold utility.
7. If possible, always install heat engines either above or below the pinch.
8. If possible, always install heat pumps across the pinch.
9. If possible, always install distillation columns either above or below the pinch.

Exercises

8.11-1. For one of the design problems that you have considered, assume a value for ΔT_{min} and include the heat effects in the separation system.

   (a) Calculate the minimum heating and cooling requirements for the process, and find the pinch temperature.

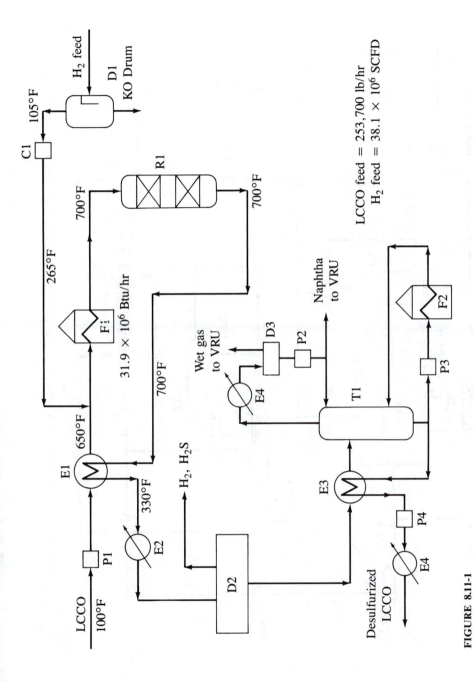

FIGURE 8.11-1
HDS process. (*From B. E. Brown Washington University Design Case Study No. 10, edited by B. D. Smith, Washington University, St Louis, Mo., 1970.*)

LCCO feed = 253,700 lb/hr
H_2 feed = 38.1 × 10⁶ SCFD

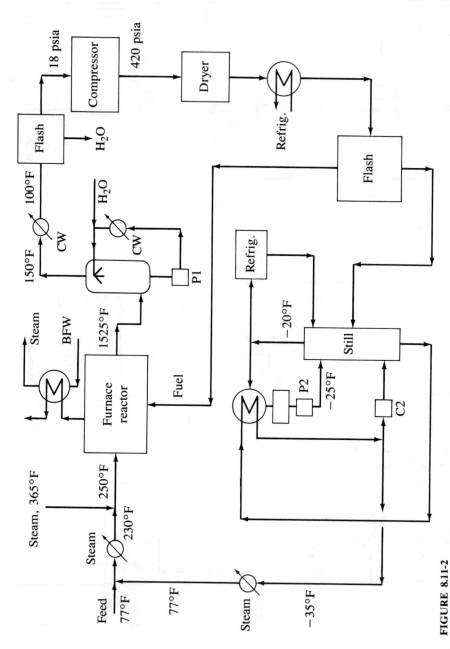

FIGURE 8.11-2
Ethylene plant. (*From W. L. Bolles, Washington University Design Case Study No. 6, edited by B. D. Smith, Washington University, St Louis, Mo., 1968.*)

(b) Do any of the distillation columns fall across the pinch? If so, pressure-shift these columns, and recalculate the minimum heating and cooling loads.

(c) Design a heat-exchanger network for above and below the pinch, and calculate the costs.

(d) How many loops cross the pinch? How can you break these loops?

8.11-2. Consider the flowsheet for the hydrosulfurization of oil, shown in Fig. 8.11-1.* From an inspection of this flowsheet, can you suggest a way of saving energy? Can you obtain a first estimate of the amount of energy that you can save? What might prevent you from realizing this energy savings? If the hydrogen is recycled, discuss the effect of improved energy integration on the optimum purge composition in this gas-recycle loop.

8.11-3. Consider the simplified flowsheet for producing ethylene from ethane† shown in Fig. 8.11-2. The reaction conditions and some features of the design problem are discussed in Exercises 5.4-8 and 6.8-7. The reactor effluent stream must be quenched to 750°F to prevent coking. From an inspection of the flowsheet, can you suggest some modifications that could be made to save energy? (In particular, consider the reactor feed and product streams.) Would energy savings be expected to affect the optimum process flows?

8.11-4. Calculate the optimum value of ΔT_{min} for the problem presented in Sec. 8.1. Assume that the overall heat-transfer coefficient for all stream matches is equal to $U = 50$ Btu/(hr·ft²·°F).

8.11-5. For the example discussed in Secs. 8.1 through 8.6, how much can we decrease the energy consumption if we reduce the ΔT from 13 to 10°F after we break the second loop (see Fig. 8.6-6)? If we consider the same problem but we break the other loop that crosses the pinch first, do we obtain the same results? Estimate the costs of the utilities and the heat exchangers for the minimum energy design, the design where we break one loop, and the design where we break both loops.

8.11-6. Consider these data:

FC_p, kW/°C	T_{in}, °C	T_{out}, °C
3	180	60
1	150	30
2	20	135
5	80	140

If $\Delta T_{min} = 10°C$, find the minimum heating load, the minimum cooling load, and the pinch temperature. Design a heat-exchanger network for the maximum energy recovery both above and below the pinch. How many loops cross the pinch? Break all the loops that cross the pinch, and restore ΔT_{min}.

---

* B. E. Brown, *A Distillate Desulfurizer*, Washington University Design Case Study No. 10, edited by B. D. Smith, Washington University, St. Louis, Mo., 1970.

† W. L. Bolles, *Ethylene Plant Design and Economics*, Washington University Design Case Study No. 6, edited by B. D. Smith, Washington University, St. Louis, Mo., 1968.

Nomenclature

A	Heat-exchanger area
C_{pi}	Heat capacity [Btu/(mol $\cdot$ °F)]
F_i	Flow rate (mol/hr)
H	Enthalpy (Btu/mol)
h_i	Film coefficient [Btu/(hr $\cdot$ ft$^2 \cdot$ °F)]
N_C	Number of cold streams
N_E	Number of exchangers
N_H	Number of hot streams
N_S	Number of streams
N_u	Number of utilities
Q_E	Excess heat supplied to a network
Q_i	Heat load (Btu/hr)
T_i	Temperature (°F)
U_i	Overall heat transfer coefficient [Btu/(hr $\cdot$ ft$^2 \cdot$ °F)]
W	Work
ΔH_v	Heat of vaporization (Btu/mol)
ΔT	Temperature difference (°F)
ΔT_{LM}	Log-mean temperature driving force (°F)

CHAPTER 9

COST
DIAGRAMS
AND THE
QUICK
SCREENING
OF PROCESS
ALTERNATIVES

Our previous synthesis and analysis enabled us to estimate the optimum design conditions (using a set of case studies) for a single process alternative. Of course, our goal is to find the best possible alternative. However, before we proceed further, we want to be certain that we understand the costs associated with the alternative we have selected. As a first step in this evaluation, we need to consider some way of summarizing the cost information.

9.1 COST DIAGRAMS

Conventional Approach

The conventional approach for summarizing cost information is to prepare a table of equipment costs that is coded to a flowsheet. Similar types of equipment, i.e., compressors, heat exchangers, distillation columns, etc., are all grouped together. A similar table is prepared for operating costs, and all the individual values for a particular type of utility, i.e., electric power, steam, etc., are summed and reported as a single entry.

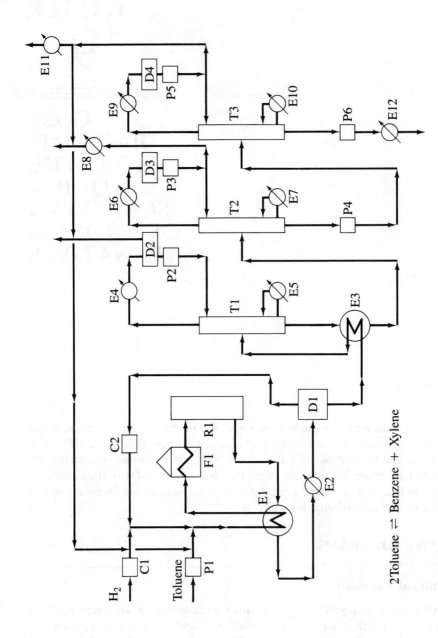

FIGURE 9.1-1
Disproportionation of toluene. (*From R. J. Hengstebeck and J. T. Banchero, Washington University Design Case Study No. 8, edited by B. D. Smith, Washington University, St Louis, Mo., 1969.*)

2Toluene ⇌ Benzene + Xylene

TABLE 9.1-1
Investment summary, $

Pumps (1949)			Towers (1969)	
P-1	5,900		T-1	25,000
P-2	1,320		T-2	37,600
P-3	1,950		T-3	35,500
P-4	1,680			
P-5	2,500		Total	98,100
P-6	980		Trays (1969)	
			T-1	5,800
Total	14,380		T-2	31,200
Pumps (1949), including spares	28,760		T-3	42,000
Pumps (1969)	50,000		Total	79,000
Exchangers (1968)			Compressors (1969)	
E-1	140,000		C-1	—
E-2	115,000		C-2	313,000
E-3	8,800			
E-4	22,000		Drums (1969)	23,650
E-5	26,000		Installed cost summary	
E-6	16,000		Pumps	185,000
E-7	9,400		Exchangers	1,140,000
E-8	4,200		Reactor	128,000
E-9	16,000		Towers (ex trays)	490,000
E-10	17,000		Trays	395,000
E-11	9,300		Compressors	751,000
E-12	6,500		Drums	130,000
			Furnace	523,000
Total	390,200			
				3,742,000
Exchangers (1969)	408,000			
Furnaces (1969)	209,000			
Reactor (1969)	29,800			

From R. J. Hengstebeck and J. T. Banchero, Washington University Design Case Study No. 8, edited by B. D. Smith, Washington University, St. Louis, Mo., 1969.

An example of a cost summary for a published case study for the disproportionation of toluene* is given in Fig. 9.1-1 and Tables 9.1-1 and 9.1-2. Figure 9.1-1 shows the flowsheet, and each piece of equipment is given a special label. Table 9.1-1 then gives the capital cost of each piece of equipment shown on Fig. 9.1-1, where all similar types of equipment are grouped together. The table also summarizes the total capital cost for each type of equipment. In Table 9.1-2 we find the total annual cost for each type of utility used as well as the annual costs for a number of factors

---

* R. J. Hengstebeck and J. T. Banchero, *Disproportionation of Toluene*, Washington University Design Case Study No. 8, edited by B. D. Smith, Washington University, St. Louis, Mo., June 26, 1969.

TABLE 9.1-2

Operating cost summary

Utilities	
Power	322
Steam	520
Fuel	333
Water	30
Total	1,205
Labor	95
Supervision	19
Taxes, insurance	166
Repairs	250
Miscellaneous	83
Payroll charges	32
Total	1850
SARE	150
Catalyst	60
Total	2060

From R. J. Hengstebeck and J. T. Banchero, Washington University Design Case Study No. 8, edited by B. D. Smith, Washington University, St. Louis, Mo., 1969.

that we have not considered as yet, such as labor, taxes, etc. Table 9.1-3 presents a somewhat different summary of this information, and it includes the costs for raw materials. One disadvantage of tables of this type is that it is difficult to visualize any type of capital versus operating cost trade-off. Another disadvantage is that we lose track of the cost of particular operations.

Process Alternatives

Our goal at this point of our process development is to make a judgment (guess) as to whether there might be a process alternative that is more profitable than our base-case design, rather than to undertake a detailed cost analysis. We expect that most of the possible alternatives will have approximately the same costs associated with labor, taxes, maintenance, repairs, etc., and therefore we neglect these factors. Similarly, we note that the costs of all the pumps and the drums are small, compared to the other items (the reactor cost is also small for this particular study), which is normally the case for most plants. Thus, we neglect the costs of all these items when we are screening alternatives. In addition, we lump the cost of the distillation columns and the trays, and then we combine these with the costs of the condensers and reboilers to develop expressions for the total costs associated with a distillation separation.

TABLE 9.1-3
Investment and operating summary

Conversion/pass, %	30
Purge gas	No
Investments, 10^6	
ISBL	3.74
OSBL	1.12
	4.86
Working capital*	1.00
	5.86
Catalyst inventory	0.06
	5.92
Operating costs, 10^6/yr	
Utilities	1.20
Labor and supervision[†]	0.15
Taxes and insurance	0.17
Repairs and miscellaneous	0.33
Catalyst	0.06
SARE	0.15
	2.06
Materials, bbl/calendar day (60°F)	
Toluene feed	3780
Products	
Benzene	1590
Xylenes	2000
H_2 feed, 10^6 SCFD	1.88
Fuel gas, 10^6 Btu/day	1700

* Principally, for a 2-week inventory of feed and products, with the products valued at cost. Note that the inventory depends on the length of the turnaround and any seasonal variation in sales, distance of the shipment, etc.
[†] Including payroll charges.
From R. J. Hengstebeck and J. T. Banchero, Washington University Design Case Study No. 8, edited by B. D. Smith, Washington University, St. Louis, Mo., 1969.

Cost Diagrams

As an alternative way of summarizing the most important processing costs, we can put all the costs on a flowsheet. The annualized, installed equipment costs (installed costs multiplied by 1/yr, see Eq. 2.5-20), are placed inside the boxes on the flowsheet, and the operating costs are attached to the stream arrows; see Douglas and Woodcock.* With this representation, we have a visual picture of how the

---

* J. M. Douglas and D. C. Woodcock, *I&EC Proc. Des. Dev.,* **27**: 970 (1985).

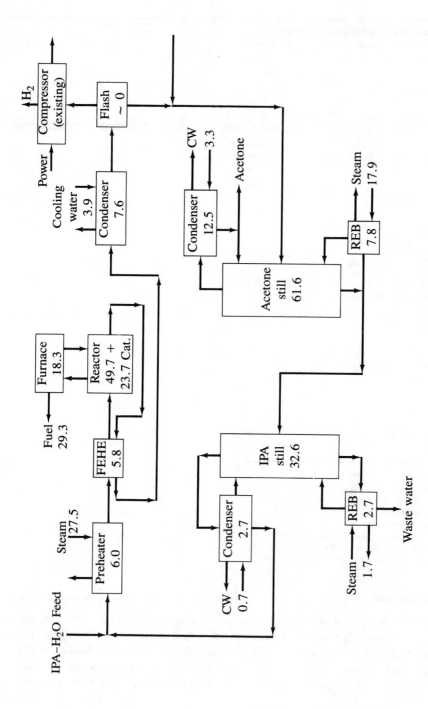

FIGURE 9.1-2
Dehydrogenation of isopropanol to produce acetone. Cost in $1000/yr. [*From J. M. Douglas and D. C. Woodcock, I&EC Proc. Des. Dev.* **24**: 470 (1985).]

costs are distributed throughout the process. Moreover, a diagram of this type can often be used to spot errors as well as to visualize potential improvements that might be obtained by modifying the flowsheet. We discuss both problems in more detail below.

If we divide each of the costs on a cost diagram by the specified production rate, we can prepare a cost diagram that indicates the dollars per pound of product which each individual piece of equipment or operating cost contributes to the total product price. This type of information is sometimes used for the design of batch processes in industry.

A cost diagram for one alternative for the dehydrogenation of isopropanol to produce acetone is shown in Fig. 9.1-2. The reaction is

$$\text{Isopropanol} \rightarrow \text{Acetone} + H_2 \tag{9.1-1}$$

Note that no pumps are included in the diagram, and we do not calculate the size or costs of either flash drums or reflux drums when we are screening alternatives. Also, since there are no side reactions, we do not include any raw-material or product cost. For this example, we are assuming that an existing compressor is available that we can use to recover the acetone leaving the flash drum with the hydrogen by-product, but we are not allowed to take any credit for the by-product stream.

Using Design Heuristics to Check the Cost Distribution

Suppose we consider the cost distribution for the two distillation operations in Fig. 9.1-2. We might assume that the common rule of thumb that the reflux ratio should be set to 20% larger than the minimum value was used to design each column. However, according to Happel and Jordan,[*] the distribution of costs should be 30% for the column, 35% for the condenser and reboiler, and 35% for the steam and cooling water at the optimum design conditions. Peters and Timmerhaus[†] present somewhat different results based on a 6.7-yr versus a 2-yr time factor, that is, 15% for the column, 10% for the condenser and reboiler, and 75% for the steam and cooling water.

Despite the fact that the two sets of results are quite different, it seems as if both distillation column designs in Fig. 9.1-2 are far from the optimum; i.e., the column cost appears to be much too large. If we increase the reflux ratio, we expect that the column cost will decrease (fewer trays, although partially balanced by an increase in the diameter) and that the costs of the condenser, reboiler, steam, and cooling water all will increase (higher vapor rates). Thus, from just an inspection of

[*] J. Happel and D. G. Jordan, *Chemical Process Economics*, 2d ed., Dekker, New York, 1975, p. 385.
[†] M. S. Peters and K. D. Timmerhaus, *Plant Design and Economics for Chemical Engineers*, 3d ed., McGraw-Hill, New York, 1980, p. 387.

the cost diagram, we expect that some additional column design studies can be justified.

Using the Cost Diagram to Infer Structural Modifications

We can also use the cost diagram to aid in generating process alternatives. In Fig. 9.1-2 if we consider the cost of the steam and the preheater (6.0 + 27.5), the fuel and the furnace (29.3 + 18.3), and the partial condenser and the cooling water (3.9 + 7.6), then we see that we are spending a large amount for heating and cooling as compared to the amount we are spending for energy integration (5.8 for the feed-effluent heat exchanger). Thus, it seems reasonable to try to make the feed-effluent heat exchanger much larger. However, as we try to increase the size of this feed-effluent exchanger, the inlet temperature to the partial condenser will approach that of the steam preheater outlet, so that the area and the capital cost will rapidly increase. Of course, we could avoid this difficulty entirely simply by eliminating the steam preheater. Hence, our inspection of the cost diagram indicates that an energy integration analysis should be undertaken.

Use of Cost Diagrams to Identify the Significant Design Variables

Numerous optimization variables exist for the flowsheet shown in Fig. 9.1-2, including the conversion, the reflux ratios in both distillation columns, the fractional recovery of acetone overhead in the product column, the fractional recovery of azeotrope overhead and water in the bottoms of the recycle column, the fractional recovery of acetone in the compressor, the approach temperature between the Dowtherm fluid leaving the furnace and the gases leaving the reactor, and the approach temperature between the steam preheater outlet and the reactor products leaving the feed-effluent heat exchanger.

Almost all of these optimization problems involve only local trade-offs. That is, the reflux ratio in either of the columns affects only the cost of that column, and the approach temperature for the feed-effluent exchanger affects only the cost of the feed-effluent exchangers, the steam preheater, and the partial condenser. However, changes in the conversion cause the recycle flow rate to change, and therefore changing the conversion affects the cost of every piece of equipment shown on the flowsheet. Thus, if the design conversion is not close to its optimum value, we can pay significant economic penalties, whereas errors in the reflux ratios are much less important; i.e., the total separation cost of either column is a relatively small fraction of the total cost of the plant.

A Systems Viewpoint

The use of cost diagrams enables us to look at the total system, whereas in Table 9.1-2 we are constrained to look at individual pieces of equipment. Similarly, the cost diagram helps us understand the interactions among various pieces of equipment. Thus, it is very helpful in screening alternatives, although the final design of the "best" alternative should be reported by using the conventional procedure.

With a cost diagram we can also break costs into gas-recycle effects, fresh feed effects, and liquid-recycle effects. We consider this approach as we look at a more complex plant in the next section.

9.2   COST DIAGRAMS FOR COMPLEX PROCESSES

The new energy integration procedure described in Chap. 8 introduces a significant amount of additional coupling and complexity into a flowsheet. This additional complexity makes it more difficult to visualize the interactions in a flowsheet. Hence, we need to find a way of simplifying the cost diagram.

Allocation Procedures

Suppose we consider our process for the hydrodealkylation of toluene (see Fig. 9.2-1). When we use the procedure described in Chap. 8 to energy-integrate the flowsheet, one of the alternative solutions we obtain is shown in Fig. 9.2-2. This flowsheet has so many interconnections that it is very difficult to gain an overall perspective of the process. However, this additional complexity is primarily caused by the addition of the heat exchangers. Hence, our first task is to remove this coupling simply by allocating the heat-exchanger costs to the individual process streams passing through each exchanger.

Allocating Heat-Exchanger Costs

Following Townsend and Linnhoff,* we allocate the annualized capital cost of the exchanger to each stream proportional to the individual film coefficient of that stream (see Eq. 8.3-7). These allocations are listed in Table 9.2-1 for the HDA process. Now the flowsheet again appears to be the same as Fig. 9.2-1 except that we have established a cost for each of the exchangers.

Lumping

We can simplify the cost diagram further by lumping costs that go together. In other words, for the purpose of evaluating process alternatives, there is no advantage to treating the annualized capital cost and the annual power cost of the

---

* D. W. Townsend and B. Linnhoff, "Surface Area Targets for Heat Exchanger Networks," Annual meeting of the Institution of Chemical Engineers, Bath, United Kingdom, April 1984.

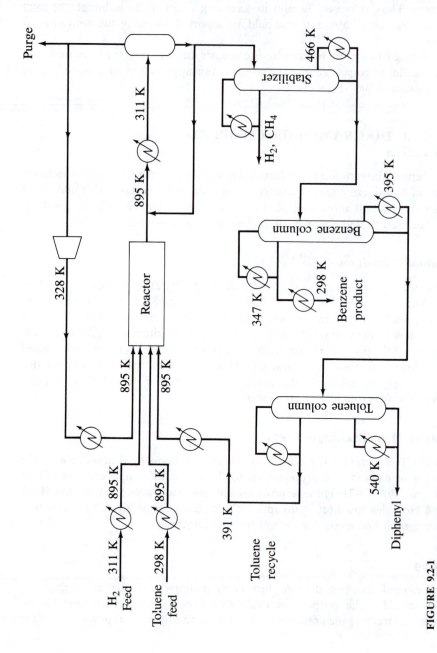

FIGURE 9.2-1
HDA process. [*From J. M. Douglas and D. C. Woodcock, I&EC Proc. Des. Dev.* **24**: 470 (1985).]

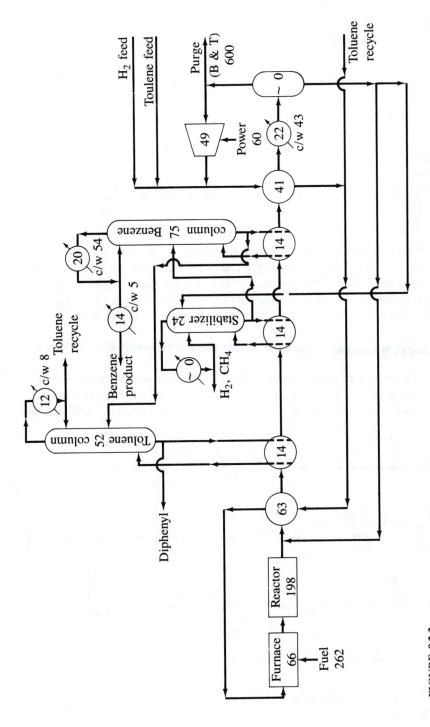

FIGURE 9.2-2

Cost diagram for energy-intergrated HDA process. [*From J. M. Douglas and D. C. Woodcock, I&EC Proc. Des. Dev.* **24**: *470 (1985).*]

TABLE 9.2-1
Allocation of heat-exchanger costs

Original cost, 10^3/yr	63	14	14	14	41
Stream	Reactor feed	Toluene reboiler	Stabilizer reboiler	Benzene reboiler	Reactor feed
h_i [kw/(m$^2 \cdot$ K)]	0.69	1.57	1.57	1.57	1.57
$1/h_i$ allocated cost	31.5	4	4	4	12
Stream	Reactor effluent	Reactor effluent	Reactor effluent	Reactor effluent	Reactor effluent
h_i [kw/(m$^2 \cdot$ K)]	0.69	0.69	0.69	0.69	0.69
$1/h_i$ allocated cost	31.5	10	10	10	29

From J. M. Douglas and D. C. Woodcock, *I&EC Proc. Des. Dev.*, **24**: 970 (1985), with permission of the American Chemical Society.

compressor separately. Hence, we combine these values into a total annual compression cost. Similarly, we lump the capital and operating costs for both the furnace and partial condenser, and we combine the column, reboiler, condenser, steam, and cooling-water costs for each distillation column into a single separator cost value. With this approach we obtain the values shown in Fig. 9.2-3.

Allocation to Process Streams

We can gain even more insight into the nature of the interactions in the process if we now allocate the costs to the processing of the fresh feed, the gas-recycle stream, and the liquid-recycle stream. We base the allocation on the flow rates and the

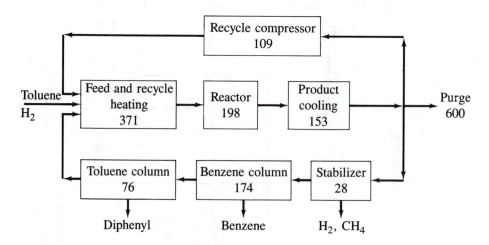

FIGURE 9.2-3
A simplified cost diagram for the HDA process. [*From J. M. Douglas and D. C. Woodcock, I&EC Proc. Des. Dev.* **24**: *470 (1985).*]

TABLE 9.2-2
Reactor cost allocation

Stream	kg · mol/hr	Cost, 10^3/yr
Gas recycle	3371	158
Liquid recycle	91	4
Toluene feed, 273 $\}$ H_2 feed, 496	769	36
	4231	198

From J. M. Douglas and D. C. Woodcock, *I&EC Proc. Des. Dev.*, **24**: 970 (1985), with permission of the American Chemical Society.

fractional heat loads of the various streams. The reactor allocations are given in Table 9.2-2, and the heating and cooling allocations are given in Table 9.2-3. The results of this allocation procedure are shown in Fig. 9.2-4. The raw-material cost listed is the value in excess of the stoichiometric requirements for the reaction, i.e., the cost we might be able to reduce by looking for a new process alternative.

Now we can clearly see that the most expensive costs are the excess hydrogen and toluene that we feed to the process. Thus, any alternative that will reduce the selectivity losses or the purge losses should be considered. The next most expensive costs are associated with the gas-recycle flow, which are considerably larger than the liquid-recycle or fresh feed costs.

Thus, the cost diagram provides a useful tool for rank-ordering an evaluation of process alternatives. It is also useful for the gross screening of alternatives, as we show in the next section.

TABLE 9.2-3
Heating and cooling cost allocation

Stream	Product cooler cost Heat load, GJ/hr	10^3/yr	Reactor feed heaters Heat load, GJ/hr	10^3/yr
Gas recycle	39.2	86	53.3	172
Liquid recycle	6.1	13	12.1	39
Fresh feed	25.5	56	49.6	160
	70.8	155	115.0	371

From J. M. Douglas and D. C. Woodcock, *I&EC Proc. Des. Dev.*, **24**: 970 (1985), with permission of the American Chemical Society.

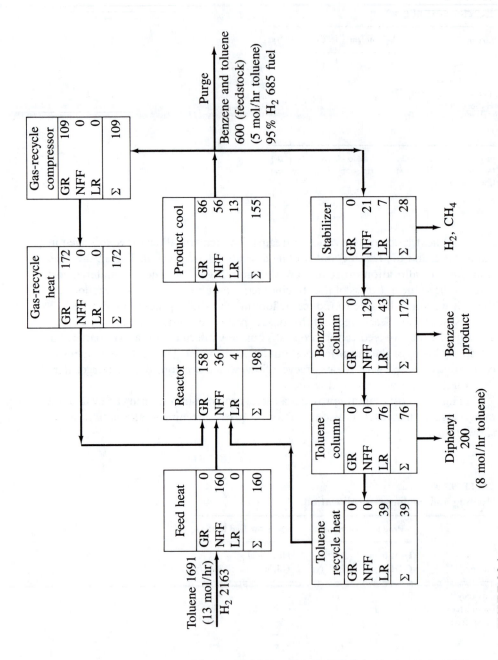

FIGURE 9.2-4
Allocating costs to fresh fuel and recycle flows. [*From J. M. Douglas and D. C. Woodcock, I&EC Proc. Des. Dev.* **24**: 470 (1985).]

9.3 QUICK SCREENING OF PROCESS ALTERNATIVES

The systematic procedure for developing a process design that we discussed earlier can also be used to generate a list of process alternatives. All that we need to do is to keep a list of each decision that we make as we proceed through the base-case design. Then as we change these decisions, we generate process alternatives. We want to make some estimate of the economic importance of each alternative, rather than repeat the design for each case, in order to minimize our design effort. We use the cost diagram as a tool in making these estimates.

Design Decisions

The decisions we made to generate the base-case design for the HDA process are listed in Table 9.3-1. We can proceed through this list decision by decision and try to evaluate the savings associated with changing any of the decisions. An alternate procedure would be to use the cost diagram to identify the largest costs and then to identify the decisions that have the greatest impact on these costs.

TABLE 9.3-1
Process alternatives for the HDA process

Level-2 decisions: Input-output structure
1. Do not purify the hydrogen feed stream.
2. Recover, rather than recycle, diphenyl so that there are three product streams (purge, benzene product, diphenyl by-product).
3. Use a gas recycle and purge stream.

Level-3 decisions: Recycle structure
1. Use a single reactor.
2. Use a gas (H_2 and CH_4) and a liquid (toluene) recycle stream.
3. Use a 5/1 H_2-to-aromatics ratio to prevent coking—assuming this to be a design constraint (although it could be formulated as an optimization problem).
4. A gas-recycle compressor is needed.
5. Operate the reactor adiabatically.
6. Do not consider equilibrium effects.

Level-4a decisions: Vapor recovery system
1. If a vapor recovery system is used, place it on the flash vapor (if recycle benzene is lost to by-product) or the purge stream (if there is no loss caused by benzene recycle).
2. Do not use a vapor recovery system.

Level-4b decisions: Liquid separation system
1. Make all separations by distillation.
2. Direct sequence of simple columns is used—probably use of complex columns should be considered.
3. Remove the light ends in a stabilizer.
4. Send the light ends to fuel—no vapor recovery system.

Level-5 decisions: Energy integration. There are numerous alternatives.

From J. M. Douglas and D. C. Woodcock, *I&EC Proc. Des. Dev.*, **24**: 970 (1985), with permission of the American Chemical Society.

For our HDA process (Fig. 9.2-4), the largest cost items correspond to the use of the excess raw materials. The decisions that affect these costs are made in level 2, the input-output structure of the flowsheet, so we want to start at the beginning of our decision list in any event. The next most important costs are those associated with the gas-recycle stream, and these correspond to the level 3 decisions. So again we find that we want to proceed sequentially through the decision list.

Purification of the Hydrogen Feed Stream

The hydrogen feed stream contains methane as an impurity, so that we might be able to cut some costs by removing the methane from the feed; i.e., we decrease the amount of inerts that pass through the process. However, methane is also produced as a by-product in the reactor. Hence, there seems to be little incentive for purifying the methane feed stream. On the other hand, if we should decide to purify the hydrogen-recycle stream, then we could feed the process through this recycle separation system. We defer consideration of this alternative until later in this section.

Recycling of Diphenyl

Since diphenyl is produced by a reversible reaction, we can recycle all the diphenyl and let the diphenyl build up in the process until it reaches its equilibrium level. If we adopt this approach, we eliminate our selectivity losses, i.e., the 8 mol/hr of toluene that gets converted to diphenyl (the purge losses of toluene and benzene are not affected). Of course, we lose the fuel credit of the diphenyl.

We can use the cost diagram to estimate the raw-material savings:

$$\text{Raw Matl. Savings} = 1691(\tfrac{8}{13}) - 200$$

$$= 840 \ (\times \ \$10^3/\text{yr}) \tag{9.3-1}$$

which is a significant value compared to the other costs. In addition, we save the cost of the toluene column which was used to separate the recycle toluene from the diphenyl. Again from the cost diagram we see that

$$\text{Savings for Toluene Column} = 76 \ (\times \ \$10^3/\text{yr}) \tag{9.3-2}$$

which is also fairly large.

We should recognize that this calculation is only qualitatively correct because the toluene column reboiler was integrated with the reactor effluent stream (see Fig. 9.2-2). Thus, the flows in the heat-exchange and quench systems will change. However, this "gross" screening of alternatives will help us decide on the priorities for undertaking more rigorous calculations.

Recycle of diphenyl also incurs cost penalties that are associated with oversizing all the equipment in the liquid-recycle loop to accommodate the increased flow rate. From Hougen and Watson* we find that the equilibrium constant is

$$K = 0.24 \text{ at } 685°C \qquad (9.3\text{-}3)$$

The stream flows for our base-case design are as follows:

Diphenyl	D
H_2	1547
CH_4	2320
Toluene	91
Benzene	273
	$\overline{273 + D}$

Thus

$$K = \frac{(D)(H_2)}{(\text{Benzene})^2} = \frac{D(1547)}{273^2} = 0.24 \qquad (9.3\text{-}4)$$

and $$D = 12 \text{ mol/hr} \qquad (9.3\text{-}5)$$

Assuming that the costs are proportional to the increased flow and using the cost diagram to estimate the liquid-recycle costs, we find that

$$\text{Increased Liquid-Recycle Cost} = \tfrac{12}{91}(4 + 13 + 7 + 43 + 39)$$

$$= 14 \, (\times \, \$10^3/\text{yr}) \qquad (9.3\text{-}6)$$

The increased liquid-recycle flow also requires that we increase the gas-recycle flow, since the hydrogen-to-aromatics ratio at the reactor inlet is specified as 5/1. Thus, the additional aromatics flow of 12 mol/hr requires an additional hydrogen flow of 5(12) = 60 mol/hr. However, the gas-recycle stream contains only 40% hydrogen, so that the total recycle flow must be increased by 150 mol/hr. Again, using the cost diagram to estimate the gas-recycle costs for a base-case gas-recycle flow of 3371 mol/hr gives

$$\tfrac{150}{3371}(158 + 86 + 109 + 172) = 24 \, (\times \, \$10^3/\text{yr}) \qquad (9.3\text{-}7)$$

Hence, the total cost savings from recycling the diphenyl is approximately

$$\text{Tot. Savings} = 840 + 76 - 14 - 24 = 878 \, (\times \, \$10^3/\text{yr}) \qquad (9.3\text{-}8)$$

This alternative leads to significant savings and should be evaluated in greater detail. Note that if we recycle the diphenyl, we expect that the optimum conversion will change dramatically (up to about 98%) because the economic trade-offs will change. Hence, it is necessary to repeat the calculations for this case.

---

* O. A. Hougen and K. M. Watson, *Chemical Process Principles*, III: *Kinetics and Catalysis*, Wiley, New York, 1947, p. 875.

Purification of the Gas-Recycle Stream

The cost of the excess hydrogen fed to the process is also very large. Most of this excess hydrogen leaves the process with the purge stream (40% H_2), and the potential savings from reducing the hydrogen loss from the cost diagram is about

$$H_2 \text{ Savings} = 2163 - 685 = 1.478 \ (\times \$10^3/\text{yr}) \tag{9.3-9}$$

If we install a hydrogen recovery unit, such as a membrane separator, we change both the overall and the recycle material balances. Suppose we purify the recycle stream to 95% H_2, the same as the feed composition. Estimates of the new process flows are shown in Fig. 9.3-1. The reduction in the gas-recycle flow from 3371 to 1628 leads to

$$\text{Gas-Recycle Saving} = (1 - \tfrac{1628}{3371})(172 + 109 + 158 + 86)$$

$$= 271 \ (\times \$10^3/\text{yr}) \tag{9.3-10}$$

Similarly the reduction of the hydrogen feed rate saves

$$H_2 \text{ Feed Savings} = \left(1 - \frac{287.4 + 273}{496 + 273}\right)(160 + 36 + 56)$$

$$= 68 \ (\times \$10^3/\text{yr}) \tag{9.3-11}$$

(Note that we do not include the stabilizer and the benzene column fresh feed processing costs in this calculation because they are not affected by the hydrogen feed flow rate.)

The total savings from purifying the hydrogen-recycle stream (neglecting the fuel value of methane and the loss of some hydrogen in the purge) are

$$\text{Potential Savings} = 1478 + 271 + 68$$

$$= 1817 \ (\times \$10^3/\text{yr}) \tag{9.3-12}$$

However, we must pay for the membrane separator from these savings. Unfortunately, there is no cost correlation available for membrane separators, and

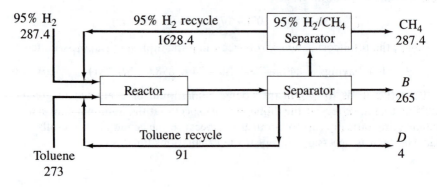

FIGURE 9.3-1
Purifying the recycle stream. [*From J. M. Douglas and D. C. Woodcock, I&EC Proc. Des. Dev.* **24**: 470 (*1985*).]

therefore we cannot estimate the incremental cost. Nevertheless, our quick estimates indicate that we should explore this possibility in more detail.

Also note that the large recycle flow of methane (60% CH_4 in the recycle gas) acts as a heat carrier (diluent). This additional flow limits the exit temperature from the adiabatic reactor to below 1300°F. If we remove most of the methane from the recycle stream, as shown in Fig. 9.3-1, we might exceed our reactor temperature constraint. If this should prove to be the case, we would change the design of the hydrogen purification unit to allow more methane to be recycled. Of course, this would also require that we adjust our estimates of the savings.

Other Decisions

We can use this same procedure to obtain a "gross" estimate of the economic impact of changing our other decisions. In particular, the heating and cooling costs are significant, so we want to consider other energy integration alternatives. Similarly, we might want to consider the use of complex distillation columns (sidestream, sidestream strippers, etc.) in an attempt to reduce the costs. The estimates are not very accurate, but they help to decide what additional design calculations can be justified.

Incentive for Changing Process Constraints

The problem statement imposes a constraint on the hydrogen-to-aromatics ratio at the reactor inlet of 5/1 in order to prevent coke formation. However, there is probably some uncertainty associated with this constraint; i.e., it is unlikely that a chemist would undertake numerous experiments to determine an exact coking limit for a wide range of conversions, reactor temperatures and pressures, etc. Instead, the chemist would probably set the ratio to a sufficiently high value that coking was never encountered (or occurred very slowly).

It is desirable to estimate the economic incentive for reducing this ratio in order to justify the cost of additional experiments. This justification needs to be accomplished sufficiently early in the development of a project that the chemist's apparatus or the pilot plant has not yet been dismantled. Again, we can use the cost diagram in making these economic estimates.

If we reduce the hydrogen-to-aromatics ratio for the base-case design to 3/1 (Rase* suggests a 2/1 ratio), the gas-recycle flow is approximately cut in half. Then the savings are

$$\text{Savings from } H_2/\text{Toluene} = 0.5 (158 + 86 + 109 + 172)$$

$$= 262 (\times \$10^3/\text{yr}) \qquad (9.3\text{-}13)$$

---

* H. F. Rase, *Chemical Reactor Design for Process Plants*, vol. 2: *Case Studies and Design Data*, Wiley, New York 1977, p. 360.

The operability (coking) of the plant must be verified experimentally before any detailed design studies are undertaken. However, now we have a justification for carrying out additional experimental work. (Coking data for plant scale application are extremely difficult to obtain in a laboratory experiment, and a considerable amount of judgment is required to interpret the results.)

Revising the Base-Case Design

Once we have determined these screening estimates of the process alternatives, we want to use our design procedure again to obtain an improved estimate of the total processing costs. We continue to use shortcut calculations for those revisions until we are certain that the project seems to be profitable; i.e., we must add the costs for pumps, reflux drums, storage, control, safety, pollution, labor, maintenance, etc. Once all these other factors have been included in the analysis, we repeat all the design calculations, using more rigorous procedures and a computer-aided design program such as FLOWTRAN, PROCESS, DESIGN 2000, ASPEN, etc. However, our goal at this point is to find the best process alternative, or the best few process alternatives.

Of course, to compare two alternatives, we must determine the optimum design conditions for each. We have used a case study approach to estimate the optimum design conditions, but there is also an approximate optimization procedure that we could use. This method is described in Chap. 10.

9.4   HDA PROCESS

Our gross screening procedure indicates that there is a large economic incentive for recycling the diphenyl and for recovering some of the hydrogen from the purge stream. There does not seem to be a design procedure or a cost correlation available in the literature for a membrane separation system, but we can evaluate the savings from recycling the diphenyl, using the procedures we described previously. Of course, when we compare process alternatives, we want to compare them at their optimum values, and so we want to improve the estimate we obtained by using our gross screening calculations.

Recycling Diphenyl—Optimum Design

The reactions for the HDA process are

$$\text{Toluene} + H_2 \rightarrow \text{Benzene} + CH_4$$
$$2\text{Benzene} \rightleftharpoons \text{Diphenyl} + H_2 \qquad (9.4\text{-}1)$$

If we recycle the diphenyl so that it builds up to its equilibrium level, we effectively have only one reaction taking place. Our heuristic for the optimum conversion for a single reaction is approximately $x = 0.98$; we trade only reactor costs against

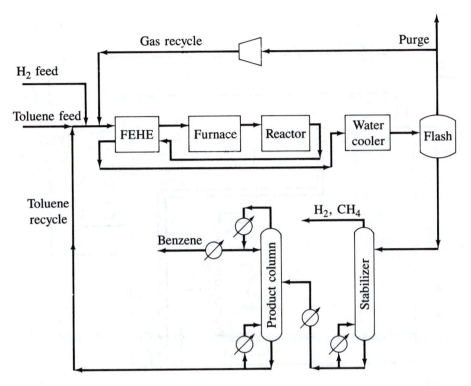

FIGURE 9.4-1
HDA process with diphenyl recycled—alternative 1. [*From D. L. Terrill and J. M. Douglas, I&EC Research,* **26**: *685 (1987), with permission of the American Chemical Society.*]

recycle costs. However, we want to check this heuristic because conversion normally is such an important design variable.

Sensitivity of the Optimum Design to Energy Integration

Several alternative heat-exchanger networks for the HDA process with diphenyl recycle were developed by Terrill and Douglas* and are shown in Figs. 9.4-1 through 9.4-5. The optimum costs for these alternatives as well as the approximate procedure presented in Sec. 8.3 are given in Table 9.4-1. The best alternative for the case with recycle has a cost of 3.57×10^6/yr compared to 4.73×10^6/yr for the case of diphenyl recovery, which is a significant change. The optimum conversion and purge composition for the case with recycle are $x = 0.977$ and $y_{pH} = 0.293$ as

---

* D. L. Terrill and J. M. Douglas, *I&EC Research,* **26**: 685 (1987).

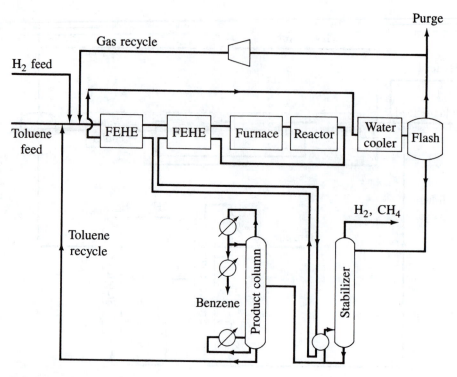

FIGURE 9.4-2
HDA process with diphenyl recycled—alternative 2. [*From D. L. Terrill and J. M. Douglas, I&EC Research,* **26**: *685 (1987), with permission of the American Chemical Society.*]

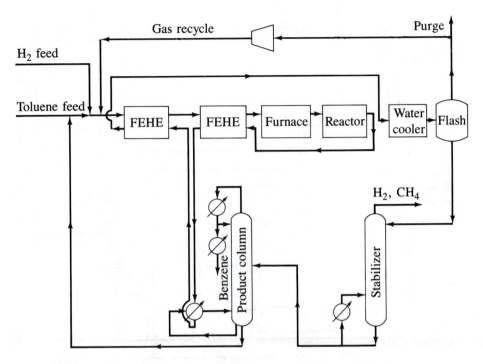

FIGURE 9.4-3
HDA process with diphenyl recycled—alternative 3. [*From D. L. Terrill and J. M. Douglas, I&EC Research,* **26**: *685 (1987), with permission of the American Chemical Society.*]

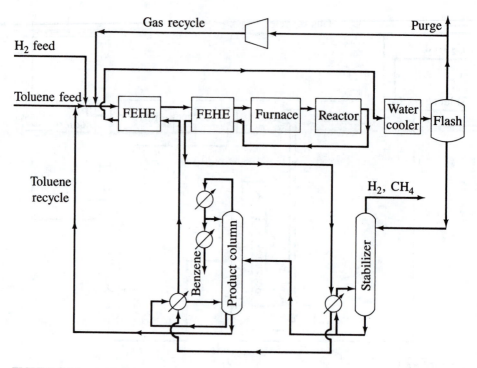

FIGURE 9.4-4
HDA process with diphenyl recycled—alternative 4. [*From D. L. Terrill and J. M. Douglas, I&EC Research,* **26**: *685 (1987), with permission of the American Chemical Society.*]

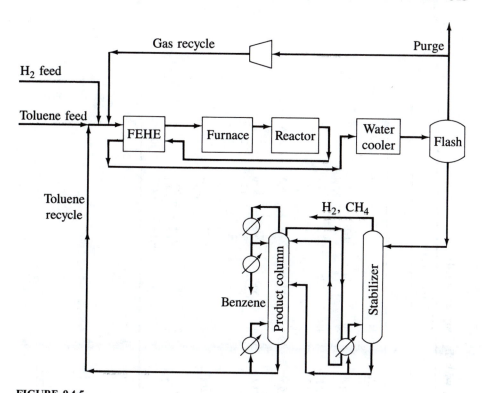

FIGURE 9.4-5
HDA process with diphenyl recycled—alternative 5. [*From D. L. Terrill and J. M. Douglas, I&EC Research,* **26**: *685 (1987), with permission of the American Chemical Society.*]

TABLE 9.4-1
Optimization results for the HDA process with diphenyl recycled

	Base case	Alternative 1	Alternative 2	Alternative 3	Alternative 4	Alternative 5	Approximate model
TAC (10^6/yr)	5.19	3.83	3.71	3.60	3.57	4.09	3.68
Annualized capital cost (10^6/yr)	1.09	1.30	1.26	1.26	1.27	1.50	1.30
Annualized operating cost (10^6/yr)	4.10	2.53	2.44	2.35	2.30	2.59	2.38
Conversion (%)	75.0	97.3	97.0	97.6	97.7	98.2	97.7
H_2 composition in gas recycle (%)	53.0	29.4	29.7	29.3	29.3	29.2	29.6
FEHE energy recovery (%), cold stream basis	85.4	94.2	93.4	89.9	85.6	90.6	81.3
ΔT_{min} (K)	10	36	32	10	9	12	26
Number of units	7	7	8	8	8	7	10
Stabilizer column Fractional loss of benzene (%)	0.5	0.2	0.2	0.2	0.2	0.4	0.2
Product column Fractional recovery (%)	99.0	99.0	97.8	99.1	99.1	99.1	99.1
Reflux ratio	1.2	1.2	1.3	1.4	1.3	1.1	1.3
Pressure (kPa)	101	101	101	101	101	586	101

From J. M. Douglas and D. C. Woodcock, *I&EC Proc. Des. Dev.*, **24**: 970 (1985), with permission of the American Chemical Society.

compared to $x = 0.675$ and $y_{pH} = 0.323$ for the process with a diphenyl by-product.

If we had used the original base-case flows (that is, $x = 0.75$, $y_{pH} = 0.4$), the cost would have been $\$4.92 \times 10^6$/yr, which demonstrates the need for optimization. We also note that the optimum design conditions are relatively insensitive to the structure of the heat-exchanger network, providing that we obtain close to the maximum energy recovery.

9.5 SUMMARY, EXERCISE, AND NOMENCLATURE

Summary

A cost diagram is prepared by writing the annualized capital cost of each significant piece of equipment inside the equipment box on the flowsheet and by attaching the values of the significant operating costs to the stream arrows. The cost diagram is useful for checking heuristics, for inferring the desirability of structural changes in the flowsheet, and for helping to identify the significant design variables. The cost diagram is also useful for obtaining very quick, but not very accurate, estimates of costs associated with various process alternatives.

Exercise

9.5-1. For any of the processes that you have designed, develop a complete list of alternatives for the input-output structure (level 2) and the recycle structure (level 3). Also develop a cost diagram for the process and discuss the economic trade-offs. Describe how the economic trade-offs for the alternatives differ from those of the base-case design, and use the cost diagram to estimate the savings (or losses) associated with one of the alternatives.

Nomenclature

Benzene	Benzene flow
D	Diphenyl flow
H_2	Hydrogen flow
K	Equilibrium constant

PART
III

OTHER
DESIGN
TOOLS AND
APPLICATIONS

CHAPTER 10

PRELIMINARY
PROCESS
OPTIMIZATION

Numerous texts discuss rigorous optimization analysis. However, at this stage of our designs, the goal is to compare two or more process alternatives when each is "close" to its optimum design conditions. Thus, we defer an exact optimization until we have selected the best alternative and until we are certain that we will proceed with a final design; i.e., we only develop a sufficient amount of accuracy to be able to make the next decision in a sequence of decisions.

In this chapter we describe an approximate optimization analysis that has been developed by Fisher, Doherty, and Douglas.* The objectives are to determine the *dominant* economic trade-offs for each design variable, to rank-order the importance of the design variables, and to estimate the incentive for optimizing a design (i.e., is our initial guess good enough, or should we try to improve it?). Thus, we want to get close to the optimum without necessarily determining the exact value of the optimum. We again use the shortcut design and cost models for our analysis.

---

* W. R. Fisher, M. F. Doherty, and J. M. Douglas, "Evaluating Significant Economic Trade-offs for Process Design and Steady-State Control Optimization Problems," *AIChE J.*, **31**: 1538 (1985).

10.1 DESIGN VARIABLES AND ECONOMIC TRADE-OFFS

Much of the optimization literature spends a great deal of time discussing the local versus global optima, the issue of convexity, constrained optimizations, etc. Hence, we need to understand the basic nature of design optimizations before we undertake any analysis. We begin by considering two simple examples, and then we attempt to generalize the results.

> **Example 10.1-1   Reactor example.** Process design optimizations are always charac-
> terized by economic trade-offs. For example, Levenspiel* describes the design of an
> oversimplified process containing only a reactor. For a reaction $A \rightarrow B$, free separa-
> tion of reactant and product, and a requirement that unconverted reactant cannot be
> recycled, the total cost of the process involves an economic trade-off between very
> high raw-material cost and a low reactor cost at low conversions balanced against low
> raw-material cost and a very high reactor cost at high conversions. Thus, we write a
> total annual cost (TAC) model as
>
> $$\text{TAC} = C_f F_F + C_{VR} V_R \qquad (10.1\text{-}1)$$
>
> The constraints on the system relate the production rate P to the fresh feed rate
> F_F and the conversion x by
>
> $$P = F_F x \qquad (10.1\text{-}2)$$
>
> and the design equation for the reactor (we assume a first-order, isothermal reaction
> in a CSTR) is
>
> $$V_R = \frac{F_F x}{k\rho(1 - x)} \qquad (10.1\text{-}3)$$
>
> All design problems have this same structure, i.e., a cost model and equality
> constraints. We could try to solve the problem as a constained optimization, but it is
> simpler merely to use Eqs. 10.1-2 and 10.1-3 to eliminate the variables F_F and x from
> the problem. After this elimination our cost model becomes
>
> $$\text{TAC} = \frac{C_F P}{x} + \frac{C_{VR} P}{k\rho(1 - x)} \qquad (10.1\text{-}4)$$
>
> The raw-material cost monotonically decreases from an unbounded value at
> $x = 0$ to its smallest possible value when $x = 1$. Also the reactor cost increases
> monotonically from its smallest possible value when $x = 0$ to an unbounded value
> when $x = 1$. Thus, there must be an optimum value of x which minimizes the cost.
> This optimum must correspond to a global minimum, and the minimum can never be
> at a constraint. Using the values given by Levenspiel, we obtain the results shown in
> Fig. 10.1-1.
>
> Design problems normally are characterized by this same type of behavior, i.e.,
> a balance between monotonically increasing and monotonically decreasing cost

---

* O. Levenspiel, *Chemical Reaction Engineering*, 2d ed., Wiley, New York, 1972, p. 130.

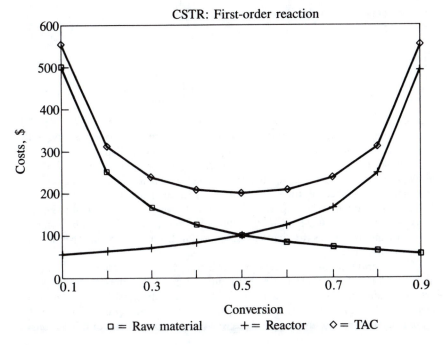

FIGURE 10.1-1
Reactor design.

functions. Moreover, they usually seem to be convex and to have a global optimum. The economic-potential graphs used for the development of a base-case design and an estimate of the optimum processing conditions that were presented in Chaps. 6 through 8 also demonstrate this type of behavior.

Example 10.1-2   Minimum reflux ratio in a distillation column—Constant-volatility systems. A cost model for a distillation column in terms of the number of trays and the vapor rate can be written as (see Appendices A.2, A.3, and A.4)

$$\text{TAC} = C_1 N^{0.8} V^{0.5} + C_2 V^{0.65} + C_3 V \qquad (10.1\text{-}5)$$

where C_1 includes the annualized cost of the column shell and trays, C_2 includes the annualized cost of the condenser and reboiler, and C_3 includes the annual cost of the cooling water and steam.

From a material balance we know that

$$V = (R + 1)D = \left(\frac{R_m R}{R_m} + 1\right)D \qquad (10.1\text{-}6)$$

For a binary mixture with a saturated-liquid feed,

$$R_m \sim \frac{1}{(\alpha - 1)x_F} \qquad (10.1\text{-}7)$$

and assuming complete recovery of the light component,

$$D \sim Fx_F \tag{10.1-8}$$

Thus,

$$V = \left[\frac{R/R_m}{(\alpha - 1)x_F} + 1 \right] Fx_F \tag{10.1-9}$$

For multicomponent mixtures, we use the results of Glinos and Malone (see Sec. 7.3) for R_m along with the appropriate material balance. Thus, we can always write the vapor rate in terms of R/R_m, the feed compositions, α's, desired spits, etc.

When R is close to R_m, Gilliland's correlation, or the approximation $N \approx 2N_m$, gives poor results. However, for binary systems we can use Smoker's equation to calculate the number of trays, while for multicomponent mixtures we can use Underwood's equation (see Appendix A.2). In both cases, the number of theoretical trays becomes unbounded as R/R_m approaches unity, and the number of trays decreases monotonically to Fenske's solution for the minimum number if R/R_m becomes unbounded.

The various terms in the cost model are shown in Fig. 10.1-2 for the values given in Table 10.1-1. Again there is a global minimum. Note that the column costs also exhibit a minimum; i.e., the height decreases but the diameter increases as R/R_m increases. From Fig. 10.1-2 we see that the costs increase rapidly if we fall below the optimum, but increase only very slowly as we exceed the optimum. For this reason, we

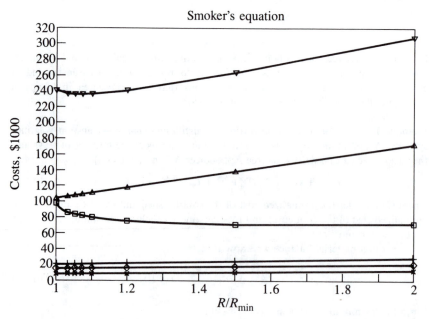

$\square$ = Column $\quad +$ = Condensor $\quad \diamondsuit$ = Rebailer $\quad \triangle$ = Steam $\quad \times$ = Cool $\quad \triangledown$ = TAC

FIGURE 10.1-2
Optimum reflux ratio.

TABLE 10.1-1
Optimum reflux ratio

Feed rate = 200 mol/hr, $x_F = 0.50$, $x_D = 0.99$, $f_D = 0.995$,
$\alpha = 2.0$, $E_0 = 0.5$, $H_0 = 15$ ft, $\Delta H_v = 13{,}300$ Btu/mol,
$U_c = 100$ Btu/(hr $\cdot$ ft$^2 \cdot$ °F), $T_b = 177$°F, $P_T = 15$ psia,
$M_G = 92$, M&S = 850

Design equations:

$$R_m = \frac{1}{(\alpha - 1)x_F}$$

$$V = (R + 1)D$$

Smoker's equation to calculate theoretical trays = N_T:

$$N = \frac{N_T}{E_0}$$

$$H_T = 2(N - 1) + H_0$$

$$D_T = 0.0164(\sqrt{V})[379 M_G(T_b + 460)(14.7)/(520 P_T)]^{1/4}$$

$$\text{Column Cost} = \frac{\text{M\&S}}{280}(3.28)(101.9)D_T^{1.06}H_T^{0.802}$$

$$\text{Condenser} = \frac{\text{M\&S}}{280}(3.29)(101.3)\left(\frac{\Delta H_v}{U\,\Delta T_m}\right)^{0.65}V^{0.65}$$

$$\text{Reboiler} = \frac{\text{M\&S}}{280}(3.29)(101.3)\left(\frac{\Delta H_v}{11{,}250}\right)^{0.65}V^{0.65}$$

$$\text{Steam} = \left(\frac{3}{1000}\right)\left(\frac{\Delta H_v}{933}\right)(8150)V$$

$$\text{Cool Water} = \left(\frac{0.06}{1000}\right)\left(\frac{1}{8.34}\right)\left(\frac{\Delta H_v}{30}\right)(8150)V$$

often let $R/R_m = 1.2$, to ensure that we are above the optimum, but that we are still within the neighborhood of the optimum.

Recycle and Product Distribution versus Unit Optimizations

In the examples above we considered optimum design problems for single-process units. However, the same type of behavior is encountered when we consider recycle optimization problems. We define a recycle optimization to be the optimization of a design variable that affects the capital and operating costs of all the equipment in a recycle loop. In contrast, unit optimizations concern design variables that affect only a single piece of equipment or a few adjacent pieces of equipment. Obviously, we expect that recycle optimizations and those that affect the product distribution will be more important than unit optimizations.

To illustrate recycle and unit optimizations, we return to the HDA process.

Example 10.1-3 Economic trade-offs in the HDA process

(a) *Conversion.* Normally it is possible to correlate the product distribution for a process in terms of the conversion of the limiting reactant (although in some cases the reaction temperature and the molar ratio of reactants are also important). For a fixed benzene production rate in the HDA process (see Fig. 8.10-2), the reactor cost and the selectivity losses (from conversion of excess toluene feed to diphenyl by-product) increase monotonically as the conversion increases. However, the cost of recovering and recycling the unconverted toluene decreases monotonically as the conversion increases. The selection of the conversion dramatically affects the flows in both the liquid and the gas (a 5/1 hydrogen-to-toluene ratio is required at the reactor inlet) recycle loops and hence affects the design of each piece of equipment in the flowsheet. Therefore, we classify this as a recycle optimization problem. There are no rules of thumb that can be used to estimate the optimum conversion for complex reactions, because the optimum depends on the selectivity losses, which are very different for various reaction systems.

(b) *Purge composition.* Whenever there is a "light" reactant and either a "light" impurity in a feed stream or a "light" by-product formed (where "light" implies having a lower boiling point than propylene), it is conventional to use a gas recycle and purge stream to remove the nonreactants from the process. As the reactant composition in the purge stream increases, the raw-material cost of the reactant will increase monotonically. However, as the reactant composition in the purge decreases, the gas-recycle flow rate and all the costs associated with equipment in the gas-recycle loop will increase monotonically to an unbounded value. Therefore, the selection of the composition of the light reactant (hydrogen) in the purge stream (or the excess hydrogen feed to the process) corresponds to a recycle optimization. Again, there are no rules of thumb available for estimating the optimum purge composition.

(c) *Molar ratio of reactor feeds.* As the molar ratio of reactants (H_2/T) approaches the stoichiometric requirement for the reaction, the cost of equipment in the vapor-recycle loop is minimized. To prevent coking and the production of undesired by-products, however, a large excess of hydrogen is required ($\geq 5/1$). Although the selection of this molar ratio corresponds to a recycle optimization, this design variable is often very difficult to incorporate into a process economic model, because of the unknown coking kinetics. Hence, it is often treated as a design constraint in order to avoid the optimization analysis.

(d) *Pressure of the flash drum and reactor pressure.* As the pressure of the flash drum is increased, the amount of aromatics lost in the purge stream decreases monotonically. However, as the pressure is increased, the wall thickness and the cost of all the equipment in the gas-recycle loop increases. Therefore, we classify the selection of the flash pressure as a recycle optimization.

    The pressure of the flash drum obviously is related to the reactor pressure. In some cases, changing the reactor pressure may affect the equilibrium conversion, the product distribution, or the phase of the reactants. Hence, purge losses are only one factor that might affect the optimum pressure. And the trade-offs will change if we install a vapor recovery system.

(e) *Approach temperature in heat exchangers.* There are rules of thumb available for estimating the optimum approach temperature in heat exchangers. These rules of thumb are not always valid, however, because the selection of the approach temperature can involve very different economic trade-offs for various units.

The optimization of the approach temperature for the feed-effluent heat exchanger, for example, involves a trade-off between the size of this exchanger and the size of both the furnace and the partial condenser. (Since only a few units affect the optimum approach temperature, we call this a unit optimization.) The approach temperature between the feed to the flash drum and the cooling-water inlet temperature to the partial condenser, however, involves a trade-off between the size of the partial condenser and the loss of aromatics in the purge stream as the flash temperature changes. (Again, this is a unit optimization.) The optimum ΔT's for these two exchangers differ by 2 orders of magnitude in the HDA process (1 K for the partial condenser and 100 K for the FEHE). Clearly, this discrepancy cannot be accounted for by the published heuristics. Of course, if a vapor recovery system is included in the flowsheet, the trade-offs will change.

Whenever an energy integration analysis (see Chap. 8) is performed, which is always an important consideration, the minimum ΔT at the pinch is an optimization variable that normally involves a trade-off between exchanger area (i.e., capital costs) and the utility requirements (i.e., operating costs).

(f) *Reflux ratio.* There is an optimum reflux ratio for each distillation column that balances the incremental number of plates against the combined costs of the column diameter, the condenser and reboiler costs, and the steam and cooling-water costs (see Example 10.1-2). This is a unit optimization, and we note that a rule of thumb is available for estimating optimum reflux ratios.

(g) *Fractional recoveries in distillation columns.* Since only the product composition of benzene is specified, the fractional recoveries of benzene overhead in the product column and the four splits in the stabilizer and recycle columns correspond to optimization variables. For example, the fractional recovery of benzene in the product column involves the trade-off of incremental trays in the stripping section and the cost of recycling benzene back through the reactor. We consider these trade-offs to be unit optimizations. A rule of thumb of greater than 99% recoveries is available, but a quick estimate of the optimum can also be evaluated (see Fisher, Doherty, and Douglas*).

Example 10.1-4   A simplified version of butane alkylation. We wish to illustrate some important design variables that are not encountered in the HDA process. For this purpose we consider a very simplified version of a butane alkylation process, where we assume that the only reactions are

$$C_4H_8 + i\text{-}C_4H_{10} \rightarrow i\text{-}C_8H_{18} \tag{10.1-10}$$

$$C_4H_8 + i\text{-}C_8H_{18} \rightarrow C_{12}H_{26} \tag{10.1-11}$$

and we assume that the feed streams are pure C_4H_8 and $i\text{-}C_4H_{10}$. A simplified flowsheet is shown in Fig. 10.1-3.

Now we assume that $E_1 < E_2$ and that the reaction kinetics are indicated by the stoichiometry. The economic trade-offs for this example are then as follows:

(a) *Conversion.* The product distribution is degraded as the conversion of C_4H_8 increases, and there is also an economic trade-off between high reactor cost at high conversion and large recycle costs at low conversions. This is a recycle trade-off.

---

* W. R. Fisher, M. F. Doherty, and J. M. Douglas, "Short-Cut Calculations of Optimal Recovery Fractions for Distillation Columns," *I&EC Proc. Des. Dev.,* **24**: 955 (1985).

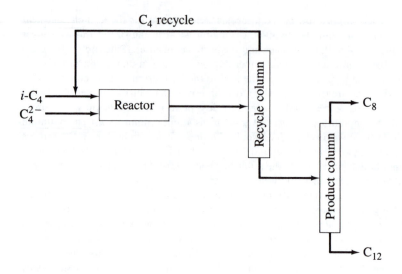

FIGURE 10.1-3
Simplified flowsheet for butane alkylation.

    (b) *Reactor temperature.* High temperatures correspond to large selectivity losses, but small reactors, whereas the opposite is true at low temperatures. The reactor temperature is a product distribution optimization problem.

    (c) *Molar ratio of reactants.* Large i-C_4H_{10}/C_4H_8 ratios decrease the selectivity losses but lead to large recycle costs of i-C_4H_{10}, and vice versa. Again we obtain a recycle optimization.

    (d) *Reflux ratios.* There is an optimum reflux ratio for each column. This is a unit optimization.

    (e) *Fractional recoveries.* There are two optimum fractional recoveries in the first column and one in the product column (the product composition is assumed to be fixed). Even though the fractional recovery of i-C_8 overhead in the first tower involves a trade-off between incremental trays in the rectifying section and recycle of i-C_8 back through the reactor, we often classify this as a unit optimization problem because we expect that the optimum value of the recycle flow of i-C_8 will be quite small (i.e., we expect that greater than 99% recoveries of i-C_8 are warranted).

Significant Design Variables

The most significant optimization variables involve product distribution or the recycle trade-offs. They include all the design variables that affect the process flow rates (conversion, purge composition, molar ratios of reactants, and possibly the reactor temperature and pressure). Unfortunately, there are no rules of thumb to select any of these variables. Thus, an optimization analysis of some type is rquired to fix the process flow rates.

We expect that these optimizations will usually correspond to a global optimum and that normally the optimum will not be at a constraint (the exception corresponds to coking constraints). The case-study approach for evaluating the economic potential that was described in Chaps. 5 through 8 can be used to verify this behavior, if necessary. The case studies also indicate the sensitivity, i.e., "flatness," of the optimum, which is always information that we desire.

Limitations of the Optimization Analysis

For the purpose of screening alternatives, we are only attempting to get in the neighborhood of the optimum design conditions. Thus, we use shortcut design and cost models. We also assume that equipment sizes are continuous and that we are not in a region where the materials of construction change as we change the reactor temperature. Our initial goals are to screen out unprofitable processes and/or to make a first evaluation as to whether a few process alternatives appear to be sufficiently profitable to warrant an additional design effort.

10.2   COST MODELS FOR PROCESS UNITS

Once the material and energy balances have been estimated for the process, we can use shortcut design procedures to calculate the equipment sizes. Then we can use Guthrie's correlations (see Appendix E.2) to calculate the installed equipment cost. We can put these installed costs on an annualized basis by using a capital charge factor, say $\frac{1}{3}$ yr, and we can calculate the utility costs. Thus, we assume that we have completed a base-case design.

To minimize the amount of computation required for process optimization, we use variable elimination and the appropriate design equations to write the annualized capital cost of each "significant" (i.e., expensive) piece of equipment and each operating cost in terms of the process flow rates. Next we use the approximate material balances described in Chaps. 5 and 6 to relate all the process flows to the significant design variables. Several examples of cost models of this type are presented here.

Heat Exchangers

Guthrie (see Appendix E.2) indicates that the installed cost of a heat exchanger can be written as

$$C_A = C_{A, BC}\left(\frac{A}{A_{BC}}\right)^{0.65} \tag{10.2-1}$$

For a capital charge factor of $\frac{1}{3}$ yr in our cost model, Eq. 2.5-20 includes a factor of $1\ \text{yr}^{-1}$, so that all quantities are on an annualized basis. The heat-exchanger area can normally be calculated from the equation

$$Q = FC_p\,\Delta t = UA\,\Delta T_m \tag{10.2-2}$$

For constant values of C_p and U, we can use Eq. 10.2-2 to eliminate A from Eq. 10.2-1 to obtain

$$C_A = C_{A,BC}\left(\frac{F \, \Delta t \, \Delta T_{mBC}}{\Delta T_m \, F_{BC} \, \Delta t_{BC}}\right)^{0.65} \tag{10.2-3}$$

or, if the stream temperatures are fixed,

$$C_A = C_{A,BC}\left(\frac{F}{F_{BC}}\right)^{0.65} \tag{10.2-4}$$

Thus, we have a simple model for heat-exchanger costs in terms of the flows.

Heat-Exchanger Utilities

For cooling water we can write

$$C_{CW} = C_{CW,BC}\left(\frac{F_{CW}}{F_{CW,BC}}\right) \tag{10.2-5}$$

Then from a heat balance we find

$$FC_p \, \Delta t = F_{CW} C_{p,CW} \, \Delta t_{CW} \tag{10.2-6}$$

and thus we obtain

$$C_{CW} = C_{CW,BC}\left(\frac{F \, \Delta t}{F_{BC} \, \Delta t_{BC}}\right) \tag{10.2-7}$$

which relates the cooling-water cost to the flow and temperatures. Again, for fixed temperatures

$$C_{CW} = C_{CW,BC}\left(\frac{F}{F_{BC}}\right) \tag{10.2-8}$$

The results for steam are similar:

$$C_{STM} = C_{STM,BC}\frac{W_S}{W_{S,BC}} \tag{10.2-9}$$

and

$$Q = FC_p \, \Delta t = W_S \, \Delta H_S \tag{10.2-10}$$

So

$$C_{STM} = C_{STM,BC}\left(\frac{F \, \Delta t}{F_{BC} \, \Delta t_{BC}}\right) \tag{10.2-11}$$

or, for fixed temperatures,

$$C_{STM} = C_{STM,BC}\left(\frac{F}{F_{BC}}\right) \tag{10.2-12}$$

Isothermal Plug Flow Reactor

For a first-order isothermal reaction in a tubular reactor, the design equation is

$$V = \frac{F}{k\rho} \ln \frac{1}{1-x} \tag{10.2-13}$$

The installed cost of this reactor can be written as

$$C_R = C_{R,BC} \left(\frac{V_R}{V_{R,BC}} \right)^{0.63} \tag{10.2-14}$$

We can relate the cost of the reactor for any conversion to the cost of the reactor at base-case conditions as follows:

$$C_R = C_{R,BC} \left[\frac{F \ln (1-x) k_{BC}}{F_{BC} \ln (1-x)_{BC} k} \right]^{0.63} \tag{10.2-15}$$

Furnaces

Available cost models for direct fired heaters relate the installed cost to the furnace heat duty only. For example,

$$C_{FN} = C_{FN,BC} \left(\frac{Q_F}{Q_{F,BC}} \right)^{0.78} \tag{10.2-16}$$

The (sensible) heat duty for the furnace is

$$Q_F = FC_p \, \Delta t \tag{10.2-17}$$

so that our cost model becomes

$$C_{FN} = C_{FN,BC} \left(\frac{F \, \Delta t}{F_{BC} \, \Delta t_{BC}} \right)^{0.78} \quad \text{or} \quad C_F = C_{F,BC} \left(\frac{F}{F_{BC}} \right)^{0.78} \tag{10.2-18}$$

Compressors

The installed cost for a compressor (comp) can be related to the required brake horsepower (B_{hp} = power/efficiency) by

$$C_{comp} = C_{comp,BC} \left(\frac{B_{hp}}{B_{hp,BC}} \right)^{0.93} \tag{10.2-19}$$

The power required for isentropic compression of an ideal-gas stream is

$$\text{Power} = \frac{3.03 \times 10^{-5}}{\gamma} \frac{P_{in} F_V}{60\rho} \left[\left(\frac{P_{in}}{P_{out}} \right)^{\gamma} - 1 \right] \tag{10.2-20}$$

where

$$\gamma = \left(\frac{C_p}{C_V} - 1 \right) \frac{C_p}{C_V} \tag{10.2-21}$$

If the gas composition is constant and the inlet and outlet pressures are roughly constant for a fixed flowsheet, then the cost model becomes

$$C_{comp} = C_{comp, BC} \left(\frac{F_V}{F_{V,BC}} \right)^{0.93} \tag{10.2-22}$$

For gas-recycle compressors, both the vapor flow rate and composition may vary (if the purge composition is optimized). In this case, the ratio of heat capacities γ may also be included in the cost model.

Distillation Columns

The installed cost of a distillation column shell (trays or packing) can be written as

$$C_{sh} = C_{sh, BC} \left(\frac{N}{N_{BC}} \right)^{0.862} \left(\frac{Dia}{Dia_{BC}} \right)^{1.066} \tag{10.2-23}$$

The column diameter varies as the square root of the column vapor rate, so we can write

$$C_{sh} = C_{sh, BC} \left(\frac{N}{N_{BC}} \right)^{0.802} \left(\frac{V}{V_{BC}} \right)^{0.533} \tag{10.2-24}$$

The vapor rate in the column is given by

$$V = (R + 1)D \tag{10.2-25}$$

For reasonably sharp splits with the light component taken overhead, the distillate flow rate is approximately $x_F F$. If the outlet composition and reflux ratio are not optimized, the number of trays for the required separation is essentially constant. For this case, our model becomes

$$C_{sh} = C_{sh, BC} \left\{ \frac{(1.2R_m + 1)x_F F}{[(1.2R_m + 1)x_F F]_{BC}} \right\}^{0.533} \tag{10.2-26}$$

If the outlet compositions are optimized but the reflux ratio is fixed (at, say, 1.2 times the minimum), then the cost model is

$$C_{sh} = C_{sh, BC} \left(\frac{\ln SF}{\ln SF_{BC}} \right)^{0.802} \left(\frac{V}{V_{BC}} \right)^{0.533} \tag{10.2-27}$$

where the separation factor SF is

$$SF = \left(\frac{x_D}{1 - x_D} \right) \left(\frac{1 - x_B}{x_B} \right) \tag{10.2-28}$$

If we also wish to optimize the reflux ratio, we can use the approximate design model of Jafarey, Douglas, and McAvoy*:

$$N = \frac{\ln \beta}{\ln \{\alpha/[1 + (\alpha - 1)/(R/R_m)]^{0.5}\}} \tag{10.2-29}$$

where

$$\beta = \frac{(R/R_m - 1 + \alpha x_F)(R/R_m - x_F)}{(R/R_m - 1)^2} \tag{10.2-30}$$

although N should be corrected so that $N \approx 2N_m$ when $R/R_m = 1.2$.

The installed cost of the column reboiler and condenser can be written as

$$C_R = C_{R,BC}\left(\frac{A_R}{A_{R,BC}}\right)^{0.65} = C_{R,BC}\left(\frac{V}{V_{BC}}\right)^{0.65} \tag{10.2-31}$$

$$C_C = C_{C,BC}\left(\frac{A_C}{A_{C,BC}}\right)^{0.65} = C_{C,BC}\left(\frac{V}{V_{BC}}\right)^{0.65} \tag{10.2-32}$$

Similarly, the operating costs for steam and cooling water can be written as

$$C_{STM} = C_{STM,BC}\left(\frac{V}{V_{BC}}\right) \tag{10.2-33}$$

$$C_{CW} = C_{CW,BC}\left(\frac{V}{V_{BC}}\right) \tag{10.2-34}$$

The vapor rate appearing in these expressions is given by Eq. 10.2-25 and the material balance for a perfect split. The reflux ratio in Eq. 10.2-25 can be calculated by using Underwood's equations or the approximations of Glinos and Malone (see Appendix A.2). We can relate the feed composition in these expressions to the extent of reactions by using simple material balances.

Total Annual Cost

Once the costs have been written in terms of the stream flows, we can use the simplified material balances illustrated in Chaps. 5 and 6 to relate the flows to the design variables. Hence, we can obtain simple cost models in terms of the design variables. A model for the total annual cost of the process is then simply the summation of the individual capital and operating costs. We use these models in our approximate optimization procedure.

---

* A. Jafarey, J. M. Douglas, and T. J. McAvoy, "Short-Cut Techniques for Distillation Column Design and Control," *I&EC Proc. Des. Dev.*, **18**: 121, 197 (1979).

10.3   A COST MODEL FOR A SIMPLE PROCESS

To illustrate the use of cost models in our approximate optimization analysis, we consider a particular example described by Fisher, Doherty, and Douglas.* A flowsheet for the simple reaction system

$$A \rightarrow P \rightarrow W \tag{10.3-1}$$

is shown in Fig. 10.3-1. Component P represents the desired product, and W is a waste by-product. The kinetics of both reactions are first-order with activation energies $E_1 < E_2$. The relative volatilities are such that $\alpha_A > \alpha_P > \alpha_W$, and we assume that the direct column sequence is favorable. The product stream flow rate and composition are specified, but the composition of the waste stream corresponds to a design optimization variable. The other optimization variables we wish to consider are the reactor conversion and temperature as well as the reflux ratio for the product column. Several other design variables are available for this process, which we have fixed using rules of thumb to simplify the analysis.

Process Flows and Stream Costs

We assume that a feed stream containing pure A is available and that all A fed to the process is recycled to extinction. That is, for the material balance calculations we assume a perfect split between A and P in the recycle column (although a

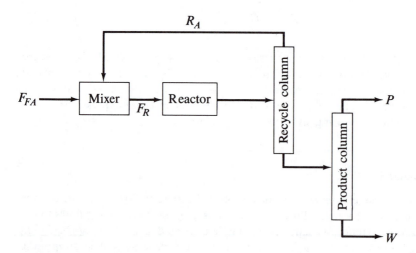

FIGURE 10.3-1
Flowsheet for the reaction system $A \rightarrow P \rightarrow W$. [*From W. R. Fisher, M. F. Doherty, and J. M. Douglas, AIChE J., 31: 1538 (1985).*]

---

* W. R. Fisher, M. F. Doherty, and J. Douglas, *AIChE J.,* **31**: 1538 (1985).

different assumption is used for the column design calculations). The desired flow of the product stream is P, and the amount of product contained in this stream to obtain a product purity x_D is

$$P_P = x_D P = 0.999P \tag{10.3-2}$$

If we let W_P be the amount of product lost in the bottoms of the product column, then the fractional recovery of the product f_P is

$$f_P = \frac{P_P}{P_P + W_P} \tag{10.3-3}$$

We also define the selectivity S as

$$S = \frac{\text{Moles of } P \text{ in Reactor Outlet}}{\text{Moles of } A \text{ Converted}} \tag{10.3-4}$$

From an overall material balance, we know that the fresh feed rate of A must be the sum of the flows of the exit streams

$$F_{F,A} = P + W \tag{10.3-5}$$

and we are assuming that no A leaves in either the product or the waste streams. Thus

$$S = \frac{P_P + W_P}{P + W} = \frac{P_P + W_P}{F_{F,A}} \tag{10.3-6}$$

We can combine these equations to obtain

$$F_{F,A} = \frac{P_P + W_P}{S} = \frac{P_P}{f_P S} = \frac{Px_D}{f_P S} \tag{10.3-7}$$

and

$$W = F_{F,A} - P = P\left(\frac{x_D}{f_P S} - 1\right) \tag{10.3-8}$$

Stream Costs

Assume that the stream costs are

$$\text{Product} = (\$20/\text{mol}) \, (P \text{ mol/hr}) \, (8150 \text{ hr/yr}) \tag{10.3-9}$$

$$\text{Fresh Feed} = (\$15.50/\text{mol})[(Px_D/f_P S) \text{ mol/hr}](8150 \text{ hr/yr}) \tag{10.3-10}$$

$$\text{By-product Value} = (\$1/\text{mol})[P(x_D/f_P S - 1) \text{ mol/hr}](8150 \text{ hr/yr})$$
$$\tag{10.3-11}$$

Selectivity and Reactor Model

We assume that the rate expressions for first-order reactions in an isothermal tubular reactor are

$$\frac{dC_A}{dt} = -k_1 C_A \quad \text{and} \quad \frac{dC_B}{dt} = k_1 C_A - k_2 C_B \qquad (10.3\text{-}12)$$

where

$$k_1 = 5.35 \times 10^{10} \exp\left(\frac{-32,500}{RT}\right) \text{hr}^{-1}$$

$$(10.3\text{-}13)$$

$$k_2 = 4.61 \times 10^{17} \exp\left(\frac{-52,500}{RT}\right) \text{hr}^{-1}$$

If we solve the rate equations, we find that

$$S = \frac{1}{x}\left(\frac{k_1}{k_1 - k_2}\right)[(1 - x)^{k_2/k_1} - (1 - x)] \qquad (10.3\text{-}14)$$

and

$$V_R = \frac{F_R}{k_1 \rho_m} \ln \frac{1}{1 - x} \qquad (10.3\text{-}15)$$

From a recycle balance we obtain

$$F_R = \frac{F_F}{x} \qquad (10.3\text{-}16)$$

and we assume that the reactor cost is given by

$$C_R = C_{R,BC}\left(\frac{V_R}{V_{R,BC}}\right)^{0.63} \qquad (10.3\text{-}17)$$

Recycle Column

We are not interested in optimizing the design of the recycle column (the reflux ratio and fractional recoveries) in this case study, but we want to include its cost in the economic model. We assume that the design reflux ratio is 1.2 times the minimum and that the theoretical number of trays is about twice the minimum:

$$N = 2N_m = \frac{2 \ln \text{SF}}{E_0 \ln \alpha_{AP}} \qquad (10.3\text{-}18)$$

An approximate expression for the minimum reflux ratio is given by Glinos and Malone* as

$$R_m = \left(\frac{\alpha_{P,W}}{\alpha_{A,W} - \alpha_{P,W}}\right)\left(\frac{x_{F,A} + x_{F,P}}{x_{F,A}}\right) + \frac{x_{F,W}}{x_{F,A}(\alpha_{A,W} - 1)} \tag{10.3-19}$$

where

$$x_{F,A} = 1 - x \qquad x_{F,P} = xS \qquad x_{F,W} = (1 - S)x \tag{10.3-20}$$

From Sec. 10.2, the appropriate cost model for this case (we use a fixed number of trays) is

$$C_{sh1} = C_{sh1,BC}\left(\frac{V_1}{V_{1,BC}}\right)^{0.533} \tag{10.3-21}$$

The column vapor rate is calculated by

$$V_1 = (1.2R_m + 1)D$$

$$= (1.2R_m + 1)\left[\frac{(1 - x)F_{F,A}}{x}\right] \tag{10.3-22}$$

The recycle column condenser and reboiler capital costs are given by

$$C_{C1} = C_{C1,BC}\left(\frac{V_1}{V_{1,BC}}\right)^{0.65} \tag{10.3-23}$$

$$C_{R1} = C_{R1,BC}\left(\frac{V_1}{V_{1,BC}}\right)^{0.65} \tag{10.3-24}$$

The associated operating costs are, for cooling water and steam,

$$C_{CW1} = C_{CW1,BC}\left(\frac{V_1}{V_{1,BC}}\right) \tag{10.3-25}$$

$$C_{STM1} = C_{STM1,BC}\left(\frac{V_1}{V_{1,BC}}\right) \tag{10.3-26}$$

Product Column. The desired economic model for the product column shell is also of the form

$$C_{sh2} = C_{sh2,BC}\left(\frac{N}{N_{BC}}\right)^{0.802}\left(\frac{V_2}{V_{2,BC}}\right)^{0.533} \tag{10.3-27}$$

---

* K. Glinos and M. F. Malone, *I&EC Proc. Des. Dev.*, **23**: 764 (1984).

In this case, we use the design model of Jafarey, Douglas, and McAvoy[†] to estimate the optimum reflux ratio and the product recovery fraction. Thus, we let

$$N = \frac{\ln \beta S}{\ln\{\alpha/[1 + (\alpha - 1)/(R/R_m)]^{0.5}\}} \tag{A.2-47}$$

where

$$SF = \left(\frac{x_D}{1 - x_D}\right)\left(\frac{1 - x_B}{x_B}\right) \tag{10.3-28}$$

and

$$\beta = \frac{(R/R_m - x_F)(R/R_m - 1 + \alpha x_F)}{(R/R_m - 1)^2} \tag{A.2-48}$$

The distillate composition for the product column is fixed, and the bottoms composition is calculated from

$$x_{B,P} = \frac{(1 - f_P)x_{D,P}S}{x_{D,P} - f_P S} \tag{10.3-29}$$

Similarly, the vapor rate becomes

$$V_2 = (R + 1)D$$

$$= \left[\frac{R/R_m}{(\alpha_{P,W} - 1)S} + 1\right]P \tag{10.3-30}$$

The condenser, reboiler, cooling-water, and steam costs, respectively, are given by

$$C_{C2} = C_{C2,BC}\left(\frac{V_2}{V_{2,BC}}\right)^{0.65} \tag{10.3-31}$$

$$C_{R2} = C_{R2,BC}\left(\frac{V_2}{V_{2,BC}}\right)^{0.65} \tag{10.3-32}$$

$$C_{CW2} = C_{CW2,BC}\left(\frac{V_2}{V_{2,BC}}\right) \tag{10.3-33}$$

$$C_{STM2} = C_{STM2,BC}\left(\frac{V_2}{V_{2,BC}}\right) \tag{10.3-34}$$

Summary

A base-case design and a set of cost calculations are presented in Table 10.3-1. The cost functions are developed from the base-case conditions $x = 0.8$, $T_r = 90°F$, $f_P = 0.995$, and $R = 1.2R_m$. We use this process cost model to describe our simplified optimization procedure.

[†] A. Jafarey, J. M. Douglas, and T. J. McAvoy, *I&EC Proc. Des. Dev.*, **18**: 121, 197 (1979).

TABLE 10.3-1
Base-case calculations

Let $P = 100$ mol/hr, $x_D = 0.999$, $x = 0.8$, $T_R = 90°F$, $f_P = 0.995$, $R/R_m = 1.2$, and $\rho_m = 0.8$ mol/ft^3.
From Eq. 10.3-13,

$$k_1 = 5.35 \times 10^{10} \exp \frac{-32,500}{1.987(460 + 90)} = 0.390 \text{ hr}^{-1}$$

$$k_2 = 4.61 \times 10^{17} \exp \frac{-52,500}{1.987(460 + 90)} = 0.03789 \text{ hr}^{-1}$$

From Eq. 10.3-14,

$$S = \frac{1}{0.8}\left(\frac{0.390}{0.390 - 0.03789}\right)[(1 - 0.8)^{0.39/0.03789} - (1 - 0.8)] = 0.907$$

Product Flow: $P = 100$

Eq. 10.3-7:   Fresh Feed $= F_{FA} = \dfrac{100(0.999)}{0.995(0.907)} = 110.67$

Eq. 10.3-8:   Waste Flow $= 110.67 - 100 = 10.67$
Eq. 10.3-9:   Prod. Value $= 11.5(100)(8150) = \$9.3725 \times 10^6/\text{yr}$
Eq. 10.3-10:   Feed Cost $= 8.5(110.67)(8150) = \$7.666 \times 10^6/\text{yr}$
Eq. 10.3-11:   By-product Value $= 1(10.67)(8150) = \$0.087 \times 10^6/\text{yr}$

Eq. 10.3-16:   Flow to React. $= \dfrac{110.67}{0.8} = 138.3$

Eq. 10.3-15:   React. Volume $= \dfrac{138.3}{0.39(0.8)} \ln \dfrac{1}{1 - 0.8} = 713.4$ ft^3

$$\frac{L_R}{D_R} = 6 \qquad V_R = \left(\frac{\pi D_R^2}{4}\right)L_R = \left(\frac{6\pi}{4}\right)D_R^3$$

$$D_R = [(2/3\pi)(713.4)]^{1/3} = 5.32 \text{ ft}$$

$$L_R = 6(5.32) = 31.9 \text{ ft}$$

Reactor cost—Guthrie correlation for a pressure vessel:

$$\text{Ann. React. Cost} = \left(\frac{792}{280}\right)(101.9)(5.32)^{1.066}(31.9)^{0.802}\left(\frac{3.18}{3}\right) = \$87,500/\text{yr}$$

Recycle column

Assume recover 0.995 of A overhead and 0.997 of P in the bottoms, $E_0 = 0.5$, $\alpha_{AP} = 2$, $\alpha_{PW} = 2$, $M_G = 60$.

$$\text{Separation Factor} = \frac{f_D}{1 - f_D}\frac{f_B}{1 - f_B} = \frac{0.995}{0.005}\frac{0.997}{0.003} = 66,134$$

Eq. 10.3-18: $N_1 = \dfrac{2}{0.5}\dfrac{\ln 66,134}{\ln 2} = 64.1$

Eq. A.3-2:   Tower Height $H_1 = 2.3N_1 = 147.3$
Eq. 10.3-20: $x_{FA} = 1 - x = 1 - 0.8 = 0.2$ $\qquad x_{F,P} = xS = 0.8(0.907) = 0.726$
$\qquad x_{F,W} = (1 - S)x = (1 = 0.907)0.8 = 0.0744$

TABLE 10.3-1

Base-case calculations (*continued*)

Eq. 10.3-19: $R_{m_1} = \left(\dfrac{2}{4-2}\right)\left(\dfrac{0.2 + 0.726}{0.2}\right) + \dfrac{0.0744}{0.2(4-1)} = 4.75$

$\qquad R_1 = 1.2R_m = 1.2(4.75) = 5.70$

$\qquad$ Recycle flow of $A = \dfrac{F_F(1-x)}{x} = \dfrac{110.67(1-0.8)}{0.8} = 27.67 = D$

Eq. 10.3-22: $V_1 = (5.7 + 1)27.67 = 185.5$ mol/hr

Eq. A.3-15: Diameter $= 0.0164(185.5)^{1/2}\left[379(60)\left(\dfrac{150 + 460}{520}\right)\right]^{1/4} = 2.85$

Appendix D.2:

Cost of Column $= \left(\dfrac{792}{280}\right)(101.9)(2.85)^{0.802}(147.3)^{1.066}\dfrac{3.18}{3}$

$\qquad\qquad\qquad = \$239,100/\text{yr}$

Condenser: Assume $\Delta H_A = 13,300$ Btu/mol, $\Delta H_p = 14,400$ Btu/mol, and $T_{bA} = 150°$F. Then

$$\Delta T_{LM} = \dfrac{120 - 90}{\ln\left[(150 - 90)/(150 - 120)\right]} = 43.2$$

$$Q_C = 13,300V_1 = 13,300(185.5) = 2.467 \times 10^6 \text{ Btu/hr}$$

$$\text{Area} = \dfrac{Q_c}{U\,\Delta T_m} = \dfrac{2.467 \times 10^6}{100(43.2)} = 571 \text{ ft}^2$$

Appendix E.2: Cost $= \left(\dfrac{792}{280}\right)(101.3)(571^{0.65})\left(\dfrac{3.29}{3}\right) = \$58,500/\text{yr}$

Eq. A.3-18: Cool-Water Cost $= \left(\dfrac{\$0.04}{1000 \text{ gal}}\right)\left(\dfrac{1 \text{ gal}}{8.34 \text{ lb}}\right)\left(\dfrac{13,300}{30}\right)(185.5)8150$

$\qquad\qquad\qquad = \$3200/\text{yr}$

Reboiler

$$Q_R = 14,400(185.5) = 2.671 \times 10^6 \text{ Btu/hr}$$

Eq. A.3-23: $\qquad A_R = \dfrac{Q_R}{11,250} = \dfrac{2.671 \times 10^6}{11,250} = 237 \text{ ft}^2$

Appendix E.2: $\qquad$ Cost $= \left(\dfrac{792}{280}\right)(101.3)(237^{0.65})\left(\dfrac{3.29}{3}\right) = \$32,700/\text{yr}$

Eq. A.3-25: $\qquad$ Steam Cost $= \left(\dfrac{\$2.80}{1000 \text{ lb}}\right)\left(\dfrac{14,400}{933}\right)(185.5)8150 = \$65,300/\text{yr}$

Product column

$\alpha_{PW} = 2$, $\Delta H_W = 15,500$ Btu/mol, $T_{b,P} = 200°$F, $M_G = 60$, and $x_{D,P} = 0.999$. Thus

Eq. 10.3-31: $x_{B,P} = \dfrac{(1 - 0.995)(0.999)(0.907)}{0.999 - 0.995(0.907)} = 0.0469$

TABLE 10.3-1
Base-case calculations   (*continued*)

$$x_F = \frac{x_{D,P}P}{x_{D,P}P + W} = \frac{0.999(100)}{0.999(100) + 10.67} = 0.903$$

Eq. 10.3-29: $\text{SF} = \dfrac{0.999}{0.001}\dfrac{1 - 0.0469}{0.0469} = 20{,}300$

Eq. A.2-48: $\beta = \dfrac{(R/R_m - x_F)(R/R_m - 1 + \alpha x_F)}{(R/R_m - 1)^2}$

$$= \frac{(1.2 - 0.903)[1.2 - 1 + 2(0.903)]}{(1.2 - 1)^2} = 14.89$$

Eq. A.2-48: $N_T = \dfrac{1}{0.5}\dfrac{\ln \beta(\text{SF})}{\ln [2/\sqrt{1 + (2 - 1)/1.2}]} = \dfrac{12.61}{0.5(0.39)} = 64.67$

Eq. A.2-23: $N = \dfrac{2}{0.5}\dfrac{\ln \text{SF}}{\ln \alpha} = 4\dfrac{\ln 20{,}300}{\ln 2} = 57.23$

Correction factor for using Eq. A.2-47 $= \dfrac{57.23}{64.67} = 0.8844$

$$N = 0.8844(64.67) = 57.23$$

Eq. A.3-2: Height $= 2.3(57.23) = 131.6$

$$R_m = \frac{1}{(\alpha - 1)x_F} = \frac{1}{(2 - 1)(0.903)} = 1.107$$

$$R = \left(\frac{R}{R_m}\right)R_m = 1.2(1.107) = 1.32$$

Eq. 10.3-22: $V_2 = (R + 1)D = (1.32 + 1)100 = 232.9$

Eq. A.3-15: $\text{Diameter} = 0.0164\sqrt{232.9}\left[379(60)\dfrac{200 + 400}{520}\right]^{1/4} = 3.24 \text{ ft}$

Appendix D.2: $\text{Column Cost} = \left(\dfrac{792}{280}\right)(101.9)(3.24^{0.802})(131.6^{1.066})\left(\dfrac{3.18}{3}\right)$

$$= \$427{,}500/\text{yr}$$

Condenser—Assume $\Delta H_p = 14{,}400$ and $\Delta H_W = 15{,}500$ Btu/mol. Then

$$\Delta T_m = \frac{120 - 90}{\ln [(200 - 90)/(200 - 120)]} = 94.2°\text{F}$$

$$Q_C = 14{,}400(232.9) = 3.354 \times 10^6 \text{ Btu/hr}$$

$$A_c = \frac{3.354 \times 10^6}{100(94.2)} = 356 \text{ ft}^2$$

$$\text{Cost} = \left(\frac{792}{280}\right)(101.3)(356)^{0.65}\left(\frac{3.29}{3}\right) = \$219{,}600/\text{yr}$$

$$\text{Cool-Water Cost} = \left(\frac{0.04}{1000}\right)\left(\frac{1}{8.34}\right)\left(\frac{14{,}400}{30}\right)(232.9)8150 = \$4400/\text{yr}$$

TABLE 10.3-1
Base-case calculations   (*continued*)

Reboiler

$$Q_R = 15,500(232.9) = 3.60 \times 10^6 \text{ Btu/hr}$$

$$A_R = \frac{3.610 \times 10^6}{11,250} = 320 \text{ ft}^2$$

$$\text{Cost} = \left(\frac{792}{280}\right)(101.3)(320)^{0.65}\left(\frac{3.29}{3}\right) = \$32,700/\text{yr}$$

$$\text{Steam Cost} = \left(\frac{4.00}{1000}\right)\left(\frac{19,500}{933}\right)(232.9)(8150) = \$88,300/\text{yr}$$

Tot. Cap. Cost = React. + Col.1 + Cond. 1 + Reb. 1 + Col. 2 + Cond. 2 + Reb. 2

$$= 29,200 + 79,700 + 29,400 + 11,000 + 142,500 + 73,200 + 10,900 = \$365,900/\text{yr}$$

Tot. Util. Cost = Coolant 1 + Steam 1 + Coolant 2 + Steam 2

$$= 3200 + 65,200 + 4400 + 88,300 = \$161,200/\text{yr}$$

Profit = Prod. − Feed + By-product − Tot. Cap. − Tot. Util

$$= 8,372,500 - 7,666,800 + 86,900 - 365,900 - 161,200$$

$$= \$1,265,500/\text{yr}$$

$$\text{Excess Feed} = 8.5\left(\frac{Px_D}{f_D S} - P\right)8150 = 8.5(100)\left[\frac{0.999}{0.995(0.907)} - 1\right]8150 = \$739,800/\text{yr}$$

Excess total cost, not including the stoichiometric feed requirement:

TAC = Excess Feed − By-product + Tot. Cap. + Tot. Util.

TAC = 739,800 − 86,900 + 365,900 + 161,200 = \$1,180,000/yr

10.4   APPROXIMATE OPTIMIZATION ANALYSIS

In a conventional optimization analysis, the gradient for each design variable is equal to zero. For our simple example we would require that

$$\frac{\partial \text{TAC}}{\partial x} = \frac{\partial \text{TAC}}{\partial T_R} = \frac{\partial \text{TAC}}{\partial f_P} = \frac{\partial \text{TAC}}{\partial (R/R_m)} = 0 \qquad (10.4\text{-}1)$$

However, for screening calculations we prefer to simplify the analysis. In particular, we would like to identify the dominant trade-offs for each design variable, the most important design variables, and the incentive for optimization. We discuss each below.

Dominant Trade-offs for Each Design Variable

After we develop a base-case design, we can change the conversion slightly and then calculate the incremental cost for each item in Table 10.3-1 divided by the incremental change that we made in the conversion. These results are shown in Table 10.4-1.

In the first column of this table, we see that as we increase the conversion, we produce more by-product (the waste cost decreases because the fuel value increases), but we are required to supply more reactant. The reactor cost increases, but the costs of the recycle column decrease because we recycle less reactant. The product column costs increase because the reflux ratio must be increased (since we are feeding more by-product to the column).

However, when we compare the positive costs in column 1, we note that the feed cost is much more important than the reactor cost and the costs of the product column. We use our trick of neglecting all costs that are an order of magnitude smaller than the largest cost, which indicates that we can neglect the effect of changes in conversion on the reactor and the product column costs. When we compare the negative costs, we see that the costs of the column shell, the condenser, and the steam in the recycle column are important. Hence, in subsequent optimization calculations we can neglect all but these largest-cost terms.

Now if we return to the base-case condition, change the temperature slightly, and then calculate the incremental cost of each item in Table 10.4-1 divided by the temperature change that we made, we obtain the results shown in the second column of Table 10.4-1. The increase in temperature causes more by-product to be formed, so we require more feed to make our desired amount of product, but we obtain a fuel credit for the by-product (i.e., the waste cost decreases). The reactor

TABLE 10.4-1
Gradients

TAC	$\dfrac{\partial \text{TAC}}{\partial x} \times 10^6$	$\dfrac{\partial \text{TAC}}{\partial T_R} \times 10^6$	$\dfrac{\partial \text{TAC}}{\partial f_p} \times 10^6$	$\dfrac{\partial \text{TAC}}{\partial (R/R_m)} \times 10^6$
Excess feed	2.8139	0.0249	-7.6976	0
By-product	-0.331	-0.0029	$+0.9056$	0
Reactor	0.041	-0.0009	-0.0182	0
Column 1	-0.078	0	-0.0426	0
Condenser 1	-0.023	0	-0.0126	0
Coolant 1	-0.006	0	-0.0032	0
Reboiler 1	-0.013	0.0001	-0.0072	0
Steam 1	-0.119	0.0002	-0.0656	0
Column 2	0.0687	0.0006	2.6268	-0.1166
Condenser 2	0.015	0.0001	-0.0419	0.0347
Coolant 2	0.001	0	-0.0025	0.0021
Reboiler 2	0.001	0	-0.0040	0.0033
Steam 2	0.0332	0.0003	-0.0909	0.0754

cost decreases, the cost of the recycle column increases (i.e., we need a larger reflux ratio because the reactant is more dilute), and the cost of the product column also increases (i.e., the product is more dilute).

Examining the positive values in column 2, we see that the effects of temperature changes on both of the column costs are negligible, except for the shell of the recycle column (the feed composition decreases so that more trays are required). Also, the change in feed cost is fairly small. The fuel credit of the waste stream and the reactor cost are both important. Thus, only three (or four) cost terms need to be considered for temperature changes.

Similar effects are observed for changes in the fractional recovery of the product overhead in the product column and for changes in the reflux ratio in the product column (see the last two columns of Table 10.4-1). Using the same order-of-magnitude arguments for the positive and negative terms separately, we find that we do not need to calculate all the processing costs. Hence, we can significantly simplify an optimization analysis by considering only the dominant costs in each trade-off. In many cases we can eliminate 75% of the calculations, although for the sake of illustration we retain most of the marginal terms in subsequent calculations.

Rank-ordering the Design Variables—The Most Important Design Variables

There is no way of comparing the various columns in Table 10.4-1 because they have different units, i.e., the first column is in ($/yr)/conversion whereas the second is in ($/yr)/°F. So we would like to find some way of putting each of the calculations on the same basis. To do this, we introduce scale factors, where the scale factor for each design variable is the maximum range for that variable. For example, we expect that the optimum value of R/R_m will be in the range

$$1 < \frac{R}{R_m} < 1.3 \tag{10.4-2}$$

so that we use a scale factor of 0.3. Similarly, we expect the optimum fractional recovery will be in the range

$$0.99 < f_P < 1.0 \tag{10.4-3}$$

so we use a scale factor of 0.01. The optimum conversion will be greater than zero and may be bounded above by the equilibrium conversion, the conversion corresponding to the maximum yield, or $x = 1$. Depending on how much information we have or how much effort we are willing to exert, we can select an appropriate range and a scale factor. We ask the chemist for a best guess of the range of temperatures (or pressures, if applicable) that are reasonable to consider, and we use this range for the scale factor.

Table 10.4-2 shows the results for the dominant cost terms when we add the *absolute values* of all the terms in each column and then multiply the sums by the

TABLE 10.4-2
Scaled gradients

	Scale factor			
	1.0	60	0.01	0.3
	$\dfrac{\partial TAC}{\partial x}\Delta x$	$\dfrac{\partial TAC}{\partial T_R}\Delta T_R$	$\dfrac{\partial TAC}{\partial f_p}\Delta f_P$	$\dfrac{\partial TAC}{\partial (R/R_m)}\Delta\left(\dfrac{R}{R_m}\right)$
Feed	2.8139	1.4982	−0.0769	0
By-product	−0.3310	−0.1763	+0.0090	0
Reactor	0.0412	−0.0545	0	0
Column 1	−0.0780	0.0047	−0.0004	0
Condenser 1	−0.0232	0.0014	−0.0001	0
Steam 1	−0.1195	0.0072	−0.0006	0
Column 2	0.0687	0.0366	0.0263	−0.0350
Condenser 2	0.0153	0.0366	0	0.0104
Reboiler 2	0.0015	0.0177	0	0.0010
Steam 2	0.0332	0.0177	−0.0009	0.0226
r_j	3.545	1.807	0.116	0.069
p_j	0.678	0.745	0.386	0.004

appropriate scale factors. Fisher, Doherty and Douglas[*] called this result the *rank-order parameter*

$$r_j = \sum \left|\frac{\partial TAC}{\partial y_i}\right| \Delta y_j \qquad (10.4\text{-}4)$$

This rank-order parameter indicates whether we are trading large positive incremental costs for large negative incremental costs or small positive incremental costs for small negative incremental costs. The units of each of the rank-order parameters are $/yr.

The rank-order parameters in Table 10.4-2 indicate that the optimization of the reactor conversion and the temperature are an order of magnitude more important than the optimization of the fractional recovery, which is about an order of magnitude more important than the optimization of the reflux ratio in the product column. The results imply that we will pay only a small penalty if we neglect the optimization of the reflux ratio. They also imply that we can obtain reasonable estimates if we simply optimize the reactor conversion and temperature —and then, if we have time available, optimize the fractional recovery. Again, we find that we can simplify the problem.

---

[*] W. R. Fisher, M. F. Doherty, and J. Douglas, *AIChE J.*, **31**: 1538 (1985).

Proximity Parameter—Incentive for Optimization

Initially we said that the criterion for an optimum is that the gradient of each design variable be equal to zero, but we have not used this information yet. The gradient is simply the sum of all the terms in each column of Table 10.4-2. By inspection we see that the positive terms might balance the negative terms for the reflux ratio, but the positive and negative terms are far out of balance for conversion and reactor temperature. Hence, we expect that our base-case value of reflux ratio is close to the optimum, but the base-case conversion and temperature are not.

To develop a quantity which characterizes the incentive for optimization, Fisher, Doherty and Douglas defined a proximity parameter as

$$p_j = \frac{|(\sum \partial \text{TAC}/\partial y_j)/\Delta y_j|}{\sum |\partial \text{TAC}/\partial y_j|/\Delta y_j} \tag{10.4-5}$$

This expression is merely the absolute value of the scaled gradient divided by the summation of the absolute values of the scaled components of the gradient. The proximity parameter is equal to zero at the optimum because the gradient is equal to zero. As we move far away from the optimum, the proximity parameter usually approaches unity because the total cost becomes asymptotic to either the total positive or the total negative cost components of the gradient (see Fig. 10.4-1).

The proximity parameters are shown in Table 10.4-2 for the base-case design. Figures 10.4-2 through 10.4-5 show plots of the total annual cost and the proximity parameters for each of the design variables. From these graphs we note that we are in the region where the optimum is fairly flat whenever the proximity parameter is

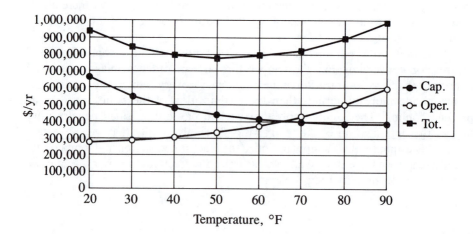

FIGURE 10.4-1
Cost behavior.

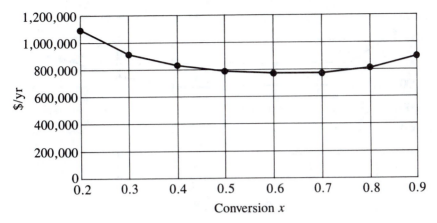

FIGURE 10.4-2a
Total excess cost versus conversion x.

less than 0.3. Hence, Fisher, Doherty, and Douglas proposed a heuristic for initial screenings:

> Whenever the proximity parameter is less than 0.3, there is little incentive to optimize.   (10.4-6)

Of course, unless each of the significant design variables has a proximity parameter less than 0.3, we must optimize all the variables. Also, as we proceed toward a final design, we might require that the proximity parameters be less than 0.2 or 0.1.

From Table 10.4-2 we see that the proximity parameter for reflux ratio is very small, but none of the other variables are close to their optimum values.

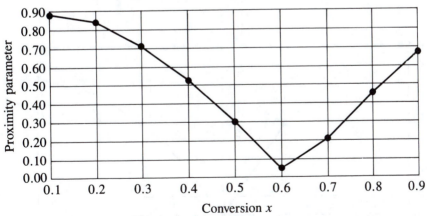

This graph was drawn from equally spaced points on the
x-axis, and therefore does not go to zero as it should.

FIGURE 10.4-2b
Proximity parameter x.

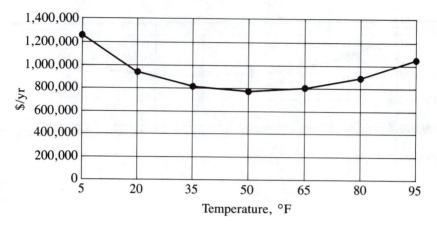

FIGURE 10.4-3a
Total excess cost versus temperature.

Categorizing Design Problems

The rank-order parameters and the proximity parameters for our example are given in Table 10.4-2. We can categorize each design variable as follows:

1. Conversion—important design variable, far from the optimum
2. Temperature—important design variable, far from the optimum
3. Fractional recovery—less important design variables, fairly close to the optimum
4. Reflux ratio—unimportant design variable, close to the optimum

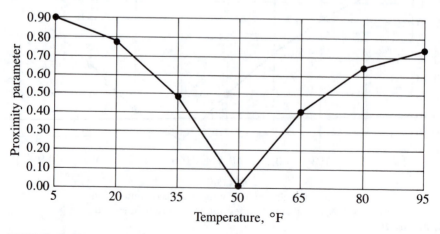

FIGURE 10.4-3b
Proximity parameter, temperature.

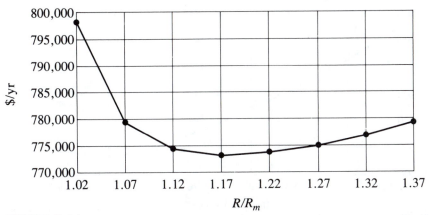

FIGURE 10.4-4a
Total excess cost versus R/R_m.

Comparisons

To illustrate the errors involved in an approximate optimization analysis, we compared the results of several case studies. In our simplified model, we deleted the cost of the condenser in column 1 and the cost of the reboiler in column 2 from the previously simplified list given in Table 10.4-2. Thus, we considered only the eight most expensive pieces of equipment as representative of the dominant costs (see Table 10.4-3).

Table 10.4-4 gives the results for the base-case design, and case 1 is the result for a rigorous optimization when all the costs are included in the analysis. We note that the optimization leads to a 73.3 % reduction in the costs. In case 2 we minimize

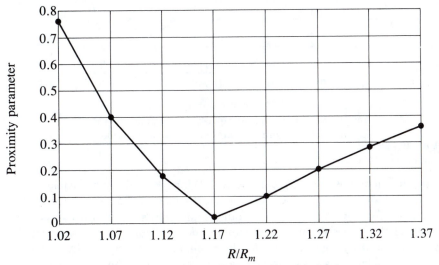

This graph was drawn from equally spaced points on the
x-axis, and therefore does not go to zero as it should.

FIGURE 10.4-4b
Proximity parameter R/R_m.

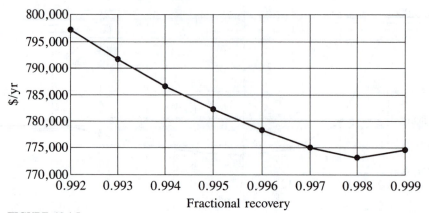

FIGURE 10.4-5a

Total excess cost versus fractional recovery.

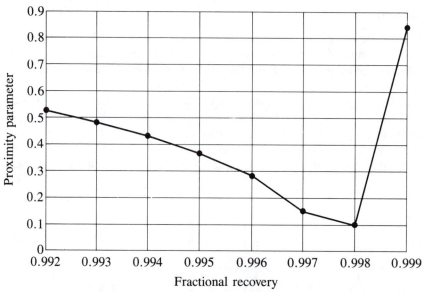

This graph was drawn from equally spaced points on the
x-axis, and therefore does not go to zero as it should.

FIGURE 10.4-5b

Proximity parameter, fractional recovery.

only the dominant cost terms (see Table 10.4-3) for the four design variables, and
then we add the costs for the other items. In cases 3 and 4 we repeat the
optimization for the dominant costs, but fix the values of the least important
variables at the values given by the common design heuristics. We see that the
errors introduced by considering only the dominant cost terms or by fixing the
values of the design variables with small rank-order functions are quite small. Thus,
at the conceptual stage of a process design, we use the rank-order function as a
guideline for determining how many design variables we need to consider.

TABLE 10.4-3
Dominant costs—scaled gradients

	$\dfrac{\partial\text{TAC}}{\partial x}\Delta x$	$\dfrac{\partial\text{TAC}}{\partial T_R}\Delta T_R$	$\dfrac{\partial\text{TAC}}{\partial f_p}\Delta f_p$	$\dfrac{\partial\text{TAC}}{\partial(R/R_m)}\Delta\left(\dfrac{R}{R_m}\right)$
Feed	2.8139	1.4988	−0.0769	0
By-product	−0.3310	−0.1763	0.0090	0
Reactor	0.0412	−0.0545	0	0
Column 1	−0.0780	−0.0047	−0.0004	0
Steam 1	−0.1195	0.0072	−0.0006	0
Column 2	0.0687	0.0366	0.0263	−0.0350
Condenser 2	0.0153	0.0082	0	0.0104
Steam 2	0.0332	0.0677	−0.0009	0.0226

TABLE 10.4-4
Results of optimization studies

Design variable	Base-case	Case 1	Case 2	Case 3	Case 4
x	0.8	0.62	0.59	0.59	0.59
T_R, °F	90	51	51	51	51
R/R_m	1.2	1.185	1.19	1.2	1.2
f_P	0.995	0.998	0.997	0.998	0.995
Dominant costs $\times 10^6$			716	718	723
TAC $\times 10^6$	1.195	772	775	777	782
ΔTAC, %	55	0	0.003	0.006	1.295

Case 1 = rigorous optimization including all the costs.
Case 2 = optimization of the dominant costs for all the design variables.
Case 3 = optimization of the dominant costs with $R/R_m = 1.2$.
Case 4 = optimization of the dominant costs with $R/R_m = 1.2$ and $f_p = 0.995$.

10.5 SUMMARY, EXERCISES, AND NOMENCLATURE

Summary

Normally only a few cost terms dominate the economic trade-offs for each optimization variable. By identifying these dominant trade-offs, we can greatly simplify an optimization analysis.

We can further simplify the optimization analysis by determining the rank-order function for each design variable y_i,

$$r_j = \sum \left|\frac{\partial\text{TAC}}{\partial y_j}\right| \Delta y_j \tag{10.5-1}$$

where the Δy_j are scale factors based in the range of a design variable within which we expect to observe the optimum. We neglect to optimize any design variables whose rank-order functions are an order of magnitude smaller than the largest rank-order function. Normally, the significant design variables are those that affect the product distribution or the process flows.

We can estimate whether an initial guess of a design variable is close to its optimum value by calculating a proximity parameter

$$p_j = \frac{|\sum \partial TAC/\partial y_j|}{\sum |\partial TAC/\partial y_j|} \tag{10.5-2}$$

Experience indicates that if $p_j < 0.3$, we are in the region where the optimum is relatively flat, although we can require a tighter tolerance.

Exercises

10.5-1. For any process design that you have considered, calculate the rank-order functions and the proximity parameters. Use the results to identify the dominant economic trade-offs. Discuss these optimization problems.

10.5-2. Consider the example given in Sec. 10.3, except consider two first-order, parallel reactions, instead of consecutive reactions. (Also, neglect the optimization of the reflux ratio and the fractional recovery in both columns.) Calculate the rank-order functions and the proximity parameters, and find the optimum design conditions. Plot the proximity parameters versus the design variables.

10.5-3. Consider the example given in Sec. 10.3, but consider the indirect column sequence rather than the direct sequence. (Neglect the optimization of the fractional recovery and the reflux ratios in both columns.) Calculate the rank-order functions and the proximity parameters. Find the optimum design conditions. How do the results for the direct and the indirect sqeuences compare? What do the sequencing heuristics indicate at the optimum flows for each case?

10.5-4. For the cyclohexane process described in Exercises 5.4-7 and 6.8-6, how many design variables are encountered at level 3? Calculate the rank-order function to determine the relative importance of these variables? Also, estimate the optimum design conditions at level 3.

Nomenclature

A, A_{BC}	Heat-exchanger area and base-case value (ft^2)
$B_{hp}, B_{hp, BC}$	Brake horsepower of compressor and base-case value (hp)
C_A, C_P	Concentrations of components A and P (mol/ft^3)
$C_A, C_{A, BC}$	Annualized cost of heat exchanger and base-case value ($/yr)
$C_C, C_{C, BC}$	Cost of condenser and base-case value ($/yr)
$C_{comp}, C_{comp, BC}$	Cost of compressor and base-case value ($/yr)
$C_{CW}, C_{CW, BC}$	Cost of cooling water and base-case value ($/yr)
C_f	Raw-material cost ($/mol)
$C_{FN}, C_{FN, BC}$	Cost of furnace and base-case value ($/yr)
C_P	Heat capacity [Btu/(mol·°F)]
$C_R, C_{R, BC}$	Cost of reactor ($/yr)
$C_R, C_{R, BC}$	Cost of reboiler and base-case value ($/yr)
$C_{sh}, C_{sh, BC}$	Cost of distillation column shell and base-case ($/yr)

$C_{STM}, C_{STM,BC}$	Cost of steam and base-case value ($/yr)
C_V	Specific heat at constant volume [Btu/(mol·°F)]
C_{VR}	Annualized reactor cost [$/(ft^3·yr)]
C_1, C_2, C_3	Cost coefficients
D	Distillate flow rate (mol/hr)
E_0	Overall plate efficiency
f_D	Fraction recovery of the light-key overhead
F	Flow rate (mol/hr)
$F_{CW}, F_{CW,BC}$	Cooling-water flow rate and base-case value (mol/hr)
F_F	Fresh feed rate (mol/hr)
F_R	Flow to reactor (mol/hr)
F_V	Feed rate to compressor (mol/hr)
H_0	Height of column sump and vapor disengaging space (ft)
k	Reaction rate constant (hr^{-1})
k_1, k_2	Reaction rate constants (hr^{-1})
M_G	Molecular weight of distillate
M&S	Marshall and Swift index (see *Chemical Engineering*)
N	Number of trays
N_T	Number of theoretical trays
p_j	Proximity parameter
P	Production rate (mol/hr)
P_{in}, P_{out}	Inlet and outlet pressures for a gas compressor (psia)
P_P	Flow of desired product (mol/hr)
P_T	Column pressure (psia)
Q	Heat duty (Btu/hr)
$Q_F, Q_{F,BC}$	Furnace heat duty and base-case value (Btu/hr)
r_j	Rank-order function ($/yr)
R	Reflux ratio
R_A	Recycle flow (mol/hr)
R_m	Minimum reflux ratio
S	Selectivity
SF	Separation factor
TAC	Total annual cost ($/yr)
T_D	Distillate temperature (°F)
U	Overall heat-transfer coefficient [Btu/(hr·ft^2·°F)]
U_C	Condenser overall heat-transfer coefficient [Btu/(hr·ft^2·°F)]
V	Vapor rate (mol/hr)
V_R	Reactor volume (ft^3)
W	Total flow of waste stream (mol/hr)
W_p	Amount of product in the waste stream (mol/hr)
$W_S, W_{S,BC}$	Flow rate of steam and base-case value (lb/hr)
x, x_{BC}	Conversion and base-case value
x_D	Distillate composition
x_F	Feed composition of light key
y_j	Design variable

Greek symbols

ΔH_v	Heat of vaporization of light key (Btu/mol)
Δt	Temperature change (°F)
ΔT_{LM}	Log-mean temperature driving force (°F)
ΔH_S	Heat of vaporization of steam (Btu/lb)
α	Relative volatility
β	See Eq. 9.2-30
γ	$(C_P/C_V - 1)/(C_P/C_V)$
ρ	Density (mol/ft^3)

CHAPTER
11

PROCESS RETROFITS

In all previous discussions we assumed that we are designing a new process. However, exactly the same techniques are useful for retrofitting a process. By retrofitting, we normally mean making minor changes in the interconnections between process equipment, the replacement of one or more pieces of equipment by some other equipment, or the change in the sizes of one or more pieces of equipment in an existing process. The first type of change involves structural modifications of the flowsheet, whereas with the last two types of change the flowsheet remains the same.

Some examples of where we might want to retrofit a process are to increase the production capacity (debottleneck a process), efficiently process new raw-material feedstocks when they are cheaper, utilize new process technologies, or reduce operating costs, because the optimum operation conditions have changed since the plant was originally built. (This changing economic environment is the reason that most companies did not usually attempt to optimize processes in the past.) The last goal is valid for a very large number of existing plants, and therefore there is a great interest in retrofit procedures. We describe a systematic procedure for process retrofits in this chapter.

11.1 A SYSTEMATIC PROCEDURE FOR PROCESS RETROFITS

The systematic procedure for process retrofits discussed here was deveolped by Fisher, Doherty, and Douglas.* The analysis is limited to single-product, continuous processes, which are the same restrictions of our other design methods. Moreover, many of the steps in the procedure are identical to the steps that we have completed previously.

The retrofit procedure also proceeds through a hierarchy of decisions:

1. Estimate an upper bound on the incentive for retrofitting.
2. Estimate the economic incentive for replacing the existing plant by using the same process flowsheet.
3. Estimate the economic incentive for replacing the existing plant with a better process alternative.
4. Estimate the incremental investment costs and savings in operating costs associated with changing the existing process.
5. Refine the retrofit calculations.

Clearly, the first four steps in this procedure focus on the same type of screening calculations that we considered earlier, and for this reason we describe the retrofit procedure before we consider the use of simulators to refine design or retrofit calculations.

We discuss each of the levels in the hierarchy above in more detail below.

Estimating an Upper Bound of the Incentive for Retrofitting

Our goal is to reduce the operating cost of an existing process, and so our first step is to prepare an operating cost diagram. That is, we prepare a cost diagram of the type described in Chap 9, except that we do not include any of the capital costs. From this operating cost diagram, we can see the total costs of the raw materials lost in the form of by-products or lost in waste streams. Similarly, we can see all the energy costs that are supplied to the process. An example of a diagram of this type for the HDA process was given in Fig. 8.10-8.

For a large number of processes, such as the HDA process, the costs of the raw-material losses will exceed the energy costs, which means that we will need to examine the possibility of changing the process flows. If we change the flows, we will also need to change the heat-exchanger network. Thus, normally we must consider the behavior of the total process.

---

* W. R. Fisher, M. F. Doherty and J. M. Douglas, "Screening of Process Retrofit Alternatives," *I&EC Research*, in press.

Estimate the Economic Incentive for Replacing the Existing Plant

One of our options is to completely replace the existing plant. This option will correspond to the largest capital investment. Also, if we do not change the process, this calculation will indicate the profitability position of one of our competitors if they built a process just like ours.

We can use the hierarchical decision procedure described in Chaps. 5 through 8 to estimate the processing costs. With the shortcut material and energy balances, equipment design procedures, and cost models, it should be possible to complete a design in about 2 days by hand (or 1 to 3 hr if software is available). One major advantage of using this hierarchical decision procedure is that it simplifies the task of generating a complete list of process alternatives to be considered.

Estimating the Economic Incentive for Replacing the Existing Plant by a Better Process Alternative

Estimating the profit potential of the "best" possible process alternative will indicate whether one of our competitors could drive us out of business by building a process that yields "shutdown" economics better than ours. We can use the gross-screening procedures described in Chap. 9 to evaluate all the process alternatives, to see whether a better process exists. If we find a better process, then we refine the calculations, using the techniques described in Chaps. 5 through 8 to estimate the optimum design conditions for each alternative.

This identification of improved alternatives also indicates the changes in the flowsheet that we should consider when we examine the retrofitting of our existing process. The savings in both raw-material and energy costs that correspond to these structural changes in the flowsheet can be evaluated. Thus, there is a systematic way of selecting structural modifications for more detailed evaluations.

Estimating the Incremental Investment Costs and Savings in Operating Costs Associated with Changing the Existing Process

Our initial retrofit analysis should focus on screening calculations to see whether a more detailed retrofit study can be justified. Of course, if we eliminate a piece of process equipment in our retrofit analysis, we still must continue to pay for that equipment (unless it has been totally depreciated). Thus, the retrofit study should focus on incremental annualized investment costs and incremental savings in operating costs.

A systematic way of screening retrofit opportunities is to

1. Eliminate the process heat exchangers.
2. Identify the significant operating variables.
3. Identify the equipment that constrains changes in the significant operating variables.

4. Remove the equipment constraints by adding excess capacity until the incremental annualized investment costs balance the incremental savings in operating costs.

5. Evaluate the optimum energy integration for the new process flows, and check the sensitivity of the optimization to energy integration changes.

6. Retrofit the heat-exchanger network.

ELIMINATE THE PROCESS HEAT EXCHANGERS. The process flows and product distribution usually dominate the process economics, and therefore we initially focus on estimating the optimum flows. To accomplish this goal and to avoid encountering constraints on heat-exchange equipment, we initially neglect the heat exchangers (although we include the operating costs for fuel, steam, and cooling water in the analysis). Once we have removed other equipment constraints and have estimated new values for the optimum flows, we will resolve the energy integration problem and retrofit the heat exchangers.

IDENTIFY THE SIGNIFICANT OPERATING VARIABLES. The number of operating variables is equal to the number of design degrees of freedom minus the number of equipment sizes that have been fixed by the design. However, some of these operating variables must be fixed so that the process constraints (e.g., production rate, product purity, molar ratio of reactants at the reactor inlet, reactor inlet or outlet temperatures, etc.) can be satisfied. The remaining operating variables should be fixed by an optimization analysis.

We can determine which of the operating variables are the most important to optimize by calculating the rank-order function (see Chap. 10), except for this analysis we include only the operating costs

$$r_j = \sum_{i=1}^{m} \left| \frac{\partial \text{TOC}_i}{\partial x_j} \right| \Delta x_{j,\,\text{max}} \tag{11.1-1}$$

Any operating variable whose rank-order function is an order of magnitude smaller than the largest rank-order function is then dropped from further consideration during the initial screening. Thus, we obtain a set of significant operating variables.

IDENTIFY THE EQUIPMENT THAT CONSTRAINS THE SIGNIFICANT OPERATING VARIABLES. If we calculate a proximity parameter (except that we retain the sign of the gradient)

$$p_j = \frac{\sum\limits_{i=1}^{m} (\partial \text{TOC}_i / \partial x_j)}{\sum\limits_{i=1}^{m} |\partial \text{TOC}_i / \partial x_j|} \tag{11.1-2}$$

we can gain an indication of the incentive for optimizing the significant operating variables; i.e., unless $p_j < 0.3$ for all the operating variables, an optimization is justified (see Chap. 10). Then we change the significant operating variables in a direction opposite to the sign of the proximity parameter; i.e., if the gradient is positive, we decrease the operating variable to look for a minimum. As we change the operating variables in these directions, normally we encounter an equipment constraint.

Note that the operating costs often lie on an equipment constraint even for the optimum design of a new plant. That is, the minimum of the total operating costs for a design (for the type of plants we are considering) normally is at lower values of the operating variables than the minimum of the total annual cost (which includes equipment as well as operating costs); see Fig. 11.1-1. Thus, after the equipment sizes have been fixed and the plant has been built, if we try to move toward the lowest possible operating cost, we will encounter an equipment constraint. The addition of existing equipment capacity to remove this constraint during the design of a process by definition will cost more than the savings in operating costs, which is why we fixed the design at the minimum total annual cost. However, the operating costs change over the life of a process, so that incremental capital costs may be justified for a process retrofit.

REMOVING EQUIPMENT CONSTRAINTS. Once we encounter an equipment constraint, we add capacity until the incremental annualized capital cost just balances the incremental savings in operating costs. Thus, we determine the new optimum retrofit trade-offs for the significant design variables. Of course, if we encounter constraints in more than one piece of equipment, we merely add capacity to both units until the incremental capital costs balance the incremental savings in operating costs. (If time permits and the retrofit economics appear promising, we might also consider the optimization of the other operating variables that had smaller rank-order functions.)

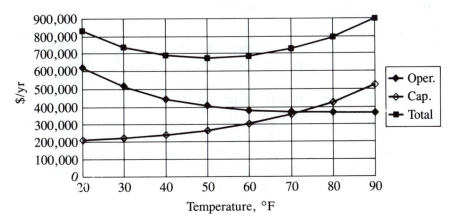

FIGURE 11.1-1
Costs versus temperature.

ENERGY INTEGRATION. From the analysis above, we obtain new estimates of the optimum process flows, assuming that no constraints in heat-exchange equipment are important. However, the new process flows correspond to a new energy integration problem. Hence, we use the procedures described in Chap. 8 to solve the heat-exchanger network design problem for the new flows. Since the optimum flows depend on recycle costs, which depend on the heat-exchanger network used, it might be necessary to iterate a few times to find a new optimum.

RETROFIT THE HEAT-EXCHANGER NETWORK. Of course, we can decrease the retrofit investment by using as much of the existing heat-exchanger equipment as possible. Thus, we readjust the flows and heat loads in an attempt to estimate the optimum retrofit conditions. We are still screening the retrofit opportunity, so that our goal is to get into the region where the optimum is relatively flat, rather than to find the exact optimum.

Refining the Retrofit Calculations

If the economic incentive for retrofitting a process is large, we must refine our screening calculations, just as we must refine the screening calculations that we used for process design. The use of computer-aided design programs to accomplish this task is discussed in Chap. 12. Before we consider this refinement, however, we present an example of our retrofit procedure.

11.2   HDA PROCESS

As an illustration of our retrofit procedure, we consider the retrofitting of an HDA process. The original optimum design variables, as taken from McKetta,* are listed in Table 11.2-1 for the flowsheet shown in Fig. 11.2-1. The steps in the systematic procedure are discussed below.

1.   Prepare an Operating Cost Diagram

An operating cost diagram is shown in Fig. 11.2-2. From this diagram we see that raw-materials costs are much more important than energy costs. Thus, we want to modify the process flows in our retrofit analysis.

2.   Design a New Plant Using the Same Process

An optimized design for the HDA process with diphenyl removed as a by-product was discussed in Sec. 8.11. The smallest total annual cost was 4.73×10^6/yr, which included an annualized capital cost of 1.47×10^6/yr and an operating cost of 3.26×10^6/yr. This optimum operating cost is lower than the values on our cost diagram, 4.62/yr, but the investment is large.

---

* J. J. McKetta, *Encyclopedia of Chemical Processing and Design*, vol. 4, Dekker, New York, 1977, p. 182.

TABLE 11.2-1

Optimum design variables for existing plant

Design variables	Value
Reactor conversion	0.75
H_2 purge composition	0.46
Inlet temperature to partial condenser	428 K
Outlet temperature from partial condenser	311 K
Benzene recovery in product column	0.99
Reflux ratio in product column	1.2
Toluene recovery in recycle column	0.986
Diphenyl recovery in recycle column	0.807
Recycle ratio in recycle column	1.0

From J. J. McKetta, *Encyclopedia of Chemical Processing and Design*, vol. 4, Dekker, New York, 1977, p. 182.

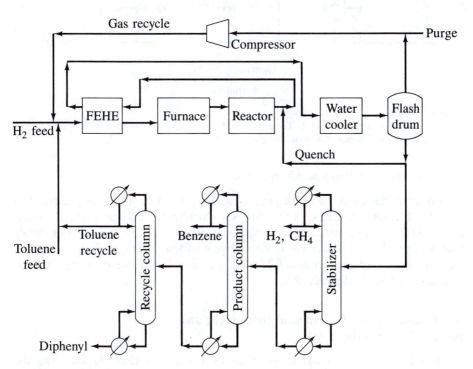

FIGURE 11.2-1
HDA process flowsheet.

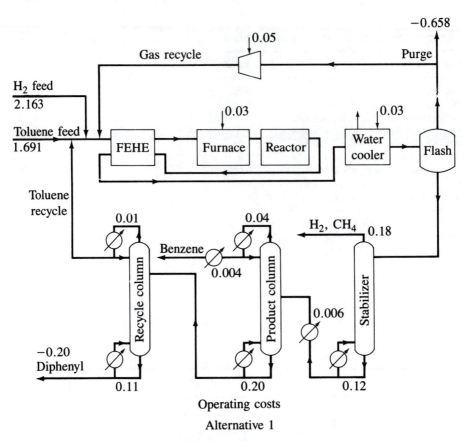

Operating costs

Alternative 1

FIGURE 11.2-2
Operating cost diagram. [*From D. L. Terrill and J. M. Douglas, I&EC Research,* **26:** *685 (1987), with permission from the American Chemical Society.*]

3. Find the Best Process Alternatives

An optimum design of the HDA process with diphenyl recycled was presented in Sec. 9.4. The total cost was 3.57×10^6/yr, which includes an annualized capital cost of 1.27×10^6/yr and an operating cost of 2.30×10^6/yr. We could probably find an even better alternative if we recovered some hydrogen from the purge stream, as we discussed in Sec. 9.3. Also, we should evaluate the other process alternatives discussed in Sec. 9.3.

4. Estimating the Incremental Investment and Savings in Operating Costs

There are at least two alternatives to consider, i.e., removing and recycling the diphenyl. To illustrate the procedure, we consider only the case where diphenyl is recovered as a by-product (although this might not be the best solution). We follow the procedure presented in Sec. 11.1.

TABLE 11.2-2
Significant operating variables

f_i	$\partial f_i/\partial x$	$\partial f_i/\partial y_{PH}$	$\partial f_i/\partial F_{cw}$	$\partial f_i/R_{PC}$
Furnace fuel	−1134	−1156	0	0
Compressor power	−125	−201	0	0
Product column utilities	0	0	0	154
Benz loss in prod. col.	0	0	0	−154
Purge losses	−311	7135	−0.5	0
Selectivity losses	5577	0	0	0
Recycle col. utilities	−382	0	0	0
Cooling water part. cond.	0	0	0	0
x_j	0.75	0.46	150	1.27
Δx_j	0.1	0.1	50	0.1
r_j	750	850	50	30
p_j	0.48	0.77	0	0

From W. R. Fisher, M. F. Doherty, and J. M. Douglas, "Screening Process Retrofit Alternatives," *I&EC Research*, in press, 1987, with permission from the American Chemical Society.

4a. ELIMINATE THE PROCESS HEAT EXCHANGERS. We eliminate the feed-effluent heat exchanger shown in Fig. 11.2-1, although we will repeat an energy integration analysis after we have revised the process flows.

4b. IDENTIFY THE SIGNIFICANT OPERATING VARIABLES. Values of the rank-order function and of proximity parameters are given in Table 11.2-2 for the largest operating costs. The results, as we might expect, show that the conversion and the purge composition are the most important operating variables (i.e., their rank-order functions are an order of magnitude larger than the values for the other two variables). Also, the large, positive values of the proximity parameters indicate that we would like to decrease the values of the conversion and purge composition. Of course, smaller values of conversion and purge composition correspond to larger recycle flows. We might not be able to obtain these larger recycle flows because of one or more equipment constraints.

4c. IDENTIFY THE EQUIPMENT CONSTRAINTS. The equipment most likely to constrain the recycle flows includes the gas-recycle compressor, the recycle distillation column, the furnace, and the partial condenser. However, we do not consider constraints in the heat-exchanger equipment (we defer that consideration until we energy-integrate the process again). Also, in the original design,* there was a considerable amount of excess capacity in the recycle column.

---

* J. J. McKetta, *Encyclopedia of Chemical Processing and Design*, vol. 4, Dekker, New York, 1977, p. 182.

Thus, for our problem, the gas-recycle compressor constrains both of the operating variables, Lower purge compositions correspond to higher gas-recycle flows, which are constrained by the compressor. Lower conversions correspond to higher liquid-recycle flows. However, because of the 5/1 hydrogen-to-aromatic ratio requirement at the reactor inlet, an increase in the toluene-recycle flow must be accompanied by an even larger increase in the gas-recycle flow. Hence, the gas-recycle compressor also, indirectly, constrains an increase in the liquid-recycle flow. Because the gas compressor capacity constrains both of the significant operating variables, we consider this retrofit problem first. The dominant operating cost trade-offs in Table 11.2-1 for both significant operating variables involve raw-materials losses balanced against fuel costs for the furnace, so that energy integration will also be very important.

The cooling-water flow rate to the partial condenser and the reflux ratio in the product column are less important operating variables. Moreover, these variables can be adjusted to obtain their optimum values with the existing equipment. that is, $p_j = 0$ for each. The cooling-water flow rate primarily trades utility costs for purge losses, and the only retrofit policy possible to reduce these costs would be to increase the area of the partial condenser before the flash drum. The reflux ratio in the product column trades product losses in the bottom (which get recycled) for increased utility costs to increase the reflux ratio. Possible retrofit policies would be to use energy integration to reduce the utilities requirements or to increase the number of trays in the stripping section to improve the product recovery (which we consider to be impractical).

4d. REMOVE THE EQUIPMENT CONSTRAINTS. Now that we have identified the equipment constraints, we want to remove those constraints. Since the gas compressor constrains both of the significant design variables, we consider this constraint first. We merely add compressor capacity, and the incremental cost of this capacity, until the incremental annualized investment balances the incremental savings in the operating costs; see Fig. 11.2-3. From this graph we see that the incremental savings are balanced by the incremental investment when we install a new compressor with 56 % of the capacity of the original unit. The incremental capital cost is \$55,000/yr, but the savings in operating costs are \$480,000/yr. Of course, the large increase in the process flows will also exceed the furnace capacity, but we consider the energy integration later.

We could also add another heat exchanger in series with the existing partial condenser and then trade the incremental capital cost for the savings in purge losses. The results of this calculation are shown in Fig. 11.2-4. At the optimum conditions, the new exchanger should have 1.5 times the area of the original unit. The incremental capital cost is \$75,000/yr, and the savings in the purge losses are \$410,000/yr.

The new values of the optimization variables are shown in Table 11.2-3. Obviously, the changes in the most significant design variables cause the process flows to change, which will change the energy integration required.

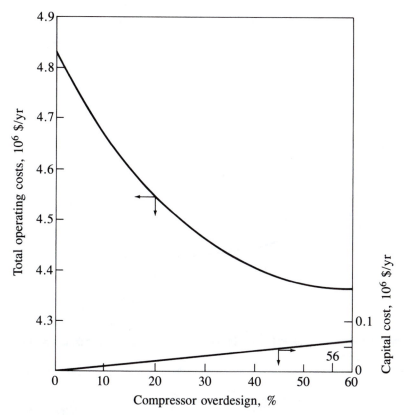

FIGURE 11.2-3
Estimation of the optimum retrofit policy for the gas-recycle compressor. (*From W. R. Fisher, M. F. Doherty, and J. M. Douglas, "Screening Process Retrofit Alternative," I&EC Research, in press, 1987, with permission from the American Chemical Society.*)

4e. ENERGY-INTEGRATE THE RETROFITTED PROCESS. The T-H diagram for the original process is shown in Fig. 11.2-5. The wide separation between the two curves indicates that there was a significant incentive for improving the energy integration. However, the T-H diagram for the partially retrofitted process is given in Fig. 11.2-6. Now, we see that the changes in the process flow rates have significantly increased the incentive for energy integration.

The evaluation of various heat-exchanger network alternatives was discussed in Sec. 8.11. The results indicated that all the network alternatives had about the same costs. Thus, we begin our energy integration retrofit analysis merely by adding exchanger capacity in series with the existing feed-effluent heat exchanger. The incremental capital cost and savings in operating costs are shown in Fig. 11.2-7. At the optimum conditions, we need a new exchanger with 68% of the area of the original unit. The capital cost is $160,000/yr, while the savings are $250,000/yr.

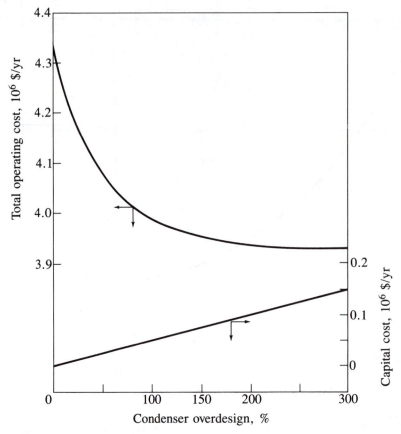

FIGURE 11.2-4
Estimation of the optimum retrofit policy for the partial condenser preceding the flash drum. (*From W. R. Fisher, M. F. Doherty, and J. M. Douglas, "Screening Process Retrofit Alternative," I&EC Research, in press, 1987, with permission from the American Chemical Society.*)

TABLE 11.2-3
Optimum operating conditions

Design variable	Original value	Retrofit value
Reactor conversion	0.75	0.69
Purge composition	0.46	0.35
Cooling water to part. cond.		
(% of design value)	150	120
Reflux ratio recy. col.	1.27	1.36
Furnace duty (10^6 BTU/hr)	11.9	11.9

From W. R. Fisher, M. F. Doherty, and J. M. Douglas, "Screening Process Retrofit Alternatives," *I&EC Research*, in press, 1987, with permission from the American Chemical Society.

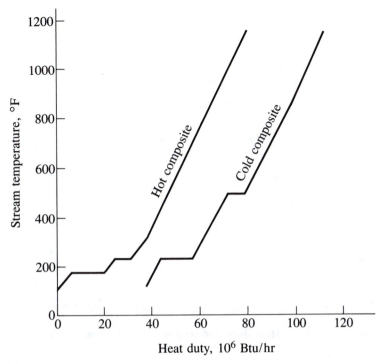

FIGURE 11.2-5
Temperature-enthalpy diagram for the HDA process at base-case operating conditions. (*From W. R. Fisher, M. F. Doherty, and J. M. Douglas, "Screening Process Retrofit Alternative," I&EC Research, in press, 1987, with permission from the American Chemical Society.*)

4f. RETROFIT THE HEAT-EXCHANGER NETWORK. With this optimum retrofit of the feed-effluent heat exchanger, the furnace load is decreased to within 10% of its design value. One way that we can resolve this discrepancy is by adding area to the feed-effluent exchanger, which according to Fig. 11.2-7 will not lead to a large increase in costs. Another alternative would be to go back and to restrict the process flows so that the existing furnace capacity is adequate. Similarly, we could add a steam heater before the feed-effluent heat exchanger, or, better, we could look for more complex networks (pressure-shifting the columns, etc.) that would satisfy our requirements.

Discussion of the Retrofit Analysis

Our retrofit analysis indicated that we can save $1,140,000/yr with an annualized investment of $290,000/yr. We could decrease the investment required by deciding not to retrofit the partial condenser. Now we can compare these savings and investment to those corresponding to building a new plant. Of course, we still might not have determined the best retrofitted process because we have not

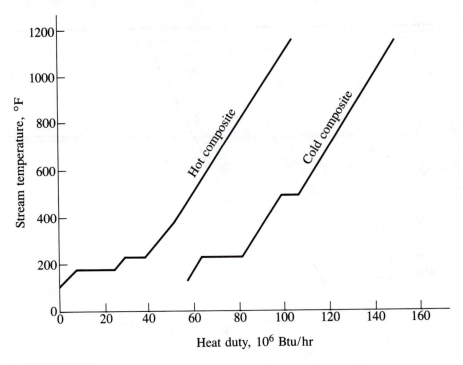

FIGURE 11.2-6
Temperature-enthalpy diagram for the HDA process with retrofitted equipment and new optimum flows. (*From W. R. Fisher, M. F. Doherty, and J. M. Douglas, "Screening Process Retrofit Alternatives," I&EC Research, in press, 1987, with permission from the American Chemical Society.*)

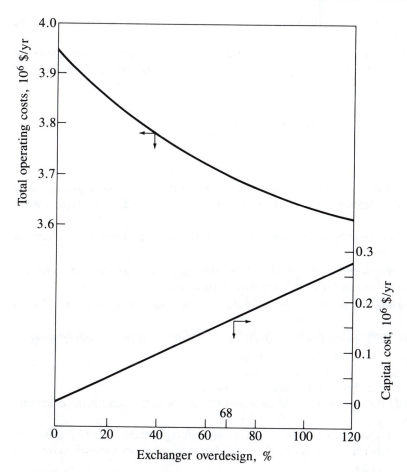

FIGURE 11.2-7
Estimation of the optimum retrofit policy for the feed-effluent heat exchanger. (*From W. R. Fisher, M. F. Doherty, and J. M. Douglas, "Screening Process Retrofit Alternative," I&EC Research, in press, 1987, with permission from the American Chemical Society.*)

considered the recycle of diphenyl or the recovery of any of the hydrogen in the purge stream.

Once we decide on a retrofit policy, we must also refine our screening calculations. The use of computer-aided design tools is discussed in the next chapter.

11.3 SUMMARY AND EXERCISES

Summary

We have described a systematic procedure for retrofitting processes that uses many of the methods described earlier. The retrofit analysis proceeds through this series of steps:
1. Estimate an upper bound on the incentive for retrofitting—prepare an operating cost diagram.
2. Estimate the incentive of replacing the existing plant with an identical system— use shortcut calculations and generate a list of process alternatives.
3. Estimate the incentive of replacing the existing plant with the best process alternative.
4. Estimate the incremental investment cost and the savings in operating costs associated with changing the existing process.
   a. Eliminate the process heat exchangers.
   b. Identify the significant operating variables.
   c. Identify the equipment that constrains changes in the significant operating variables.
   d. Remove the equipment constraints by adding excess capacity until the incremental, annualized capital cost balances the incremental savings in operating costs.
   e. Energy-integrate the process.
   f. Retrofit the heat-exchanger network.
5. Refine the retrofit calculations, if justified.
6. Find the best retrofit alternative.

Exercises

11.3-1. Retrofit the HDA process by recycling the diphenyl to extinction.

11.3-2. For any process that you have designed, look up the original case study and compare the raw-material costs, utilities costs, and M&S index with current values. How would you expect the optimum design values to change as the economic factors change? Calculate the rank-order functions and the proximity parameters, using current prices but the original case-study values of the optimum design variables. Retrofit the process.

CHAPTER 12

COMPUTER-AIDED DESIGN PROGRAMS

Our preliminary design (screening) calculations enable us to see whether more detailed design studies can be justified. In addition, they help us to select the three or four flowsheets that are the most promising and to estimate the optimum design conditions for each of these process alternatives, i.e., the compositions, temperatures, and pressures of every stream as well as the equipment sizes. However, our screening calculations were based on a variety of approximations. Hence, if our screening indicates that a more detailed study can be justified, we need to obtain more rigorous solutions to our design equations; i.e., we need to derive a set of material balances that are rigorous, we need to account for losses of reactants and products, we need to use more rigorous equipment design equations, etc. We use a computer-aided design (CAD) program for this purpose.

Numerous CAD programs are available commercially, including FLOW-TRAN, PROCESS, DESIGN 2000, ASPEN, ChemCAD, HYSIM, SPEEDUP, etc. In addition, many companies have their own proprietary programs. Since more academic institutions use FLOWTRAN than any other program, we use FLOW-TRAN in our examples. However, the other programs are fairly similar in their general structure. FLOWTRAN was developed by Monsanto Co. at the expense of 60 worker-years of effort and a cost of more than \$2 million.

12.1 GENERAL STRUCTURE OF COMPUTER-AIDED DESIGN PROGRAMS (FLOWTRAN)

Computer-aided design programs all have essentially the same components:

1. An executive system
2. A physical properties data bank
3. A thermodynamics properties package
4. A collection of design (and cost) subroutines for a variety of process units

The current trend is to add optimization routines, process dynamic capabilities, and a friendlier user interface.

The executive system reads the input data, controls the order in which the equipment subroutines are calculated, prints the results of the calculations, etc. The physical properties data bank in FLOWTRAN contains data for 180 chemical compounds, including molecular weights, critical constants, heat capacities, acentric factors, etc. These pure-component properties are used by the thermodynamics package to calculate gas and liquid densities and enthalpies, as well as vapor-liquid equilibrium relationships, for the process streams in the plant. Then the equipment design subroutines use this information on the stream properties to calculate equipment sizes (and costs) for the various process units, i.e., furnaces, heat exchangers, compressors, distillation columns, etc.

Our discussion of CAD programs will not be adequate to use these programs efficiently. Instead, our goal is to describe the design information that is required to use one of these programs, to discuss an approach for solving design problems by using these programs, and to give examples of some results. In addition, we clarify the relationship between our preliminary design estimates and CAD calculations.

To accomplish these goals, we restrict our attention to design problems where the basic FLOWTRAN routines are adequate for solving the problem. The program is much more flexible and powerful than our examples illustrate, and an interested reader should obtain a user's manual.*

Executive Routine

The executive routine reads the input data, initializes the variables used in the calculations, arranges the order of calculations to correspond to the connections of the process units, calls for the thermodynamics package as it is needed, converges the recycle calculations, prints the output results, and repeats the calculation procedure for parametric case studies. The format for the input data is given in

---

* J. D. Seader, W. D. Seider, and A. C. Pauls, *FLOWTRAN Simulation—An Introduction*, CAChE Corp., Cambridge, Mass., 1977 (available from Ulrich's Book Store, Ann Arbor, Mich.); J. Peter Clark, ed., *Exercises in Process Simulation Using FLOWTRAN*, ibid.; R. R. Hughes, *CAChE Use of FLOWTRAN on UCS*, ibid.

TABLE 12.1-1
FLOWTRAN input data

TITLE	Any title to identify the case study.
PROPS	Five numeric constants; the first constant indicates the number of components, and the next four correspond to the choices of the thermodynamic options in Table 12.1-2.
PRINT	The stream flows for the solution can be printed in a variety of units (lb · mol/hr is the default unit), the stream connections for the flowsheet can be printed, the input data can be reproduced, and/or the physical property data records can be printed.
FILE	Used to retrieve data records from private files.
RETR	Used to identify the chemical compounds used in the case study (see Appendix D).
PAIR	Used to supply liquid-phase activity coefficients when regular solution theory is not applicable.
NEW BLOCK	Used to enter new equipment subroutines.
BLOCK	Identifies the type of design equations and the stream connections following the formats in Appendix D.
PARAM	Specifies the design parameters for each equipment subroutine following the formats in Appendix D.
MOLES	Specifies the flow rates of each component in the input streams and initial estimates of recycle streams.
TEMP	Specifies the temperature of each feed stream and initial estimates of the temperature of recycle streams.
PRESS	Specifies the pressure of each feed stream and initial estimates of the recycle streams.
NOFLSH	Used to suppress a flash calculation on a recycle stream.
END CASE	Parametric case studies.
END JOB	

From J. D. Seader, W. D. Seider, and A. C. Pauls, *FLOWTRAN Simulation—An Introduction*, CAChE Corp., Cambridge, Mass., 1977.

Table 12.1-1. The statements used in other programs are slightly different, but the same type of information is required.

Physical Properties Data Bank and Thermodynamics Package

One of the most tedious tasks in process design is looking up data for the physical properties of the compounds of interest and then using these properties to compute the vapor-liquid relationships, the enthalpies of the process streams, etc. Computer-aided design programs have much of this information stored in data banks for numerous compounds, so that these routine calculations are greatly simplified. For example, FLOWTRAN contains the values shown in Table 12.1-3 for the 180 chemical compounds (see Appendix D.1). It is possible to add similar data for other compounds and to determine the constants of interest directly from experimental data. (the user's manual* should be consulted for these procedures.)

[†] J. D. Seader, W. D. Seider, and A. C. Pauls, *FLOWTRAN Simulation—An Introduction*, CAChE Corp., Cambridge, Mass., 1977.

TABLE 12.1-2
FLOWTRAN thermodynamics options

Key word	PROPS
Number of chemical species (25 maximum)	_____
Vapor pressure option	_____
1. Antoine equation	
2. Cavett equation	
Vapor fugacity option	_____
1. Ideal-gas equation	
2. Redlich-Kwong equation	
Liquid fugacity option	_____
1. Vapor pressure; Chao-Seader if supercritical	
2. Redlich-Kwong and Poynting equations; Chao-Seader if supercritical	
3. Chao-Seader equation (used with * components)	
4. Grayson-Streed equation (used with * components, $T_{max} < 900°F$)	
5. Option 2 except Prausnitz-Shair for supercritical N_2, CO, AR, O_2, NO, CH_4	
Liquid activity coefficient equation option	_____
1. Ideal solution	
2. Regular solution	
3. Wilson	
4. Van Laar	
5. Renon	
6. Renon with regular solution for unspecified pairs	

From J. D. Seader, W. D. Seider, and A. C. Pauls, *FLOWTRAN Simulation—An Introduction,* CAChE Corp., Cambridge, Mass., 1977.

    These data for pure components are adequate to predict the thermodynamic properties of the mixtures encountered in process streams, provided that the mixtures satisfy the assumptions of ideal or regular solution theory (no hydrogen bonding). For more complex mixtures it is necessary either to supply estimates of the liquid-phase activity coefficients or to supply experimental data and to use the correlation routines available in the program to estimate the activity coefficients. (Again, the user's manual* should be consulted for these procedures.)

    The options available in FLOWTRAN for estimating the thermodynamic properties of streams are given in Table 12.1-2. For the purposes of illustration, we consider only the simplest case where regular solution theory should give reasonable predictions. However, the program is not limited in this respect.

---

* J. D. Seader, W. D. Seider, and A. C. Pauls, *FLOWTRAN Simulation—An Introduction,* CAChE Corp., Cambridge, Mass., 1977.

TABLE 12.1-3
Physical property data record

Properties	Common symbol	FLOWTRAN symbol	Units
Basic			
Molar weight	MW	MW	
Normal boiling point	T_{Nbp}	NBP	°R
Critical temperature	T_c	TC	°R
Critical pressure	P_c	PC	psia
Critical compressibility	z_c	ZC	
Liquid volume constant*		VL	
Liquid volume (60°F and 14.7 psia)		LDEN60	gal/(lb·mol)
Enthalpy			
Liquid enthalpy constant*		HL	
Ideal-gas heat capacity constants	$a_1, \ldots, a_5$	CP(i)	Btu/(lb·mol·°F)
Liquid enthalpy constants		LH(i)	
Equilibria			
Solubility parameter	δ	DELTA	$(\text{cal/mL})^{1/2}$
Expansion factor*		EXPF	
Acentric factor	ω	OMEGA	
Antoine vapor pressure constants	a_1, a_2, a_3	VPA(i)	psia °F
Cavett vapor pressure constants	a_1, a_2	VPC(i)	psia, °F

* Proprietary Monsanto constants computed by PROPTY.

From J. D. Seader, W. D. Seider, and A. C. Pauls, *FLOWTRAN Simulation—An Introduction*, CAChE Corp., Cambridge, Mass., 1977.

Equipment Subroutines

Another time-consuming feature of process design is the calculation of the sizes (and costs) of the process equipment, once the stream flow rates, temperatures, and pressures have been specified. This effort is even more tedious if we must use trial-and-error procedures to determine the stream flows or temperatures because of the presence of recycle loops. However, the computer can solve these trial-and-error calculations relatively rapidly, so that it is the ideal tool to use to improve our estimates of the material balances and to solve complicated sets of equipment design equations.

Three kinds of information must be supplied to the program to use the equipment subroutines: the type of design equations we desire to solve, how the various process units are connected in the process flowsheet, and design parameters for the particular unit under consideration. A list of the equipment subroutines available in FLOWTRAN is given in Table 12.1-4, and some of these subroutines are given in Appendix D.

We note that there are no equipment subroutines available for calculating the size of reactors, for the shortcut design of plate absorbers or extraction columns, for the shortcut design of distillation columns when the relative volatility is not constant, and for the detailed design of cooler condensers (when the vapor and

TABLE 12.1-4
FLOWTRAN subroutines

	Block name	Title
Flash		
	IFLSH	Isothermal flash
	AFLSH	Adiabatic flash
	BFLSH	General-purpose flash
	KFLSH	Isothermal three-phase flash
	FLSH3	Adiabatic/isothermal three-phase flash
Distillation		
	FRAKB	Rigorous distillation (KB method)
	DISTL	Shortcut distillation (Edminster)
	DSTWU	Shortcut distillation (Winn-Underwood)
	SEPR	Constant split fraction separation
	AFRAC	Rigorous distillation/absorption (matrix method)
Absorption/stripping		
	ABSBR	Rigorous absorber/stripper
Other separation		
	EXTRC	Rigorous liquid-liquid extraction
Heat exchange		
	EXCH1	Shortcut heat exchanger
	CLCN1	Shortcut cooler condenser
	DESUP	Shortcut desuperheater
	HEATR	Heat requirements
	EXCH2	Shortcut partial/total vaporizer/condenser
	BOILR	Shortcut reboiler/intercooler
	HTR3	Three-phase heater/cooler
	EXCH3	Shortcut heat exchanger
Miscellaneous unit operations		
	ADD	Stream addition
	MIX	Stream addition with no phase change
	SPLIT	Stream split
	PUMP	Centrifugal pump size and power
	MULPY	Stream multiplication by a parameter
	GCOMP	Compressor and turbine
	PART	General-purpose stream splitter
Stream convergence		
	SCVW	Bounded Wegstein stream convergence
Control		
	CNTRL	Feedback controller
	PCVB	Multiple-parameter control block
	DSPLT	Distillate feed forward control
	RCNTL	Ratio, sum and difference feedback controller
Cost analysis		
	CAFLH	Flash drum cost
	CFLH3	
	CIFLH	
	CKFLH	
	CAFRC	Distillation column cost
	CDSTL	

TABLE 12.1-4
FLOWTRAN subroutines (*continued*)

	Block name	Title
	CFRKB	
	CABSR	Packed absorber cost
	CCLN1	Heat-exchanger cost
	CEXC1	
	CEXC2	
	CEXC3	
	CPUMP	Pump cost
	CCOMP	Compressor cost
	CTABS	Tray absorber cost
	CHETR	Heat-exchanger cost
	BPROD	By-product value
	PRODT	Product stream value
	RAWMT	Raw-material value
	PROFT	Profitability analysis
Report		
	SUMRY	Stream output editor
	TABLE	Component physical properties table
	GAMX	Liquid-activity-coefficients table
	SPRNT	Stream print block
	ASTM	Analytical distillation of a stream
	CURVE	Heating and cooling curves
Reaction		
	REACT	Chemical reactor
	AREAC	Adiabatic add/subtract reactor
	XTNT	Chemical reactor (extent of reaction model)

From J. D. Seader, W. D. Seider, and A. C. Pauls, *FLOWTRAN Simulation—An Introduction*, CAChE Corp., Cambridge, Mass., 1977.

liquid phases are not in equilibrium). It is possible to develop subroutines for these, and other, problems and to add them to FLOWTRAN, but we do not consider problems of this type.

12.2 MATERIAL BALANCE CALCULATIONS

In Chaps. 5, 6, and 7 we discussed shortcut procedures, for estimating both overall and recycle material balances. We used these shortcut calculations to help fix the structure of the flowsheet. One structure that we discussed in detail for the HDA process is shown in Fig. 12.2-1, although, as we discussed in Chap. 9, this structure probably does not correspond to the best design. However, since our goal here is to demonstrate how CAD calculations can be used to refine our approximate calculations, we develop a more detailed solution for the flowsheet given in Fig. 12.2-1.

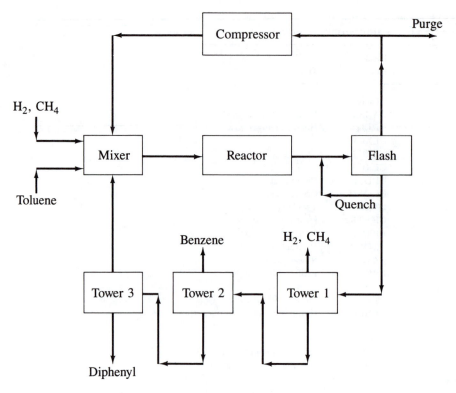

FIGURE 12.2-1
HDA process, block flowsheet.

General Approach to CAD Calculations

Once the structure of the flowsheet has been fixed, then the conventional approach to developing a design is to calculate the material balances, calculate the energy balances, calculate the equipment sizes, calculate the capital and operating costs, and evaluate the process profitability. Most of the CAD manuals imply that all these problems can be solved simultaneously, but they recommend the gradual development of a solution by breaking the complete problem down into a series of smaller problems, The reason for this evolutionary approach is that it is easy to make mistakes entering the input data, the codes have fairly large running times, and it is both expensive and tedious to make numerous runs that fail.

   Our goal here is to use a CAD program to evaluate the approximations we made in our screening calculations to verify our selection of the most promising process alternatives. We based our material balances on the complete recovery of all valuable materials, along with some other approximations for vapor-liquid splits. Since the energy balance calculations and the calculations of equipment sizes and costs all depend on the material balances, our approach is to revise the material balance calculations first.

Simplified Flowsheet for Material Balance Calculations

Our process flowsheet (without any heaters or coolers) is shown in Fig. 12.2-1. However, to perform a set of material balance calculations, we need to consider only the units where the component flows change, i.e., the reactor, the phase splitter (flash drum), the columns, and stream mixing or splitting. In addition, we consider the locations where pressure changes take place, because most of the equipment subroutines require that the pressure be specified. For our example, we assume that the feed streams are at 535 psia, that the pressures of the gas- and liquid-recycle streams are raised to this level, that the reactor operates at 500 psia, that the pressure in the flash drum is 465 psia, that the stabilizer pressure is 150 psia, and that the product and recycle column pressures are 15 psia (we should use 20 psia for final design calculations, but here we want to check our shortcut results).

Our simplified flowsheet for the process, which shows the units needed for the material balance and pressure-change calculations, is shown in Fig. 12.2-2. Note that we *must include* units where stream mixing and splitting take place (which in practice might only be a tee), and we include valves that are used to drop the pressure if a phase change might occur.

Sequential Modular CAD Programs and Stream Tearing

Many of the CAD programs that are commercially available (PROCESS, DE-SIGN 2000, ASPEN, etc.), as well as FLOWTRAN, have a sequential modular structure. That is, if we know the inputs to a process unit, the equipment subroutines will calculate the outputs. Thus, if we fix the feed flow rates of hydrogen and toluene and if we know the gas- and liquid-recycle flows for the flowsheet shown in Fig. 12.2-2, we can calculate the reactor feed conditions. Then we can calculate the reactor product stream, the split that takes place in the flash drum, the outputs from the valves and towers, the purge split, etc.

Stream Tearing

However, we do not know the gas and recycle flows initially. So if recycle loops are present in the flowsheet, we must tear the recycle streams. That is, suppose we tear both of the gas- and liquid-recycle streams (sse Fig. 12.2-3) and we guess the component flows in each of these streams as well as the temperatures and pressures. Then we can calculate all the outputs from every unit, and eventually we can calculate all the component flows in the gas- and liquid-recycle streams. If we change our guessed values until they match our calculated values, we will have obtained a converged set of material balances. Thus, the sequential modular codes include convergence blocks that will iterate on these guessed and calculated values until they agree within a tolerance specified by the user.

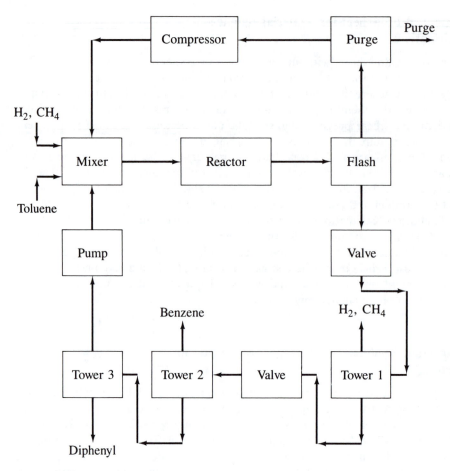

FIGURE 12.2-2
HDA process, flowsheet showing all units.

Convergence

Of course, convergence is obtained much more rapidly if the initial guesses are close to being correct. Some inexperienced users start by guessing zero values. Even though the CAD programs usually converge with zero starting values for the recycle flows, a very large number of iterations normally is required (i.e., it takes a very large number of iterations to build up large recycle flows from a zero starting value). However, we can use our shortcut material balances to supply "good" starting values.

Minimizing the Number of "Tear Streams"

After some thought, we might realize that it is very inefficient to tear both the gas- and liquid-recycle streams. If, instead, we tear the stream entering or leaving the

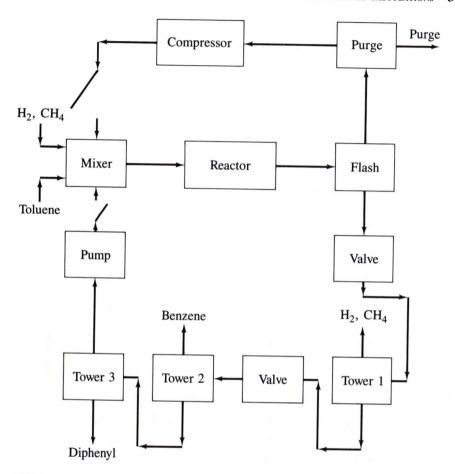

FIGURE 12.2-3
HDA process, tearing recycle streams.

reactor, we break both recycle loops (see Fig. 12.2-4). Most CAD packages include an algorithm that will indicate which streams to tear to minimize the number of convergence blocks that need to be included in the flowsheet. We can also use our shortcut calculations to obtain good estimates of the feed to the reactor or the flash drum.

Check the Physical Properties

We prefer to choose the stream entering the flash drum as the starting point for our calculations. It is essential to ensure that the physical property data provide realistic predictions for the process under consideration, and therefore it is always a good idea to initiate a CAD study by undertaking some flash calculations. In addition, specifying the feed to a flash drum in a flowsheet almost always breaks

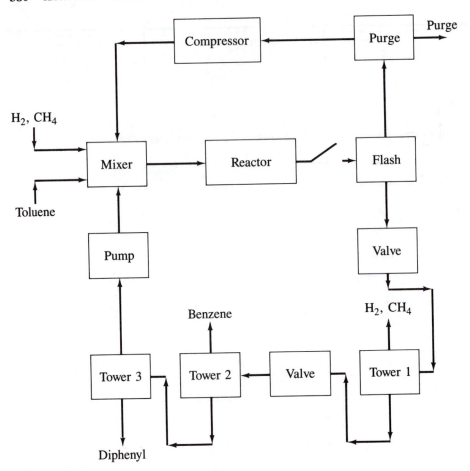

FIGURE 12.2-4
HDA process, alternate tearing scheme.

one or more recycle loops. Moreover, starting with the feed to a flash drum allows us to evaluate some of the decisions we made concerning the liquid separation system.

Other Types of CAD Programs

There are CAD packages (SPEEDUP, QUASILIN, ASCEND II) that are based on an equation-solving approach. That is, they attempt to solve all the equations describing the flowsheet simultaneously. These programs should be much more efficient if there are a large number of recycle loops, because the sequential modular programs suffer if there are many nested iterations. In addition, simultaneous modular iteration schemes have been developed (FLOWPACK and a recent modification of FLOWTRAN*), which use equipment subroutines but converge all the recycle loops simultaneously. These newer types of codes also make it possible to optimize the flowsheet at the same time as the recyle loops are being converged. Thus, improved codes should be available in the future.

Computer Information Diagram

The initial step in developing a CAD program is to translate our process flowsheet into a computer information diagram. The computer information diagram is essentially the same as the flowsheet, except that we must show all mixers and splitters and the location of our convergence block(s). In FLOWTRAN we must number each of the streams from S01 to S99 (if less that 10 streams are present, we number them from S1 to S9). Moreover, we give each unit an arbitrary name of up to six letters and numbers (starting with a letter), and we indicate the type of subroutine (see Table 12.1-4 and Appendix D) that we will use for the calculation. A computer information diagram for HDA material balances is shown in Fig. 12.2-5.

Starting Values—Approximate Material Balances

Using the approximation procedures described in Chaps. 5 and 6 (and in Appendix B for the HDA plant), we can estimate the overall and recycle material balances. For a case where the desired production rate of benzene is 265 mol/hr, the conversion is $x = 0.75$ (the selectivity is $S = 0.9694$), the purge composition is $y_{ph} = 0.4$, and the molar ratio of hydrogen to toluene is 5/1, we obtain the stream flows shown in Table 12.2-1. Then in Chap. 6 we described a shortcut procedure for flash calculations, and the revised flows are shown in Table 12.2-2.

We note from these revised flows that we do not meet the desired production rate specification (see the flash liquid flow of benzene) because so much benzene went overhead with the flash vapor. We also note that the gas-recycle flows no longer balance and that a significant amount of benzene is recycled with the gas-recycle stream. Of course, the errors are not as bad as Table 12.2-2 seems to indicate, because the benzene in the gas-recycle stream will be recycled through the reactor and be flashed again. At this point, the amount of benzene in the flash liquid will increase. We could try to solve this iterative problem by hand, but instead we use our CAD program.

It probably would be undesirable to recycle this much benzene to the reactor because a substantial fraction of this benzene would be converted to diphenyl (if we had recycled all the diphenyl, the benzene recycle would not be a problem because

---

* V. D. Lang, L. T. Biegler, and I. E. Grossmann, "Simultaneous Optimization and Heat Integration with Process Simulation," Paper no. 72b, 1986 Annual AIChE Meeting, Miami Beach, November 1986, submitted to *Computers in Chemical Engineering.*

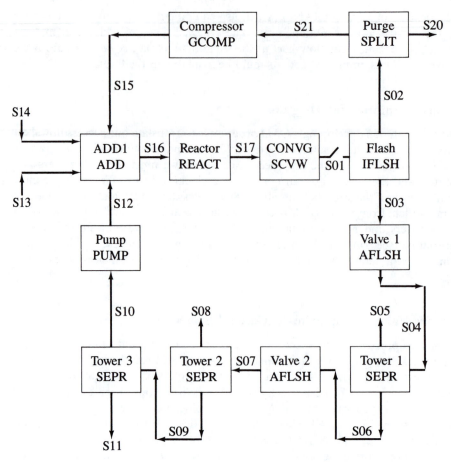

FIGURE 12.2-5
HDA process, flowtran subroutines.

there would be no selectivity loss). However, our correlation for product distribution (the selectivity data in Appendix B) is based on a pure toluene feed to the reactor, so that we cannot estimate the loss of the benzene to diphenyl because of the recycle.

If we had installed a vapor recovery system on the flash vapor stream, our estimates of the process flows would have been more accurate. However, it should be a good test of our approximate material balances to compare this worst-case condition shown in Table 12.2-2 to a rigorous CAD solution. Hence, we use the values shown in Table 12.2-2 as initial values for our CAD runs.

Flash Calculations

We start our CAD study by just considering the flash drum, i.e., we want to ensure that the physical properties are reasonable. We call the unit FLASH, and we use an

TABLE 12.2-1
Stream flows for perfect splits in flash drum–level 3

Component	H_2 feed	Toluene feed	Gas recycle	Liquid recycle	Reactor feed
H_2	467.9	0	1350.0	0	1816.9
CH_4	24.6	0	2025.0	0	2049.6
Benzene	0	0	0	0	0
Toluene	0	273.4	0	91.1	364.5
Diphenyl	0	0	0	0	0

Component	Delta reaction	Reactor exit	Flash vapor	Flash liquid	Purge	Gas recycle
H_2	−269.2	1548.7	1548.7	0	198.7	1350.0
CH_4	273.4	2323.0	2323.0	0	298.0	2025.0
Benzene	265.0	265.0	0	265.0	0	0
Toluene	−273.4	91.1	0	91.1	0	0
Diphenyl	4.2	4.2	0	4.2	0	0

IFLSH subroutine (see Appendix D.2). The computer information diagram is shown in Fig. 12.2-6. This diagram is identical to the flash unit that has been isolated from Fig. 12.2-5.

Program Input

We follow the input format given in Table 12.1-1, and as a title we choose HDA FLASH. There are five components in the feed stream (hydrogen, methane, benzene, toluene, and diphenyl) and we decide to use the Antoine equation for vapor pressure (option 1 in Table 12.1-2), the Redlich-Kwong equation for the

TABLE 12.2-2
Stream flows with a flash–level 4

Component	H_2 feed	Toluene feed	Gas recycle	Liquid recycle	Reactor feed
H_2	467.9	0	1350.0	0	1816.9
CH_4	24.6	0	2025.0	0	2049.6
Benzene	0	0	0	0	0
Toluene	0	273.4	0	91.1	364.5
Diphenyl	0	0	0	0	0

Component	Delta reaction	Reactor exit	Flash vapor	Flash liquid	Purge	Gas recycle
H_2	−269.2	1548.7	1547.0	2	198.5	1348.5
CH_4	273.4	2323.0	2312.0	11	296.6	2015.4
Benzene	265.0	265.0	29.6	235.4	3.8	25.8
Toluene	−273.4	91.1	3.6	82.4	0.5	3.1
Diphenyl	4.2	4.2	0	4.2	0	0

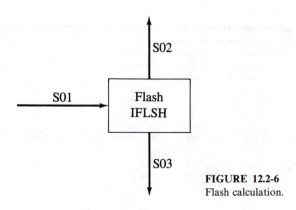

FIGURE 12.2-6
Flash calculation.

vapor fugacity (option 2 in Table 12.1-2), option 5 for the liquid fugacity, and option 2 for the liquid activity coefficient. Hence, the TITLE and PROPS statements are

TITLE   HDA   FLASH

PROPS 5 1   2 5 2

We decide to print the input program along with the results, so that we can maintain a record

PRINT   INPUT

All the components of interest are listed in the physical properties data bank (see Appendix D.1 for the component list and the allowable names to be used in the program). Note that FLOWTRAN uses biphenyl instead of diphenyl, a possible source of confusion which exists elsewhere in the literature. Therefore we can write the retrieve statement as either

RETR HYDROGEN METHANE BENZENE TOLUENE BIPHENYL

or

RETR H2 Cl BZ TOL B-P

The block statement associates the unit name that we selected, FLASH, with the equipment subroutine in FLOWTRAN, IFLSH, and identifies the input and output streams. From Appendix D.2, we see that the block statement for an IFLSH subroutine should be

BLOCK   FLASH   IFLSH   SO1   SO3   SO2

Similarly, if the flash drum temperature and pressure are 100°F and 465 psia for the first case, and we desire to print the K values (the FLOWTRAN K values will be different from estimates using the Hadden and Grayson correlations in Appendix C.1), we write

PARAM   FLASH 1   100   465   1

We use our earlier estimates of the stream flow rates (see Table 12.2-2), and we enter the flows in the same order as they appear in the RETR statement:

MOLES   SO1   1   1549   2323   265   91   4

The temperature and pressure statements for the feed stream are

TEMP   SO1   100

PRESS   SO1   465

and this completes one case.

To consider the behavior of the flash drum at other pressures, we introduce a new title statement, a new PARAM statement and a new PRESS statement, and another END CASE card. A program for operation at 100°F and both 465 and 480 psia is shown in Table 12.2-3. Note that the PARAM statement for this second case is

PARAM   FLASH   2   480   1

because we are changing only the second entry in the PARAM statement (see Appendix D.2).

Program Output

There are several parts to the computer output, including a stream table and an equipment summary, which gives the heat added or removed from the drum.

TABLE 12.2-3
Flash calculations

```
TITLE HDA FLASH DRUM
PROPS 5 1 2 5 2
PRINT INPUT
RETR HYDROGEN METHANE BENZENE TOLUENE BIPHENYL
BLOCK FLASH IFLSH S01 S03 S02
PARAM FLASH 1 100 465 1
MOLES S01 1 1549 2323 265 91.1 4.2
TEMP S01 100
PRESS S01 465
END CASE
TITLE HDA FLASH - INCREASED PRESSURE
PARAM FLASH 2 480
PRESS S01 480
END CASE
END JOB
```

TABLE 12.2-4
Flash calculation output

```
FLASH (IFLSH)  T =  100.00  F, P =   465.00  PSIA,  V/F =  0.9197  MOLS/MOL
               FEED = S01 ,  BOTTOMS = S03 ,  OVERHEAD = S02
               HEAT ADDED =  0.5821E-10  BTU/HR
               STREAM OUTPUT

                   --------MOLES/HR--------
                   TOTAL    LIQUID    VAPOR     LBS/HR   MOLE PC   K-VALUE
  1 HYDROGEN      1549.00   1.36430   1547.64   3122.78  36.5995  99.0667
  2 METHANE       2323.00   10.0976   2312.90   37265.6  54.8874  20.0037
  3 BENZENE       265.000   236.797   28.2028   20698.6  6.26140  0.010401
  4 TOLUENE       91.1000   87.4667   3.63329   8393.41  2.15250  0.003628
  5 BIPHENYL      4.20000   4.199610  .000387   647.640  0.09920  0.000008

  TOTAL MOLES/HR  4232.30   339.925   3892.37
  TOTAL LBS/HR    27366.7   42761.3   70128.0

  DEGREES F                  100.00
  PSIA                       465.00
  MOLE FRAC VAPOR            0.9197
  1000 BTU/HR               -1479.30  -4215.73  2736.43
  BTU/LB-F                   0.6052    0.4030   0.7347
  MOLE WT                    16.57     80.51    10.99
  ACTUAL LB/CUFT             1.3943    52.2530  0.8591
  ACTUAL GPM AND CFM                   65.2992  829.546
  DEGREES API                          30.90
  SP.GR. AT 60 F                       0.87129
  GPM AT 60 F                          62.7475
  BPD AND MMCFD AT 60,14.7             2151.34  35.4050
```

However, the output of primary interest is shown in Table 12.2-4. These are the same values that were used to compare the rigorous and approximate flash calculations in Table 7.1-1.

FLOWTRAN Program for Material Balance Calculations

The computer information diagram we use for the material balance calculations is given in Fig. 12.2-5. However, since it is easy to make mistakes in developing any CAD code, it is always a good idea to build the program in small pieces and to debug each of these smaller portions. Thus, we might develop separate programs for the liquid separation system, i.e., from the flash drum through the pump shown in Fig. 12.2-5, and another program for most of the gas-recycle loop, i.e., from the flash vapor stream back through the reactor. We can use our approximate stream flows given in Table 12.2-1 or 12.2-2 as starting values for the two streams.

A program for the calculation of the material balances is given in Table 12.2-5. We use SEPR blocks initially for the distillation columns, because they provide the simplest way of estimating the component flows for specified fractional recoveries, and we use the conventional rules of thumb to fix these fractional recoveries. (If necessary, we adjust these split fractions to satisfy our product purity requirement.) The feed rate to the flash drum is taken from Table 12.2-2. The PARAM statement for the SPLIT block on the purge stream requires that we specify the split fraction. We can use our approximate material balances (see Table 12.2-1 to estimate this value)

$$\text{Split Fraction for Purge Stream} = \frac{198.7 + 298.0}{1548.7 + 2323} = 0.1283 \quad (12.2\text{-}1)$$

The PARAM statement for the recycle compressor requires that we specify the exit pressure and the efficiencies, while the PARAM statement for the REACT block requires that we specify the conversion of each reaction. These conversions are related to one another by our selectivity correlation. Again, we use our approximate solution to estimate the appropriate values

$$\text{Conversion 1} = \frac{364.5 - 91.1}{364.5} = 0.75 \quad (12.2\text{-}2)$$

$$\text{Conversion 2} = \frac{273.4 - 265}{273.4} = 0.0307 \quad (12.2\text{-}3)$$

The ADD block requires that we specify the component flow rates of each feed stream (in the order in which they appear in the component list in the RETR statement). The MOLES statement for the fresh feed toluene is

$$\text{MOLES} \quad \text{S13} \quad 4 \quad 273.4$$

TABLE 12.2-5
HDA plant material balances

```
TITLE HDA PLANT - MATERIAL BALANCES - CONTROL SYSTEMS
PROPS 5 1 2 5 2
PRINT INPUT
RETR HYDROGEN METHANE BENZENE TOLUENE BIPHENYL
BLOCK FLASH IFLSH S01 S03 S02
BLOCK VALV1 AFLSH S03 6*0 S04 0
BLOCK TOWR1 SEPR S04 S06 S05
BLOCK VALV2 AFLSH S06 6*0 S07 0
BLOCK TOWR2 SEPR S07 S09 S08
BLOCK TOWR3 SEPR S09 S11 S10
BLOCK PUMP PUMP S10 S12
BLOCK PURG SPLIT S02 S20 S21 5*0
BLOCK COMP GCOMP S21 S15
BLOCK TOLFD MULPY S30 S13
BLOCK FMIX ADD S12 S13 S14 S15 3*0 S16
BLOCK REACT REACT S16 6*0 S17 0
BLOCK CONVG SCVW S17 0 0 FLASH S01 0 0
BLOCK PCONT CNTRL S08 TOLFD 1
BLOCK MCONT RCNTL S16 S16 PURG 2
PARAM FLASH 1 100 465 1
PARAM VALV1 1 150 0 1
PARAM TOWR1 1 1. .995 .003 22*0 1
PARAM VALV2 1 15 0 1
PARAM TOWR2 1 1. 1. .995 .001 21*0 0
PARAM TOWR3 1 1. 1. 1. .995 .005 20*0 0
PARAM PUMP 1 535
PARAM PURG 1 2 .128 .872 5*0
PARAM COMP 1 535 0 1 0 .8 .8
PARAM TOLFD 1 .6625
PARAM REACT 1 1265 500 0 2 4 .75 -1 1 1 -1 21*0
PARAM REACT 32 3 .03 1 0 -2 0 1 74*0
PARAM CONVG 1 3*0 2*1 6*0
PARAM PCONT 1 6 265 1 .3 0 .001 0 0
PARAM MCONT 1 4 1 5 1 .5 .05 0 .001 0 0
MOLES S01 1 1549 2323 265 91.1 4.2
MOLES S30 4 400
MOLES S14 1 467.9 24.6
TEMP S01 100
TEMP S30 100
TEMP S14 100
PRESS S01 465
PRESS S30 575
PRESS S14 575
END CASE
END JOB
```

because this stream contains only toluene, which is the fourth component in our list.

RELATIONSHIP BETWEEN OUR APPROXIMATE MATERIAL BALANCES AND THE FLOWTRAN INPUT. We used our approximate material balances to estimate the fresh feed rates to the process, the split fraction for the purge stream, the reactor conversions for the two reactions, and as a starting guess when we tore the recycle loop at the feed to the flash drum. Hence, our approximate material balances provide a good starting point for the more rigorous analysis. It is not a simple matter to make "reasonable" guesses of the values required for the FLOWTRAN input if these approximate solutions are not available.

Need for Feedback Controllers

Also note that the FLOWTRAN input requires different information from what we used for our initial design. That is, for the FLOWTRAN input we must specify both fresh feed rates, the reactor conversion (we can use our selectivity correlation to calculate the conversion of the by-product reaction), and the split for the purge stream. In contrast, in our shortcut calculations, we based the design on a specification of the production rate, the reactor conversion, the purge composition (which is equivalent to specifying the fresh feed hydrogen flow), and a constraint of maintaining a 5/1 hydrogen-to-aromatics ratio at the reactor inlet.

To have our FLOWTRAN program produce the desired production rate, we will need to add a feedback controller to adjust the toluene fresh feed rate until the benzene production rate matches the desired value. Similarly, we will need to adjust the split fraction for the purge stream until the hydrogen-to-aromatics ratio at the reactor inlet matches the desired value of 5/1. Each of these feedback controllers introduces a new level of iteration. Hence, we normally defer adding these controllers until after we have obtained a converged set of material balances. The order in which FLOWTRAN undertakes the calculations is the same as the order specified for the BLOCK statements.

Including Feedback Controllers—Nested Iterations

When we attempt to install a feedback controller that will manipulate the fresh feed rate of toluene so that the flow rate of the benzene product stream is 265.0 mol/hr, we find that the CNTRL block can only manipulate a parameter in a block statement and cannot adjust a flow rate directly. To resolve this difficulty, we can add a MULPY block to the toluene fresh feed stream and then adjust the stream multiplier; see Fig. 12.2-7.

We also want to add a feedback controller to adjust the split fraction of the purge stream, to keep the hydrogen-to-aromatics ratio equal to 5/1 at the reactor inlet. However, the RCNTL block will allow us to control the hydrogen-to-toluene flow, but not the ratio of hydrogen to the sum of the benzene and toluene. We could

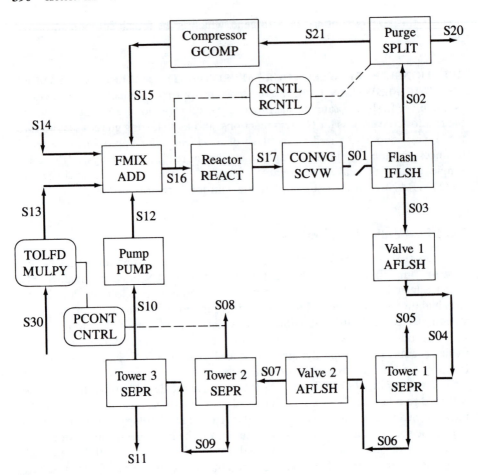

FIGURE 12.2-7
HDA control systems.

overcome this difficulty by writing the code for a new block that would sum the toluene and benzene flows at the reactor inlet and then using a ratio controller. However, since our goal here is merely to illustrate how a CAD program can be used to refine our approximate material balances and since we probably would not want to allow this much benzene to be recycled, we simply add a ratio controller for the hydrogen-to-toluene flow; see Fig. 12.2-7.

When we add these feedback controllers in FLOWTRAN, we nest the iterations required to converge the calculations. That is, suppose it takes five iterations to converge the material balances, i.e., the SCVW block. After these calculations have converged, a change is made to attempt to converge the product flow rate controller. After this change has been made, it might take another five iterations to converge the SCVW block again, and then a second change is made in an attempt to converge the product flow controller. This nested iteration procedure is continued until both the SCVW block and the CNTRL block are converged. At this point, a change is made to converge the RCNTL block. Then the whole procedure starts again until both the SCVW block and the CNTRL block are converged, and then a second change is made to attempt to converge the RCNTL block.

Clearly, nested iterations will greatly increase the computer time (and cost) required to obtain a solution. Therefore, we suggest that initially the convergence tolerances be set to 0.001 multiplied by the desired value of the property being controlled, rather than the default value of 0.00001.

Material Balances for the Base-Case Flowsheet

The stream flows calculated by the program given in Table 12.2-5 are given in Table 12.2-6. Of course, we could decrease the error in these material balances by requiring a tighter tolerance on the controller blocks. However, the results are adequate for a second-level screening of the process alternatives.

It is interesting to compare the stream flows shown in Table 12.2-6 with the approximate results we obtained in Table 12.2-2 (see Table 12.2-7). The approximations given in Table 12.2-2 correspond to a worst-case condition, where we made no attempt to correct for the fact that a significant amount of benzene leaves the flash drum with the vapor stream. Despite this error, most of the approximate stream flows are within 3 % of their more exact values, which is adequate for screening calculations. We base our error analysis on the total flows because most of the heat loads and equipment sizes depend on only the total flows.

Material Balances when Benzene Is Recovered
from the Flash Vapor Stream

If we include a vapor recovery system on the flash vapor stream (see Fig. 12.2-8), we obtain the results given in Table 12.2-8. Now if we compare the stream flows in Table 12.2-8 with the results of the approximate balances given in Table 12.2-1, we see that the errors are quite small, i.e., within a 2 to 6 % error. Errors of this magnitude can easily be tolerated when we are screening alternatives.

Discussion of the Approximate Material Balances

Hopefully we have demonstrated the magnitude of the errors that can be expected from using our very simple, approximate material balances. We use these approximations initially to see whether the process has any chance of being profitable or whether the project should be terminated. If further design effort can be justified, we use the approximations to identify promising alternatives. However, we would always use a set of more rigorous material balances obtained by using a CAD program for final design calculations. Our improved balances are still not rigorous

TABLE 12.2-6
HDA plant material balances

HDA PLANT - MATERIAL BALANCES - CONTROL SYSTEMS

STREAM NAME:

	S01 LBMOL/HR	S02 LBMOL/HR	S03 LBMOL/HR	S04 LBMOL/HR	S05 LBMOL/HR
1 HYDROGEN	1577.14	1575.66	1.48202	1.48202	1.48202
2 METHANE	2399.24	2388.12	11.1190	11.1190	11.0634
3 BENZENE	296.602	29.7555	266.847	266.847	0.80054
4 TOLUENE	92.8057	3.47663	89.3291	89.3291	0.0
5 BIPHENYL	4.61002	0.00040	4.60963	4.60963	0.0
TOTAL LBMOL/HR	4370.40	3997.01	373.386	373.386	13.3460
TOTAL LB/HR	74096.5	44131.3	29965.3	29965.3	242.996
1000 BTU/HR	71543.00	2808.79	-4621.20	-4621.20	16.24
DEGREES F	1265.00	100.00	100.00	98.89	142.49
PSIA	500.000	465.000	465.000	150.000	150.000
DENSITY, LB/FT3	0.4543	0.8637	52.2572	0.0000	0.4292
MOLE FRAC VAPOR	1.0000	1.0000	0.0000	0.0202	1.0000

STREAM NAME:

	S06 LBMOL/HR	S07 LBMOL/HR	S08 LBMOL/HR	S09 LBMOL/HR	S10 LBMOL/HR
2 METHANE	0.05560	0.05560	0.05560	0.0	0.0
3 BENZENE	266.046	266.046	264.716	1.33023	1.33023
4 TOLUENE	89.3291	89.3291	0.08933	89.2398	88.7936
5 BIPHENYL	4.60963	4.60963	0.0	4.60963	0.02305
TOTAL LBMOL/HR	360.040	360.040	264.861	95.1796	90.1469
TOTAL LB/HR	29722.3	29722.3	20685.5	9036.72	8288.36
1000 BTU/HR	-794.97	-794.97	-2663.12	-814.03	-752.64
DEGREES F	370.21	193.69	168.83	234.37	231.20
PSIA	150.000	15.000	15.000	15.000	15.000
DENSITY, LB/FT3	42.6642	0.0000	51.0575	49.2994	48.6827
MOLE FRAC VAPOR	0.0000	0.5458	0.0000	0.0000	0.0000

STREAM NAME:

	S11 LBMOL/HR	S12 LBMOL/HR	S13 LBMOL/HR	S14 LBMOL/HR	S15 LBMOL/HR
1 HYDROGEN	0.0	0.0	0.0	467.900	1383.40
2 METHANE	0.0	0.0	0.0	24.6000	2096.72
3 BENZENE	0.0	1.33023	0.0	0.0	26.1247
4 TOLUENE	0.44620	88.7936	279.375	0.0	3.05241
5 BIPHENYL	4.58658	0.02305	0.0	0.0	0.00035
TOTAL LBMOL/HR	5.03278	90.1469	279.375	492.500	3509.30
TOTAL LB/HR	748.361	8288.36	25740.0	1337.92	38746.4
1000 BTU/HR	-14.09	-752.64	-3821.65	343.96	3157.50
DEGREES F	405.12	-231.20	100.00	100.00	124.65
PSIA	15.000	535.000	575.000	575.000	535.000
DENSITY, LB/FT3	54.2040	48.6827	53.1678	0.2543	0.9497
MOLE FRAC VAPOR	0.0000	0.0000	0.0000	1.0000	1.0000

STREAM NAME:

	S16 LBMOL/HR	S17 LBMOL/HR	S20 LBMOL/HR	S21 LBMOL/HR	S30 LBMOL/HR
1 HYDROGEN	1851.30	1577.47	192.261	1383.40	0.0
2 METHANE	2121.32	2399.74	291.398	2096.72	0.0
3 BENZENE	27.4550	296.695	3.63076	26.1247	0.0
4 TOLUENE	371.221	92.8054	0.42422	3.05241	0.0
5 BIPHENYL	0.02340	4.61146	0.00005	0.00035	400.000
TOTAL LBMOL/HR	4371.32	4371.32	487.714	3509.30	400.000
TOTAL LB/HR	74112.6	74112.6	5384.89	38746.4	36853.6
1000 BTU/HR	-1072.83	71719.65	342.73	2466.06	-5471.70
DEGREES F	126.11	1265.00	100.00	100.00	100.00
PSIA	535.000	500.000	465.000	465.000	575.000
DENSITY, LB/FT3	0.0000	0.4543	0.8637	0.8637	53.1678
MOLE FRAC VAPOR	0.9111	1.0000	1.0000	1.0000	0.0000

TABLE 12.2-7
Stream compositions and flows

Component composition	H_2 feed	Toluene feed	Gas recycle	Liquid recycle	Reactor in
			Approximate		
H_2	0.95	0	0.40	0	0.429
CH_4	0.05	0	0.60	0	0.484
Benzene	0	0	0	0	0
Toluene	0	1.000	0	1.00	0.086
Diphenyl	0	0	0	0	0
Total flow	492.5	273.4	3375	91.1	4231

Component composition	Reactor out	Flash vapor	Flash liquid	Purge	Gas recycle
H_2	0.366	0.396	0.006	0.397	0.397
CH_4	0.549	0.595	0.032	0.594	0.594
Benzene	0.063	0.008	0.703	0.008	0.008
Toluene	0.021	0.001	0.246	0.001	0.01
Diphenyl	0.001	0	0.012	0	0
Total flow	4232	3887	335	499	3393

FLOWTRAN

Component composition	H_2 feed	Toluene feed	Gas recycle	Liquid recycle	Reactor in
H_2	0.95	0	0.394	0	0.423
CH_4	0.05	0	0.598	0	0.485
Benzene	0	0	0.007	0.15	0.006
Toluene	0	1.00	0.001	0.985	0.085
Diphenyl	0	0	0	0	0
Total flow	492.5	279.3	3509	90.1	4371
Error in flow, %	0	2.1	3.8	1.1	3.2

Component composition	Reactor out	Flash vapor	Flash liquid	Purge	Gas recycle
H_2	0.361	0.394	0.004	0.394	0.397
CH_4	0.549	0.600	0.030	0.597	0.594
Benzene	0.068	0.007	0.715	0.007	0.008
Toluene	0.021	0.001	0.239	0.001	0.001
Diphenyl	0.001	0	0.012	0	0
Total flow	4371	3997	373.4	487.7	3393
Error in flow, %	3.2	2.8	10.2	2.3	

because we are not certain that we can design distillation columns that have exactly the split fractions that we have assumed.

An inspection of our approximate and exact results indicates that the major error in our approximate calculations is caused by the flash drum; i.e., our shortcut calculations are based on a perfect vapor-liquid split, so that we did not properly account for the benzene and toluene leaving in the flash vapor and the

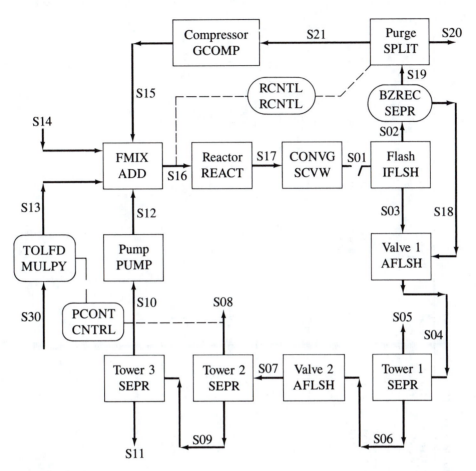

FIGURE 12.2-8
HDA process with benzene recovery.

hydrogen and methane dissolved in the flash liquid. However, our shortcut flash calculations did correctly indicate the magnitude of this error, although it would have been tedious to converge the flash calculations by hand. Despite the error, the shortcut calculations are sufficiently accurate for screening purposes.

A second source of error in the approximate balances was due to the assumption of complete recoveries. The losses depend on the column sequence we choose, as well as the design of these columns, which is the reason why we neglected the losses in our initial calculations. However, the losses are only a small source of error in our screening calculations.

Now that we have developed an improved set of material balances, it is still necessary to check the approximations we made to calculate the energy balances, the equipment sizes, and the equipment costs. We briefly discuss these problems in the remainder of this chapter.

TABLE 12.2-8
Stream compositions and flows, with benzene recovery

Component composition	H$_2$ feed	Toluene feed	Gas recycle	Liquid recycle	Reactor in	Reactor out	Flash vapor	Flash liquid	Purge
				Approximate					
H$_2$	0.95	0	0.40	0	0.429	0.366	0.40	0	0.40
CH$_4$	0.05	0	0.60	0	0.484	0.549	0.60	0	0.60
Benzene	0	0	0	0	0	0.063	0	0.735	0
Toluene	0	1.00	0	1.00	0.086	0.022	0	0.253	0
Diphenyl	0	0	0	0	0	0.001	0	0.012	0
Total flow	492.5	273.4	3375	91.1	4231	4232	3872	360.3	496.7
			1.6	1.1					
				FLOWTRAN					
H$_2$	0.95	0	0.409	0	0.437	0.372	0.409	0.008	0.409
CH$_4$	0.05	0	0.591	0	0.475	0.541	0.591	0.044	0.591
Benzene	0	0	0	0.014	0	0.037	0	0.699	0
Toluene	0	1.00	0	0.985	0.087	0.022	0	0.239	0
Diphenyl	0	0	0	0	0	0.001	0	0.011	0
Total flow	492.5	274.4	3321	92.1	4181	4181	3800	382.2	478.6
Error in flow, %	0	0.4	1.6	1.1	1.2	1.2	1.9	5.7	3.7

12.3   COMPLETE PLANT SIMULATION

Once we have completed a set of material balance calculations, we can work on developing a simulation of the complete plant. Again, it is advisable for a beginner to put the program together in small chunks and to debug these small chunks individually, rather than attempting to immediately solve the complete problem. We use the results of our material balance calculations to fix the flows for these small subproblems.

Some examples of the types of studies that normally are useful to undertake are discussed below. Then we discuss a program for the complete HDA process.

Distillation Column Calculations

Our material balance calculations were based on split (SEPR) blocks to describe the behavior of the distillation columns. With these blocks we merely specify the fraction of each component taken overhead, so the results will not necessarily correspond to the distribution of the nonkeys that we would obtain with more rigorous models of distillation columns. Hence, as the next step in the development of a complete simulation, we might develop a simple program for just the distillation section, where we replace the use of split blocks by Fenske-Underwood-Gilliland routines (or their equivalent); see Fig. 12.3-1. The Fenske-Underwood-Gilliland calculations (called DSTWU blocks in FLOWTRAN) will also give us an estimate for the column designs that we can use to check our shortcut calculations.

The Fenske-Underwood-Gilliland design procedure is limited to systems having a constant relative volatility. Thus, the calculations might be seriously in error for nonideal mixtures where there is hydrogen bonding and where the activity coefficients are large. Similarly, the assumption of a constant relative volatility is valid only when all the components in a mixture have essentially the same heat of vaporization (i.e., are close boilers). Thus, there are often cases where we want to check the accuracy of the Fenske-Underwood-Gilliland estimates by using rigorous column models.

Most of the rigorous, tray-by-tray programs are simulation programs. That is, the number of trays above and below the feed must be specified, as well as the reflux ratio, and then the subroutine calculates the splits. However, when we are designing a column, we are trying to determine the number of trays that we need to accomplish a desired split. Therefore, normally we need to use an iterative procedure when simulation subroutines are used for design.

Experience indicates that these tray-by-tray subroutines are often slow to converge (i.e., they have the longest running times and require the most iterations of all the subroutines available). For this reason, if we include rigorous column models in a program that is also using iteration to close the material balances calculations, the computing costs are often excessive. Hence, in sequential modular simulators, we usually use the tray-by-tray routines in separate programs; i.e., we use the results from rigorous distillation calculations to adjust the split fractions in simple models in a material balance program or complete plant simulation.

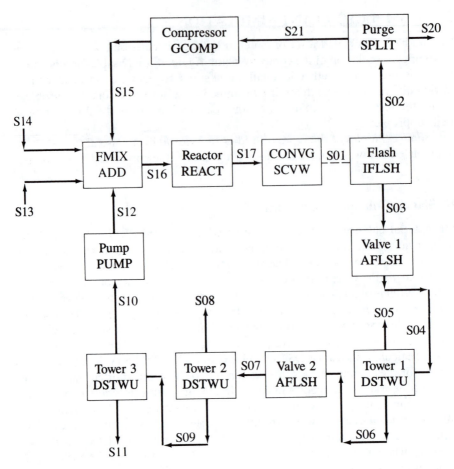

FIGURE 12.3-1
HDA process with column models.

Another approach that we could use is to develop simulation programs that avoid including rigorous models for distillation columns as part of any material balance recycle loop. That is, any time we have a distillation column in a material balance recycle loop, we can double the flow rate of this stream (using a MULPY block in FLOWTRAN), then split the stream in half (using a SPLIT block in FLOWTRAN), and then use a SEPR block to connect one of the exit streams from the splitter to the recycle portion of the flowsheet and a rigorous model on the other half of the stream. Since we can control the order in which the simulator calls the subroutines, we can use this approach to completely converge the material balances before we call the rigorous distillation routine. A simple example of this type is shown in Fig. 12.3-2.

(*a*) Flowsheet

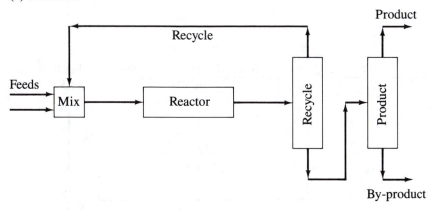

(*b*) Computer information diagram

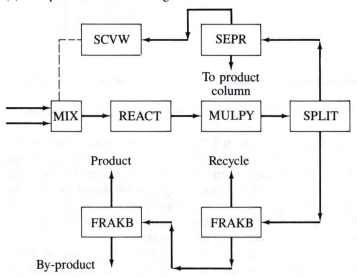

FIGURE 12.3-2
A simple plant.

Energy Balances and Heat Exchangers

In the material balance computer information diagram in Fig. 12.2-7, we neglected the quench stream that uses flash liquid to reduce the temperature of the reactor effluent to 1150°F. An inspection of the flowsheet with the quench stream included indicates that the quench stream merely provides a recycle loop around the flash drum. Thus, if we make a material balance from the reactor effluent to the flash vapor and the pressure reduction valve before the stabilizer, i.e., if we include the quench-recycle loop completely within this balance, then the process flow rates will

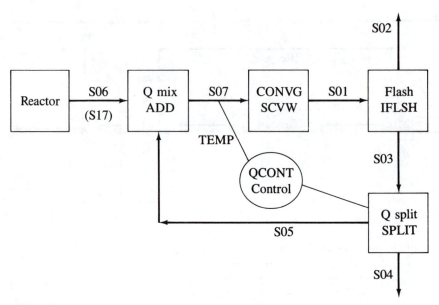

FIGURE 12.3-3
Quench calculation.

not change. However, the heat duty of the partial condenser will depend on the quench flow rate.

To calculate the quench flow rate, and the load on the partial condenser, we must adjust the flow rate of the quench stream to decrease the reactor exit temperature to 1150°F. Thus, we need to install another controller that will solve the problem iteratively; see Fig. 12.3-3. Again, a beginner is advised to solve this problem separately and to make certain that convergence is obtained before attempting to add an iteration loop to a large program. We use our shortcut calculations as a starting point to converge the calculations.

CAD programs also make it fairly easy to generate the temperature-enthalpy curves for each process stream that are needed for the energy integration analysis. In particular, when there is a phase change in a stream containing a mixture, the temperature-enthalpy calculations are tedious to undertake by hand. In the initial simulation of a complete plant, we would include only heaters and coolers on the streams and then design the heat-exchanger network using the procedure described in Chap. 8. However, a procedure for incorporating the heat-exchanger design procedure into a sequential modular simulator has been presented by Lang, Biegler, and Grossmann.*

---

* V. D. Lang, L. T. Biegler, and I. E. Grossmann, "Simultaneous Optimization and Heat Integration with Process Simulation," Paper no. 72b, 1986 Annual AIChE Meeting, Miami Beach, November 1986, submitted to *Computers in Chemical Engineering*.

Complete Plant Simulation

After we have used simple studies to make certain that our simulator subroutines will converge and will give the correct predictions, we can put these subroutines together to generate a plant simulation. Table 12.3-1 gives a program for the HDA process that contains a feed-effluent heat exchanger (see Fig. 12.3-4). Since we have already calculated all the process flows, we can tear as many recycle streams as we desire in this flowsheet, and we do not need to include any controllers. With this approach, we can calculate the equipment sizes and the required utility flows with a minimum of computational cost. Of course, if we want to change the processing conditions (i.e., the values of the design variables), we must include the controllers in the program.

Cost Models, Process Profitability, and Optimization

Many simulators, including FLOWTRAN, include equipment cost correlations. To use these correlations, normally it is necessary to include a factor that accounts for inflation, such as the Marshall and Swift index that we used in our models. By supplying the unit costs for cooling water, steam, etc., it is also possible to calculate the costs for the utility streams. The capital and operating costs along with the cost of labor, maintenance and repairs, taxes and insurance, etc. (see Chap. 2), can be combined to obtain an estimate of the profitability of the process.

Each simulator run corresponds to a single set of design variables. One way that we could estimate the optimum design conditions is to make a set of case-study runs that correspond to the range of the design variables where our shortcut design calculations indicated that the optimum was fairly flat. Many simulators also include optimization routines that can be used to find the optimum.

Remember that costs change over the years, and therefore the optimum design conditions will change. Thus, at the end of the 3-yr period, or so, that it is required to build a plant, the optimum design might be different from the final design that is approved for construction. For this reason, some thought needs to be given to the flexibility of the process to meet changing conditions in the economic environment.

What Remains to Be Done

Once a set of CAD calculations has been used to verify the selection of the best process alternative, the conceptual design effort has been completed. However, it is still necessary to develop a control system for the process, to consider the safety aspects of the process, and to add a significant amount of detail associated with the final design. Safety and control problems that are discovered might require the basic flowsheet to be changed again. For this reason, our initial CAD studies should focus primarily on finding the best process alternative and the cost penalties associated with other alternatives. Some additional discussion of safety, control, etc., is given in Sec. 13.3.

TABLE 12.3-1
HDA process

```
TITLE HDA FLOWSHEET
PROPS 5 1 2 5 2
PRINT INPUT
RETR HYDROGEN METHANE BENZENE TOLUENE BIPHENYL
BLOCK FLASH IFLSH S01 S03 S02
BLOCK QSPLIT SPLIT S03 S04 S05 5*0
BLOCK VALV1 AFLSH S05 6*0 S06 0
BLOCK TOWR1 DISTL S06 S08 S07
BLOCK VALV2 AFLSH S08 6*0 S09 0
BLOCK TOWR2 DISTL S09 S11 S10
BLOCK PCOOL HEATR S10 S40
BLOCK TOWR3 DISTL S11 S13 S12
BLOCK PUMP PUMP S12 S14
BLOCK PURGE SPLIT S02 S30 S31 5*0
BLOCK GCOMP GCOMP S31 S17
BLOCK FMIX ADD S14 S15 S16 S17 3*0 S18
BLOCK FEHE EXCH3 S18 S23 S19 S24
BLOCK FURN HEATR S19 S20
BLOCK REACT REACT S20 6*0 S21 0
BLOCK QMIX ADD S04 S21 5*0 S22
BLOCK CONDS HEATR S24 S25
PARAM FLASH 1 100 465 0
PARAM QSPLIT 1 2 .2805 .7195
PARAM VALV1 1 150 0 0
PARAM TOWR1 1 .7 6 5 2*150 .0357 .0202 0 1
PARAM VALV2 1 15 0 0
PARAM TOWR2 1 1.8 24 12 2*15 .735 .5458 0 0
PARAM PCOOL 1 100 3*0 1 0
PARAM TOWR3 1 .1 4 3 2*15 .9472 0 0 0
PARAM PUMP 1 535
PARAM PURGE 1 2 .122 .878 5*0
PARAM GCOMP 1 535 0 1 0 .8 .8
PARAM FEHE 1 1100 20 10 15 2*0 2
PARAM FURN 1 1150 15 0 0 1 0
PARAM REACT 1 1265 500 0 2 4 .75 -1 1 1 -1 21*0
PARAM REACT 32 3 .03 1 0 -2 0 1 74*0
PARAM CONDS 1 100 5 0 0 1 0
MOLES S01 1 1577.53 2402.98 392.33 124.8 6.24776
MOLES S15 4 279.375
MOLES S16 1 467.9 24.6
MOLES S23 1 1577.53 2402.98 392.33 124.8 6.24776
TEMP S01 100
TEMP S15 100
TEMP S16 100
TEMP S23 1150
PRESS S01 465
PRESS S15 535
PRESS S16 535
PRESS S23 500
END CASE
END JOB
```

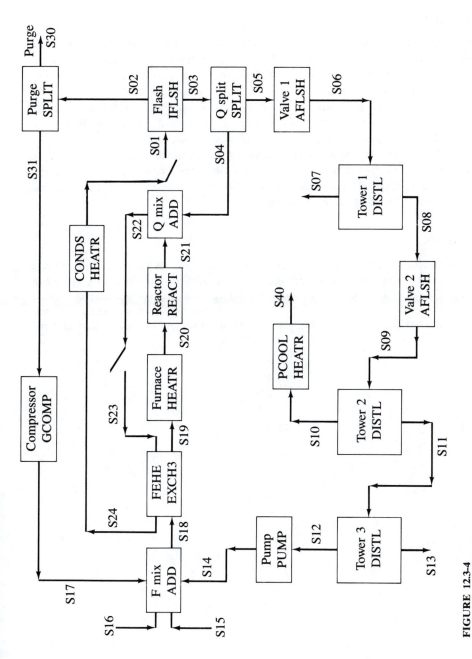

FIGURE 12.3-4
HDA process, complete plant simulation.

12.4 SUMMARY AND EXERCISES

Summary

The large CAD programs, such as FLOWTRAN, PROCESS, DESIGN 2000, ASPEN, etc., are powerful tools, but they are somewhat tedious to use. It is easy to make mistakes in the input data, and these mistakes can be costly in terms of computer time. Hence, the best approach to developing a CAD program is to consider only small portions of the plant at one time and to debug the code corresponding to this small part of the plant. Then we add another small portion, and gradually we generate a code for the complete process.

To use these programs efficiently, i.e., to minimize the number of iterations required, it is usually necessary to have good estimates for recycle flows, the splits in purge streams, reactor conversions, etc. We use the results from our shortcut calculations to provide these estimates.

Exercises

12.4-1. Develop a material balance program (using either FLOWTRAN or another CAD program that you might have available) for the HDA process with diphenyl recycled.

12.4-2. For one of the processes that you have designed, develop a rigorous material balance program. Also, develop a CAD program for the distillation sequence, and then develop a program for the complete plant. How do the rigorous calculations compare with your shortcut approximations?

CHAPTER
13

SUMMARY
OF THE
CONCEPTUAL
DESIGN
PROCEDURE
AND EXTENSIONS
OF THE
METHOD

We have described a systematic procedure for the conceptual design of a limited class of petrochemical processes, i.e., continuous, vapor-liquid processes that produce a single product. Of course, many other types of processes could be considered. Moreover, numerous other types of design studies need to be undertaken to complete a final design.

Unfortunately, it is not possible to cover all this material in a one-semester course. Petrochemical processes are selected for consideration because they are the most common. Similarly, the emphasis is placed on conceptual design because the equipment used in the process and the structure of the flowsheet are fixed at this stage of the design activity; i.e., all the other design activities depend on the results of the conceptual design.

405

The systematic procedure we used to develop a conceptual design was hierarchical. A brief review of this procedure is given in Sec. 13.1. Brief outlines of hierarchical procedures that can be used to develop conceptual designs for solids processes and batch processes are given in Sec. 13.2. Finally, some other types of design problems that need to be solved before a final design can be developed are briefly discussed in Sec. 13.3.

13.1 REVIEW OF THE HIERARCHICAL DECISION PROCEDURE FOR PETROCHEMICAL PROCESSES

To simplify the conceptual design of a process, we decompose the problem into a hierarchy of decisions. The decision levels that we consider are given in Table 13.1-1. The decisions that need to be made at each level for petrochemical processes are given in Table 13.1-2.

The input-output information required is presented in Sec. 4.1, and the heuristics that are available to help make the decisions presented in Table 13.1-2 are discussed in Chaps. 5 through 8. If no heuristics are available to make a decision, we merely make a guess. We go through the complete design procedure in this way, to generate a base-case design. We try to develop a complete design as rapidly as possible to see whether there is some reason why we should terminate all work on the project.

As we proceed through the base-case design, we keep track of the decisions we make. In addition, we prepare a cost diagram for the base-case design (see Secs. 9.1 and 9.2) as an aid in identifying the most expensive processing costs. Then we attempt to evaluate how changes in one or more of our original decisions will affect the processing costs; see Sec. 9.3. We continue to evaluate process alternatives in this way until we obtain the best process alternative.

It might be necessary to change the flowsheet corresponding to the best process alternative because of safety, start-up, controllability considerations, etc.

TABLE 13.1-1
Hierarchy of decisions

Level 1. Batch versus continuous
Level 2. Input-output structure of the flowsheet
Level 3. Recycle structure of the flowsheet
Level 4. General structure of the separation system
      a. Vapor recovery system
      b. Liquid recovery system
Level 5. Energy integration

From J. M. Douglas, *AIChE J.*, **31**, 353 (1985).

TABLE 13.1-2

Design decisions for continuous processes

Level 1:   Batch versus continuous—below we consider only continuous processes
Level 2:   Input-output structure of flowsheet
     1. "Should we purify the raw-material streams before they are fed to the reactor?" If the impurities are inert, there are no quantitative heuristics.
     2. "Should a reversible by-product be recovered or recycled to extinction?" No quantitative heuristic is available.
     3. "Do we need a gas recycle and a purge stream?" A quantitative heuristic seemed to be available before the recent invention of membrane separation processes to separate gaseous mixtures.
     4. "Is O_2 from air or H_2O a reactant that is not recovered and recycled?" (An excess amount must be specified.)
     5. "How many product streams will there be?" Reasonable heuristics seem to be available, except for the case of a reversible by-product.
Level 3:   Recycle structure
     1. "How many reactor systems are required?" The heuristics seem to be reasonable. "Is there any separation between the reactors?" Usually a decision can be made based on the chemist's data.
     2. "How many recycle streams are there?" Heuristics are available.
     3. "Should we use an excess of one reactant?" Normally the chemist's data will indicate the answer.
     4. "Is a gas-recycle compressor required?" A heuristic is available.
     5. "Should the reactor be operated adiabatically, with direct heating (or cooling), or is a diluent (heat carrier) needed?" Some calculations are needed to use the heuristic.
     6. "Do we want to shift the equilibrium conversion?" Calculations and judgment are required.
Level 4:   Separation system
     1. "What is the structure of the vapor and liquid recovery system?" Heuristics are available.
Level 4a: Vapor recovery system
     1. "What is the best location of the vapor recovery system?" A heuristic is available.
     2. "What is the best type of vapor recovery system to use?" No heuristics are available.
Level 4b: Liquid separation system
     1. "What separations can be made by distillation?" A heuristic that usually works is available.
     2. "What sequence of distillation columns should be used?" The published heuristics are limited to sharp splits of ideal mixtures for a single feed, but in many cases they do not lead to the best sequence. Thus, calculations are required.
     3. "How should the light ends be removed?" Calculations and judgment are required.
     4. "Should the light ends be vented, sent to fuel, or recycled to the vapor recovery system?" Calculations and judgment are required.
     5. "How should we accomplish the other separations?" No heuristics are available.
Level 5:   Heat-exchanger network—a design procedure is available (see Chap. 8)

From W. R. Fisher, M. F. Doherty, and J. M. Douglas, "Screening of Process Retrofits Alternatives," *I&EC Research*. in press.

Moreover, the accuracy of the preliminary design calculations needs to be refined to obtain more accurate estimates of the costs. Similarly, other equipment, such as pumps, drums, storage tanks, etc., need to be added. Thus, the conceptual design is merely a starting point for other design studies.

13.2 DESIGN OF SOLIDS PROCESSES AND BATCH PROCESSES

Changing economic conditions normally cause changes in the types of processes that we build. At present there is a growing interest in the design of batch processes for both speciality chemicals and biotechnology. The design of bioprocesses is discussed by Bailey and Ollis.* A review of their chapter 11 on product recovery operations indicates that solids processing units (crystallization, filtration, and drying) are very common. Similarly, solids processing steps are commonly encountered in polymer processes.

In this section, we present brief outlines of systematic procedures that can be used for the conceptual design of solids processes and batch processes. The procedures are arranged into a hierarchical structure, similar to the procedure presented in Sec. 13.1. However, new types of economic trade-offs are encountered, and often new types of constraints must be considered.

Solids Processes

Our discussions up to this point have been limited to vapor-liquid processes. However, some petrochemical processes include solid processing steps in order to isolate the product. For example, the separation of xylene isomers is often accomplished by using crystallization instead of distillation because of the close boiling points of the components. Similarly, the production of adipic acid includes crystallization steps. Of course, if a crystallization step is present, normally filtration (we consider the use of a centrifuge as an alternative) and drying are also required.

To include solid processing steps in our synthesis procedure, it is necessary to modify the structure of the separation system, level 4, to include liquid-solid splits. Actually, to make the procedure even more general, gas-solid splits and liquid-liquid splits should be included (e.g., gas-phase olefin production requires a gas-solid split). The process alternatives for crystallization (and/or precipitation), solid-liquid separation (filtration, centrifugation, settling, etc.), and drying, as well as the unit operation models, and cost correlations must be added.

An initial framework for a synthesis procedure for solids processes has been published by Rossiter and Douglas.† The focus of this initial work was on

---

* J. E. Bailey and D. F. Ollis, *Biochemical Engineering Fundamentals*, 2d ed., McGraw-Hill, New York, 1986.

† A. P. Rossiter and J. M. Douglas, *Chem. Eng. Res. Des.*, **64**: 175 (1986); **64**: 184 (1986).

TABLE 13.2-1
Input information for solid processes

1. Products
   a. Desired production rate and purity
   b. Desired particle size (and distribution) and bulk properties
   c. Price of product, or price versus purity
   d. Valuable by-products, if any
2. Raw materials
   a. Composition and physical state of all raw materials
   b. Price of each raw material, or price versus purity
3. Solids generation
   a. Available methods for generating solid product of desired characteristics
   b. Solubility data for product and possible impurities
   c. Reaction stoichiometry (if any) and selectivity data
4. Processing constraints (these will vary from process to process, but typically include)
   a. Product temperature constraints due to thermal instability
   b. Crystallizer (precipitator) slurry density limitations due to a decline in product quality or poor
      flow properties at high solids concentration
5. Plant and site data
   a. Cost of utilities—fuel, steam levels, cooling water, refrigeration, etc.
   b. Waste disposal facilities and costs

From A. P. Rossiter and J. M. Douglas, *Chem. Eng. Res. Des.*, **64**: 175 (1985).

moderate- to high-tonnage, continuous, inorganic processes that produce solid products from liquid and/or solids feeds. The input information required is presented in Table 13.2-1, and the decisions required are listed in Table 13.2-2. An example of the application of the procedure to a design problem has been published by Rossiter,* and an application to a retrofit study was given by Rossiter, Woodcock, and Douglas.†

Batch Processes

The design of batch processes was discussed in Sec. 4.2. The design of batch plants requires not only that we select the units to be used in the process and the interconnections between these units, but also that we decide whether we want to merge adjacent batch operations into a single vessel and/or to replace some batch units by continuous units. Hence, the design of batch processes is more difficult than the design of a continuous process.

To simplify the understanding of the design of a batch plant, we start by designing a continuous process, using the techniques presented in Chaps. 4 through

---

* A. P. Rossiter, *Chem. Eng. Proc. Des.*, **64**: 191 (1986).
† A. P. Rossiter, D. C. Woodcock, and J. M. Douglas, "Use of a Hierarchical Decision Procedure for Retrofit Studies of Solids Processes," Paper presented at the 1986 Annual AIChE Meeting, Miami Beach, November, 1986.

TABLE 13.2-2

Hierarchical desision procedure for solid processes

1. Batch versus continuous process—we consider only continuous processes
2. Input-output structure
   a. Should we purify the raw-material streams before processing, or should we process the feed impurities?
   b. Is a purge stream required?
   c. How many product streams are required?
   d. What is the economic potential (i.e., product value minus raw-material cost minus disposal cost for purge and waste)?
3. Recycle structure and crystallizer considerations (including reaction, if any)
   a. What type of crystallizer should be used?
   b. Should the product-forming reaction (if any) take place within the crystallizer, or separately?
   c. How many crystallizer effects or stages are required?
   d. How many recycle streams are required?
   e. What is the economic potential (i.e., the economic potential at level 2 minus the sum of the annualized capital and the operating cost of the crystallizer)?
4. Separation system specification—several solid-liquid separations might be needed
   a. How can the primary product be recovered?
   b. What types of solids recovery systems are required?
   c. How should the waste-solid separation be accomplished?
   d. Are any liquid-liquid separations required?
   e. Location of separation units (purge or recycle streams or both)?
   f. What is the economic potential (i.e., the economic potential at level 3 minus the separation system (annualized) capital and operating costs minus liquor loss cost minus washing annualized capital and operating costs)?
5. Product drying
   a. What type of dryer should be used?
   b. What losses can be expected?
   c. What is the economic potential (i.e., the level 4 economic potential minus the annualized capital and operating costs of the dryer)?
6. Energy systems
   a. What are the minimum heating and cooling loads?
   b. How many heat exchangers of what size are required?
   c. What is the economic potential (i.e., the level 5 economic potential minus the annualized capital and operating cost) of the heat-exchanger network?

From A. P. Rossiter and J. M. Douglas, *Chem. Eng. Res. Des.*, **64**: 175 (1985).

9. Then we use the systematic approach developed by Malone and coworkers[*] that is given in Table 13.2-3. This procedure is also hierarchical, so that a series of small problems can be considered that eventually lead to the best design.

OTHER STUDIES IN THE DESIGN OF BATCH PROCESSES. The design of batch processes is expected to take on growing importance in the future, and for this

[*] O. Iribarren and M. F. Malone, "A Systematic Procedure for Batch Process Synthesis," Paper presented at the 1985 Annual AIChE Meeting, Chicago, Ill.; C. M. Myriatheas, "Flexibility and Targets for Batch Process Designs," M.S. Thesis, University of Massachusetts, Amherst, 1986.

TABLE 13.2-3

A hierarchical procedure for the conceptual design of dedicated batch processes

1. Design a continuous process first (if possible), using the procedure described in Sec. 13.1. Use this procedure to find the best process alternative and to identify the dominant design variables. If continuous units are not available for some processing steps, start with the best guess of a flowsheet that shows each processing step individually.
2. Replace each continuous unit by a batch unit.
   a. Include only an intermediate storage tank for recycle.
   b. Calculate the optimum cycle times for each unit by minimizing the total annual cost of the complete process.
      (1) This calculation provides a bound on the cost for the case where the intermediate storage required to schedule the plant is free. (If the plant is not profitable with free storage, it will not be profitable when storage is included.)
      (2) The results will provide some measure of the economic incentive for modifying the chemist's recipe.
      (3) Normally, the cost of each operation in the optimized batch process will exceed the cost of the corresponding unit in the continuous plant.
      (4) The results are used later as a guide to merging units.
   c. Calculate the optimum design by setting the cycle times of every unit equal to each other.
      (1) This calculation provides a bound for the cost when there is a maximum equipment utilization.
      (2) However, there will be no flexibility in the design.
      (3) Again, a measure of the economic incentives for changing the chemist's recipe is obtained.
3. Consider merging adjacent batch units for the design in 2b.
   a. Merge units with similar cycle times and size factors.
   b. Compare the costs of the merged units with the costs of the comparable continuous units.
      (1) If the costs of the continuous units are cheaper, retain the continuous units.
      (2) Otherwise keep the merged batch units.
   c. Continue to merge units until the costs increase.
4. Consider the use of parallel units (or parallel merged units).
   a. The goal is to increase equipment utilization.
   b. The ratio of the cycle times must be matched to the inverse ratio of the number of units.
   c. Normally, use at most three parallel units.
5. Add the intermediate storage needed to schedule the plant and optimize the design.
6. Optimize the best flowsheet alternative including storage.
7. Check the operability of the process, using a batch simulator.

From M. F. Malone, personal communication.

reason we are including a survey of some of the previous work. Most of these studies consider fixed cycle times for the batch units, which makes them different from Malone's approach.

Ketner* developed a procedure for minimizing the capital cost (by using linear cost correlations) for single-product plants with a fixed flowsheet that contain both batch and semicontinuous units. The cycle times and size factors of the batch units were held constant, and then the trade-off that balances the batch

---

* S. Ketner, *Chem. Eng.*, **121** (Aug. 22, 1960).

equipment sizes against the continuous equipment sizes was evaluated. Loonkar and Robinson* considered the same problem, but used power-law expressions. They also extended the results to multiproduct plants.[†]

A heuristic procedure to fix the sizes of the batch units in a multiproduct plant having fixed cycle times and fixed size factors was developed by Sparrow, Forder, and Rippen.[‡] The number of units in parallel was also used as an optimization variable in this study. Introducing parallel units changes the structure of the flowsheet, and a very large number of possible solutions can be generated. Grossmann and Sargent[§] relaxed the cycle time constraint, but assumed that the cycle time was a function of the batch size.

Takamatsu, Hashimoto, and Hasebe[¶] noted that the size of each batch unit had to be determined by taking into account the schedule of the complete plant in addition to the production capacity. Both single-product and multiproduct processes with intermediate storage tanks and parallel units were considered, and the scheduling was used as an additional optimization variable. They also derived an analytical expression for the minimum volume of a storage in terms of the batch sizes entering and leaving the tank. Karami and Reklaitis[∥] developed analytical estimates and bounds for the limiting storage volume for plants composed of several collections of batch, semicontinuous, or continuous operations.

Simulation programs for batch plants have also been developed by Sparrow, Rippin, and Forder;[**] Overturf, Reklaitis, and Woods;[††] and Rippin.[‡‡] These are particularly useful for checking final designs.

Flatz[§§] presented a shortcut procedure for calculating equipment sizes for multiproduct plants, for generating process alternatives, and for estimating the optimum conditions corresponding to standard equipment sizes. This procedure most resembles Malone's approach.

13.3 OTHER SIGNIFICANT ASPECTS OF THE DESIGN PROBLEM

The goal of our conceptual design effort was to decide whether an idea for a new process was sufficiently promising from an economic point of view that a more detailed study could be justified. If the results of this study appear to be promising,

* Y. Loonkar and J. D. Robinson, *Ind. Eng. Chem. Process Des. Dev.*, **17**: 166 (1970).

† J. D. Robinson and Y. R. Loonkar, *Process Tech. International*, **11**: 861 (1972).

‡ R. E. Sparrow, G. J. Forder, and D. W. T. Rippen, *Ind. Eng. Chem. Process Des. Dev.*, **14**: 197 (1975).

§ I. E. Grossmann and R. W. H. Sargent, *Ind. Eng. Chem. Process Des. Dev.*, **18**: 343 (1979).

¶ T. Takamatsu, I. Hashimoto, and S. Hasebe, *Computers and Chem. Eng.*, **3**: 185 (1979); *Ind. Eng. Chem. Process Des. Dev.*, **21**: 431 (1982) and **23**: 40 (1984).

∥ A. I. Karami and G. V. Reklaitis, *AIChE J.*, **31**: 1516 (1985) and **31**: 1528 (1985).

** R. E. Sparrow, D. W. T. Rippin, and G. J. Forder, *The Chem. Eng.*, p. 520 (1974).

†† B. W. Overturf, G. V. Reklaitis, and J. M. Woods, *Ind. Eng. Chem. Process Des. Dev.*, **17**: 166 (1978).

‡‡ D. W. T. Rippin, *Computers and Chem. Eng.*, **7**: 137 (1983) and **7**: 463 (1983).

§§ W. Flatz, *Chem. Eng.*, p. 71 (Feb. 25, 1980) and p. 105 (July 13, 1981).

it is common practice to improve the accuracy of the calculations by using one of the CAD programs, such as PROCESS, DESIGN 2000, ASPEN, etc., that we briefly discussed in Chap. 12. However, many other aspects of the total design problem still remain to be considered:

1. Environmental constraints
2. Control of the process
3. Start-up, shutdown, and coping with equipment failures
4. Safety
5. Site location and plant layout
6. Piping and instrumentation diagrams
7. Final design of equipment
8. Planning for construction

Each item on the list normally introduces new costs, and these additional costs may make the process unprofitable. Hence, it is important to try to estimate when large new costs may be incurred. Unfortunately, there is not sufficient time in a one-semester course to cover all these topics, but a brief discussion of some, as well as some references that may of be interest, are given below. The discussion emphasizes the factors to be considered just after the conceptual design has been completed.

Environmental Constraints

The problems associated with the release of chemicals into the environment have received so much attention in recent years that almost everyone is aware of the importance of environmental constraints. Hence, it is essential to consider the processing costs necessary to meet any environmental requirements. At the conceptual stage of a process design, we include a rough estimate of these costs by associating a pollution treatment cost with all the streams that leave the process as waste streams.

That is, suppose we estimate the annualized, installed cost of a pollution treatment facility and add the operating costs of this facility. Next we allocate the total annualized cost of this facility to all the process streams that it is expected to handle, where this allocation is based on the amount of each stream that is handled and the biological oxygen demand of the materials in that stream. If this information is available, then we can relate the pollution treatment costs to each of the waste streams leaving a particular process in a plant complex. Moreover, as experience is accumulated, we should be able to provide fairly close estimates of pollution treatment costs. These are the costs we look for when we are developing a conceptual design.

Of course, if we underestimate these costs, our conceptual design results may be very misleading. Thus, it is essential to consult an environmental expert in the company at the beginning of a conceptual design study. Similarly, after the

TABLE 13.3-1

A hierarchical approach to control system synthesis

I.   Steady-state considerations. If we can identify and eliminate control problems by using steady-state models (which are much simpler than the dynamic models), we can minimize our design effort.
  A. Identify the significant disturbances.
    1. Those that affect the process constraints.
    2. Those that affect the operating costs.
    3. If disturbances do not have a significant effect on either 1 or 2 above, ignore them from further consideration—this simplifies the problem.
  B. Make certain that the manipulative variables available in the flowsheet are adequate (both in number and sensitivity) to be able to satisfy the process constraints and to optimize the operating variables over the complete (reasonable) range of the disturbances.
    1. If the number of manipulative values is not adequate, the process is not controllable.
    2. To restore controllability, we can
      a. Modify the flowsheet to introduce more manipulative variables.
      b. Modify the equipment designs so that some constraints never become active over the complete range of the disturbances.
      c. Neglect the least important optimization variables.
  C. See whether any equipment constraints are encountered that prevent the changes in the manipulative variables from satisfying the process constraints or optimizing the operating variables over the complete (reasonable) range of the disturbances.
    1. If the process constraints cannot be satisfied, the constrained equipment must be overdesigned, to restore the operability of the process.
    2. If the process is operable when there are equipment constraints, the savings in operating costs by introducing equipment overdesign in order to remove equipment constraints might be economically justified.
  D. Use heuristics to select the controlled variables such that the steady-state behavior of the process will be close to the optimum steady-state performance (see W. R., Fisher, M. F. Doherty, and J. M. Douglas, *Proceedings of the American Control Conference*, p. 293, Boston, June 1985).
  E. Select pairings of the manipulative and controlled variables for single-loop controllers.
    1. Criteria.
      a. High sensitivities.
      b. Small dead times, i.e., close together on the flowsheet.

conceptual design has been completed, it is essential to consult the environmental expert again, i.e., after the best flowsheet has been determined and better estimates of the process flows have been obtained.

Also it may be necessary to design a new pollution treatment facility for the process, and time must be allowed so that the construction of this facility matches that of the process. Similarly, we could undertake a conceptual design for a new pollution treatment facility and then develop a more detailed design later in the development of a project.

Process Control

The conceptual design and even a fairly rigorous optimum design using a CAD program are normally based on the assumptions that the connections between the process and its environment remain constant. However, the demand for the product normally changes with time, the compositions of the feed streams will also

2. Evaluate pairings.
   a. Relative gain array.
   b. Singular value decomposition.
3. Eliminate pairings with large interactions.
4. Several alternative control systems may be developed.

II. Normal dynamic response—small perturbations and linear process dynamics.
   A. Requirements to build a dynamic model.
      1. All equipment capacities must be specified, i.e., the holdup in the tubes and the shells of each heat exchanger, the holdup on the trays in a distillation column, etc.
      2. The sizes of reflux drums, column sumps, flash drums, intermediate storage vessels, etc., must be specified.
   B. Assume perfect level control in any unit where there are two-phase mixtures.
   C. Evaluate the stability of the uncontrolled and controlled processes.
   D. Use linear dynamic models to evaluate the steady-state plant control systems having the fewest interactions.
      1. Use the difference between the total operating cost of the optimum steady-state control response and the dynamic response of the controlled plant as a performance measure to compare control system alternatives for an assumed pattern of disturbances—check the sensitivity of the results to the disturbance pattern.
      2. Evaluate the robustness of the control system.
      3. If the dynamic response is not satisfactory,
         a. Change the control system.
         b. Modify the flowsheet.
   E. Design the level controllers, and recheck the performance.

III. Abnormal dynamic operation—large perturbations and nonlinear dynamic response.
   A. Start-up and shutdown.
      1. Normally, a flowsheet showing all intermediate storage is used as a starting point.
      2. The flowsheet should be checked and modified to correspond to the start-up strategy.
      3. The control systems required for plant start-up and shutdown are different from the controls used for normal operation.
   B. Failures.
      1. A failure analysis of the flowsheet needs to be undertaken.
      2. Special control systems to handle failures might be needed.

IV. Implementation of the control.
   A. Should distributed control be used? How?
   B. What kind of computer control–human interface is required?

From W. R. Fisher, M. F. Doherty, and J. M. Douglas, *Chem. Eng. Res. Des.,* **63**: 353 (1985).

fluctuate, the cooling-water temperature returned from the cooling towers changes from day to night and from summer to winter, the composition and the heating value of the fuel supply will vary, the pressure and temperatures of the steam supply will fluctuate, etc. Thus, as the connections between the process and its environment change, these changes will disturb the behavior of the process. The purpose of a control system is to ensure that the process will operate "satisfactorily," despite the fact that these disturbances occur.

A hierarchical approach to synthesizing control systems for complete processes has been proposed by Fisher, Doherty, and Douglas.* The steps in the hierarchy are listed in Table 13.3-1. If the process can not be controlled, if start-up

---

* W. R. Fisher, M. F. Doherty, and J. M. Douglas, *Chem. Eng. Res. Dev.,* **63**: 353 (1985).

is very difficult, or if the process becomes unsafe because of a failure in one or more pieces of equipment, then it may be necessary to change the flowsheet or even to abandon the project. Since flowsheet modifications normally are very expensive, it is desirable to identify any potential control problems as early as possible in the development of a design. From Table 13.3-1 we see that a steady-state control study can be undertaken as soon as a conceptual design has been completed (i.e., we need to have a flowsheet available).

Start-up Considerations

The conceptual design produces a small number of process flowsheets that should be considered further. Normally, these flowsheets are not complete because they do not include all the minor equipment needed to operate the plant, i.e., we would need to add intermediate storage, pumps, reflux drums, column sumps, etc. In addition, it is usually necessary to add special equipment to be able to start up the plant easily. For example, if we consider an adiabatic, exothermic reactor with a feed-effluent heat exchanger (see Fig. 13.3-1) where the reactor exit temperature is sufficiently high that there is an adequate temperature driving force at the reactor inlet, then the process can operate satisfactorily at steady-state conditions. However, there is no way to start up the process, and a start-up furnace must be installed to initially supply heat to the reactants. Similarly, a special piping system is often installed that makes it possible to fill reflux drums and reboilers with the appropriate materials (purchased materials, if necessary).

Thus, a complete flowsheet should be developed fairly early in the life of a project, and some consideration should be given to developing a start-up strategy. By making a preliminary evaluation of a start-up strategy early, it is often possible to identify changes in the flowsheet and the design that might be required. With this approach we can avoid the very large costs associated with oversights that can occur late in the development of a design project. A preliminary start-up evaluation of this type also simplifies the more detailed start-up study that must be undertaken during the construction of the plant.

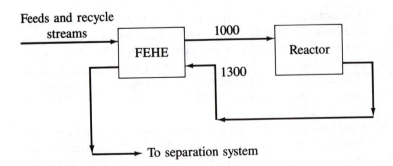

FIGURE 13.3-1
Feed-effluent heat exchanger.

Safety

Safety studies should be undertaken throughout the life of a project. The initial study must be carried out by the chemist, who needs to recognize the nature of the materials being handled. In particular, it is necessary to know which raw-material, intermediate, and product components are flammable, unstable, toxic, corrosive, highly reactive, especially sensitive to impurities, etc., to be able to handle these materials safely in the laboratory.

The properties of the materials also need to be considered during the conceptual design. If at all possible, we prefer that the design correspond to conditions outside the explosive limits for any stream in the process. However, we also need to examine whether changes in operating pressures, temperatures, compositions, etc., will cause a stream to move into the explosive range.

Similarly, if the presence of corrosive components means that special materials of construction will be needed for the equipment that processes those components, we might want to modify the design to minimize the amount of expensive equipment required. For highly reactive materials or situations where the reaction rates are very sensitive to changes in impurities or process parameters, we also might want to modify the design.

When the intermediate storage units, or other units having capacitances, are added to the flowsheet, we want to minimize the inventory of any hazardous materials and to consider special safety systems that will ensure safe operation. As the design proceeds, we must add pressure relief systems on the process vessels, to be certain to avoid hazardous operations. In other words, we want to be able to predict what might happen if something goes wrong with the operation of the plant, and we want to be absolutely sure that we have a safe situation when something does go wrong.

An excellent set of guidelines for a safety evaluation was developed by Batelle Laboratories for the AIChE.* Table 13.3-2 decribes the potential hazards, possible initiating events, the propagation and amelioration actions, and the consequences of accidents. Figure 13.3-2 shows a flowsheet that describes the steps in a hazard evaluation. Note that this figure indicates that design changes might be required, and, if possible, we want to identify these design modifications as early in the life of a project as we can. Several checklists that are useful for safety studies are included in the manual.

Site Location

Site location also has an impact on conceptual design because the utilities available on a site, e.g., cooling-water temperatures, will depend on the geographical location. Similarly, the costs of raw materials will reflect the transportation costs,

---

* *Guidelines for Hazard Evaluation Procedures*, prepared by the Battelle Columbus Division for the Center for Chemical Process Safety, AIChE, New York, 1985.

TABLE 13.3-2
Hazards and accidents

| Hazards | Initiating events or upsets | Intermediate events (system and operator responses to upsets) | | Accident consequences |
		Propagating	Ameliorative	
Significant inventories of	Machinery and equipment malfunctions	Process parameter deviations	Safety system responses	Fires
(a) Flammable materials	(a) Pumps, valves	(a) Pressure	(a) Relief valves	Explosions
(b) Combustible materials	(b) Instruments, sensors	(b) Temperature	(b) Backup utilities	Impacts
(c) Unstable materials		(c) Flow rate	(c) Backup components	
(d) Toxic materials		(d) Concentration	(d) Backup systems	
(e) Extremely hot or cold materials		(e) Phase/state change		
(f) Inerting gases (methane carbon monoxide)				
Highly reactive	Containment failures	Containment failures	Mitigation system responses	Dispersion of toxic materials
(a) Reagents	(a) Pipes	(a) Pipes	(a) Vents	Dispersion of highly reactive materials
(b) Products	(b) Vessels	(b) Vessels	(b) Dikes	
(c) Intermediate products	(c) Storage tanks	(c) Storage tanks	(c) Flares	
(d) By-products	(d) Gaskets	(d) Gaskets, bellows, etc.	(d) Sprinklers	
		(e) Input/output or venting		

Reaction rates especially sensitive to
(a) Impurities
(b) Process parameters

Human errors
(a) Operations
(b) Maintenance
(c) Testing

Loss of utilities
(a) Electricity
(b) Water
(c) Air
(d) Steam

External events
(a) Floods
(b) Earthquakes
(c) Electric storms
(d) High winds
(e) High-velocity impacts
(f) Vandalism

Method or information errors
(a) As designed
(b) As communicated

Material releases
(a) Combustibles
(b) Explosive materials
(c) Toxic materials
(d) Reactive materials

Ignition/Explosion

Operator errors
(a) Omission
(b) Commission
(c) Diagnosis/decision making

External events
(a) Delayed warning
(b) Unwarned

Method or information failure
(a) Amount
(b) Usefulness
(c) Timeliness

Control responses operator responses
(a) Planned
(b) Ad hoc

Contingency operations
(a) Alarms
(b) Emergency procedures
(c) Personnel safety equipment
(d) Evacuations
(e) Security

External events
(a) Early detection
(b) Early warning

Information flow
(a) Routing
(b) Methods
(c) Timing

From *Guidelines for Hazard Evaluation Procedures*, prepared by the Battelle Columbus Division for the Center for Chemical Process Safety, AIChE, New York, 1985.

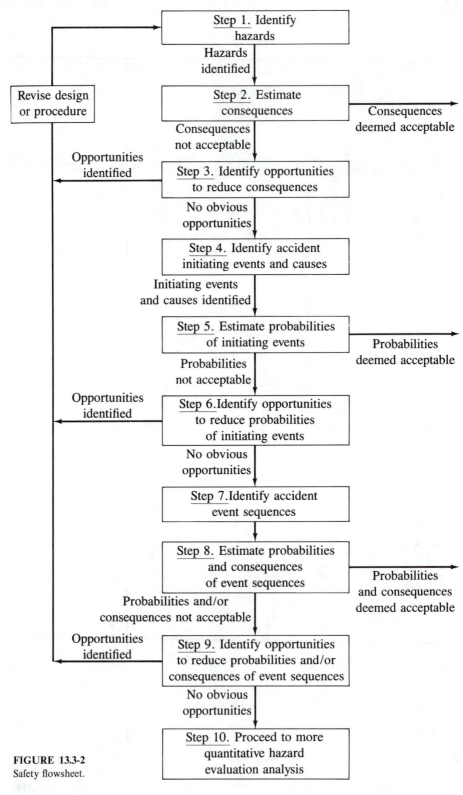

FIGURE 13.3-2
Safety flowsheet.

420

depending on where these materials are produced. Moreover, the site location may depend on the major uses of the product.

For example, most plants that produce benzene are located on the Gulf Coast of Texas. The benzene product is often used to produce ethylbenzene, by reacting the benzene with ethylene. Often the ethylbenzene is dehydrogenated to make styrene. The Gulf Coast is chosen for all these processes because there is a large supply of toluene (from crude oil) and a large supply of ethylene (from ethane cracking plants). Both ethane and ethylene would be quite expensive to transport elsewhere in the country. Styrene is also produced on the Gulf Coast, to minimize transportation costs.

However, most of the plants that produce polystyrene are located in the northeast. Most of the fabricators who use polystyrene as a raw material are located in the northeast, so that the polystyrene plants are located close to these fabricators. Thus, the overall costs are minimized when styrene monomer is shipped, but not the other materials.

Other Design Problems

The other design problems, e.g., final equipment design, piping and instrumentation diagrams, plant layout, project engineering, etc., are considered to be well beyond the scope of this text. All these problem areas are very important to the success of the commercialization of a project, and each area poses many new challenges. An understanding of the process, however, is essential to developing successful solutions in each area, and that basic understanding is most closely related to the conceptual design.

PART
IV

APPENDIXES

APPENDIX
A

SHORTCUT
PROCEDURES
FOR EQUIPMENT
DESIGN

Normally we use shortcut equipment-design procedures when we screen process alternatives. We want to focus on the most expensive pieces of processing equipment during this screening activity, and therefore we usually focus on gas absorbers, distillation columns, heat exchangers and furnaces, gas compressors (and refrigeration systems), and reactors. Some useful shortcut models for most of these units are presented in this appendix.

We do not include a discussion of reactors because they are so dependent on the reaction chemistry. Thus, one of the many texts on reactor design needs to be consulted to develop a reactor model. Also, note that our list of shortcut methods is not complete, and other models are available in the many texts on unit operations and design.

A.1 NUMBER OF TRAYS FOR A GAS ABSORBER

The design of plate gas absorbers and that of distillation columns have many similarities. Therefore, we decribe the shortcut procedures for finding the number of trays required for each type of unit, and then we present a procedure that can be used for the design (i.e., length and diameter) of both types of units.

Shortcut Procedures for the Number of Theoretical Plates in a Gas Absorber

In general, we expect that both the gas and liquid flow rates in a gas absorber will change as the solute is transferred from the gas to the liquid stream. For cases where the solvent is nonvolatile and the carrier gas is noncondensible, we often use molar ratios to describe the compositions; i.e., the liquid compositions are given in terms of moles of solute per mole of solvent, and similarly for the gas stream. The relationships between the total gas and liquid flows, G and L, respectively, and the carrier flows G_S and L_S are

$$G_S = G(1 - y) \qquad L_S = L(1 - x) \qquad \text{(A.1-1)}$$

Similarly, the relationship between molar ratios and mole fractions is

$$Y = \frac{y}{1 - y} \qquad X = \frac{x}{1 - x} \qquad \text{(A.1-2)}$$

We denote the composition of both the gas and the liquid leaving a tray by the tray number.

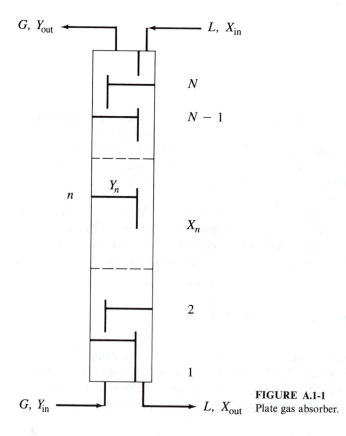

FIGURE A.1-1   Plate gas absorber.

A material balance around tray n in Fig. A.1-1 gives

$$G_S(Y_{n-1} - Y_n) = L_S(X_n - X_{n+1}) \tag{A.1-3}$$

and we see that the operating line will always be linear on a $Y = X$ diagram.

According to the definition of a theoretical plate, we assume that equilibrium is established on each plate. For moderate pressures P_T of, say, less than 50 to 100 psig, where we can consider that the gas mixtures are ideal, the equilibrium relationship can be written as

$$P_T y = \gamma P^\circ x \tag{A.1-4}$$

or

$$y = \frac{Y}{1+Y} = \frac{\gamma P^\circ}{P_T} x = \frac{\gamma P^\circ}{P_T} \frac{X}{1+X} = m \frac{X}{1+X} \tag{A.1-5}$$

In general, the activity coefficient γ will change with both composition and temperature, and the vapor pressure P° will change with temperature throughout the column (the temperature will vary because as the solute condenses in the solvent, it gives up its heat of vaporization). Hence, the equilibrium relationship is nonlinear on a Y-versus-X diagram. Because of this nonlinearity, normally we need to employ some graphical or numerical procedure to solve the design equations.

KREMSER EQUATION FOR PLATE TOWERS. For dilute mixtures the problem becomes much simpler, and we can obtain an analytical solution of the design equations. The result is called the *Kremser equation,*[*] and it can be written as

$$\frac{y_{in} - mx_{in}}{y_{out} - mx_{in}} = \frac{A^{N+1} - 1}{A - 1} \tag{A.1-6}$$

where

$$A = \frac{L}{mG} \tag{A.1-7}$$

or

$$N + 1 = \ln\left[1 + (A-1)\left(\frac{y_{in} - mx_{in}}{y_{out} - mx_{in}}\right)\right]\frac{1}{\ln A} \tag{A.1-8}$$

A graph of this expression is shown in Fig. A.1-2.

Back-of-the-Envelope Approximation. Order-of-magnitude arguments can be used to simplify the equation (see Sec. 3.3)

$$N \approx \ln\left[(A-1)\left(\frac{y_{in}}{y_{out}}\right)\right]/(A-1) \tag{A.1-9}$$

or, if $L/(mG) \approx 1.4$,

$$N + 2 \approx 6 \log \frac{y_{in}}{y_{out}} \tag{A.1-10}$$

---

[*] A. Kremser, *Natl. Petrol. News,* **22**(21): 42 (1930).

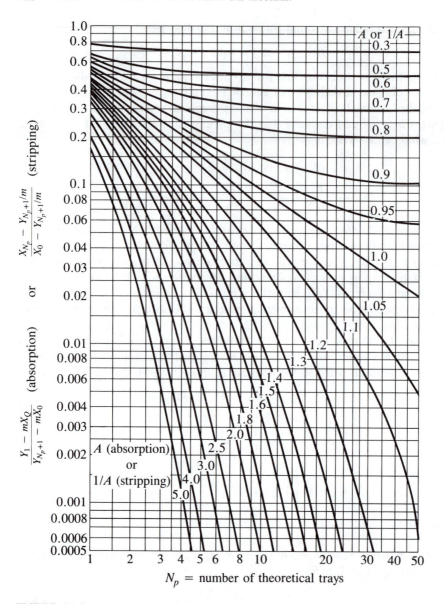

FIGURE A.1-2

Number of theoretical stages for countercurrent cascades, with Henry's-law equilibrium and constant absorption or stripping factors. [*After Hachmuth and Vance, Chem. Eng. Progr.,* **48**: *523, 570, 617* (*1952*).]

Applicability of the Kremser Equation. We can use our approximation procedures to derive a criterion for evaluating the applicability of the Kremser equation:

$$Y_{\text{in}}\left\{1 + \left[2(A_{21} - 2A_{12}) + \frac{\Delta H_v^2}{RC_p T_L^2}\right]\frac{G_S}{L_S}\right\} \leq 0.1 \qquad (A.1\text{-}11)$$

where A_{12} and A_{21} are the Margules constants of the solute and solvent, respectively, at infinite dilution.

We obtain this expression by first writing a material balance similar to Eq. A.1-3 around the top of the tower

$$G_S(Y_{n-1} - Y_{\text{out}}) = L_S(X_n - X_{\text{in}}) \qquad (A.1\text{-}12)$$

Also, we arrange the equilibrium relationship given by Eq. A.1-5 to obtain

$$Y = \left(\frac{\gamma P^\circ}{P_T}\right)\left(\frac{1 + Y}{1 + X}\right)X = mX \qquad (A.1\text{-}13)$$

If m is essentially constant, then both the operating and equilibrium lines will be linear and we can write the Kremser equation in terms of molar ratios and molar flow rates. Hence, we expect the Kremser equation to be valid whenever m is essentially constant.

To test the constancy of m, we expand it in a Taylor series around the condition of infinite dilution:

$$Y = 0 \qquad X = 0 \qquad T_L = \text{Inlet Liquid Temp.}$$

and at these conditions we let

$$m_0 = \frac{\gamma_0 P_0^\circ}{P_T} \qquad (A.1\text{-}14)$$

The Taylor series expansion gives the result

$$m = m_0 + Y\left(\frac{\partial m}{\partial Y}\bigg|_0\right) + X\left(\frac{\partial m}{\partial X}\bigg|_0\right) + (T - T_L)\left(\frac{\partial m}{\partial T}\bigg|_0\right) \qquad (A.1\text{-}15)$$

From the definition of m,

$$\frac{\partial m}{\partial Y}\bigg|_0 = \left[\frac{\gamma P^\circ}{P_T}\left(\frac{1}{1 + X}\right)\right]\bigg|_0 = m_0 \qquad (A.1\text{-}16)$$

Similarly,

$$\frac{\partial m}{\partial X}\bigg|_0 = \left[\frac{\gamma P^\circ}{P_T}\frac{-1}{(1 + X)^2}\right]\bigg|_0 + \left[\frac{P^\circ}{P_T}\left(\frac{1 + Y}{1 + X}\right)\right]\bigg|_0\left(\frac{\partial \gamma}{\partial x}\frac{\partial x}{\partial X}\right)\bigg|_0 \qquad (A.1\text{-}17)$$

If we assume that the activity coefficient is given by the Margules equation

$$\ln \gamma = x_2^2[A_{12} + 2x_1(A_{21} - A_{12})] \qquad (A.1\text{-}18)$$

and let

$$\gamma_0 = \exp A_{12} \qquad (A.1\text{-}19)$$

then we find that

$$\left.\frac{\partial m}{\partial X}\right|_0 = -m_0 - 2m_0(2A_{12} - A_{21}) \tag{A.1-20}$$

Both γ and P° depend on temperature, but for most substances the temperature dependence of the activity coefficient is much less than that for vapor pressure; that is, $d(\ln \gamma)/dT$ is proportional to the heat of mixing, whereas $d(\ln P^\circ)/dT$ is proportional to the heat of vaporization. Hence, to simplify our analysis, we neglect the temperature dependence of γ. Over moderate temperature ranges, we assume that the vapor pressure dependence on temperature is given by the Clausius-Clapeyron equation

$$\frac{P^\circ}{P_0^\circ} = \exp\left[-\frac{\Delta H_V}{R}\left(\frac{1}{T} - \frac{1}{T_L}\right)\right] \tag{A.1-21}$$

and we find that

$$\left.\frac{\partial m}{\partial T}\right|_0 = m_0 \frac{\Delta H_V}{RT_L^2} \tag{A.1-22}$$

With these approximations, Eq. A.1-15 becomes

$$m \approx m_0\left[1 + Y - X - 2(2A_{12} - A_{21})X + \frac{\Delta H_V}{RT_L^2}(T - T_L)\right] \tag{A.1-23}$$

We would like to obtain a criterion that would indicate the maximum possible difference between m and m_0. For this reason, we write Eq. A.1-23 as

$$m = m_0\left[1 + Y_{in} + 2(A_{21} - 2A_{12})X_{out} + \frac{\Delta H_V}{RT_L^2}(T_{out} - T_L)\right] \tag{A.1-24}$$

However, if we expect to design the absorber for 99% recoveries, then

$$G_S Y_{in} \approx L_S X_{out} \tag{A.1-25}$$

Similarly, if we assume that the heat released when the solute condenses appears only as a sensible heat change of the liquid, then

$$G_S Y_{in} \Delta H_V = C_p L_S(T_{out} - T_L) \tag{A.1-26}$$

Thus, Eq. A.1-24 can be written as

$$\frac{m}{m_0} = 1 + Y_{in}\left\{1 + \left[2(A_{21} - 2A_{12}) + \frac{\Delta H_V^2}{RC_p T_L^2}\right]\frac{G_S}{L_S}\right\} \tag{A.1-27}$$

Of course, if the second term is very small, say, less than 0.1, then $m \approx m_0$ and the Kremser equation will adequately describe the system. Hence, a criterion for the applicability of the Kremser equation is

$$Y_{in}\left\{1 + \left[2(A_{21} - 2A_{12}) + \frac{\Delta H_V^2}{RC_p T_L^2}\right]\frac{G_S}{L_S}\right\} \le 0.1 \tag{A.1-11}$$

Multicomponent Absorption

For multicomponent absorption with dilute solutions, we can use the Kremser equation to describe each component. The slope of the equilibrium line for each component will be different, and the absorption factors will be different:

$$m_j = \frac{\gamma_j P_j^\circ}{P_T} \qquad A_j = \frac{L}{m_j G} \qquad \text{(A.1-28)}$$

We estimate the number of trays required based on the recovery of a key component; i.e., normally we want to recover 99% or so of some component:

$$N + 1 = \ln\left[1 + (A_K - 1)\left(\frac{y_{K,\text{in}} - m_K x_{K,\text{in}}}{y_{K,\text{out}} - m_K x_{K,\text{in}}}\right)\right]\frac{1}{\ln A_K} \qquad \text{(A.1-29)}$$

where K refers to the key. Then the recoveries of the other components can be determined

$$\frac{y_{j,\text{in}} - m_j x_{j,\text{in}}}{y_{j,\text{out}} - m_j x_{j,\text{in}}} = \frac{A_j^{N+1} - 1}{A_j - 1} \qquad \text{(A.1-30)}$$

Of course, the design problem becomes much more difficult if heat effects are important.

COLBURN'S METHOD. Colburn* has presented shortcut procedures that can be used even when the equilibrium and operating lines are curved. He developed these expressions by using order-of-magnitude arguments to simplify the design equations for packed columns, and then he wrote the expressions for plate towers by analogy. In addition, he lists the results for stripping, distillation, and extraction as well as absorption. His results are given in Table A.1-1.

OTHER PROCEDURES. Several other shortcut procedures for estimating the number of theoretical trays have been presented in the literature.† In general, these other procedures are more tedious to apply, but they give more accurate predictions. Also a number of computer programs are available for tray-by-tray solutions for absorber problems.

---

* A. P. Colburn, "Simplified Calculation of Diffusional Processes," *Ind. Eng. Chem.*, **33**: 459 (1941).

† G. Horton and W. B. Franklin, "Calculation of Absorber Performance and Design," *Ind. Eng. Chem.*, **32**: 1384 (1940); W. C. Edmister, "Design for Hydrocarbon Absorption and Stripping," *Ind. Eng. Chem.*, **35**: 837 (1943); B. D. Smith, *Design of Equilibrium Staged Processes*, McGraw-Hill, New York, 1963, chap. 8; T. K. Sherwood, R. L. Pigford, and C. R. Wilke, *Mass Transfer*, McGraw-Hill, New York, 1975, chap. 9; and U. V. Stockar and C. R. Wilke, "Rigorous and Short-Cut Design Calculations for Gas Absorption Involving Large Heat Effects. 2. Rapid Short-Cut Design Procedure for Packed Gas Absorbers," *Ind. Eng. Chem. Fund.*, **16**: 94 (1977).

TABLE A.1-1

Colburn's shortcut equations

Theoretical plates:
$$N_P = \frac{\log[(1-P)M + P]}{\log(1/P)}$$

Transfer units:
$$N_T = \frac{2.3 \log[(1-P)M + P]}{1-P}$$

$$R = \frac{L_m}{G_m}$$

	N_P	N_T	P	M
Absorption				
Case 1, constant m/R	N_P	N_{OG}	$\dfrac{m}{R}$	$\dfrac{y_1 - mx_2}{y_2 - mx_2}$
Case 2, varying m/R	N_P	$N_{OG} - 1.15 \log \dfrac{1-y_2}{1-y_1}$	$\dfrac{m_1}{R_2}$	$\dfrac{y_2 - m_2 x_2}{Y_2 - m_2 X_2}\left(\dfrac{1 - m_2/R_2}{1 - y_1^*/y_1}\right)$
	N_P	$N_{OG} + 1.15 \log \dfrac{1+Y_1}{1+Y_2}$	$\dfrac{m_2}{R_2}$	$\dfrac{Y_1 - m_2 X_2}{Y_2 - m_2 X_2}\left(\dfrac{1 - m_2/R_2}{1 - Y_1^*/Y_1}\right)$
Desorption				
Case 1, constant R/m	N_P	N_{OL}	$\dfrac{R}{m}$	$\dfrac{x_1 - y_2/m}{x_2 - y_1/m}$
Case 2, varying R/m	N_P	$N_{OL} - 1.15 \log \dfrac{1-x_2}{1-x_1}$	$\dfrac{R_2}{m_2}$	$\dfrac{x_2 - y_2/m_2}{X_2 - Y_2/m_2}\left(\dfrac{1 - R_2/m_2}{1 - x_1^*/x_1}\right)$
	N_P	$N_{OL} + 1.15 \log \dfrac{1+X_1}{1+X_2}$	$\dfrac{R_2}{m_2}$	$\dfrac{X_1 - Y_2/m_2}{X_2 - Y_2/m_2}\left(\dfrac{1 - R_2/m_2}{1 - X_1^*/X_1}\right)$

Distillation, enriching†

Case 1, constant m/R	N_P	N_{OG}	$\dfrac{m}{R}$	$\dfrac{y_1 - mx_2}{y_2 - mx_2}$
Case 2, varying m/R	N_P	N_{OG}	$\dfrac{m_2}{R_2}$	$\left(\dfrac{y_1 - m_2 x_2}{y_2 - m_2 x_2}\right)\left(\dfrac{1 - m_2/R_2}{1 - y_1^*/y_1}\right)$
Multicomponent mixture	N_P	N_{OG}	$\dfrac{K_{av}}{R}$	$\dfrac{y_1 - m_2 x_2}{y_2 - m_2 x_2}$

Distillation, stripping, closed steam‡

Case 1, constant R/m	N_P	$N_{OL} + \dfrac{2.3\log(m/R)}{1 - R/m}$	$\dfrac{R}{m}$	$\dfrac{x_1 - x_2/m}{x_2 - x_2/m}$
Case 2, varying R/m	N_P	$N_{OL} + \dfrac{2.3\log(m_2/R_2)}{1 - R_2/m_2}$	$\dfrac{R_2}{m_2}$	$\left(\dfrac{x_1 - x_2/m_2}{x_1 - x_2/m_2}\right)\left(\dfrac{1 - R_2/m_2}{1 - x_1^*/x_1}\right)$
Multicomponent mixtures	N_P	$N_{OL} + \dfrac{2.3\log(m_2/R_2)}{1 - R_1/m_2}$	$\dfrac{K_{av}}{R}$	$\dfrac{x_1 - x_2/m_2}{x_2 - x_2/m_2}$

Distillation, stripping, open steam‡

Case 1, constant R/m	N_P	N_{OL}	$\dfrac{R}{m}$	$\dfrac{x_1}{x_2}$
Case 2, varying R/m	N_P	N_{OL}	$\dfrac{R_2}{m_2}$	$\left(\dfrac{x_1}{x_2}\right)\left(\dfrac{1 - R_2/m_2}{1 - x_1^*/x_1}\right)$

Extraction, stripping

Case 1, constant R/m	N_P	N_{OL}	$\dfrac{R}{m}$	$\dfrac{w_1 - v_2/m}{w_2 - v_2/m}$
Case 2, varying R/m	N_P	$N_{OL} - \theta$	$\dfrac{R_2}{m_2}$	$\left(\dfrac{w_1 - v_2/m_2}{w_2 - v_2/m_1}\right)\left(\dfrac{1 - R_2/m_2}{1 - w_1^*/w_1}\right)$

(*Continued*)

TABLE A.1-1 (*Continued*)

	N_P	N_T	P	M
		Extraction, enriching		
Case 1, constant m/R	N_P	N_{OG}	$\dfrac{m}{R}$	$\dfrac{v_1 - mw_2}{v_2 - mw_2}$
Case 2, varying m/R	N_P	$N_{OG} - \theta'$	$\dfrac{m_2}{R_2}$	$\left(\dfrac{v_1 - m_2 w_2}{v_2 - m_2 w_2}\right)\left(\dfrac{1 - m_2/R_2}{1 - v_1^*/v_1^*}\right)$

Note: Equations for varying m/R or R/m are approximate and hold best for large values of M (say, 20 or larger); $x_1^* = x_{1\,\text{equil}}$; $y_1^* = y_{1\,\text{equil}}$. From A. P. Colburn, *Ind. Eng. Chem.*, **33**: 459 (1941).

† Concentrations and m are based on high boiler or "heavy key."

‡ Concentrations and m are based on low boiler or "light key."

$$\theta = 1.15 \log\left(\frac{1 - w_2}{1 - w_1}\right) - 1.15(1 - n)\log\left[\frac{1 - (1 - n)w_2}{1 - (1 - n)w_1}\right]$$

$$\theta' = 1.15 \log\left(\frac{1 - v_2}{1 - v_1}\right) - 1.15(1 - n)\log\left[\frac{1 - (1 - n)v_2}{1 - (1 - n)v_1}\right]$$

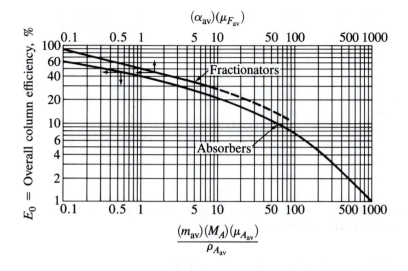

$$\alpha \quad = \text{Relative volatility of key components}$$

μ_F = Molal average viscosity of feed, cP

μ_A = Molal average viscosity of liquid, cP

M_A = Average molecular weight of liquid

ρ_A = Liquid density, lb/ft^3

m = Mole fraction of solute in gas in equilibrium with liquid/mole fraction of solute in liquid

av = At average column temperature and pressure

FIGURE A.1-3
O'Connell's correlation for overall tray efficiencies. [*From H. E. O'Connell, Trans. AIChE,* **42**: *751(1946).*]

Actual Plates—Plate Efficiencies

Unfortunately, our assumption of equilibrium on each tray is seldom met in practice, so we must find a relationship between the number of theoretical trays required for a separation and the number of actual trays. This relationship involves the plate efficiency of the column. There are a variety of ways to predict plate efficiencies, and these range from very simple, quick estimates to very detailed, computational procedures.*

For preliminary process designs, we normally use the simplest method available (although it is not very accurate), which corresponds to O'Connell's

---

* R. H. Perry and C. H. Chilton, *Chemical Engineer's Handbook*, 5th ed., McGraw-Hill, New York, 1973, p. 18–13.

correlation*(see Fig. A.1-3). If we fit a straight line to the graph for gas absorbers in the range from 30 to 60% efficiencies, we find that

$$E_0 = \frac{0.377}{(mM_L\mu_L/\rho_L)^{0.209}}$$
(A.1-31)

Of course, at later stages in a process development, we would use more rigorous procedures for calculating the plate efficiency.

Once we have estimated an overall plate efficiency, the number of actual trays required is

$$N_{\text{act}} = \frac{N}{E_0}$$
(A.1-32)

To complete the design, we must find the tower height, the tower diameter, and the cost. We discuss these calculations in Sec. A.3.

A.2 DISTILLATION COLUMNS: NUMBER OF TRAYS

A variety of design procedures for distillation columns, ranging from shortcut methods to rigorous tray-by-tray calculations, are described in numerous books. Space limitations preclude a detailed review of these various techniques. Instead, we present only a brief summary of some of these procedures and a short discussion of some related material.

Overhead and Bottom Compositions

In many cases either the overhead or the bottoms composition from a distillation column will be fixed by product purity specifications. If this is not the case, then the end compositions represent optimization problems, and they depend on the process economics. For preliminary designs, we use the rule of thumb that we desire 99.5% recoveries of the light key in the overhead and 99.5% of the heavy key in the bottoms. Also, we assume that all the components lighter than the light key are taken overhead and that all components heavier than the heavy key leave with the bottoms.

Column Pressure

The operating pressure for a distillation column normally is fixed by the economic desirability of using a condenser supplied with cooling water from the cooling towers. We want to ensure that the condenser will be sufficiently large to condense

---

*H. E. O'Connell, "Plate Efficiency of Fractionating Columns and Absorbers," *Trans. AIChE*, **42**: 751 (1946).

the vapors even on hot summer days, so we assume that cooling water is available at 90°F. Also, the design of the cooling towers is based on the assumption that all the streams returned to the tower have temperatures of 120°F or less, because the scaling and corrosive characteristics of water become very pronounced above that temperature (i.e., calcium and magnesium salts are deposited on the exchanger walls).

If the bubble point of the overhead mixture from the column is greater than 130°F, which allows a temperature driving force of 10°F at the condenser outlet, then we can use a total condenser and operate the column at atmospheric pressure. (Actually, we would operate the column at an excess of ambient pressure, say 25 psig, so that any leakage would be toward the outside of the column.) However, if the condensing temperature of the overhead mixture is less than 100°F, we prefer to increase the column pressure until we can condense the overhead at 115°F or so. (We reduce the exit temperature of the cooling water in order to keep the tower pressure as low as possible rather than using a refrigerated condenser.)

In some cases, e.g., if hydrogen, methane, or similar light materials are present in the overhead, we use a partial condenser. The product is taken as a vapor from the flash drum, and the column pressure is set sufficiently high to obtain an adequate supply of reflux. Also, for a few situations, e.g., demethanizers, deethanizers, and cryogenic systems, we are forced to use both high pressures and refrigerated condensers. In cases where the overhead bubble point exceeds 267°F, we operate the condenser at atmospheric pressure, but we generate steam instead of using cooling water.

If a reasonable operating pressure will condense the overhead between 100 and 130°F, we next check the bubble point of the bottoms stream at this pressure, to estimate the temperature of the steam required for the reboiler. Depending on whether this temperature is slighty above (that is, 25 to 70°F) the temperature of a steam supply in the plant, for example, 25 psia = 267°F, 115 psia = 338°F, 420 psia = 450°F, 1000 psia = 545°F (although a pressure this high might not be available), we might decide to reduce the column pressure somewhat to accommodate a lower-pressure steam supply to the reboiler. For bubble points above 400 to 500°F, depending on the steam supply available in the plant, we might operate the distillation column under vacuum conditions or use a furnace as a reboiler.

For most distillation column design problems, the selection of the operating pressure for stand-alone columns (i.e., not energy-integrated with the rest of the process) is a relatively simple and straightforward task; the primary consideration is the condensation temperature of the overhead stream. However, in some cases the costs of high-pressure steam must be balanced against lower temperatures of the overhead, or refrigerated condensers are required. For these situations the use of heat integration techniques* often prove to be particularly advantageous.

---

* C. S. Robinson and E. R. Gilliland, *Elements of Fractional Distillation*, 4th ed., McGraw-Hill, New York, 1950, p. 167.

Estimating Bubble Points and Dew Points

For preliminary designs, we would like to have a shortcut procedure for estimating the bubble point and the dew point of mixtures. For most hydrocarbon mixtures (those that do not exhibit hydrogen bonding) we write the equilibrium expression as

$$y_i = K_i x_i \qquad (A.2\text{-}1)$$

Since

$$\sum y_i = 1 = \sum K_i x_i \qquad (A.2\text{-}2)$$

we can also write Eq. A.2-1 as

$$y_i = \frac{K_i x_i}{\sum K_i x_i} \qquad (A.2\text{-}3)$$

Now if we divide both the numerator and the denominator by the K value of the heavy key K_{HK} and define the relative volatility as

$$\alpha_i = \frac{K_i}{K_{\text{HK}}} \qquad (A.2\text{-}4)$$

then Eq. A.2-3 becomes

$$y_i = \frac{\alpha_i x_i}{\sum \alpha_i x_i}$$

BUBBLE POINTS. To estimate the bubble point for a specified liquid composition x_i (which we estimate from the column material balances), we simply calculate

$$\frac{y_i}{x_i} = K_i = \frac{\alpha_i}{\sum \alpha_i x_i} \qquad (A.2\text{-}5)$$

and then use one of the K value correlations, such as that given in Appendix C.1, to find the temperature at the specified column pressure. Normally, we calculate the K value for the light key to estimate the temperature.

DEW POINT. A simple expression that can be used to estimate the dew point can be derived in a similar fashion:

$$\frac{y_i}{x_i} = K_i = \frac{1/x_i}{\sum y_i/\alpha_i} \qquad (A.2\text{-}6)$$

Once we estimate K_i for the light key, we use a correlation to find the temperature.

Relative Volatility

Once we have estimated the bubble points of the overhead and bottoms streams (we estimate the dew point of the overhead if we use a partial condenser) at the operating pressure of the column, we can calculate the relative volatility of the components with respect to the heavy key at both the top and the bottom. For cases where the relative volatility is reasonably constant, we can use a variety of methods to estimate the number of theoretical trays required for the separation. The constant value of α used in these calculations is normally taken to be the geometric mean of the top and bottom values:

$$\alpha_{av} = \sqrt{\alpha_{top}\alpha_{bottom}} \tag{A.2-7}$$

Expensive columns (which are those we want to design most accurately) normally correspond to close boiling materials, and for close boilers α_{top} is usually close to α_{bottom}, and both are in the range from 1 to 2. Whenever the boiling points of the overhead and bottoms are widely separated, α_{top} is often much greater than α_{bottom}. The assumption of constant α is not valid for these systems, and the assumption of constant molar overflow is not usually valid. However, since the splits are easy for these cases, normally there are not many trays and errors in the preliminary designs do not have a great impact on the total processing costs. Thus, in many cases we can obtain a reasonable first estimate by using the smallest value of α in the shortcut design equations.

A criterion that can be used to estimate the effect of variations in α is presented later in this appendix.

Estimating the Number of Theoretical Trays for Sharp Splits in Simple Columns

Initially we consider the case of sharp splits between the light key and the heavy key (i.e., no components between the keys) and simple columns (i.e., one overhead and one bottoms stream). Both empirical and analytical procedures are discussed. Then a shortcut procedure for sloppy splits is presented. Shortcut procedures for evaluating column sequencing are presented in Sec. A.4, complex columns are considered in Sec. A.5, and the energy integration of stand-alone column sequences is discussed in Sec. A.6.

Fenske-Underwood-Gilliland Procedure for Estimating the Number of Theoretical Trays

One of the most commonly used procedures for obtaining quick estimates of the number of theoretical trays required for a distillation separation is called the Fenske-Underwood-Gilliland procedure.

GILLILAND'S CORRELATION. Gilliland* developed an empirical correlation for the number of theoretical trays in terms of the minimum number of trays at total reflux N_m, the minimum reflux ratio R_m, and the actual reflux ratio R; see

---

* C. S. Robinson and E. R. Gilliland, "Elements of Fractional Distillation," 4th ed., McGraw-Hill, New York, 1950, p. 347.

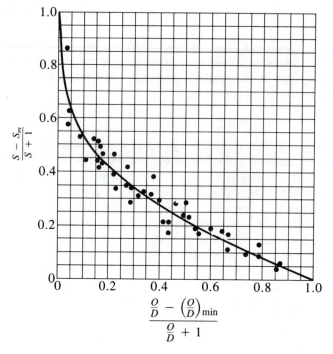

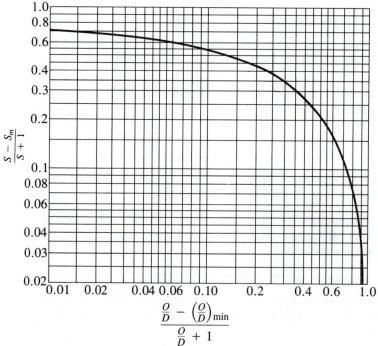

FIGURE A.2-1
Gilliland correlation. (*From C. S. Robinson and E. R. Gilliland, Elements of Fractional Distillation, 4th ed., McGraw Hill, New York, 1950, p. 349.*)

Fig. A.2-1. A simple equation for Gilliland's graphical correlation was developed by Eduljee *

$$\frac{N - N_m}{N + 1} = 0.75\left[1 - \left(\frac{R - R_m}{R + 1}\right)^{0.5688}\right]$$ (A.2-8)

Hence, we can calculate N after we have estimated N_m, R_m, and R.

FENSKE'S EQUATION FOR THE MINIMUM TRAYS AT TOTAL REFLUX N_m. For constant-α systems, Fenske derived an expression for the minimum number of trays at total reflux. The result for binary separations is

$$N_m = \frac{\ln\{[x_D/(1 - x_D)][(1 - x_W)/x_W]\}}{\ln \alpha} = \frac{\ln SF}{\ln \alpha}$$ (A.2-9)

where the composition ratio in the numerator is often called the *separation factor* SF. The analysis is valid for multicomponent mixtures, and the separation factor in Eq. A.2-9 for these systems can be written as

$$SF = \left(\frac{x_{D,LK}}{x_{D,HK}}\right)\left(\frac{x_{W,HK}}{x_{W,LK}}\right) = \left(\frac{d_{LK}}{d_{HK}}\right)\left(\frac{w_{HK}}{w_{LK}}\right) = \left(\frac{r_{LK}}{1 - r_{LK}}\right)\left(\frac{r_{HK}}{1 - r_{HK}}\right)$$ (A.2-10)

where x_i = mole fraction, d_i = distillate flow, w_i = bottoms flow, r_i = fractional recovery, LK = light key, and HK = heavy key.

UNDERWOOD'S EQUATION FOR THE MINIMUM REFLUX RATIO R_m. For a binary system with constant α, the minimum reflux ratio corresponds to the common intersection of the operating line in the rectifying section of the tower

$$y = \frac{R_m}{R_m + 1} x + \frac{x_D}{R_m + 1}$$ (A.2-11)

the equilibrium curve

$$y = \frac{\alpha x}{1 + (\alpha - 1)x}$$ (A.2-12)

and the q line representing the feed quality

$$y = \frac{q}{q - 1} x - \frac{z_F}{q - 1}$$ (A.2-13)

where q is the heat required to vaporize 1 mol of feed divided by the heat of vaporization. Using these expressions to eliminate y and x, we find that

$$\frac{R_m z_F + q x_D}{R_m(1 - z_F) + q(1 - x_D)} = \frac{\alpha[x_D(q - 1) + z_F(R_m + 1)]}{(R_m + 1)(1 - z_F) + (q - 1)(1 - x_D)}$$ (A.2-14)

---

* H. E. Eduljee, "Equations Replace Gillilands' Plot," *Hydrocarbon Proc.*, **54**(9): 120 (September 1975).

For the case of a *saturated-liquid feed* where $q = 1$, this expression becomes

$$R_m = \left(\frac{1}{\alpha - 1}\right)\left[\frac{x_D}{x_F} - \frac{\alpha(1 - x_D)}{1 - x_F}\right] \tag{A.2-15}$$

Moreover, for high-purity separations where $x_D \sim 1.0$ with reasonably well-balanced columns (where $0.2 < x_F < 0.8$) and moderate values of α, we see that

$$R_m \approx \frac{1}{(\alpha - 1)x_F} \tag{A.2-16}$$

Similarly, for a saturated-vapor feed where $q = 0$, Eq. A.2-14 reduces to

$$R_m = \left(\frac{1}{\alpha - 1}\right)\left[\frac{\alpha x_D}{y_F} - \left(\frac{1 - x_D}{1 - y_F}\right)\right] - 1 \tag{A.2-17}$$

and with the same simplifying assumptions, we note that

$$R_m \approx \frac{1}{(\alpha - 1)y_F} - 1 \tag{A.2-18}$$

Underwood's binary equation for the minimum reflux ratio given above also provides a conservative (although often a very conservative) estimate for the minimum reflux ratio for multicomponent mixtures:

$$R_m = \left(\frac{1}{\alpha - 1}\right)\left(\frac{x_{D,LK}}{x_{F,LK}} - \frac{\alpha x_{D,HK}}{x_{F,HK}}\right) \tag{A.2-19}$$

However, an exact solution for multicomponent systems with constant α can be obtained by using another of Underwood's equations. For the case of sharp separations, first we solve the equation below for the value of θ between the α values for the light and heavy keys:

$$\sum_{j=1}^{n} \frac{\alpha_j x_{F,j}}{\alpha_j - \theta} = 1 - q \tag{A.2-20}$$

Next, we substitute this value of θ into the equation

$$R_m + 1 = \sum_{j=1}^{n} \frac{\alpha_j x_{D,j}}{\alpha_j - \theta} \tag{A.2-21}$$

An exact solution of this pair of equations requires a trial-and-error procedure. Some approximate solutions of these equations are presented later in this appendix.

OPERATING REFLUX RATIO R. As the reflux ratio is increased above the minimum, the number of trays required for a given separation decreases, so that the capital cost of the column decreases. However, increasing the reflux ratio will increase the vapor rate in the tower, and, as we show later, higher vapor rates correspond to more expensive condensers and reboilers, along with higher cooling-water and steam costs. Therefore, there is an optimum reflux ratio for any specified separation.

However, experience has shown that the value of the optimum reflux ratio normally falls in the range $1.03 < R/R_m < 1.3$. Furthermore, the slope of the cost curve is very steep below the optimum, but relatively flat above the optimum. Hence, it is common practice to select an operating reflux ratio somewhat above the optimum, and for first estimates we often use the rule-of-thumb value:

Rule of thumb:

The operating reflux ratio is chosen so that $R/R_m = 1.2$ (A2-22)

SIMPLIFIED APPROXIMATION OF GILLILAND'S CORRELATION. From a number of case studies, Gilliland* noted that when $R/R_m \sim 1.2$,

$$N \sim 2N_m = 2\frac{\ln \text{SF}}{\ln \alpha}$$ (A.2-23)

This result provides a very simple procedure for estimating the number of theoretical trays required for a specified separation.

MODIFICATIONS OF THE FENSKE-UNDERWOOD-GILLILAND PROCEDURE. Instead of Fenske's equation for the minimum number of trays at total reflux, some investigators use Winn's equation.[†] This approach leads to good predictions over wider ranges of the relative volatility than Fenske's equation. Similarly, many investigators prefer to use Erbar and Maddox's correlation[‡] for the number of theoretical plates instead of Gilliland's. All these techniques give quite good predictions, providing that the value of α is relatively constant.

Approximate Expressions for the Minimum Reflux Ratios

Glinos and Malone[§] have developed approximate expressions for the minimum reflux ratio that can be used instead of Underwood's equations, Eqs. A.2-20 and A.2-21. Their results for ternary mixtures are given in Table A.2-1, and the expressions for four-component mixtures are given in Table A.2-2. They have also proposed a set of lumping rules for the case of more than four components.

---

* C. S. Robinson and E. R. Gilliland, *Elements of Fractional Distillation*, 4th ed., McGraw-Hill, New York, 1950, p. 350.

† F. W. Winn, "New Relative Volatility Method for Distillation Calculations," *Petrol. Refiner*, **37**(5): 216 (1958).

‡ J. H. Erbar and R. N. Maddox, *Petrol. Refiner*, **40**(5): 183 (1961).

§ K. Glinos and M. F. Malone, *I&EC Proc. Des. Dev.*, **23**: 764 (1984).

TABLE A.2-1
Minimum reflux ratios for ternary mixtures

A/BC:
$$R_m = \frac{\alpha_{BC}(x_{AF} + x_{BF})}{f x_{AF}(\alpha_{AC} - \alpha_{BC})} + \frac{x_{CF}}{f x_{AF}(\alpha_{AC} - 1)}$$

where $f = 1 + \frac{1}{100} x_{BF}$.

AB/C:
$$R_m = \frac{(x_{BF} + x_{CF})/(\alpha_{BC} - 1) + x_{AF}/(\alpha_{AC} - 1)}{(x_{AF} + x_{BF})(1 + x_{AF} x_{CF})}$$

From K. Glinos and M. F. Malone, *I&EC Proc. Des. Dev.*, **23**: 764 (1984).

Analytical Procedures for the Number of Theoretical Trays—Ideal Mixtures

In addition to the shortcut procedures for calculating the number of trays, there are some analytical procedures for the case of ideal mixtures.

SMOKER'S EQUATION FOR BINARY SEPARATIONS. Smoker* developed an exact analytical solution for the case of binary mixtures. The equilibrium relationship is written as

$$y = \frac{\alpha x}{1 + (\alpha - 1)x} \tag{A.2-24}$$

Then for the rectifying section we calculate the number of trays, using the algorithm below:

$$m = \frac{R}{R + 1} \tag{A.2-25}$$

$$b = \frac{x_D}{R + 1} \tag{A.2-26}$$

$$x_0 = x_D \tag{A.2-27}$$

$$k = \frac{-[m + b(\alpha - 1) - \alpha] \pm \sqrt{[m + b(\alpha - 1) - \alpha]^2 - 4bm(\alpha - 1)}}{2m(\alpha - 1)}$$

where $0 < k < 1$ $\tag{A.2-28}$

$$c = 1 + (\alpha - 1)k \tag{A.2-29}$$

$$x'_D = x_D - k \tag{A.2-30}$$

$$x'_n = x_n - k \tag{A.2-31}$$

$$N_R = \left(\ln \left\{ \frac{x'_D}{x'_n} \left[\frac{1 - mc(\alpha - 1)x'_n/(\alpha - mc^2)}{1 - mc(\alpha - 1)x'_D/(\alpha - mc^2)} \right] \right\} \right) \left(\ln \frac{\alpha}{mc^2} \right)^{-1} \tag{A.2-32}$$

---

* E. H. Smoker, *Trans. AIChE*, **34**: 165 (1938).

TABLE A.2-2
Minimum reflux ratios for four-component mixtures

A/BCD: $\quad R_m = \dfrac{\alpha_B(x_A + x_B)}{x_A(\alpha_A - \alpha_B)} + \dfrac{\alpha_C x_C}{x_A(\alpha_A - \alpha_C)} + \dfrac{x_D}{x_A(\alpha_A - 1)}$

AB/CD: $\quad R_m = \dfrac{\dfrac{\alpha_C x_A}{\alpha_A - \alpha_C} + \dfrac{\alpha_C(x_B + x_C)}{\alpha_B - \alpha_C}}{(x_A + x_B)[1 + x_A(x_C + x_D)]} + x_D \dfrac{\dfrac{x_A}{\alpha_A - 1} + \dfrac{x_B}{\alpha_B - 1}}{(x_A + x_B)^2}$

ABC/D: $\quad R_m = \dfrac{\dfrac{x_A}{\alpha_A - 1} + \dfrac{x_B}{\alpha_B - 1} + \dfrac{x_C + x_D}{\alpha_C - 1}}{(x_A + x_B + x_C)[1 + x_D(x_A + x_B)]}$

ABC/BCD: $\quad R_m = \dfrac{1}{\alpha_A x_A + \alpha_B x_B + \alpha_C x_C + x_D - 1}$

From K. Glinos and M. F. Malone, *I&EC Proc. Des. Dev.*, **23**: 764 (1984).

Similarly, for the stripping section we modify the expressions above and let

$$m = \frac{Rx_F + qx_D - R + qx_B}{(R + 1)x_F + (q - 1)x_D - (R + q)x_B} \tag{A.2-33}$$

$$b = \frac{(x_F - x_D)x_B}{(R + 1)x_F + (q - 1)x_D - (R + q)x_B} \tag{A.2-34}$$

Here k is the same quadratic equation as for the rectifying section

$$x'_D = x_n - k \tag{A.2-35}$$

$$x'_m = x_B - k \tag{A.2-36}$$

and N_S is the same expression as for N_R. For the case of saturated-liquid feed, $x_n = x_F$; otherwise, x_n represents the intersection of the q line and the rectifying line.

Approximate Solution of Smoker's Equation

An approximate solution of Smoker's equation has been developed by Jafarey, Douglas, and McAvoy.[*] The result for the rectifying section is

$$
\begin{aligned}
N_R &= \ln\left\{\left(\frac{1 - x_F}{1 - x_D}\right)\left[\frac{R(\alpha - 1)x_D - 1}{R(\alpha - 1)x_F - 1}\right]\right\}\left[\ln\alpha\left(\frac{R}{R + 1}\right)\right]^{-1} \\
&= \ln\left\{\left(\frac{x_D}{1 - x_D}\right)\left(\frac{1 - x_F}{x_F}\right)\left[\frac{1 - 1/R(\alpha - 1)x_D}{1 - 1/R(\alpha - 1)x_F}\right]\right\}\left(\ln\frac{\alpha R}{R + 1}\right)^{-1} \quad \text{(A.2-37)}
\end{aligned}
$$

---

[*] A. Jafarey, J. M. Douglas, and T. J. McAvoy, *I&EC Proc. Des. Dev.*, **18**: 197 (1979).

This expression resembles Fenske's equation at total reflux, except that there are reflux correction terms in the numerator and denominator. As R becomes very large, Eq. A.2-37 reduces to Fenske's equation.

If we use the simple approximation for the minimum reflux rate given by

$$R_m = \frac{1}{(\alpha - 1)x_F} \tag{A.2-38}$$

we can write Eq. A.2-37 as

$$N_R = \frac{\ln\left[\left(\dfrac{x_D}{1 - x_D}\right)\left(\dfrac{1 - x_F}{x_F}\right)\left(\dfrac{R/R_m - x_F/x_D}{R/R_m - 1}\right)\right]}{\ln\left[\alpha R/(R + 1)\right]} \tag{A.2-39}$$

Now we see that the number of trays becomes unbounded as R approaches R_m. Hence, Eq. A.2-39 has the correct limiting values.

The equations for the stripping column are developed in the same way, and the results are

$$N_S = \frac{\ln\left\{\left(\dfrac{x_n}{x_W}\right)\left[\dfrac{R_S(\alpha - 1)(1 - x_W) - \alpha}{R_S(\alpha - 1)(1 - x_n) - \alpha}\right]\right\}}{\ln\left[\alpha(R_S - 1)/R_S\right]} \tag{A.2-40}$$

where

$$R_S = \frac{R(x_F - x_W) + q(x_D - x_W)}{x_D - x_F} \tag{A.2-41}$$

For the case of a saturated-liquid feed and a sharp separation, this becomes

$$N_S = \frac{\ln\left\{\left(\dfrac{1 - x_W}{x_W}\right)\left(\dfrac{x_F}{1 - x_F}\right)\left[\dfrac{R(\alpha - 1)x_F + \alpha x_F - 1}{R(\alpha - 1)x_F - 1}\right]\right\}}{\ln\left\{\alpha[(R + 1)x_F/(Rx_F + 1)]\right\}} \tag{A.2-42}$$

This expression also reduces to Fenske's equation as R becomes very large and predicts that N_S becomes unbounded as R approaches R_m.

SIMPLIFIED APPROXIMATE SOLUTION. We expect that the approximate solution presented above will be quite conservative, because the assumption used to obtain Eq. A.2-37 is equivalent to drawing the upper end of the rectifying line through the point $y_D = x_D = 1.0$ (that is, we set $x_D = 1.0$ in Eq. A.2-28 to simplify this expression). Thus, the operating line is moved closer to the equilibrium line, and more trays will be required.

For expensive columns with numerous trays, we expect that most of the trays will be located near the ends of a McCabe-Thiele diagram. In these regions both the operating lines and the equilibrium lines will be essentially straight, and the Kremser equation should provide a reasonable approximation for the system. The denominator terms in Eqs. A.2-37 and A.2-42 correspond to the absorption and stripping factors for the two ends of the column, that is, $A = L/(mV) = \alpha R/(R + 1)$ for the rectifying section. For normal column operation, i.e., when $R/R_m \sim 1.2$, we expect these terms to be more significant than the reflux correction terms in the numerator. Therefore, we simplify our solutions by writing

$$\left(\frac{\alpha R}{R + 1}\right)^{N_R} = \left(\frac{x_D}{1 - x_D}\right)\left(\frac{1 - x_F}{x_F}\right) \tag{A.2-43}$$

and

$$\left(\frac{\alpha(R + 1)x_F}{Rx_F + 1}\right)^{N_R} = \left(\frac{1 - x_W}{x_W}\right)\left(\frac{x_F}{1 - x_F}\right) \tag{A.2-44}$$

Multiplying these two solutions and assuming that $N_R \sim N_S = N/2$, we obtain

$$\alpha^N\left(\frac{Rx_F}{Rx_F + 1}\right)^{N_R} = \left(\frac{x_D}{1 - x_D}\right)\left(\frac{1 - x_W}{x_W}\right) = \text{SF} \tag{A.2-45}$$

or

$$N = \frac{\ln \text{SF}}{\ln\left(\alpha/\sqrt{1 + 1/Rx_F}\right)} \tag{A.2-46}$$

This simple expression normally gives predictions that are within a few trays of the exact solution.

Behavior Near Minimum Reflux

If we retain the reflux correction terms in Eqs. A.2-42 and A.2-45 but assume that $N_R = N_S = N/2$, then we can derive the equation

$$N = \frac{\ln\left\{[x_D/(1 - x_D)][(1 - x_W)/x_W]\beta\right\}}{\ln\left[\alpha/(1 + 1/Rx_F)^{1/2}\right]} \tag{A.2-47}$$

where

$$\beta = \frac{(R/R_m - x_F)(R/R_m - 1 + \alpha x_F)}{(R/R_m - 1)^2} \tag{A.2-48}$$

We can use this result to estimate the sensitivity of the change in the number of plates as we decrease the reflux ratio below $R/R_m = 1.2$. This result is expected to be conservative. We can "tune" the approximation by using Eq. A.2-47 to calculate N when $R/R_m = 1.2$, and then we compare the result to Eq. A.2-23. If we introduce a correction factor into Eq. 2-47 to make the results agree when $R/R_m = 1.2$, we will probably obtain a more accurate estimate.

UNDERWOOD'S EQUATION FOR MULTICOMPONENT SYSTEMS. Underwood has also proposed an expression for calculating the number of trays for multicomponent mixtures. At the operating vapor rate of the column, we solve for the values of θ and θ' satisfying the expressions*

Rectifying section:

$$V = \sum \frac{\alpha_i D x_{iD}}{\alpha_i - \theta} \tag{A.2-49}$$

Stripping:

$$-V' = \sum \frac{\alpha_i B x_{iB}}{\alpha_i - \theta'} \tag{A.2-50}$$

Then we use these values to calculate N_R and N_S from the expressions for the rectifying section

$$\left(\frac{\theta_k}{\theta_j}\right)^{N_R} = \frac{\sum \alpha x_{iF}/(\alpha_i - \theta_j)}{\sum \alpha_i x_{iF}/(\alpha_i - \theta_k)} \tag{A.2-51}$$

and the stripping section

$$\left(\frac{\theta'_k}{\theta'_j}\right)^{N_S} = \frac{\sum \alpha_i x_{iF}/(\alpha_i - \theta'_k)}{\sum \alpha x_{iF}/(\alpha_i - \theta'_j)} \tag{A.2-52}$$

A Criterion for Constant Relative Volatility

Most of the short-cut procedures for estimating the number of theoretical trays require that the relative volatility be constant. To develop a criterion for constant α, first we evaluate the relative volatilities of the light key with respect to the heavy key at the top and the bottom of the column:

$$\alpha_T = \left(\frac{K_1}{K_2}\right)_{\text{top}} \qquad \alpha_B = \left(\frac{K_1}{K_2}\right)_{\text{bottom}} \tag{A.2-53}$$

Then it is common practice to estimate an average volatility as the geometric mean

$$\alpha_{GM} = \sqrt{\alpha_T \alpha_B} \tag{A.2-54}$$

RELATIONSHIP BETWEEN THE GEOMETRIC MEAN AND THE ARITHMETIC MEAN. To simplify the analysis, we write the average volatility in terms of the arithmetic mean rather than the geometric mean. The relationship between these quantities can be established by letting

$$\alpha_B = \alpha_T(1 + \epsilon) \tag{A.2-55}$$

and then using a Taylor series expansion of the geometric mean to obtain

$$\alpha_{GM} = \sqrt{\alpha_T^2(1 + \epsilon)} = \alpha_T(1 + \tfrac{1}{2}\epsilon - \tfrac{1}{8}\epsilon^2 + \cdots) \tag{A.2-56}$$

---

* See C. J. King, *Separation Processes*, McGraw-Hill, New York, 1971, p. 427.

Since the arithmetic mean is

$$\alpha_{av} = \frac{\alpha_T + \alpha_B}{2} = \alpha_T(1 + \tfrac{1}{2}\epsilon) \qquad \text{(A.2-57)}$$

we see that Eq. A.2-56 will be within 10% of the arithmetic mean if

$$\tfrac{1}{8}\epsilon^2 \leq 0.1(\tfrac{1}{2}\epsilon) \qquad \text{(A.2-58)}$$

or $$\epsilon \leq 0.4 \qquad \text{(A.2-59)}$$

Hence, a criterion for the two kinds of means to be approximately the same is

$$\frac{\alpha_B - \alpha_T}{\alpha_T} \leq 0.4 \qquad \text{(A.2-60)}$$

EFFECT OF VARIATIONS IN α ON THE COLUMN DESIGN. Gilliland's correlation predicts that the number of trays required to achieve a particular separation is approximately equal to twice the number of trays at total reflux:

$$N \sim 2 \frac{\ln\{[x_D/(1 - x_D)][(1 - x_W)/x_W]\}}{\ln \alpha} = \frac{2 \ln SF}{\ln \alpha} \qquad \text{(A.2-61)}$$

If we let

$$\alpha = \alpha_{av}(1 + \phi) \qquad \text{A.2-62}$$

then $$N = \frac{2 \ln SF}{\ln \alpha_{av}(1 + \phi)} = \frac{2 \ln SF}{\ln \alpha_{av} + \ln (1 + \phi)} = \frac{2N_m}{1 + \ln (1 + \phi)/\ln \alpha_{av}} \qquad \text{(A.2-63)}$$

However, for small changes in α we can use Taylor series expansions to write

$$\ln (1 + \phi) \sim \phi \qquad \text{(A.2-64)}$$

$$\frac{1}{1 + \phi/\ln \alpha_{av}} \sim 1 - \frac{\phi}{\ln \alpha_{av}} \qquad \text{(A.2-65)}$$

so that Eq. A.2-68 becomes

$$N \sim 2N_m\left(1 - \frac{\phi}{\ln \alpha_{av}}\right) \qquad \text{(A.2-66)}$$

Thus, variations in α will introduce less than a 10% error in the column design if we require that

$$\frac{\phi}{\ln \alpha_{av}} \leq 0.1 \qquad \text{(A.2-67)}$$

After substituting Eq. A.2-62 with $\alpha = \alpha_T$ and the definition of α_{av}, Eq. A.2-67 becomes

$$\frac{\alpha_T - \alpha_B}{\alpha_T + \alpha_B} \leq 0.1 \ln \frac{\alpha_T + \alpha_B}{2} \qquad \text{(A.2-68)}$$

which provides a simple criterion for evaluating the effect of variations in α on the column design.

McCabe-Thiele Diagrams—Ideal and Nonideal Binary Separations

The McCabe-Thiele procedure often is used to calculate the number of theoretical trays for binary mixtures, particularly when the relative volatility is not constant. In addition, by introducing fictitious molecular weights, the McCabe-Thiele method can be used to describe heat effects caused by large differences in the heats of vaporization of the keys. To account for heat of mixing effects, it is necessary to use the Ponchon-Savarit technique, a method decribed by Knight and Doherty,[*] or a computer program that solves the tray-by-tray material and energy balances. The details of the Ponchon-Savarit procedure or computer routines are available in texts on distillation.

Sloppy Splits

Most of the research on column design and shortcut procedures has been limited to sharp splits in simple columns. However, the optimum fractional recovery for any stream leaving a column, except for a product stream, represents an economic trade-off between adding more trays in the end of the column (i.e., the top or the bottom) where the stream of interest leaves the column balanced against either a decrease in the loss of materials in waste or fuel streams or a decreased cost to recycle the material through the reactor system. Shortcut procedures for estimating the optimum recoveries have been proposed by Fisher, Doherty, and Douglas.[†] In addition, a case study that illustrates the desirability of making a sloppy split (i.e., not a high fractional recovery) has been presented by Ingleby, Rossiter, and Douglas.[‡]

A procedure for estimating the minimum reflux ratio for sloppy splits of ternary mixtures has been published by Glinos and Malone[§] and is presented below. They noted that Underwood's equation (Eq. A.2-21) for the minimum reflux ratio can be written as a linear function of the overhead composition of the light key x_{AD}, providing that only A and B leave overhead. If we merely flash the feed, we obtain a lower bound on the distillate composition x_{AD}, which corresponds to $R_m = 0$.

Then when we add a rectifying section and continue to increase the reflux ratio, we will increase the purity of the overhead as a linear function of the reflux

---

[*] J. R. Knight and M. F. Doherty, *I&EC Fund.*, **25**: 279 (1986).

[†] W. R. Fisher, M. F. Doherty, and J. Douglas, *I&EC Proc. Des. Dev.*, **24**: 955 (1985).

[‡] S. Ingleby, A. P. Rossiter, and J. Douglas, *Chem. Eng. Res. Des.*, **64**: 241 (1986).

[§] K. Glinos and M. F. Malone, *I&EC Proc. Des. Dev.*, **23**: 764 (1984).

ratio until we reach a point where we obtain a sharp AB/BC split. The value of the minimum reflux ratio for this AB/BC split is

$$R_m = \frac{1}{\sum \alpha_i x_{i,F} - 1} \tag{A.2-69}$$

At this condition the fraction of component B in the feed that is recovered overhead is

$$\text{Fraction of } B \text{ Recovered Overhead} = f_B = \frac{\alpha_{AC} - 1}{\alpha_{BC} - 1} \tag{A.2-70}$$

So we can calculate the composition of A overhead x_{AD}. Then if we merely draw a straight line connecting these points, we can estimate the value of R_m for any value of x_{AD} that falls between the values calculated above; see Fig. A.2-2.

   If we continue to increase the reflux ratio, we continue to increase the purity of the overhead as a linear function of the reflux ratio until we obtain a sharp A/BC split. We can estimate the minimum reflux ratio for this case where $\chi_{AD} = 1$, using the approximate results given in Table A.2-1. Then we merely draw a straight line connecting the values for the AB/BC and A/BC splits (see Fig. A.2-1), and we can estimate the values of R_m corresponding to any other value of χ_{AD}. The figure also gives a plot of the ratio of the fractional recoveries of B, f_B, to the fractional recovery of A, f_A, overhead, but these results are nonlinear.

   Glinos and Malone also considered the cost of sloppy splits in the bottom of a column. Again, the results for a simple flash, the AB/BC split, Eqs. A.2-69 and A.2-70, and an AB/C split, Table A.2-1, provide three points that define two straight lines. The graph is simpler to construct if the reboil ratio S is used instead of the reflux ratio (see Fig. A.2-2), but the ideas are exactly the same.

Actual Trays—Plate Efficiency

A simple, but not very accurate, technique for estimating overall plate efficiencies is to use O'Connell's correlation for fractionators; see Fig. A.1-3. If we write an equation for the curve in the range from 30 to 90% efficiencies, we obtain

$$E_0 = \frac{0.4983}{(\alpha\mu_F)^{0.252}} \sim \frac{0.5}{(\alpha\mu_F)^{1/4}} \tag{A.2-71}$$

Also, if we consider columns having saturated-liquid feeds, and if we use the result that the viscosity of most liquids at their normal boiling points is 0.3 cP,* then

$$E_0 \sim \frac{0.5}{(0.3\alpha)^{1/4}} \tag{A.2-72}$$

---

* R. H. Perry and C. H. Chilton, *Chemical Engineer's Handbook*, 5th ed., McGraw-Hill, New York, 1973, p. 3-246.

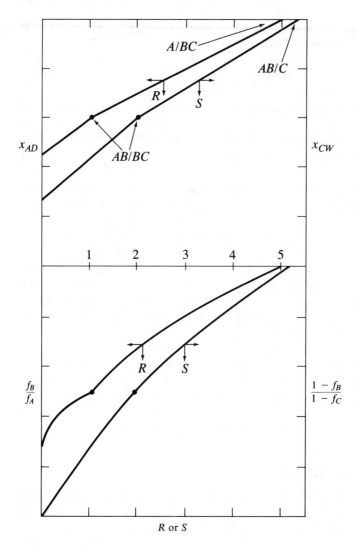

FIGURE A.2-2
Sloppy splits. Distillate or bottoms composition fixes reflux or reboil ratio and ratio of fractional recoveries. Dark circles indicate the critical condition for a sharp split. [*From K. Glinos and M. F. Malone, I&EC Proc. Des. Dev.,* **23**; *764(1984), with permission of the American Chemical Society.*]

Costly separations correspond to low values of α, but for a range of α from 1.3 to 3, the efficiency predicted by Eq. A.2-70 only changes from 63 to 51 %. Hence, the overall efficiency is relatively insensitive to α, and we can obtain a conservative design estimate for atmospheric columns by assuming that

$$E_0 = 0.5 \tag{A.2-73}$$

The number of actual trays in the column is then

$$N_{act} = \frac{N}{E_0} \qquad (A.2\text{-}74)$$

This value is needed to calculate the tower height, as we discuss now.

A.3   DESIGN OF GAS ABSORBERS AND DISTILLATION COLUMNS

Guthrie's correlations (see Appendix E.1) give expressions for the cost of plate columns in terms of the column height and the column diameter. Hence, we need to develop a procedure for calculating the height and diameter. Both the gas absorbers and distillation columns are described by the same models with the exceptions noted below.

Tower Height

In Secs. A.1 and A.2 we developed expressions for the number of plates in a gas absorber and in distillation columns. The most common tray spacing used for plate towers is 2 ft. Hence, for our initial designs, we assume a 2-ft spacing between trays. Then the height corresponding to the trays is $2(N/E_0 - 1)$.

It is common practice to include some additional space, say 5 to 10 ft, at the top of the tower as a vapor-liquid disengaging space (we may even install demisters at the top of the tower to prevent the carryover of liquid droplets). Similarly, it is common practice to include space at the bottom of the tower for a liquid sump. The liquid stream leaving an absorber or a distillation column often is fed through a heat exchanger and then to a distillation column. If we want to ensure that the feed to another distillation tower is not interrupted, we maintain a 5-min or so supply of liquid in the bottom of the tower. The height of the sump required to hold this 5-min supply of liquid obviously will depend on the liquid flow rate and the column diameter; but to simplify our preliminary calculations we assume that an additional 5 to 10 ft at each end should be adequate.

With this simplification, we can write that the tower height is

$$H = \frac{2N}{E_0} + H_0 \qquad (A.3\text{-}1)$$

where H_0 includes the space at the ends of the column for vapor disengagement and the liquid sump. This expression is commonly used for tower designs.

As an approximation, we could express the additional space at the ends of the tower as a percentage of the height that contains trays. For our studies we include a 15% allowance for this excess space, and we write Eq. A.3-1 as

$$H = \frac{2(1.15)N}{E_0} = \frac{2.3N}{E_0} \qquad (A.3\text{-}2)$$

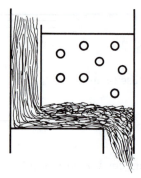

FIGURE A.3-1
Entrainment.

This should be a reasonable estimate for large towers with many trays, but it probably will be too low for smaller towers. We use Eqs. A.3-1 and A.3-2 for both gas absorbers and distillation columns.

Flooding Velocity and Tower Diameter

The tower diameter usually is selected so that the vapor velocity is between 60 and 80% of the flooding velocity. The early study of Souders and Brown* assumed that flooding was caused by the entrainment of droplets carried along with the gas; see Fig. A.3-1. When a small droplet in the tower is stationary, its weight will be balanced by the drag force exerted by the fluid

$$\tfrac{4}{3}\pi R^3(\rho_L - \rho_G)g = C_D(\tfrac{1}{2}\rho_G v^2)\pi R^2 \tag{A.3-3}$$

For turbulent flow, i.e., large particle Reynolds numbers, the drag coefficient is constant,[†] $C_D = 0.44$, so that the velocity is

$$v = \sqrt{\frac{8gR}{3(0.44)}}\sqrt{\frac{\rho_L - \rho_G}{\rho_G}} \tag{A.3-4}$$

We can use this expression to estimate the vapor velocity such that all droplets larger than a given size will fall back onto the tray.

Except for systems operating near their critical point, $\rho_G \ll \rho_L$. Hence, we can write Eq. A.3-4 as

$$v = K_0\sqrt{\rho_L/\rho_G} \tag{A.3-5}$$

Also, the densities of most liquids are about the same, so we can write

$$v\sqrt{\rho_G} = F = \text{Constant} \tag{A.3-6}$$

This constant F factor provides a quick means for estimating flooding velocities.

---

* M. Souders, Jr., and G. G. Brown, "Design of Fractionating Columns," *Ind. Eng. Chem.*, **26**: 98 (1934).

† R. H. Perry and C. H. Chilton, *Chemical Engineer's Handbook*, McGraw-Hill, New York, 1973, p. 5-62.

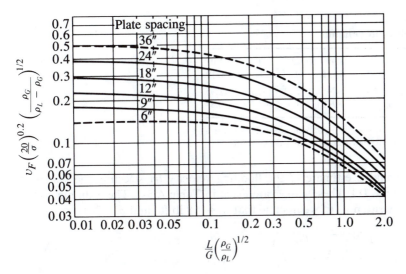

FIGURE A.3-2
Flooding limits for bubble-cap and perforated plates. L/G=liquid-to-gas mass ratio at point of consideration. [*J. R. Fair, Petro. Chem. Eng.*, 33(10): 45 (*September 1961*).]

GAS ABSORBERS—FLOODING VELOCITY. A more rigorous flooding correlation developed by Fair* is shown in Fig. A.3-2. Note that the density groups appearing in this correlation are the same as those obtained above. If we write the group appearing on the abscissa in terms of molar flow rates and insert our rule of thumb for a gas absorber that $L_S/(mG_S) = 1.4$, we see that

$$\frac{L}{S}\left(\frac{\rho_G}{\rho_L}\right)^{1/2} = \frac{L_S}{G_S}\frac{M_L}{M_G}\left(\frac{\rho_G}{\rho_L}\right)^{1/2} = 1.4m\frac{M_L}{P_T}\left(\frac{\rho_G}{M_G}\right)^{1/2} = 1.4\left(\frac{\gamma P^\circ}{\rho_L}\right)\frac{M_L}{M_G}\left(\frac{\rho_G}{\rho_L}\right)^{1/2} \tag{A.3-7}$$

Normally, absorbers are designed to operate at temperatures well below the normal boiling points of the solute, so $P^\circ/P_T \leq 0.5$ or so. The liquid phase activity coefficient often is in the range from 1 to 10; and if we consider cases where the solute is being recovered from an air stream into water at standard conditions, we find that

$$\frac{L}{G}\left(\frac{\rho_G}{\rho_L}\right)^{1/2} \sim 1.4(10)(0.5)\left(\frac{18}{29}\right)\sqrt{\frac{29/379}{62.4}} \sim 0.03\left(\frac{\gamma P^\circ}{P_T}\right) \sim 0.15 \tag{A.3-8}$$

However, we see from Fig. A.3-2 that the ordinate does not change very much for $(L/G)(\rho_G/\rho_L)^{1/2}$ in the range from 0.01 to 0.2. Thus, we should be able to

---

* J. R. Fair; see R. H. Perry and C. H. Chilton, *Chemical Engineer's Handbook*, 5th ed., McGraw-Hill, New York, 1973, p. 18-6.

TABLE A.3-1
Vapor velocities in columns

Tray spacing, ft	3.0	2.0	1.5	1.0
Ordinate	0.425	0.33	0.26	0.19
Flooding, $v_F\sqrt{\rho_G} = F_F$	3.24	2.51	1.98	1.45
60% of flooding, F	1.94	1.51	1.19	0.87

J. R. Fair; see R. H. Parry and C. H. Chilton, "Chemical Engineer's Handbook," 5th ed., McGraw-Hill, New York, 1973, p. 18-6.

obtain a reasonable estimate of the flooding velocity by selecting the values at $(L/G)(\rho_G/\rho_L)^{1/2} = 0.1$. These values are listed in Table A.3-1. The ordinate on the graph can be written as (if we assume that $\rho_L \sim 58$ lb/ft³ and $\sigma = 20$)

$$v_F\left(\frac{20}{\sigma}\right)^{0.2}\left(\frac{\rho_G}{\rho_L - \rho_G}\right)^{1/2} \sim v_F\sqrt{\frac{\rho_G}{\rho_L}} \sim \frac{v_F\sqrt{\rho_G}}{\sqrt{58}} = \frac{F_F}{\sqrt{58}} \tag{A.3-9}$$

Then if we consider that we desire to operate at 60% of the flooding velocity, we obtain the other values listed in Table A.3-1. Most of our preliminary designs are for towers with 2-ft tray spacings, and we assume that

$$F = 1.5 = v\sqrt{\rho_G} = v\sqrt{M_G\rho_m} \tag{A.3-10}$$

The same result is obtained for distillation columns, and so we use this expression for both gas absorbers and distillation columns.

Tower Cross-Sectional Area and Diameter

To find the tower diameter, first we convert the molar flow rate of vapor to a volumetric flow $Q = (V \text{ mol/hr})/(\rho_m \text{ mol/ft}^3)$, and then we divide by the velocity:

$$A_T = \frac{V/\rho_m}{1.5(3600)/\sqrt{M_G\rho_m}} = \frac{V\sqrt{M_G/\rho_m}}{1.5(3600)} \tag{A.3-11}$$

Actually, the cross-sectional area must be larger than this value because about 12% of the area is taken up by the downcomers.* Hence, we write

$$A_T = \frac{1}{0.88(1.5)3600} V\sqrt{M_G/\rho_m} = 2.1 \times 10^{-4} V\left(\frac{M_G}{\rho_m}\right)^{1/2} \tag{A.3-12}$$

Once we have estimated the column area, it is a simple matter to calculate the diameter

$$D_T = \left(\frac{4A_T}{\pi}\right)^{1/2} \tag{A.3-13}$$

---

* See B. D. Smith, "Design of Equilibrium Staged Processes," McGraw-Hill, New York, 1963, p. 486.

or

$$D_T = 0.0164\sqrt{V}\left(\frac{M_G}{\rho_m}\right)^{1/4} \tag{A.3-14}$$

The diameter is relatively insensitive to changes in the operating temperature and pressure.

Since the vapor rate in a distillation column is different above and below the feed tray, we need to evaluate the tower diameter for stills at these two locations. If the two values are not too different, we design the tower to correspond to the larger diameter. Assuming ideal-gas behavior and saturated-liquid feeds, we can write that

$$D_T = 0.0164\sqrt{V}\left[379M_G\left(\frac{T_b}{520}\right)\left(\frac{14.7}{P}\right)\right]^{1/4} \tag{A.3-15}$$

so that the largest value of D_T will correspond to the key component for which $M_G T_b$ is largest. When the two estimates of the tower diameter are significantly different, we build a tower, called a *swedge column*, in two sections having the appropriate diameters. We prefer to avoid these swedge columns, however, as they are more expensive to build.

Limitations on the Design Conditions

It is undesirable to build very tall and skinny towers because the towers will bend, and might buckle, in strong winds. A design guideline often used is that the column height should be less than about 175 ft, but a better design guideline is that the height-to-diameter ratio should be less than 20 to 30.

Plate towers are seldom used if the tower diameter is less than 1.5 ft, area = 1.77 ft^2, because it is not possible to get inside the tower (through manholes placed every five trays or so) to clean the tower. Instead, packed towers are used if the diameter is small. Either plate or packed towers are used if the column diameter is less than 4.5 ft, area = 15.9 ft^2, although the tray spacing normally is less than 2 ft in this range. Similarly, if the required height is above 190 ft or so, we might redesign the tower with a smaller tray spacing.

Shortcut Cost Procedure for Column Trays and the Shell

Once we have calculated the tower height and diameter, we can use Guthrie's correlations to estimate the tower cost; i.e., we cost the shell as a pressure vessel, and then we add the cost of the trays. As a quick approximation, we might assume that the cost of the trays is about 20% the cost of the column shell (assuming that everything is carbon steel).

Distillation Column Auxiliaries

For distillation columns, we also must design the condenser and reboiler. In addition, we must find the cooling-water and steam requirements.

COLUMN CONDENSER AND COOLING WATER. We consider the case of a total condenser, which is the most common situation. Then the condenser heat duty is the heat required to completely condense the vapor passing overhead. With cooling water available at 90°F and being returned at 120°F, heat balances give

$$Q_C = \Delta H_V \, V = U_C A_C \frac{120 - 90}{\ln\left[(T_b - 90)/(T_b - 120)\right]} = w_C C_P (120 - 90) \quad \text{(A.3-16)}$$

We normally assume that an overall heat-transfer coefficient for the condenser $U_C = 100$ Btu/(hr·ft²·°F) gives reasonable results. Hence, the required heat-transfer area for the condenser is

$$A_C = \left[\frac{\Delta H_V}{30(100)} \ln \frac{T_b - 90}{T_b - 120}\right] V \quad \text{(A.3-17)}$$

and the required flow of cooling water is

$$w_C = \left(\frac{\Delta H_V}{30}\right) V \quad \text{(A.3-18)}$$

If cooling water costs $\$C_W/10^3$ gal, the annual cost of cooling water will be

$$\text{Ann. Cost} = \left(\frac{\$0.01 C_W}{1000 \text{ gal}}\right)\left(\frac{1 \text{ gal}}{8.34 \text{ lb}}\right)\left(\frac{\Delta H_V}{30} V \frac{\text{lb}}{\text{hr}}\right)\left(8150 \frac{\text{hr}}{\text{yr}}\right)$$

$$= 3.26 \times 10^{-4} C_W \, \Delta H_V \, V \quad \text{(A.3-19)}$$

Similarly, we can find the capital cost of the condenser by using Guthrie's correlation (see Sec. E.1). To develop a shortcut model, we consider an exchanger with 1000 ft² of area, so that the purchased cost is $8900. The cost exponent is 0.65, and we consider a floating-head exchanger, carbon-steel construction, an operating pressure of less than 150 psig, and an installation factor of 3.29. Thus, the installed cost is

$$C_A = \left(\frac{\text{M\&S}}{280}\right)(3.29)(8900)\left(\frac{A}{1000}\right)^{0.65}$$

or

$$C_A = \left(\frac{\text{M\&S}}{280}\right)(328 A^{0.65}) \quad \text{(A.3-20)}$$

After substituting Eq. A.3-17 into this expression, we find that the installed cost can be written as

$$C_C = \left(\frac{\text{M\&S}}{280}\right)(328)\left(\frac{\Delta H_V}{3000} \ln \frac{T_b - 90}{T_b - 120}\right)^{0.65} V^{0.65} \quad \text{(A.3-21)}$$

REBOILER AND STEAM SUPPLY. If we use steam to supply the heat to produce $\bar{V}$ mol/hr of vapor at the bottom of the tower, a heat balance gives

$$Q_R = \Delta H_{\bar{V}} \bar{V} = U_R A_R \Delta T_m = W_S \Delta H_S \qquad \text{(A.3-22)}$$

The temperature driving force in the reboiler must be constrained to be less than about 30 to 45°F, to prevent film boiling. We expect to obtain a very high value of the overall heat-transfer coefficient in the reboiler because we have heat transfer between a condensing vapor and a boiling liquid. Thus, we expect that there will be a limiting heat flux in the reboiler, and we assume that

$$U_R \Delta T_m \sim 11{,}250 \text{ Btu/(hr} \cdot \text{ft}^2) \qquad \text{(A.3-23)}$$

With this approximation the required heat-transfer area is

$$A_R = \frac{\Delta H_{\bar{V}}}{11{,}250} \bar{V} \qquad \text{(A.3-24)}$$

and the required steam supply is

$$W_S = \frac{\Delta H_{\bar{V}}}{\Delta H_S} \bar{V} \qquad \text{(A.3-25)}$$

For the case of 25-psi steam, where $T_S = 267°F$, $\Delta H_S = 933$ Btu/lb, and the cost is $\$C_S/10^3$ lb, the annual steam cost is

$$\text{Ann. Cost} = \frac{\$C_S}{1000 \text{ lb}} \left(\frac{\Delta H_{\bar{V}}}{933} \bar{V} \frac{\text{lb}}{\text{hr}} \right) \left(8150 \frac{\text{hr}}{\text{yr}} \right)$$
$$= 8.74 \times 10^{-3} C_S \Delta H_{\bar{V}} \bar{V} \qquad \text{(A.3-26)}$$

The installed capital cost of the reboiler can be estimated from Eq. A.3-20, and we find that

$$C_R = \left(\frac{\text{M\&S}}{280} \right)(328) \left(\frac{\Delta H_V}{11{,}250} \right)^{0.65} \bar{V}^{0.65} \qquad \text{(A.3-27)}$$

The vapor rate in the bottom of the tower depends on the quality of the feed q

$$\bar{V} = V + F(q - 1) \qquad \text{(A.3-28)}$$

where
$$q = \frac{H_G - H_F}{H_G - H_L} \qquad \text{(A.3-29)}$$

which is the heat required to convert 1 mol of feed to a saturated vapor divided by the molal latent heat of vaporization.

Summary of the Cost Model

If we combine the expressions above, we can write a model for the total annual cost (TAC) of a distillation separation (i.e., a capital charge factor of $\frac{1}{3}$ yr is used to annualize the installed equipment cost) in terms of the design variables:

TAC = Column Shell and Trays + Condenser + Reboiler
    + Cooling Water + Steam

$$= \frac{1}{3}\left(\frac{M\&S}{280}\right)(120 D_T H^{0.8})(2.18 + F_C)$$

$$+ \frac{1}{3}\left(\frac{M\&S}{280}\right)(101.3)(2.29 + F_C)(A_C^{0.65} + A_R^{0.65}) + \text{Steam} + \text{Cooling Water}$$

$$= \frac{M\&S}{3(280)}(101.9)(2.18 + F_C)\left\{0.0164\sqrt{V}\left[379 M_G\left(\frac{T_b}{520}\right)\left(\frac{14.7}{P}\right)\right]^{1/4}\right\}$$

$$\times \left(\frac{2N}{E_0} + H_0\right)^{0.8}$$

$$+ \frac{M\&S}{3(280)}(101.3)(2.29 + F_C)\left(\frac{\Delta H_V}{3000}\ln\frac{T_b - 90}{T_b - 120}\right)^{0.65} V^{0.65}$$

$$+ \frac{M\&S}{3(280)}(101.3)(2.29 + F_C)\left(\frac{\Delta H_V}{11,250}\right)^{0.65} V^{0.65}$$

$$+ 3.26 \times 10^{-4} C_W \Delta H_V V + 8.74 \times 10^{-3} C_S \Delta H_{\bar{V}} \bar{V} \qquad \text{(A.3-30)}$$

COLBURN'S COST MODEL. An alternate cost model was developed by Colburn.[*] The form of the cost model is

$$\text{TAC} = (C_1 N + C_2 + C_3)V \qquad \text{(A.3-31)}$$

where $C_1 N$ is the cost of the column shell, C_2 includes the cost of the condenser and reboiler, and C_3 includes the cost of the steam and cooling water. An example that illustrates the application of the procedure is given in Happel and Jordan.[†] Using Taylor series expansions, we could reduce Eq. A.3-30 to Eq. A.3-31.

Packed Versus Plate Towers

For preliminary process design we are concerned primarily with the most expensive pieces of equipment. We expect that the expensive towers will have large vapor rates and diameters greater than 4.5 ft. Hence, normally the expensive towers

---

[*] A. P. Colburn "Collected Papers on the Teaching of Chemical Engineering," ASEE Summer School for the Teaching of Chemical Engineering, Pennsylvania State University, 1936.

[†] J. Happel and D. G. Jordan, *Chemical Process Economies*, 2d ed., Dekker, New York, 1975, p. 385.

will be plate columns. Tray towers are preferred when the solvent flows are small, i.e., packing would not be wetted uniformly, and where internal cooling is desired.

But packed towers are used when the materials are highly corrosive or foam badly, when low-pressure drops are required, and if the tower diameter is less than 2 ft. The height of a packed tower is estimated by calculating the number of transfer units and the height of a transfer unit. Often, shortcut procedures such as those given in Table A.1-1 will give reasonable predictions for the number of transfer units. The height of a transfer unit can be predicted by a variety of correlations that have been reviewed by Fair.[*] However, to obtain a quick estimate of the tower height, Jordan[†] recommends letting

$$H_{OG} = 2.0 \text{ ft} \tag{A.3-32}$$

This estimate should be reasonable for packing sizes from 0.75 to 1.5 in. and for gas flow rates in the range of 200 to 1400 lb/(hr·ft²).

A.4   DISTILLATION COLUMN SEQUENCING

In Sec. 7.3 we discussed the use of heuristics and computer codes for selecting the best sequence of distillation columns. However, as an alternative approach, we can use our order-of-magnitude arguments to simplify the problem. By simplifying problems of this type, we often improve our understanding of the most important features that dominate the design. The procedure we present here is due to Malone et al.[‡]

Cost Model

For the sake of simplicity, we consider ternary mixtures, and we want to determine whether it is better to use the direct sequence (i.e., split A from B and C and then split B and C) or the indirect sequence (i.e., split C from A and B and then split A and B); see Fig. 7.3-3. We base our decision on the total separation costs of the two sequences, and so we need a cost model for a distillation separation. We use Guthrie's correlations (see Sec. E.1) to estimate the capital costs of the column, condenser, and reboiler, and then we add the cost of the steam and cooling water. Thus, we write the cost of a single column as

$$\text{TAC} = C_0 N_T^{0.8} V^{0.5} + C_1 V^{0.65} + C_2 V \tag{A.4-1}$$

where C_0 is the annualized cost coefficient for the column shell, $N_T =$ number of theoretical trays, $V =$ vapor rate, C_1 is the annualized cost coefficient for both the

---

[*] R. H. Perry and C. H. Chilton, *Chemical Engineer's Handbook*, 5th ed., McGraw-Hill, New York, 1973, sec. 18.

[†] D. G. Jordan, *Chemical Process Development*, vol. 2, Interscience, New York, 1968, p. 452.

[‡] M. F. Malone, K. Glinos, F. E. Marquez, and J. M. Douglas, *AIChE J.*, **31**: 683 (1985).

condenser and reboiler, and C_2 is the annual cost coefficient of the steam and the cooling water. If a base-case design is available, we can use a Taylor series expansion to write Eq. A.4-1 as

$$\text{TAC} = K_0 + K_1 N_T + K_2 V \qquad (A.4\text{-}2)$$

This is the cost model that we use to evaluate column sequences. Now if we consider the difference in cost between the direct and indirect sequences, using Eq. A.4-2, we obtain

$$\Delta\text{TAC} = K_1(N_{AB}^D - N_{AB}^I + N_{BC}^D - N_{BC}^I)$$
$$+ K_2(V_{AB}^D - V_{AB}^I + V_{BC}^D - V_{BC}^I) \qquad (A.4\text{-}3)$$

NUMBER OF PLATES. According to Gilliland's correlation (see Eq. A.2-23), when $R \sim 1.2R_{\min}$,

$$N_T \sim 2N_{\min} \qquad (A.2\text{-}23)$$

where the minimum number of plates is given by Fenske's equation. If we write $N_{\min}$ in terms of fractional recovery of the light key overhead r_i and the fractional recovery of the heavy key in the bottoms r_j, then for the A/B split,

$$N_{AB} = \frac{2 \ln \{[r_A/(1 - r_A)][r_B/(1 - r_B)]\}}{\ln a_{AB}} \qquad (A.4\text{-}4)$$

However, the fractional recoveries for the A/B split will be the same for both the direct and indirect sequences (Fig. 7.3-3)

$$N_{AB}^D = N_{AB}^I \qquad (A.4\text{-}5)$$

The same result will be obtained for the B/C split. Thus, normally Eq. A.4-3 reduces to

$$\Delta\text{TAC} = K_2(V_{AB}^D - V_{AB}^I + V_{BC}^D - V_{BC}^I) \qquad (A.4\text{-}6)$$

This result corresponds to the well-known heuristic

Select the column sequence based on the smallest total vapor load.   (A.4-7)

LIMITATIONS OF THE VAPOR LOAD HEURISTIC. Of course, we see from Eq. A.4-3 that this heuristic will be valid only if the cost coefficient K_1 is the same for each column. However, if C were a corrosive component, both columns in the direct sequence (Fig. 7.3-3) would have to be made from expensive materials, whereas only one column in the indirect sequence would require corrosion protection. Thus, the cost coefficients would not be the same. Similarly, if the columns operate at different pressure levels in the direct and indirect sequences, their cost coefficients will be different. Thus, the vapor-load heuristic is not always valid. Also note that Fenske's equation is limited to mixtures having constant relative volatilities.

Vapor Rate

The vapor rate in each column can be related to the reflux and distillate rates by a material balance

$$V_i = (R_i + 1)D_i \tag{A.4-8}$$

If we use the rule of thumb

$$R_i = 1.2R_{m,i} \tag{A.4-9}$$

along with the approximate material balances that correspond to perfect splits (rather than greater than 99% recoveries)

$$D_i x_{i,D} = F x_{i,F} \tag{A.4-10}$$

then it is a simple matter to estimate the vapor flows given a knowledge of the minimum reflux ratios.

The common procedure for calculating reflux ratios for constant-volatility, multicomponent mixtures is to solve Underwood's equations

$$\frac{\alpha_{AC} x_{AF}}{\alpha_{AC} - \theta} + \frac{\alpha_{BC} x_{BF}}{\alpha_{BC} - \theta} + \frac{x_{CF}}{1 - \theta} = 1 - q \tag{A.4-11}$$

and

$$R_m = \frac{\alpha_{AC} x_{AD}}{\alpha_{AC} - \theta} + \frac{\alpha_{BC} x_{BD}}{\alpha_{BC} - \theta} + \frac{x_{CD}}{1 - \theta} \tag{A.4-12}$$

For the direct sequence, we solve Eq. A.4-11 for the value of θ in the range $\alpha_{AC} < \theta < \alpha_{BC}$ and substitute this result into Eq. A.4-12 to find R_m; for the indirect sequence we determine the θ in the range $\alpha_{BC} < \theta < 1$ and again use Eq. A.4-12 to find R_m. The minimum reflux ratios for the remaining binary columns can be found by using the shortcut approximation of Underwood's equation for binary mixtures

$$R_m = \frac{1}{(\alpha - 1)x_F} \tag{A.2-16}$$

where we use the appropriate α and feed composition for the last column in either the direct or indirect sequence (see Fig. 7.3-3), i.e.,

$$R_{m,B/C}^D = \frac{1 - x_{AF}}{(\alpha_{BC} - 1)x_{BF}} \qquad R_{m,A/B} = \frac{1 - x_{CF}}{(\alpha_{AB} - 1)x_{AF}} \tag{A.4-13}$$

Instead of using Underwood's equations for the ternary mixtures (Eqs. A.4-11 and A.4-12), however, we prefer to use the approximate expressions developed by Glinos and Malone so that we can obtain an explicit expression for the vapor rate; see Table A.2-1. Their results indicate that for

A/BC:
$$R_m = \frac{\alpha_{BC}(x_{AF} + x_{BF})}{f x_{AF}(\alpha_{AC} - \alpha_{BC})} + \frac{x_{CF}}{f x_{AF}(\alpha_{AC} - 1)} \tag{A.4-14}$$

where
$$f = 1 + \tfrac{1}{100}x_{BF} \tag{A.4-15}$$

and for

AB/C:
$$R_m = \frac{(x_{BF} + x_{CF})/(\alpha_{BC} - 1) + x_{AF}/(\alpha_{AC} - 1)}{(x_{AF} + x_{BF})(1 + x_{AF}x_{CF})} \qquad \text{(A.4-16)}$$

Now, if we combine Eqs. A.4-6 and A.4-8 through A.4-16, we obtain

$$\frac{\Delta TAC}{K_2 F} = \frac{\Delta V}{F} = 1.2\left\{\left(\frac{x_{BF} + x_{CF}}{\alpha_{BC} - 1}\right)\left(\frac{x_{AF}x_{CF}}{1 + x_{AF}x_{CF}}\right)\right.$$

$$\left. + \left(\frac{1}{\alpha_{AC} - 1}\right)\left[\frac{x_{CF} - fx_{AF} + x_{AF}x_{CF}}{f(1 + x_{AF}x_{CF})}\right]^2 - \frac{\alpha_{BC}(x_{AF} + x_{BF})f - 1}{(\alpha_{AC} - \alpha_{BC})f}\right\} - x_{AF} \qquad \text{(A.4-17)}$$

Whenever this result is negative, we prefer the indirect sequence.

Discussion

The result, Eq. A.4-17, provides us with an approximate, explicit expression that we can use to determine the best column sequence as a function of the feed composition to the distillation train. Thus, we can use this result in place of the heuristics; see Sec. 7.3. Malone et al.* present a comparison of this expression to the published heuristics as well as discuss the sensitivity of the solution. That is, they describe the conditions under which the wrong choice of the column sequence corresponds to large cost penalties. This type of sensitivity information is often more useful for preliminary designs than obtaining a rigorous solution for a particular case.

Using Bounding Arguments

In another study, Glinos and Malone† plotted the line where the total vapor rate for the direct sequence was equal to the total vapor rate for the indirect sequence on the plane of the feed compositions; see Fig. A.4-1. The relative volatilities are shown in parentheses, and Underwood's equations for sharp splits were used for the calculations. From an inspection of these diagrams, they proposed these simple criteria:

Use direct sequence when
$$\frac{x_A}{x_A + x_C} > \mu = \frac{\alpha_{AB} - 1}{\alpha_{AC} - 1} \qquad \text{(A.4-18)}$$

Use the indirect sequence if
$$\frac{x_A}{x_A + x_C} < \lambda = \frac{1}{\alpha_{AC} + 1} \qquad \text{(A.4-19)}$$

Calculate the vapor rate if
$$\mu > \frac{x_A}{x_A + x_C} > \lambda \qquad \text{(A.4-20)}$$

Thus, they use bounds to develop very simple expressions for some cases.

---

* M. F. Malone, K. Glinos, F. E. Marquez, and J. Douglas, *AIChE J.*, **31**: 683 (1985).

† K. Glinos and M. F. Malone, "*Optimality Regions for Complex Column Alternatives in Distillation Systems*," paper submitted to *Chem. Eng. Res. Des.*, 1987.

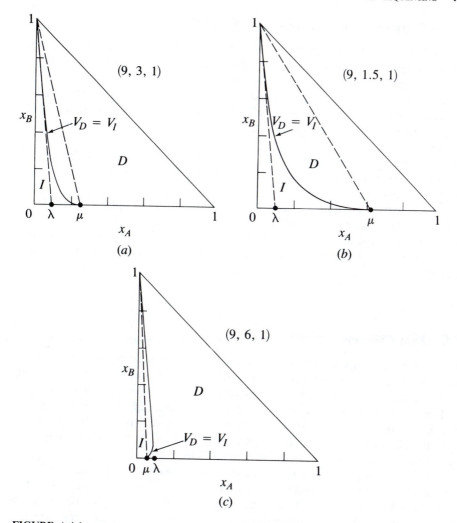

FIGURE A.4-1
Bounds for the direct and indirect sequences. (*From K. Glinos and M. F. Malone, "Optimality Regions for Complex Column Alternatives in Distillation Systems," submitted to Chem. Eng. Res. Dev., 1987.*)

It is interesting that Eq. A.4-18 agrees with the heuristic to select the direct sequence when the amount of the lowest boiler is small. Also, Eq. A.4-19 is a similar result for the indirect sequence when the heaviest component is a large fraction of the feed. However, for a case where the volatilities are (9, 3, 1) and the feed compositions are $x_A = x_B = 0.15$ and $x_C = 0.7$, Fig. A.4-1 shows that the direct sequence is still favored. Similarly, we see from Fig. A.4-1 that the region where the indirect sequence is best shrinks as the A/B split becomes more difficult than the B/C split, which contradicts the common heuristic "do the easiest splits first and leave the difficult splits until last." This result again shows the danger of using heuristics.

A.5   COMPLEX DISTILLATION COLUMNS

Normally we do not consider the use of complex distillation columns in our initial design, except for the use of pasteurization columns to remove light ends from a product stream (see Sec. 7.3). Instead, we first look for the best sequences of simple columns, and we evaluate the profitability of the process. If additional design effort can be justified, we examine the possibility of replacing two adjacent columns in the simple sequence by complex columns, to see whether we can reduce the separation costs. Remember that if we can reduce the recycle costs of a reactant for a reaction system where there are significant by-product losses, then we might be able to translate these separation system savings to raw-materials savings if we reoptimize the process flows.

Most texts about unit operations do not include very complete discussions of complex distillation columns. Thus, a brief introduction is presented here for sidestream columns, sidestream strippers and rectifiers, and prefractionators and Petlyuk columns. These results are taken from various papers by Malone and coworkers.

Sidestream Columns

We can sometimes use a single sidestream column to replace two columns in either the direct or the indirect sequence; see Fig. A.5-1. If the indirect sequence is favored, we say that the "primary" separation corresponds to the AB/C split, and we replace the two-column sequence by a single sidestream above the feed; see Fig. A.5-2. However, if the direct sequence is favored, we say that the "primary" separation corresponds to the A/BC split, and we replace the two-column sequence by a single column with a sidestream below the feed.

For a sidestream above the feed (or below it), we note from Fig. A.5-2 that there are only three column sections compared to the four sections available in the indirect (or direct) sequence. We also note that the A/B (or B/C) split takes place in only the upper (or lower), single-column section. Hence, this "secondary" separation is limited by the vapor rate required for the primary separation.

Moreover, since we have only three column sections, we can no longer achieve any purity that we desire. The recoveries of B and C or the composition of the bottoms can be specified arbitrarily, and the composition of A in the distillate can be fixed at any value. However, there is a maximum concentration of B (and a minimum concentration of A) that will be obtained in a sidestream above the feed even if an infinite number of trays is used in the upper section. Glinos and Malone[*] showed that this minimum concentration of A and the maximum concentration of

---

[*] K. Glinos and M. F. Malone, *I&EC Proc. Des. Dev.*, **24**: 1087 (1985) and **23**: 764 (1984).

(*a*) Direct

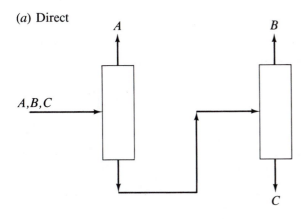

(*b*) Indirect

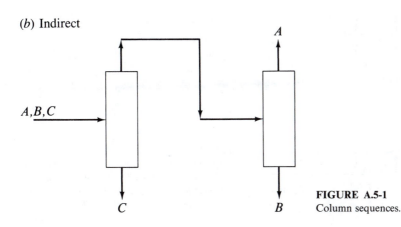

FIGURE A.5-1
Column sequences.

B can be estimated by using the expressions below for the case of a high-purity overhead

$$(x_{AS,\,min})^2 - \left[1 + \frac{x_{BF}}{R_2(x_{AF} + x_{BF})} + \frac{1}{R_2(\alpha_{AB} - 1)}\right]x_{AS,\,min} + \frac{x_{AF}/(1 - x_{CF})}{R_2(\alpha_{AB} - 1)} = 0$$

(A.5-1)

$$x_{BS,\,max} = 1 - x_{AS,\,min} - x_{CS}$$

(A.5-2)

where R_2 is the reflux ratio that corresponds to the primary split (AB/C) at the feed plate

$$R_2 = \frac{V_2}{D + S}$$

(A.5-3)

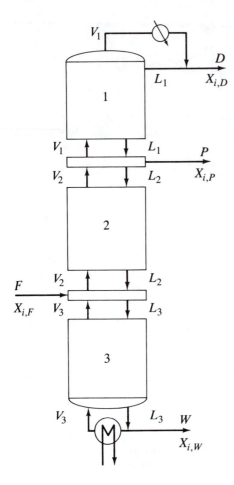

FIGURE A.5-2

Sidestream above the feed. (*From K. Glinos and M. F. Malone, "Optimality Regions for Complex Column Alternatives in Distillation Systems," submitted to Chem. Eng. Res. Dev., 1987.*)

The value of $R_{2,\min}$ can be estimated by using Underwood's equations or the approximate expressions of Glinos and Malone* (see Table A.2-1) for the AB/C split

$$R_{2,\min} = \frac{x_{AF}/(\alpha_{AC} - 1) + (x_{BF} + x_{CF})/(\alpha_{BC} - 1)}{(x_{AF} + x_{BF})(1 + x_{AF}x_{CF})} \qquad (A.5\text{-}4)$$

The approximate expression is usually within 4% of the exact result, which is adequate for screening purposes.

DESIGN OF COLUMNS WITH SIDESTREAMS ABOVE THE FEED. If we have a case where a high purity of the sidestream is not required (e.g., suppose that we

---

* K. Glinos and M. F. Malone, *I&EC Proc. Des. Dev.*, **24**: 1087 (1985).

plan to recycle this stream back to a reactor and that the impurities do not affect the product distribution), we can select $x_{A,W} = 0$; $x_{C,D} = 0$; either the overhead purity of A, x_{AD}, or the fractional recovery of A overhead (but not both); the purity and the fractional recovery of C in the bottoms; and the composition of the lightest component of A in the sidestream, x_{AS}. Next we calculate $R_{2,\min}$, using Eq. A.5-4, and then we let $R_2 = 1.2R_{2,\min}$. Now, we can evaluate $x_{AS,\min}$ by using Eq. A.5-1, and we see whether our specification of x_{AS} exceeds this value. If it does not, we must change our specification for x_{AS}.

Assuming that the desired purity can be obtained, Glinos and Malone were able to prove that a column with a sidestream above the feed would always be cheaper than the corresponding two columns for the indirect sequence. About 10% fewer trays were required, and vapor savings up to 30% can be obtained. The largest savings correspond to balanced volatilities, although the sidestream purity decreases. For low values of x_{AF}, high-purity sidestreams can be obtained, but the vapor savings decrease.

PASTEURIZATION COLUMNS. For the special case of sidestream columns where the desired product is the intermediate boiler and there is a waste or fuel by-product that is either much lighter than the product (so we recover the product as a sidestream above the feed) or else is much heavier than the product (so we recover the product as a sidestream below the feed), Glinos and Malone* find that they can obtain very simple solutions. For the case where $\alpha_{AC} \gg \alpha_{BC}$ and the sidestream is above the feed, the minimum amount of impurity in the product sidestream will be

$$x_{AS,\min} = \frac{x_{AF}(\alpha_{BC} - 1)}{(\alpha_{BC}x_{BF} + x_{CF})(\alpha_{AB} - 1)} \tag{A.5-5}$$

and the minimum vapor rate is given by

$$\frac{V_{\min}}{F} = \frac{\alpha_{AB}x_{BF} + x_{CF}}{\alpha_{BC} - 1} + x_{AF} \tag{A.5-6}$$

From Eq. A.5-5 we see that if $\alpha_{AB} \gg 1$ and/or if x_{AF} is small, then we can obtain a product with high purity as a sidestream.

The corresponding results for sidestreams below the feed are

$$x_{CS,\min} = \frac{\alpha_{AB}(x_{AF} + x_{BF})}{\alpha_{AB}x_{AF} + x_{BF}} \tag{A.5-7}$$

$$R_m = \frac{x_{AF} + x_{BF}}{x_{AF}(\alpha_{BC} - 1)} \tag{A.5-8}$$

$$\frac{S}{F} = \frac{x_{BF} + x_{CF} - x_{CF}/x_{CW}}{1 - x_{CS}/x_{CW}} \sim \frac{x_{BF}}{1 - x_{CS}} \tag{A.5-9}$$

---

* K. Glinos and M. F. Malone, *I&EC Proc. Des. Dev.*, **24**: 1087 (1985) and **23**: 764 (1984).

Thus, we have fairly simple expressions available for the preliminary designs of pasteurization columns.

Sidestream Strippers and Sidestream Rectifiers

If the purity requirements of a sidestream column are not satisfactory, we can add a column section, as either a sidestream stripper or a sidestream rectifier (see Figs. A.5-3 and A.5-4). Now we again have four column sections, so that any specified recoveries and purities can be obtained.

SIDESTREAM STRIPPERS. For the configuration shown in Fig. A.5-3, we use the specifications and overall material balances to fix the external flows. Then, to

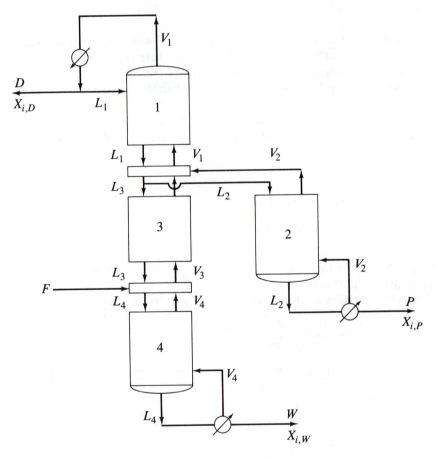

FIGURE A.5-3
Sidestream stripper. (*From K. Glinos and M. F. Malone, "Optimality Regions for Complex Column Alternatives in Distillation Systems," submitted to Chem. Eng. Res. Dev., 1987.*)

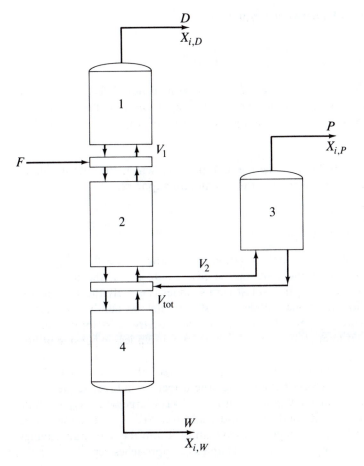

FIGURE A.5-4
Sidestream rectifier. (*From K. Glinos and M. F. Malone, "Optimality Regions for Complex Column Alternatives in Distillation Systems," submitted to Chem. Eng. Res. Dev., 1987.*)

estimate the internal flows V_1 and V_3, Glinos and Malone* developed the expressions

$$V_{3,\min} = \left(1 - q + \frac{x_{CF}}{\theta_2 - 1}\right)F \tag{A.5-10}$$

and

$$V_{1,\min} = \left[\frac{x_{AF}x_{CF}\theta_2}{\alpha_{BC}(\theta_2 - 1) - x_{CF}\theta_2/(\alpha_{BC} - 1)}\right]F \tag{A.5-11}$$

---

* K. Glinos and M. F. Malone, *I&EC Proc. Des. Dev.*, **24**: 1087 (1985) and **23**: 764 (1984).

where θ_2 is the root of Underwood's equation

$$\frac{\alpha_{AC} x_{AF}}{\alpha_{AC} - \theta} + \frac{\alpha_{BC} x_{BF}}{\alpha_{BC} - \theta} + \frac{x_{CF}}{1 - \theta} = 1 - q \qquad \text{(A.5-12)}$$

in the range $\alpha_{BC} > \theta_2 > 1$. For the case of saturated-liquid feeds, they also developed approximate solutions for Underwood's equation so that no iteration is required. Also, a procedure for estimating the number of trays was developed by Glinos.*

SIDESTREAM RECTIFIERS. For the configuration shown in Fig. A.5-4, Glinos developed an expression for estimating the minimum vapor rate

$$V_R = \left[\frac{\alpha_{AC} x_{AF}}{\alpha_{AC} - \theta_1} + \frac{\alpha_{AC} x_{AF} x_{BF} \theta_1}{x_{AF} \theta_1 (\alpha_{AC} - 1) - \alpha_A + \theta_1} + 1 - q \right] F \qquad \text{(A.5-13)}$$

where θ_1 is the root of Underwood's equation in the range $\alpha_{AC} > \theta_1 > \alpha_{BC}$.

Glinos and Malone found that for the case where $q = 1$, the total vapor generated for the two reboilers in a sidestream stripper is exactly the same as the vapor requirement for the single reboiler in a sidestream rectifier. The operating costs may differ, however, since the reboiler of the sidestream stripping section operates at a different temperature and therefore may use a less expensive utility.

POTENTIAL SAVINGS. Glinos and Malone[†] proved that these sidestream columns require less total vapor than either the direct or indirect sequence. In addition, they showed that the vapor savings are always large when x_{BF} is small, which agrees with the results of Tedder and Rudd (see Sec. 7.3). The maximum savings possible is 50%, independent of the volatilities, and the maximum savings is obtained when $x_{AF} = (\alpha_{AB} - 1)/(\alpha_{AC} - 1)$ and x_{BF} approaches zero.

Thermally Coupled Columns—Petlyuk Columns

For a ternary mixture A, B, C, the easiest split possible is between A and C, letting B distribute between the top and bottom of the column. Then either we can split the A/B overhead and the B/C bottoms stream in two additional columns, or we can combine these columns into a single column (see Fig. A.5-5). With this arrangement, we need only one condenser and one reboiler for both columns; i.e., the columns are thermally coupled since the reboiler from the downstream column supplies the vapor for both columns and the condenser on the downstream column

---

* K. Glinos, "A Global Approach to the Preliminary Design and Synthesis of Distillation Trains," Ph.D. Thesis, University of Massachussetts, Amherst, 1984.

† K. Glinos and M. F. Malone, "*Optimality Regions for Complex Column Alternatives in Distillation Systems,*" paper submitted to *Chem. Eng. Res. Des.*, 1986.

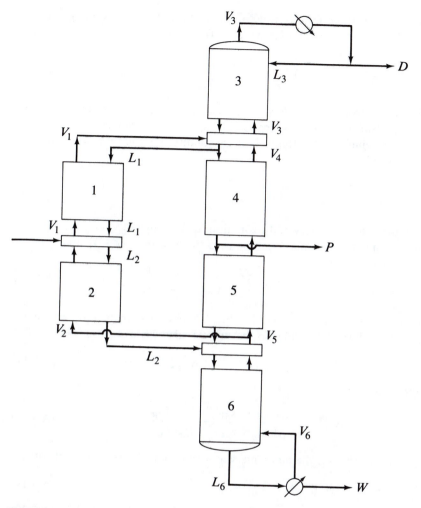

FIGURE A.5-5
Petlyuk column. (*From K. Glinos and M. F. Malone, "Optimality Regions for Complex Column Alternatives in Distillation Systems," submitted to Chem. Eng. Res. Dev., 1987.*)

provides the reflux for both columns. This configuration is often called a *Petlyuk column.*

If we make a sharp A/C split in the first column and fix the product flows to match our specifications, then two internal flows must be fixed to be able to calculate all the other flows. If we decrease L_3, we will eventually reach a limiting condition at either the upper feed or the lower feed of the downstream column, depending on which feed condition is controlling. For the case were we make a sharp AB/BC split or a sloppy A/BC split in the prefractionator, and the upper feed point is controlling, Glinos and Malone* showed that (assuming that the prefractionator is also at its limiting condition)

$$V_{min}^U = \left(\frac{\alpha_{AC} x_{AF}}{\alpha_{AC} - \theta_1} \right) F \tag{A.5-14}$$

where θ_1 is the root of Underwood's equation, Eq. A.5-12, in the range $\alpha_{AC} > \theta > \alpha_{BC}$. But for either a sharp AB/BC split or a sloppy AB/C split in the prefractionator (which is operating at limiting conditions), and the lower feed point controls, the result is

$$V_{6,min}^L = \left(\frac{x_{CF}}{\theta_2 - 1} \right) F \tag{A.5-15}$$

where θ_2 is the root of Underwood's equation in the range $\alpha_{BC} > \theta_2 > 1$.

If $q = 1$, then mass balances can be used to show that $V_3 = V_6$ and

$$V_{6,min} = \left\{ \max \left[\frac{\alpha_{AC} x_{AF}}{\alpha_{AC} - \theta_1}, \frac{x_{CF}}{\theta_2 - 1} \right] \right\} F \tag{A.5-16}$$

Glinos and Malone also noted that Eq. A.5-14 corresponds to the minimum vapor rate for a sharp A/BC split, so that we can also write

$$V_{min}^{A/BC} = (R_{min}^{A/BC} + 1) x_{AF} F \tag{A.5-17}$$

where, approximately,

$$R_{min}^{A/BC} = \frac{\alpha_{AC}(x_{AF} + x_{BF})}{x_{AF}(\alpha_{AC} - \alpha_{BC})} + \frac{x_{CF}}{x_{AF}(\alpha_{AC} - 1)} \tag{A.5-18}$$

Similarly, Eq. A.5-16 gives the minimum vapor rate for a sharp AB/C split, and so

$$V_{min}^{AB/C} = (R_{min}^{AB/C} + 1)(x_{AF} + x_{BF}) F \tag{A.5-19}$$

and $\qquad V_{6,min} = \{ \max[x_{FF}(R_{min}^{A/BC} + 1), (x_{AF} + x_{BF})(R_{min}^{AB/C} + 1)] \} F \qquad$ (A.5-20)

Thus, simple approximate expressions that are useful for conceptual designs are available.

FLEXIBILITY IN THE DESIGN. Even though the vapor rate $V_{6,min}$ establishes the minimum reboiler duty of the separation system, it is still possible to choose a value for V_1 independently (providing that the choice for V_1 does not change which feed to the downstream column is controlling). To understand this flexibility associated with the design, we consider a case where the lower feed is controlling,

---

* K. Glinos and M. F. Malone, *I&EC Proc. Des. Dev.*, **24**: 1087 (1985) and **23**: 764 (1984).

$V_{min}^L > V_{min}^4$, and a sharp AB/BC split is accomplished in sections 1 and 2. For this case the fraction of B recovered overhead in the prefractionator is[*]

$$f_B = \frac{\alpha_{BC} - 1}{\alpha_{AC} - 1} \tag{A.5-21}$$

and the amount of B fed to the downstream column at the upper feed location is $f_B x_{BF} F$. Column sections 3 and 4 operate at a condition above the minimum, so that they can handle larger amounts of B. Thus, we could have designed for a sloppy AB/C split and taken more B overhead. However, there is an upper bound on the fractional recovery of B overhead in the prefractionator, say $f_{B,max}$, which corresponds to the situation where the minimum vapor rate in column sections 3 and 4 becomes equal to $V_{6,min}$. We write these bounds as

$$V_{1,min}^{AB/BC} < V_{1,min}^L < V_{1,min}^{AB/C} \tag{A.5-22}$$

where
$$V_{1,min}^L = \frac{\alpha_{AC} x_{AF}}{\alpha_{AC} - \theta_2} + \frac{\alpha_{BC} x_{BF} f_B}{\alpha_{BC} - \theta_2} \tag{A.5-23}$$

where θ_2 is again the root of Underwood's equation in the range $\alpha_{BC} > \theta_2 > 1$.

Similarly, if the upper column controls, then column sections 5 and 6 can handle more B than corresponds to a sharp split in the prefractionator. That is, we can perform a sloppy A/BC split in the prefractionator, but now we encounter a lower bound on the fraction of B taken overhead $f_{B,max}$. Hence, we can write

$$V_{1,min}^{AB/BC} < V_{1,min}^4 < V_{1,min}^{A/BC} \tag{A.5-24}$$

where $V_{1,min}$ is now evaluated by using Eq. A.5-23, except that θ_2 is replaced by θ_1. Fidkowski and Krolikowski[†] developed an expression for this bound on the fractional recovery of B:

$$F_{B,max} = \frac{(\alpha_{AC} - \alpha_{BC})V_{3,min} - \alpha_{AC} x_{AF}^F}{(\alpha_{AC} - 1)V_{3,min} - (\alpha_{AC} x_{AF} + x_{BF})F} \tag{A.5-25}$$

The results above indicate that a Petlyuk column can handle a range of feed compositions without changing the reboiler duty, but merely by changing the flows in column sections 1, 2, 4, and 5. This flexibility both for the limiting conditions and for operating conditions has also been reported by others.[‡]

---

[*] K. Glinos and M. F. Malone, *I&EC Proc. Des. Dev.*, **24**: 1087 (1985).

[†] Z. Fidkowski and L. Krolikowski, *AIChE J.*, **32**: 537 (1986).

[‡] W. J. Stupin and F. J. Lockhart, "Thermally Coupled Distillation Columns—A Case Study," 64th Annual AIChE Meeting, San Francisco, 1971; and T. L. Wayburn and J. D. Seader, in *Proc. 2d Intl. FOCAPD Conf.*, A. W. Westerberg and H. Chien, eds., CACLE Corp. c/o Prof. B. Carnahan, Department of Chemical Engineering, University of Michigan, Ann Arbor, Mich., 48109, p. 765, June 1983.

BEST DESIGN CONDITIONS. The smallest of all the possible minimum vapor rates in the prefractionator is $V_{1,\text{min}}^{AB/BC}$. In column sections 4 and 5, the vapor rates are given by $V_4 = V_5 = V_3 - V_1$. For $q = 1$, $V_3 = V_6$, and so V_3 is fixed by Eq. A.5-16. Thus, the column diameter in section 4 will decrease as V_1 increases, but will never reach V_3 (which fixes the diameter of section 3). We prefer to build columns that have the same diameter in each section for reasons of costs, so we prefer to make V_4 as large as possible (although it will be less than V_3). This largest value of V_4 corresponds to the vapor rate $V_{1,\text{min}}^{AB/BC}$ in the prefractionator. Glinos and Malone* showed that this value is given by the expression

$$V_{1,\text{min}}^{AB/BC} = \left(\frac{\alpha_{AC}x_{AF} + \alpha_{BC}x_{BF} + x_{CF}}{\alpha_{AC} - 1}\right)F \tag{A.5-26}$$

PERFORMANCE OF PETLYUK COLUMNS. Glinos and Malone[†] were able to show that the maximum vapor savings were again 50%, independent of the relative volatilities, and that they occur when $x_A = (\alpha_{AB} - 1)/(\alpha_{AC} - 1)$ and $x_B \to 0$, which are the same results as for a sidestream stripper. However, they also found that large savings could be obtained in many cases when x_B was large, especially when the A/B split was not much easier than the B/C split.

Prefractionators

If we add a condenser and a reboiler to the first column in a Petlyuk configuration, i.e., we remove the thermal coupling, we call the result a *prefractionator*; see Fig. A.5-6. We again look for a sharp A/C split in the prefractionator, and we let B distribute between the top and bottom of this column. The minimum vapor rate for the downstream column depends on whether the upper or lower feed is controlling. Since both feeds to the downstream column are essentially binary mixtures and are saturated liquids, it is simple to develop expressions to estimate their values. Hence,

$$V_{6,\text{min}}^4 = \frac{\alpha_{AC}x_{AF}}{\alpha_{AC} - \alpha_{BC}} + \frac{\alpha_{BC}(\alpha_{BC} - 1)x_{BF}}{(\alpha_{AC} - 1)(\alpha_{AC} - \alpha_{BC})} \tag{A.5-27}$$

and

$$V_{6,\text{min}} = \frac{x_{CF}}{\alpha_{BC} - 1} + \frac{\alpha_{BC}(\alpha_{AC} - \alpha_{BC})x_{BF}}{(\alpha_{AC} - 1)(\alpha_{BC} - 1)} \tag{A.5-28}$$

The corresponding result for the prefractionator is given by Eq. A.5-26, and the fractional recovery of B overhead in the prefractionator is given by Eq. A.5-21. The

---

* K. Glinos and M. F. Malone, *I&EC Proc. Des. Dev.*, **24**: 1087 (1985).

[†] K. Glinos and M. F. Malone, "*Optimality Regions for Complex Column Alternatives in Distillation Systems*," paper submitted to *Chem. Eng. Res. Des.*, 1987.

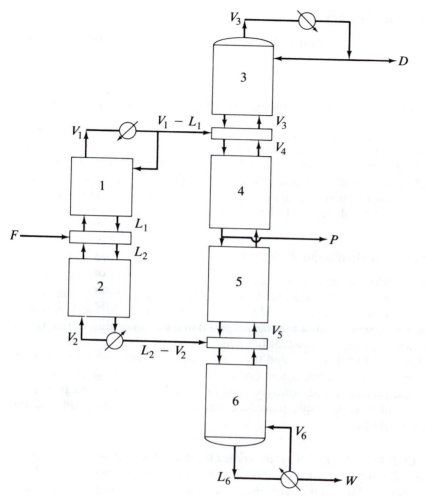

FIGURE A.5-6
Prefractionator. (*From K. Glinos and M. F. Malone, "Optimality Regions for Complex Column Alternatives in Distillation Systems," submitted to Chem. Eng. Res. Dev., 1987.*)

total vapor generated by both reboilers is then

$$V_{\text{min, tot}} = V_{1,\text{min}} + V_{6,\text{min}} \tag{A.5-29}$$

PERFORMANCE OF PREFRACTIONATORS. The maximum possible savings with a prefractionator are never as large as with a Petlyuk column. If the upper feed controls, the maximum fractional savings are $(\alpha_{AC} - \alpha_{BC})/(\alpha_{AC} - 1)$, which occurs when $x_B \to 1$. When the lower feed controls, the maximum fractional savings are $(\alpha_{BC} - 1)/(\alpha_{AC} - 1)$ and are obtained when $x_C \to 0$. Hence, the prefractionator savings may be large for large values of x_B.

Heuristics for Complex Columns

Glinos and Malone* summarized the results of their studies by developing a set of heuristics. These heuristics are presented in Sec. 7.3.

A.6 ENERGY INTEGRATION OF DISTILLATION COLUMNS

Distillation separations consume large amounts of energy. Hence, there is an economic incentive to make the energy consumption as small as possible. The basic idea behind energy integration (see Chap. 8) is to use the heat that must be removed from hot streams which need to be cooled to add heat to streams which need to be heated. Obviously, we would like to achieve as much energy integration as can be economically justified when we design distillation columns.

Heat Effects in Distillation Columns

For most distillation separations the feed enters as a saturated liquid, and the two product streams are also saturated liquids. Expensive separations correspond to relatively close boiling mixtures, and for these cases an energy balance can be used to show that approximately all the heat supplied to the reboiler must be removed in the condenser. Thus, in most distillation separations, heat is supplied to the reboiler at a high temperature, and this same amount of heat is removed from the condenser at a lower temperature. For this reason we say that heat is *degraded* across a temperature range, which is equal to the difference in boiling points. We would like to have a simple procedure for estimating the heat load and this temperature range.

SHORTCUT ESTIMATES OF THE TEMPERATURE RANGE. Shortcut expressions for ΔT and $Q \Delta T$ have been developed by Glinos, Malone, and Douglas[†] and they are presented here. Over narrow temperature ranges we can use the Clausius-Clapeyron equation to relate the vapor pressure to temperature

$$P_i^\circ = P_{i,0} \exp\left[\frac{-\Delta H_i}{R}\left(\frac{1}{T} - \frac{1}{T_0} \right) \right]$$ (A.6-1)

For low-pressure systems, the K value is given by

$$K_i = \frac{P_i^\circ}{P_{\text{tot}}}$$ (A.6-2)

---

[*] K. Glinos and M. F. Malone, "*Optimality Regions for Complex Column Alternatives in Distillation Systems,*" paper submitted to *Chem. Eng. Res. Des.*, 1987.

[†] K. Glinos, M. F. Malone, and J. M. Douglas, *AIChE J.*, **31**: 1039 (1985).

However, using our shortcut procedure for bubble point calculations, Eq. A.2-5, we can also write

$$K_i = \frac{\alpha_i}{\sum \alpha_i x_i} \tag{A.2-5}$$

For ideal, close-boiling mixtures, the heats of vaporization of all the components are about the same, so that we can replace ΔH_i by an average ΔH. Thus, if we combine Eqs. A.6-1, A.6-2, and A.2-5 and we let

$$\sigma_k = \sum \alpha_i x_{i,k} \tag{A.6-3}$$

we can obtain an expression relating the temperatures at two points in a column

$$\frac{1}{T_2} - \frac{1}{T_1} = \frac{R}{\Delta H} \ln \frac{\sigma_2}{\sigma_1} \tag{A.6-4}$$

Using this result, we can relate the temperature of the distillate and the bottoms to the feed temperature

$$T_D = \left(-\frac{R}{\Delta H} \ln \frac{\sigma_F}{\sigma_D} + \frac{1}{T_F} \right)^{-1} \tag{A.6-5}$$

$$T_B = \left(\frac{R}{\Delta H} \ln \frac{\sigma_B}{\sigma_F} + \frac{1}{T_F} \right)^{-1} \tag{A.6-6}$$

Then, by combining these expressions and assuming that $\Delta H \gg RT_F$, we can write an expression for the temperature drop across the column

$$\Delta T = \frac{RT_F^2}{\Delta H} \ln \frac{\sigma_D}{\sigma_B} \tag{A.6-7}$$

HEAT INPUT TO THE COLUMN. The heat supplied to the reboiler and removed by the condenser is merely the heat of vaporization ΔH multiplied by the vapor rate V (note that we are assuming saturated-liquid feeds)

$$Q = \Delta H V \tag{A.6-8}$$

By combining Eqs. A.6-7 and A.6-8 we find that

$$Q \, \Delta T = R V T_F^2 \ln \frac{\sigma_D}{\sigma_B} \tag{A.6-9}$$

Since

$$V = (R + 1)D = (1.2R_m + 1)D \tag{A.6-10}$$

we see that $Q \, \Delta T$ is approximately a constant for a given separation task, i.e., it depends primarily on the compositions.

Multieffect Distillation

Multieffect distillation has been discussed numerous places in the literature,* although it is not widely used. We consider multieffect columns because they both provide a simple example of the energy integration of distillation columns and illustrate the effect of the pressure shifting of columns. As we increase the pressure in a column, we increase both the overhead and bottoms temperatures, so that in many cases we can make energy matches that otherwise would not be possible. Our discussion of multieffect distillation is not complete, and an interested reader should consult the literature. Our goal here is to illustrate some effects encountered in the energy integration of columns.

Suppose we consider the distillation of a binary mixture, but we split the feed roughly in half, raise the pressure of one of the streams, and send each stream to a separate distillation column; see Fig. A.6-1. If our pressure shifting was such that the overhead temperature in the high-pressure column is greater than the reboiler temperature in the low-pressure column, then we can combine the condenser of the high-pressure column with the reboiler of the low-pressure column. Hence, we only need to supply steam to one reboiler in the high-pressure column, and we only need to supply cooling water to the condenser in the low-pressure column; i.e., the heat that must be removed to condense the overhead of the high-pressure column can be used to supply the heat needed in the reboiler of the low-pressure column.

Since only one-half of the feed is supplied to the high-pressure column, the distillate flow from this column will be one-half of the value for the case of a single column. Also, from Eq. A.6-10 the vapor rate will be cut in half, and from Eq. A.6-8 the reboiler heat duty will be cut in half. Thus, we can accomplish the same separation with a multieffect column configuration (see Fig. A.6-1) as we can in a conventional column, but we require only half as much steam and cooling water.

Of course, when we use a multieffect configuration, we must use two separate columns. Hence, multieffect systems will be of interest only when the energy savings are adequate to pay for the higher investment. In addition, however, we must supply the heat to the reboiler of the high-pressure column at a higher temperature than we would need for a single column. The condenser temperature for the multieffect and single columns will be the same, but we degrade the heat required for the multieffect system over a larger temperature range.

To estimate this temperature range, we remember that Eq. A.6-9 indicates that $Q \, \Delta T$ is essentially a constant. Hence, if we cut the heat load in half, the temperature range will double. Our shortcut procedures enable us to estimate all the quantities involved for a particular system.

A LOWER BOUND ON UTILITY CONSUMPTION. As we introduce additional effects in our distillation separation, we continue to decrease the energy requirements, although the temperature level through which the energy is degraded will

---

* C. S. Robinson and E. R. Gilliland, *Elements of Fractional Distillation*, 4th ed., McGraw-Hill, New York, 1950; and C. J. King, *Separation Processes*, 2d ed., McGraw-Hill, New York, 1980.

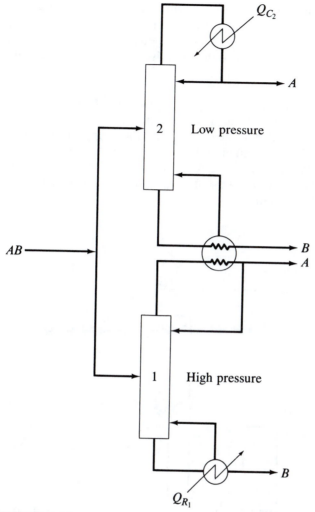

FIGURE A.6-1
Multieffect column. [*From M. J. Andrecovich and A. W. Westerburg, AIChE J.,* **31**: *363 (1985).*]

continue to increase. Andrecovich and Westerberg[*] showed that there was a very simple procedure for estimating a lower bound for the utility consumption. By defining ΔT_{avail} as the difference between the temperatures of the highest hot utility available and the lowest cold utility, the minimum utility required is simply

$$Q_{min} = \frac{Q\,\Delta T}{\Delta T_{avail}} \tag{A.6-11}$$

where $Q\,\Delta T$ is calculated for a single column by using Eq. A.6-9.

---

* M. J. Andrecovich and A. W. Westerberg, *AIChE J.,* **31**: 363 (1985).

Of course, we must allow for a 10°F, or so, temperature driving force between energy-integrated condensers and reboilers, and we can change both Q and ΔT by only integral values as we add effects, so that our bound is not quite correct. However, it does provide an indication of the incentive for undertaking a more detailed analysis.

T-Q **DIAGRAMS.** In the discussion of energy integration in Chap. 8, we used temperature-enthalpy diagrams as an aid in understanding the behavior of the process. These diagrams are also useful when we consider the energy integration of distillation columns. The effects of multieffect distillation and the lower bound on the utility consumption are illustrated in Fig. A.6-2. We see from the diagram that we minimize the utilities consumption if we shift the pressures so that we can stack column sections on top of one another.

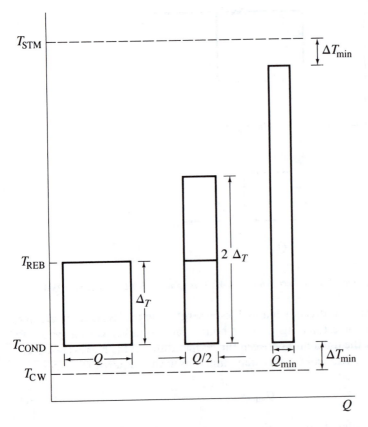

FIGURE A.6-2
Minimizing utilities by using a multieffect configuration. [*From M. J. Andrecovich and A. W. Westerburg, AIChE J.*, **31**: 363 (1985).]

Stand-Alone Column Sequences

Andrecovich and Westerberg showed that this same stacking procedure was useful for evaluating a bound for the utility consumption for column sequences. That is, for a five-component mixture, there are 14 possible sequences. We find the $(Q\,\Delta T)_{tot,i}$ for sequence 1 by adding the values of $Q\,\Delta T$ for each column in that sequence, and we do this calculation for each sequence. Hence, we can find which sequence has the smallest value of

$$(Q\,\Delta T)_{tot,i} = \sum (Q\,\Delta T)_i \tag{A.6-12}$$

Then the minimum utility bound is estimated by using the expression

$$Q_{min} = \frac{(Q\,\Delta T)_{tot,min}}{\Delta T_{avail}} \tag{A.6-13}$$

On a T-Q diagram, we divide the $Q\,\Delta T$ for each separation task into widths equal to Q_{min}, and then we stack the various pieces that we obtain between the utility levels. An illustration of this procedure* is given in Fig. A.6-3. Thus, there is an easy way of estimating the utility bounds.

Of course, as we add effects in a multieffect column, we are increasing the capital cost, even though the energy costs are decreasing. Nevertheless, the pressure shifting to allow the stacking of columns will provide a lower bound on the utility requirements, even if we forbid the use of any multieffect columns. For example, in Fig. A.6-4a we consider three columns. By pressure shifting and stacking the columns, we can reduce the utility consumption. The lowest utility bound then becomes equal to the value of Q for the column with the largest heat load (see Fig. A.6-4b)

$$Q_{min} = Q_{i,max} \tag{A.6-14}$$

We can reduce this load by replacing the column with the largest heat load $Q_{i,max}$ by a multieffect column (see Fig. A.6-4c). Normally, we prefer to pressure-shift as little as possible when we stack the columns.

Integration of Column Sequences with a Process

If heat is available from process streams that must be ejected to a cold utility, we prefer to use this heat rather than a hot utility to supply the energy to a distillation system. In this situation ΔT_{avail} depends on the nature of the composite curves (see Chap. 8). We never want to have a column straddle the process pinch, and our goal is to fit the columns in between the hot and cold composite curves (see Fig. A.6-5) or below the grand composite curve (see Chap. 8).

---

* M. J. Andrecovich and A. W. Westerberg, *AIChE J.*, **31**: 363 (1985).

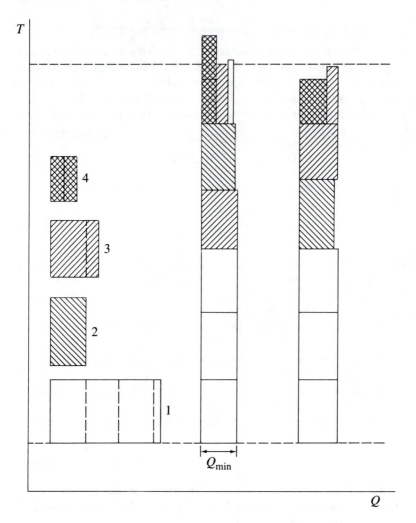

FIGURE A.6-3
Minimum utilities, multieffect configuration for four separations. [*From M. J. Andrecovich and A. W. Westerberg, AIChE J.*, **31**: *363 (1985).*]

Limitations of the Procedure

If we energy-integrate columns, then we will change the optimum reflux ratio because we do not have to pay for utilities. We expect that the optimum reflux ratio will be shifted to higher values, so that the operating reflux ratio and the vapor rate will increase. As the vapor rate is increased, Eq. A.6-8 indicates that the column heat load will increase. Hence, some iteration will be required to find the best design.

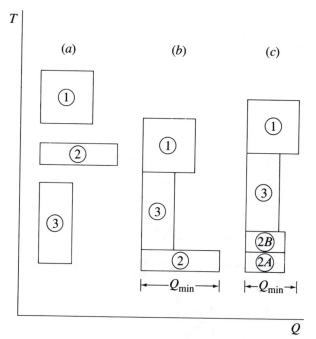

FIGURE A.6-4
Varying utilities: (*a*) Three columns; (*b*) stacked configuration; (*c*) multieffect. [*From M. J. Andrecovich and A. W. Westerberg, AIChE J., **31**: 363 (1985).*]

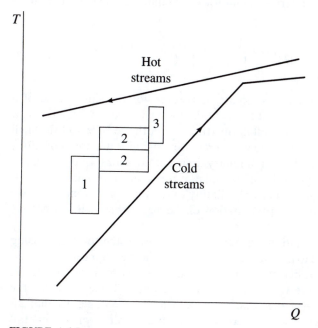

FIGURE A.6-5
T-H diagram and column stacking. [*From M. J. Andrecovich and A. W. Westerberg, AIChE J., **31**: 363 (1985).*]

A.7 HEAT-EXCHANGER DESIGN

Guthrie's correlations (see Appendix E.2) give the cost of a heat exchanger in terms of its area. The heat-transfer area of a heat exchanger is calculated from an equation such as

$$Q = UA\Delta T_m \tag{A.7-1}$$

where Q = heat duty, U = overall heat-transfer coefficient, and ΔT_m = log-mean temperature driving force.

Heat-Transfer Coefficient

The overall heat-transfer coefficient in this expression is related to the individual film coefficients by

$$\frac{1}{U} = \frac{1}{h_i} + \frac{1}{h_0}\frac{A_i}{A_0} \tag{A.7-2}$$

where we neglect the wall resistance and we include the fouling factors in the individual coefficients. Also, the individual coefficients, excluding the fouling factors, normally are given in terms of empirical correlations, such as the Dittus-Boelter equation

$$\frac{h_i D}{k} = 0.023\left(\frac{DG}{\mu}\right)^{0.8}\left(\frac{C_p\mu}{k}\right)^{1/3}\left(\frac{\mu}{\mu_w}\right)^{0.14} \tag{A.7-3}$$

Thus, to obtain an accurate estimate of U, even for the simple case of a double-pipe heat exchanger, we must evaluate the thermal conductivities, the heat capacities, and the viscosities of the fluid mixtures on both the tube and the shell sides of the exchanger. This procedure would require a significant amount of effort, which might not be warranted for preliminary design calculations. In particular, if the heat-exchanger area required for a specific unit turns out to be fairly small, so that the cost of that unit contributes only slightly to the total processing costs, we prefer to obtain only a rough approximation of the size rather than a rigorous design.

To avoid tedious calculations in our first estimates of the total processing costs, it is common practice to base the initial design of a heat exchanger directly on an overall heat-transfer coefficient, such as the values listed in Table A.7-1. An equivalent approach is to use values, such as those given in Table A.7-2, for the individual resistances. Then if the process appears to be profitable, so that an additional design effort can be justified, more accurate values of the overall coefficients can be calculated. In most of our studies we use the values given in Table A.7-1.

TABLE A.7-1
Overall heat-transfer coefficients

Reasonable estimates of overall heat-transfer coefficients (including fouling and wall resistances) to use for preliminary designs

System	U, Btu/(hr $\cdot$ ft^2 $\cdot$ °F)
Condensing vapor to boiling liquid	250
Condensing vapor to flowing liquid	150
Condensing vapor to gas	20
Liquid to liquid	50
Liquid to gas	20
Gas to gas	10
Partial condenser	30

Simplified Models for Condensers (or Feed Vaporizers)

There are design problems where we cool a vapor, condense it, and then subcool the liquid. The normal heat-exchanger design equation, Eq. A.7-1, is not valid for this case. Instead, first we calculate the heat duties for the process stream for the three sections of the exchanger (see Fig. A.7-1)

$$Q_1 = FC_{PV}(T_1 - T_C) \qquad Q_2 = F\,\Delta H_V \qquad Q_3 = FC_{PL}(T_C - T_2) \quad \text{(A.7-4)}$$

From these estimates, we can calculate the total heat duty and the required flow rate of cooling water

$$Q_T = Q_1 + Q_2 + Q_3 = w_C(1)(120 - 90) \qquad \text{(A.7-5)}$$

TABLE A.7-2
Design resistances shell-and-tube heat exchangers

Fluid	No phase change	Boiling liquid	Condensing vapor
Fixed gases	0.045	—	—
Light hydrocarbon gases	0.035	—	0.004
Aromatic liquids	0.007	0.011	0.007
Light hydrocarbon liquids	0.004	0.007	—
Chlorinated hydrocarbons	0.004	0.009	—
Steam	0.045	—	0.001
Boiler water	0.003	0.004	—
Cooling tower water	0.007	—	—

Note: Resistance includes fouling and metal wall allowances. Dimensions of resistances are hr $\cdot$ ft^2 $\cdot$ °F/Btu. From H. F. Rase and M. H. Barrow, "Project Engineering of Process Plants," Wiley, New York, 1957.

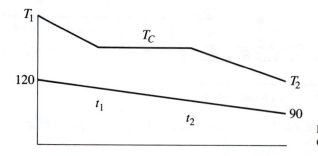

Next we calculate the cooling-water temperatures corresponding to the discontinuities in the condensation profile

$$\frac{w_c(120 - t_1)}{w_c(120 - 90)} = \frac{Q_1}{Q_1 + Q_2 + Q_3} \tag{A.7-6}$$

$$\frac{w_c(t_2 - 90)}{w_c(120 - 90)} = \frac{Q_3}{Q_1 + Q_2 + Q_3} \tag{A.7-7}$$

Once these intermediate temperatures have been evaluated, we can use the normal design equations to find the areas in each of the three sections:

$$Q_1 = U_1 A_1 \frac{T_1 - 120 - (T_c - t_1)}{\ln\left[(T_1 - 120)/(T_c - t_1)\right]} \tag{A.7-8}$$

$$Q_2 = U_2 A_2 \frac{t_1 - t_2}{\ln\left[(T_c - t_2)/(T_c - t_1)\right]} \tag{A.7-9}$$

$$Q_3 = U_3 A_3 \frac{T_c - t_2 - (T_2 - 90)}{\ln\left[(T_c - t_2)/(T_2 - 90)\right]} \tag{A.7-10}$$

Also, we can use the heat-transfer coefficients given in Table A.7-1 or A.7-2 for the three sections of the exchanger.

The problem of designing a feed vaporizer—where we first heat the stream to its boiling point, next vaporize the stream, and then superheat the vapor—is essentially the same, except that it is simpler because the temperature of the heating medium, which usually is steam, is constant throughout the exchanger. For this case we use the fractional heat duties in the three portions of the exchanger to estimate the quantity of steam condensed in the three sections.

Arithmetic Mean Versus Log-Mean Driving Force

Of course, it is simpler to use the arithmetic mean rather than the log-mean temperature as a driving force to calculate the area by using Eq. A.7-1. Thus, we would like to develop an expression that will indicate when the arithmetic mean provides a reasonable approximation of the log-mean driving force.

To develop a criterion of this type, we choose Δt_2 to be the smallest driving force at one end of the exchanger, and we write

$$\frac{\Delta t_1}{\Delta t_2} = 1 + \epsilon \qquad \text{(A.7-11)}$$

The arithmetic mean driving force is simply

$$\Delta t_{av} = \frac{\Delta t_1 + \Delta t_2}{2} = \frac{\Delta t_2}{2}(1 + 1 + \epsilon) = \Delta t_2(1 + \tfrac{1}{2}\epsilon) \qquad \text{(A.7-12)}$$

In contrast, the log-mean expression is

$$\Delta t_{ln} = \frac{\Delta t_1 - \Delta t_2}{\ln(\Delta t_1/\Delta t_2)} = \Delta t_2 \frac{\epsilon}{\ln(1 + \epsilon)} \qquad \text{(A.7-13)}$$

If ϵ is small, we can write

$$\ln(1 + \epsilon) \sim \epsilon - \tfrac{1}{2}\epsilon^2 + \tfrac{1}{3}\epsilon^3 + \cdots = \epsilon(1 - \tfrac{1}{2}\epsilon + \tfrac{1}{3}\epsilon^2 + \cdots) \qquad \text{(A.7-14)}$$

Substituting this result into Eq. A.7-13 and then using synthetic division, or another Taylor series expansion, gives

$$\Delta t_{ln} = \Delta t_2(1 + \tfrac{1}{2}\epsilon - \tfrac{1}{12}\epsilon^2 + \cdots) \qquad \text{(A.7-15)}$$

Now we see that if the ϵ^2 term is very small, the result will become identical to the arithmetic mean given by Eq. A.7-12. Hence, suppose we require that

$$\tfrac{1}{12}\epsilon^2 \leq 0.1(\tfrac{1}{2}\epsilon) \qquad \text{(A.7-16)}$$

or
$$\epsilon \leq 0.6 \qquad \text{(A.7-17)}$$

Then, from Eq. A.7-11 for the arithmetic and geometric mean driving forces to be about the same, we require that

$$\frac{\Delta t_1}{\Delta t_2} \leq 1.6 \qquad \text{(A.7-18)}$$

Multipass Exchangers

Most industrial exchangers include multiple tube passes and/or multiple shell passes. However, for conceptual designs we limit our focus to simple counter-current exchangers. The correction factors for multipass exchangers are given in most texts on heat transfer as well as that by Perry and Chilton.

Furnaces

Guthrie's correlation gives the cost of furnaces in terms of the heat absorbed by the process fluid. Thus, no design procedure is necessary for conceptual designs.

A.8   GAS COMPRESSORS

Guthrie's correlations give the cost of a gas compressor in terms of the brake horsepower bhp (see Appendix E),

$$\text{Installed Cost} = \frac{\text{M\&S}}{280}(517.5)(\text{bhp})^{0.82}(2.11 + F_C) \qquad \text{(A.8-1)}$$

Design Equations

Assuming a compressor efficiency of 0.8 gives

$$\text{bhp} = \frac{\text{hp}}{0.8} \qquad \text{(A.8-2)}$$

The horsepower is given by Eq. 6.5-1:

$$\text{hp} = \left(\frac{3.03 \times 10^{-5}}{\gamma}\right)P_{\text{in}} Q_{\text{in}}\left[\left(\frac{P_{\text{out}}}{P_{\text{in}}}\right)^{\gamma} - 1\right] \qquad \text{(A.8-3)}$$

and the gas exit temperature is

$$\frac{T_{\text{out}}}{T_{\text{in}}} = \left(\frac{P_{\text{out}}}{P_{\text{in}}}\right)^{\gamma}$$

Operating Cost

The operating costs are based on the bhp and a motor efficiency of 0.6.

Multistage Compressors

For multistage compressors, an equal compression ratio is used for each stage; see Eq. 6.5-5.

A.9   DESIGN OF REFRIGERATION SYSTEMS

Even though refrigeration cycles are discussed in numerous thermodynamic textbooks, normally it is not a trivial matter to use the basic ideas to develop a design procedure. Hence, instead of merely describing the basic ideas, we present a fairly simple design case study that illustrates the economic trade-offs encountered. In Chap. 3 we discussed the use of a gas absorber to recover a solvent from a gas stream (i.e., acetone from air), and we noted that a condensation process would be a process alternative. Hence, we choose the recovery of acetone from air as the problem to consider.

Initial Flowsheet and Screening Calculations

We want to recover up to 10.3 mol/hr of acetone from 687 mol/hr of air, where the feed stream is at ambient conditions. If we use a condensation process, our first

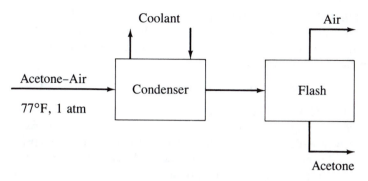

FIGURE A.9-1
Acetone condensation—refrigeration.

sketch of a flowsheet is shown in Fig. A.9-1. If we guess that we want to recover 99.5% of the acetone, so that we can directly compare the cost of our condensation process to the absorption system that we discussed in Chap. 3, then we reduce the mole fraction of the acetone from the inlet value of 0.15 to

$$y_2 = (1 - 0.995)(0.15) = 7.5 \times 10^{-5} \tag{A.9-1}$$

which is quite small. There must be an optimum fractional recovery, but we use this value for our base-case design. We also assume that the acetone leaving as the liquid stream from the phase splitter is pure, despite the fact that a small amount of air is dissolved in this acetone.

ESTIMATING THE TEMPERATURE AND PRESSURE OF THE PHASE SPLITTER. The only temperature specified in the problem statement is that of the feed stream. However, we expect that the vapor and liquid leaving the phase splitter will be in equilibrium, so that from thermodynamics we expect that the partial pressure of acetone in the flash vapor will be equal to its vapor pressure:

$$p_A = P_T y_A = P_A^\circ \tag{A.9-2}$$

ATMOSPHERIC-PRESSURE DESIGN. If we operate the condenser at atmospheric pressure $P_T = 1$ atm, from Eq. A.9-2 we find that

$$P_A^\circ = 7.5 \times 10^{-5} \tag{A.9-3}$$

If we plot the data for the vapor pressure of acetone given in Perry and Chilton's handbook[*] as P° versus $1/T$, where T is in $^\circ R$, we find that the temperature is $-128^\circ F$, which is quite a low temperature.

---

[*] R. H. Perry and C. H. Chilton, *Chemical Engineer's Handbook*, 5th ed., McGraw-Hill, New York, 1973, p. 3-49.

HIGH-PRESSURE DESIGN. An alternate approach for condensing the acetone would be to increase the pressure of the feed stream until the partial pressure of the acetone were equal to the vapor pressure. At 77°F, the vapor pressure of acetone is 0.33 atm, and from Eq. A.9-2 we find that

$$(7.5 \times 10^{-5})P_T = 0.33 \qquad (A.9-4)$$

so that $P_T = 4000$ atm. This is a very high pressure; i.e., most high-pressure processes operate in the range from 300 to 500 psia, although hydrocrackers operate at 1000 psia and low-density polyethylene processes operate at 43,000 psia. Also note that Eq. A.9-2 is no longer valid at this pressure level and that we must include fugacity correction terms above 50 psia, or so.

CHOOSING A BASE-CASE CONDITION. There must be an optimum pressure and temperature for the phase splitter; i.e., as we increase the pressure (which increases the cost of a feed compressor), we can operate the phase splitter at higher temperatures (which decreases the cost of the refrigeration system, which includes a compressor). As a first design, suppose that we operate at atmospheric pressure, which gives us one bound on our design (and places the maximum load on the design of a refrigeration system).

If we choose a 12°F approach temperature (the rule-of-thumb value is 2 or 3°F for low temperatures, but we should evaluate the optimum for a final design), we can now put temperatures on our flowsheet; see Fig. A.9-2. However, we would not want to exhaust air to the atmosphere at this low temperature (it would cause condensation and ice formation). Hence, we install a feed-effluent exchanger, which also reduces the load on the refrigeration system. If we pick an approach temperature of 10°F, we obtain the revised flowsheet shown in Fig. A.9-3. Of course, we would also not send acetone to storage at this low temperature, so that actually we would install another feed-effluent exchanger. However, since the acetone product flow is small compared to the other flows, we neglect it in our first set of calculations.

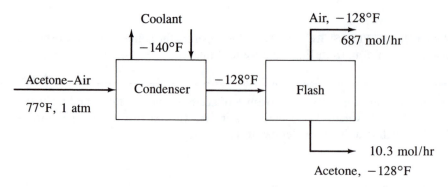

FIGURE A.9-2
Acetone condensation—refrigeration.

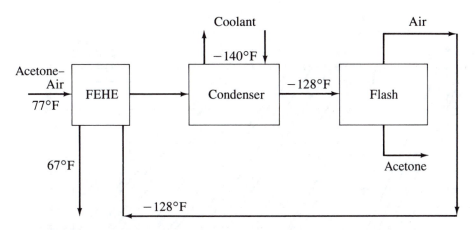

FIGURE A.9-3
Acetone condensation—refrigeration.

Planning a Refrigeration System

If we examine the graph of the pressure versus temperature for various refrigerants shown in fig. 12-30 of Perry and Chilton* (also see Fig. A.9-4), clearly only a few materials can be used at $-140°F$. Letting the lowest operating pressure in the refrigerant process be about 5 psig, so that air does not leak into the system, we see that only ethylene and Freon-14 are acceptable for the low-temperature operation.

We also require that the refrigeration system reject heat to cooling water at $90°F$, which corresponds to a refrigerant temperature of $100°F$ if we allow a $10°F$ approach temperature. Neither ethylene nor Freon-14 will reach this temperature, and therefore we must use a two-stage process; i.e., the ethylene or Freon-14 will reject heat to another refrigerant, such as propane, ammonia, etc., which will then reject the heat to the cooling water.

If we suppose that our acetone recovery system is a part of a petrochemical complex, then propane and ethylene may be readily available, and so we choose these materials for our base-case calculations. To obtain a quick check of whether these fluids are acceptable, we use the Hadden and Grayson correlations given in Appendix C.1. From these correlations, and assuming $10°F$ approach temperatures, we obtain the results given in Table A.9-1, which appear reasonable.

Modified Flowsheet

When we add the equipment for the refrigeration system to the flowsheet shown in Fig. A.9-3, we obtain the result shown in Fig. A.9-5. Note that we have considered

---

* R. H. Perry and C. H. Chilton, *Chemical Engineer's Handbook*, 5th ed., McGraw-Hill, New York, 1973, p. 12-30.

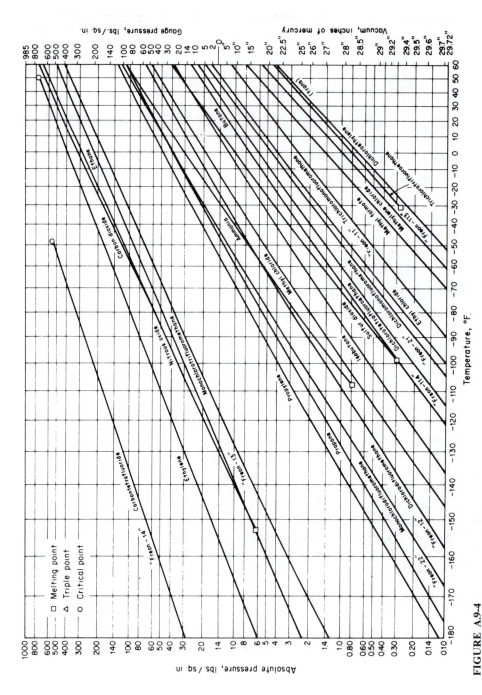

FIGURE A.9-4

Pressure-temperature relationship of refrigerants. (*E. I. du Pont de Nemours & Co., Inc. From R. H. Perry, D. W. Green , and J. O. Maloney, Perry's Chemical Engineers' Handbook, 6th ed, McGraw-Hill, New York, 1984, p. 12-26.*)

TABLE A.9-1
Refrigeration system conditions

Propane	Ethylene
$K = 1$ at 100°F when $P = 190$ psia	$K = 1$ at -32°F when $P = 250$ psia
$K = 1$ at 17 psia when $T = -42$°F	$K = 1$ at 17 psia when $T = -150$°F

the possibility of using aftercoolers after the compressors, since compression causes a temperature rise. These aftercoolers use cooling water supplied by the cooling towers.

Base-Case Design for the Refrigeration Process

In our development of a process flowsheet, we estimated the material flows of the process streams, although not for the refrigeration fluids. Moreover, we selected the

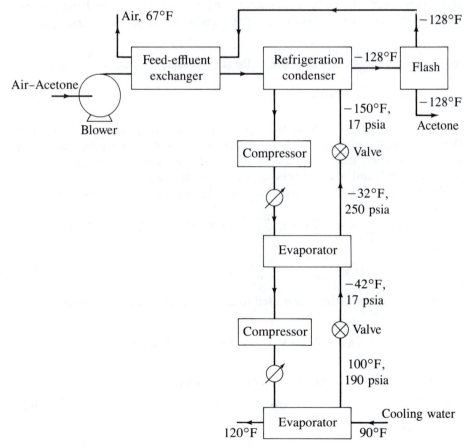

FIGURE A.9-5
Process flowsheet.

temperatures and pressures of most of the streams. Based on this information, we can now use energy balances to calculate the other stream temperatures and flows, and then we can calculate the sizes and costs of the equipment.

Obviously, our results depend on the assumptions we made for the inlet acetone composition, the fractional recovery, the operating pressures, and the approach temperatures. Thus, if our results indicate that refrigeration is an attractive recovery process, we will need to determine the sensitivity of the design to these assumptions. However, we also must consider the absorption and adsorption alternatives, and we develop a final design only for the least expensive process.

Energy Balances

FEED-EFFLUENT EXCHANGER. From an inspection of the flowsheet (Fig. A.9-5) we know the flow rate and the temperature change of the air returning through the feed-effluent heat exchanger. The heat duty of this exchanger is (the heat capacity values were taken from Perry and Chilton, pp. 3–134 and 3–126)

$$Q_{FEHE} = \left(687 \frac{mol}{hr}\right)\left(0.238 \frac{Btu}{lb \cdot {}^\circ F}\right)\left(29 \frac{lb}{mol}\right)[67 - (-128)]$$

$$= 925{,}000 \frac{Btu}{hr} \tag{A.9-5}$$

which is a fairly small heat load. We expect that acetone might start to condense on the feed, rather than the return side of this exchanger, and so we run the feed stream through the shell side. The dew point of the feed mixture can be estimated from Eq. A.9-2 and a vapor-pressure curve for acetone; $P^\circ = P_T y = 0.015$ and $T = -21^\circ F$. Then the heat required to cool the feed to the dew point (DP) is

$$Q_{DP} = \left[\left(600 \frac{lb}{hr}\right)\left(0.506 \frac{Btu}{lb}\right)\right] + 687(0.238)(29)[77 - (-21)]$$

$$= 494{,}000 \frac{Btu}{hr} \tag{A.9-6}$$

Hence, the incremental heat load used to cool the feed stream and condense some of the acetone is

$$Q_{inc} = 925{,}000 - 494{,}000 = 431{,}000 \text{ Btu/hr} \tag{A.9-7}$$

If we let x = moles of acetone condensed, where

$$x \approx 10.3 - 687 y_A \tag{A.9-8}$$

then we can write

$$Q_{inc} = [600(0.506) + 687(0.238)(29)](-21 - t_2) + (243 \text{ Btu/mol})(58)(10.3)$$
$$= 5045(-21 - t_2) \tag{A.9-9}$$

TABLE A.9-2
Incremental heat load

t, °F	$P° = y$	x	$\Delta H_C = $ 14,100x	$\Delta H_S = $ 5045$(-21 - t_2)$	Q_{inc}
-21	0.015	0	0	0	0
-40	0.0072	5.35	75,500	45,400	120,900
-80	0.0019	9.48	133,600	297,600	431,300

Now, if we select temperatures, use a vapor-pressure curve to find the vapor pressure at these temperatures, and apply Eq. A.9-2 to find the mole fraction of acetone and Eq. A.9-8 to find x, we obtain the results in Table A.9-2. From these calculations we see that the effluent of the process stream is about $-80°$F.

Moreover, we note from Table A.9-2 that the latent heat effect is a fairly small fraction of the total heat duty of this exchanger ($133.6/925 = 15\%$), so that we base the log-mean Δt on the inlet and outlet stream temperatures. The design calculations are given in Table A.9-3.

REFRIGERATION CONDENSER. Now that we have estimated the temperature of the process stream entering the refrigeration condenser, we can calculate the heat duty of the refrigeration system:

$$Q_R = [600(0.506) + 687(0.238)(29)][-80 - (-128)] + 243(58)(10.3 - 9.58)$$

$$= 242,200 + 10,200$$

$$= 252,400 \text{ Btu/hr} \tag{A.9-10}$$

Similarly, we see that the latent heat effect is only a small fraction of the total.

Refrigeration Cycles

The simplest procedure for designing the two refrigeration cycles is to use pressure-enthalpy diagrams, since the expansion through the valve takes place at constant enthalpy and the compression is approximately isentropic. We expect the estimates we obtain from these diagrams to be somewhat different from our results using the Hadden and Grayson correlations, so we repeat the analysis. The diagrams we

TABLE A.9-3
Exchanger designs

Unit	Q, 10^6 Btu/hr	Δt_1	Δt_2	Δt_m	U	A, ft^2
Feed-effluent exchanger	0.925	10	48	24.2	20	1909
Refrigeration condenser	0.252	70	22	41.4	20	304.3
Evaporator in ethylene loop	0.408	155	10	52.9	20	386
Evaporator in propane loop	0.611	15	10	12.3	20	2484

used for the calculations were published by Edmister,* but a more recent set published by Reynolds[†] is available.

To estimate the temperatures, pressures, and enthalpies of the various streams in the refrigeration loop (see Fig. A.9-8), we start our analysis at point 1 (see Fig. A.9-8) on the propane diagram (Fig. A.9-6), where the fluid is a saturated liquid at 100°F, and we read P_1 and H_1 (see Table A.9-4). Next we expand the fluid through the valve at constant enthalpy to a pressure of 17 psia to point 2, and we read temperature T_2 from the diagram. Then we evaporate the liquid at constant temperature and pressure until point 3, where it is a saturated vapor, and we find H_3 and S_3. The compression takes place at constant entropy until we reach the original pressure at point 4, and we find T_4 and H_4. If T_4 is much larger than 100°F, we install an aftercooler. By allowing a temperature rise of the cooling water of 10°F and a 10°F approach temperature, the temperature at point 5 will be 110°F, and we can find H_5. Then the vapor is condensed again. The ethylene refrigeration loop is calculated in an identical fashion by starting with a temperature 10°F higher than T_2.

Using this approach, we obtain the values given in Table A.9-4. We neglect the aftercoolers in both loops because temperatures are not excessive. The diagrams are difficult to read, so we might expect to obtain more accurate estimates by using a computer-aided design program.

Material and Energy Balances for the Refrigeration System

REFRIGERATION CONDENSER. The heat duty of the refrigeration condenser is given by Eq. A.9-10, and the enthalpy charge of ethylene is given in Table A.9-4. Hence, we can calculate the circulation rate of ethylene:

$$\text{Ethylene Flow} = w_E = \frac{Q_R}{H_3 - H_2} = \frac{252,400}{86 - (-44)} = 1941 \text{ lb/hr} \quad \text{(A.9-11)}$$

The evaporator duty in the ethylene loop is then

$$Q_{EE} = 1941(H_4 - H_1) = 1941[166 - (-44)] = 407,600 \text{ Btu/hr} \quad \text{(A.9-12)}$$

Similarly, the circulation rate of propane is

$$\text{Propane Flow} = w_p = \frac{Q_{EE}}{H_3 - H_2} = \frac{407,600}{100 - (-2)} = 3996 \text{ lb/hr} \quad \text{(A.9-13)}$$

and the heat duty of the final evaporator is

$$Q_{EP} = 3996(H_4 - H_1) = 3996[151 - (-2)] = 634,000 \text{ Btu/hr} \quad \text{(A.9-14)}$$

---

* W. C. Edmister and B. I. Lee, "Applied Hydrocarbon Thermodynamics, vol. 1," 2d ed., Gulf Publishing Co., Houston, Texas, 1984.

[†] W. C. Reynolds, *Thermodynamic Properties in SI*, Department of Mechanical Engineering, Stanford University, Stanford, Calif., 1979.

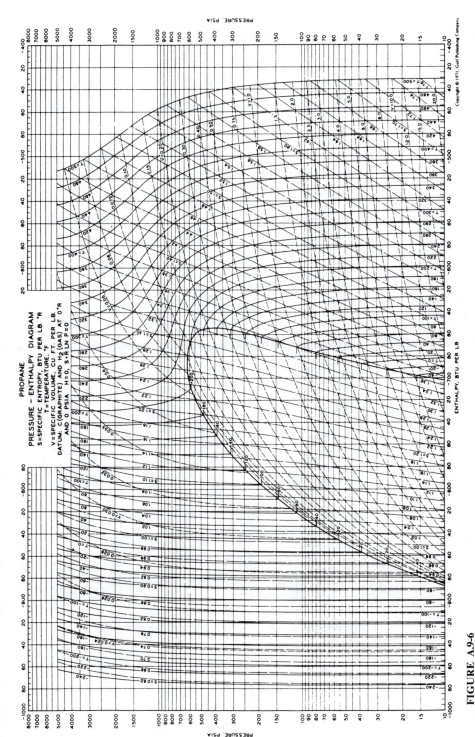

FIGURE A.9-6

Pressure-enthalpy diagram for propane. (*From W. C. Edmister and B. I. Lee, " Applied Hydrocarbon Thermodynamics, vol. 1," 2d ed., Gulf Publishing Co., Houston, Texas, 1984.*)

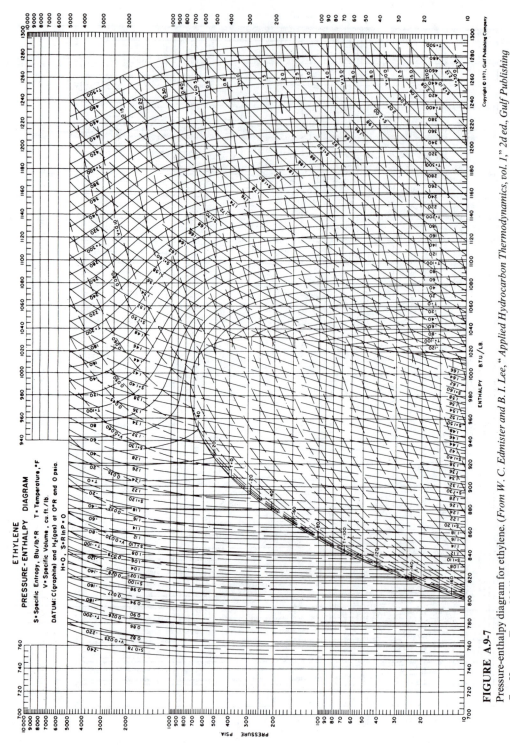

FIGURE A.9-7

Pressure-enthalpy diagram for ethylene. (*From W. C. Edmister and B. I. Lee, "Applied Hydrocarbon Thermodynamics, vol. 1," 2d ed., Gulf Publishing Co., Houston, Texas, 1984.*)

TABLE A.9-4
Refrigeration system conditions

Variable	Point 1	Point 2	Point 3	Point 4
Propane				
P, psia	190	17	17	190
T, °F	100	−40	−40	135
H, Btu/lb	−2	−2	100	151
S			1.37	1.37
Ethylene				
P	250	17	17	180
T	−30	−150	−150	115
H	−44	−44	86	166
S			1.675	1.675

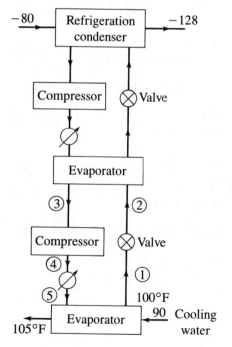

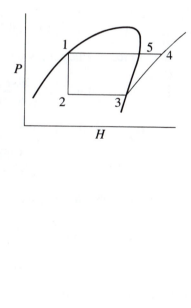

FIGURE A.9-8
Refrigeration system.

Equipment Design

HEAT EXCHANGERS. Now that the heat duties of all the heat exchangers have been estimated and the temperatures of the streams have been fixed, it is a straightforward task to calculate the size and cost of the exchangers. The results are given in Table A.9-5.

The cooling-water flow rate is determined from an energy balance

$$Q_{EP} = 611{,}400 = wC_p(120 - 90) \qquad \text{(A.9-15)}$$

so that

$$w = \frac{611{,}400}{30} = 20{,}400 \text{ lb/hr} \qquad \text{(A.9-16)}$$

If the cost of cooling water is \$0.075/1000 gal and the plant operates 8150 hr/yr, the annual cost is

$$\text{Ann. Cost} = \left(\frac{\$0.075}{1000 \text{ gal}}\right)\left(\frac{1 \text{ gal}}{8.34 \text{ lb}}\right)\left(20{,}400 \frac{\text{lb}}{\text{hr}}\right)\left(8150 \frac{\text{hr}}{\text{yr}}\right)$$

$$= \$1440/\text{yr} \qquad \text{(A.9-17)}$$

COMPRESSORS. The horsepower for the ethylene compressor is given by the expression

$$\text{hp} = \left(H_4 - H_3 \frac{\text{Btu}}{\text{lb}}\right)\left(w_E \frac{\text{lb}}{\text{hr}}\right)\left(\frac{1 \text{ hr}}{3600 \text{ s}}\right)\left(778 \frac{\text{ft} \cdot \text{lb}}{\text{Btu}}\right)\left(\frac{1 \quad \text{hp}}{550 \text{ ft} \cdot \text{lb/s}}\right)\frac{1}{E_c E_m} \qquad \text{(A.9-18)}$$

where E_c = compressor efficiency ~ 0.8 and E_m = motor efficiency ~ 0.9. Thus,

$$\text{hp} = (166 - 86)\left(\frac{1941}{3600}\right)\left(\frac{778}{550}\right)\frac{1}{(0.9)(0.8)} = 84.7 \qquad \text{(A.9-19)}$$

Similarly, the results for the propane compressor are

$$\text{hp} = (151 - 100)\left(\frac{3996}{3600}\right)\left(\frac{778}{550}\right)\frac{1}{(0.9)(0.8)} = 111.2 \qquad \text{(A.9-20)}$$

At a cost of \$0.042/kwhr, the annual power cost for the ethylene loop compressor is

$$\text{Ann. Cost} = \left(\frac{\$0.042}{\text{kwhr}}\right)\left(\frac{1 \text{ kw}}{1.341 \text{ hp}}\right)(84.7 \text{ hp})\left(8160 \frac{\text{hr}}{\text{yr}}\right)$$

$$= \$21{,}700/\text{yr} \qquad \text{(A.9-21)}$$

while that for the propane loop is

$$\text{Ann. Cost} = 21{,}700\left(\frac{111.2}{84.7}\right) = \$28{,}500/\text{yr} \qquad \text{(A.9-22)}$$

BLOWER. We also might need a blower for the feed stream. The air flow rate is $687 \text{ mol/hr} = 687(359 \text{ ft}^3/\text{mol})(537/492)/60 = 4500 \text{ ft}^3/\text{min}$, and Peters and Timmerhaus* give the purchased cost of turboblowers in terms of this volumetric flow rate. If we guess that the pressure drop is 5 psi on both the tube side and the shell side of heat exchangers, then the pressure drop of the air stream is 15 psi, so that we need a 30-psi blower. The cost is

$$\text{Cost} = 55{,}000(4)(\tfrac{870}{280}) = \$683{,}600 \qquad (A.9\text{-}23)$$

where we have assumed an installation factor of 4.0. Also, the power cost, assuming a blower efficiency of 70%, a motor efficiency of 90%, and a power cost of \$0.042/kwhr, is

$$H_p = \left(15\,\frac{\text{lb}}{\text{in.}^2}\right)\left(144\,\frac{\text{in.}^2}{\text{ft}^2}\right)\left(687\,\frac{\text{mol}}{\text{hr}}\right)\left(392\,\frac{\text{ft}^3}{\text{mol}}\right)\left(\frac{1}{3600\text{ s}}\right)\left(\frac{1\text{ hp}\cdot\text{s}}{550\text{ ft}\cdot\text{lb}}\right)\frac{1}{0.7(0.9)}$$

$$= 466.3 \qquad (A.9\text{-}24)$$

$$\text{Ann. Cost} = \left(\frac{\$0.042}{\text{kwhr}}\right)\left(\frac{1\text{ kw}}{1.342\text{ hp}}\right)(466.3\text{ hp})\left(8160\,\frac{\text{hr}}{\text{yr}}\right)$$

$$= \$119{,}200/\text{yr} \qquad (A.9\text{-}25)$$

The capital cost of the blower is very high, and we note from the cost correlation that we could cut this cost by a factor of 10 if we could use a 3-psi turboblower, i.e.,

$$\text{Cost} \sim 5100(4)(\tfrac{870}{280}) = \$63{,}400 \qquad (A.9\text{-}26)$$

Hence, we must estimate the pressure drop of these exchangers fairly carefully and look for ways of minimizing the pressure drop. If we guess that a 3-psi turboblower will be adequate, then the power costs will be

$$\text{Ann. Cost} = 119{,}200(\tfrac{3}{15}) = \$23{,}800/\text{yr} \qquad (A.9\text{-}27)$$

Other Equipment Costs

The capital cost of the heat exchangers is estimated from Guthrie's correlation, Appendix E.2, based on the values given in Table A.9-3. We assume that the installation factor is 3.29, that we use floating-head exchangers, that the pressure correction factor for the evaporators is 1.1, that we use stainless-steel (or low-Ni steel) exchangers at temperatures below $-20°\text{F}$, that the M&S ratio is $\tfrac{870}{280}$, and that the capital charge factor is $\tfrac{1}{3}$. Similarly, we consider motor-driven, reciprocating compressors with an installation factor of 3.11.

---

* M. S. Peters and K. D. Timmerhaus, "Plant Design and Economics for Chemical Engineers," 3d ed., McGraw-Hill, New York, 1980, p. 470.

TABLE A.9-5
Total processing costs

	Design capacity	Purchased cost, $	Installed cost, $	Annualized cost, $/yr	Utility	Utility flow	Annual cost, $/yr
Feed-effluent exchanger, ft^2	1909	14,000	357,800	119,300			
Refrigeration condenser, ft^2	304	4,300	109,900	36,300			
Ethylene evaporator, ft^2	386	4,900	137,700	45,900			
Propane evaporator, ft^2	2577	17,000	191,100	64,700	Water	21,133 lb/hr	1,500
Ethylene compressor, hp	84.7	21,000	203,000	67,700	Electric		21,700
Propane compressor, hp	111.2	25,000	241,600	80,500	Electric		28,500
Blower, 3 psi	4500 ft^3/min	5,100	63,400	21,100	Electric		75,500
			1,304,400	434,800			

The results of the cost calculations are given in Table A.9-5. In addition, the utility costs are listed. We see that roughly an investment of $1,304,000 ($434,800/yr) and an operating cost of $75,500/yr are needed to recover acetone valued at $881,300/yr. If we optimize the process, we might be able to improve the profitability. However, we must also complete the designs for a gas absorber and an adsorption process before we can decide which process might be best.

Shortcut Design Procedures for Refrigeration Loops

From the case study presented above, we see that the design of refrigeration processes is quite tedious. Fortunately, Shelton and Grossmann* have presented a shortcut procedure that significantly simplifies the calculations. From Fig. A.9-9 we see that the heat load for the evaporator (points 1 to 2) can be written as

$$Q_{\text{evap}} = (\Delta H_1^V - \Delta H_{12}^L)F_{RF} \qquad (A.9\text{-}28)$$

where ΔH_1^V is the molar latent heat of vaporization and ΔH_{12}^L is the molar lost heat of vaporization. Since

$$\Delta H_{12}^L = H^L(T_2, P_2) - H^L(T_1, P_1) \qquad (A.9\text{-}29)$$

if we assume that the enthalpy of a saturated liquid is independent of pressure and that the heat capacity of the liquid is constant over the temperature range, then we can write that

$$\Delta H_{12}^L = C_p^L(T_2 - T_1) \qquad (A.9\text{-}30)$$

and Eq. A.9-28 becomes

$$Q_{\text{evap}} = [\Delta H_1^V - C_p^L(T_2 - T_1)]F_{RF} \qquad (A.9\text{-}31)$$

For the isotropic compression, points 2 to 3, we can write that

$$\left(\frac{\partial H}{\partial P}\right)_s = V \qquad (A.9\text{-}32)$$

Instead of attempting to integrate this expression by using an equation of state, Shelton and Grossmann* noted that the curves for constant entropy on a Mollier diagram, where log pressure was plotted against enthalpy, are almost linear. Therefore, writing Eq. A.9-32 in terms of ln P and using the ideal gas law give

$$\left(\frac{\partial H}{\partial \ln P}\right)_s = PV = RT \qquad (A.9\text{-}33)$$

---

* M. R. Shelton and I. E. Grossmann, *Computers and Chem. Eng.*, **9**: 615 (1985).

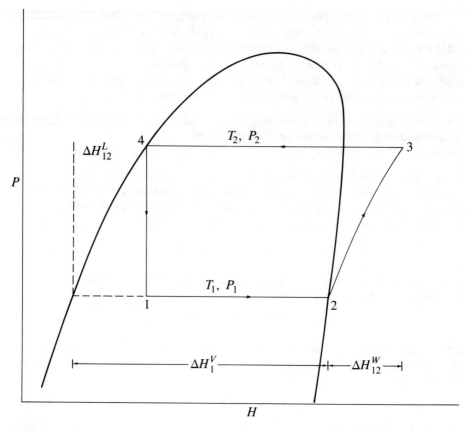

FIGURE A.9-9
Refrigeration process. [*From M. R. Shelton and I. E. Grossmann, Computers and Chem. Eng.*, **9**: 615 (*1985*).]

Evaluating T at T_2 and integrating then give

$$\Delta H_{12}^{w} = RT_2 \ln \frac{P_2}{P_1} \qquad \text{(A.9-34)}$$

We can rewrite this expression in terms of temperatures by introducing the Clausius-Clapeyron equation

$$\ln \frac{P_2}{P_1} = \frac{\Delta H_1^{V}}{R} \left(\frac{1}{T_1} - \frac{1}{T_2} \right) \qquad \text{(A.9-35)}$$

Then the power requirement for the compressor becomes

$$W_{12} = \Delta H_{12}^{w} F_{RF} = \frac{\Delta H_1^{V} (T_2 - T_1) F_{RF}}{T_1} \qquad \text{(A.9-36)}$$

The condenser duty, points 3 to 4, can be written as (see Fig. A.9-9)

$$Q_{\text{cond}} = Q_{\text{evap}} + W_{12} = (\Delta H_1^V - \Delta H_{12}^L + \Delta H_{12}^w)F_{RF} \qquad \text{(A.9-37)}$$

Substituting our previous results, we obtain

$$Q_{\text{cond}} = \left[\Delta H_1^V \frac{T_2}{T_1} - C_p^L(T_2 - T_1)\right]F_{RF} \qquad \text{(A.9-38)}$$

Summary

Thus, we have obtained simple expressions for the quantities of interest:

Evaporator: $\quad Q_{\text{evap}} = [\Delta H_1^V - C_p^L(T_2 - T_1)]F_{RF}$

Power requirement: $\quad W_{12} = \dfrac{\Delta H_1^V (T_2 - T_1)F_{RF}}{T_1}$

Condenser: $\quad Q_{\text{cond}} = \left[\Delta H_1^V \dfrac{T_2}{T_1} - C_p^L(T_2 - T_1)\right]F_{RF}$

A.10   REACTORS

A kinetic model must be used to calculate the required reactor volume. For tubular reactors, the length-to-diameter ratio must exceed 6, but preferably it should be greater than 10. The cost of a tubular reactor is estimated by using Guthrie's cost correlation for pressure vessels.

For continuous stirred tank reactors, we usually set the L/D ratio to about 1.0/1.5. Cost correlations for CSTRs are available in Peters and Timmerhaus.*

A.11   SUMMARY OF SHORTCUT EQUIPMENT-DESIGN GUIDELINES AND NOMENCLATURE FOR APPENDIX A

We have presented a number of design guidelines during our discussions. It is convenient to combine these into a single list which we can refer to when we attack other problems. The list presented below is not complete, so it always pays to ask experienced engineers if they know of any other guidelines that may apply to a particular problem.

Perhaps we should emphasize that it is dangerous to blindly use any published rule-of-thumb or design value, since any one of these generalizations is limited to certain types of applications. Therefore, it is always worth understanding the origin of the rule in order to gain some feeling for the potential limitations.

---

* M. S. Peters and K. D. Timmerhaus, *Plant Design and Economics for Chemical Engineers*, 3d ed., McGraw-Hill, New York, 1980, pp. 790-791.

Moreover, for final design calculations we always prefer to use rigorous calculation procedures because we must be as certain as we can that the final design will correspond to an operable process, i.e., the cost of modifying an existing plant to make it meet design specifications is enormous compared to the cost of making changes at the design stage. However, time pressures and budget constraints sometimes force us to take risks, so good judgment is always required.

I. *Physical Properties*

Procedures for estimating physical properties of materials are described by R. C. Reid, J. M. Prausnitz, and T. K. Sherwood, *The Properties of Gases and Liquids*, 3d ed., McGraw-Hill, New York, 1977. A condensed version of some of these methods is given in R. H. Perry and C. H. Chilton, *Chemical Engineer's Handbook*, 5th ed., McGraw-Hill, New York, 1973, p. 3–226. The *API Data Book* and the *International Critical Tables*, as well as many other handbooks, also contain much valuable information.

It is always better to base designs on actual experimental data rather than estimates obtained from correlations. However, we often use the correlations for preliminary designs when data are not available.

A simple procedure for estimating K values for vapor-liquid equilibria is presented in Sec. C.1 of Appendix C.

II. *Distillation Column Design*

Distillation columns are one of the most common unit operations, and normally the distillation costs are a large fraction of the total processing costs.

A. *Phase Equilibrium*

1. General considerations

    a. Careful attention must be given to highly nonideal mixtures. In particular, it is essential to determine whether azeotropes may be formed. An interesting discussion of nonideal effects has been presented by G. J. Pierotti, C. H. Deal, and E. L. Derr, *Ind. Eng. Chem.*, **51**: 95 (1959).

    b. If the relative volatility is 1.1 or less, extraction or some other operation may be cheaper than distillation.

    c. The presence of heat-sensitive compounds or polymerizable materials may require vacuum operation.

    d. Corrosive materials require more expensive materials of construction.

2. Shortcut design procedures

    a. Most shortcut design procedures, e.g., that of Fenske, Underwood, and Gilliland, require the assumption of constant relative volatility α.

    (1) Constant α is obtained for similar materials.

        (*a*) The heats of vaporization of the light and heavy keys are essentially the same.

        (*b*) The boiling points of the light and heavy keys are close together.

(c) Use ideal liquid mixtures.

(d) The separations are relatively difficult and are expensive.

(2) For similar materials (ideal solutions) where the boiling points are widely separated, heat effects are important.

(a) Base the preliminary design on the Kremser equation for the top and bottom sections.

(b) Use fictitious molecular weights and a McCabe-Thiele diagram.

(c) The separations normally are simple, and the columns are less expensive, so that great accuracy is not required for preliminary designs.

(3) For some nonideal liquid mixtures (widely separated boiling points).

(a) Base preliminary designs on the Kremser equation for the rectifying and stripping sections.

(b) The separations often are easy and the columns relatively inexpensive, so that great accuracy is not required.

B. *Column Material Balances*

1. Normally only one composition, which may be either the top or the bottom, is fixed by product specifications.

2. For first estimates assume between 99.0 and 99.9% recoveries of valuable components for normal separations.

   a. These estimates normally must be modified if azeotropes are present.

   b. Lower estimates often are used for high-pressure columns and columns with partial condensers or refrigerated condensers.

3. Use Fenske's equation at total reflux to estimate the splits of the nonkey components.

C. *Column Design*

1. Estimate the column pressure.

   a. Condense the overhead at 100 to 130°F.

   (1) Use of a total condenser is preferred.

   (2) Make certain that there is adequate reflux if a partial condenser is used.

   (3) Compare bottoms temperature to available steam temperatures.

   (a) We desire a 30 to 45°F Δt or more.

   (b) Readjust pressure if necessary.

2. Estimate minimum reflux rate.

   a. Use one of Underwood's equations if α is constant.

   b. Use McCabe-Thiele diagram for binary mixtures.

3. Let $R = 1.2R_m$ for first estimate.

4. Estimate number of theoretical plates—use sieve trays (or valve trays).

   a. $N \sim 2N_m$, or Gilliland's correlation.

   b. Use Kremser equation or McCabe-Thiele diagram.

5. Estimate the actual number of plates.
   a. Assume an overall plate efficiency of 50%.
      (1) It should be higher for close-boiling, ideal mixtures.
      (2) It should be lower for mixtures with large differences in their boiling points.
   b. Use O'Connell's correlation (although it is not very reliable).
      (1) Use α_{av} if α is approximately constant.
      (2) Use different estimates for the top and bottom sections if the Kremser equation is used for design.
6. Estimate the column height.
   a. Use a 2.0-ft tray spacing (a 2-ft 6-in. tray spacing for manhead locations).
   b. Add 8 to 15 ft at top for vapor disengagement.
   c. Add extra space at bottom as a liquid surge tank.
      (1) Normally use 3- to 5-min holding time based on bottoms flow.
      (2) Use 6-min holding time if bottom product is fed to a furnace.
      (3) Use 8- to 15-ft space for first designs.
   d. If column height is greater than 175 ft or if L/D is greater than about 20 to 30, special foundations are required.
      (1) Redesign with 18-in. (or 12-in.) tray spacing.
      (2) Build the rectifying and stripping sections separately, if necessary.
7. Estimate the column diameter for both the rectifying and stripping sections.
   a. Estimate the vapor rates in the top and bottom in ft^3/s.
   b. Estimate the vapor velocity.
      (1) Guess the velocity.
         (a) For atmospheric operation assume the superficial velocity is 3 ft/s and the pressure drop is 3 in.H_2O per tray.
         (b) For pressures below 100 mmHg use 6 to 8 ft/s. Be careful that the pressure drop is not so high that the bottoms temperature becomes excessive.
         (c) For high-pressure operation, the velocity should be reduced from 3 to 1 ft/s as the pressure increases.
      (2) Use an F factor of $F = 1.5$ for preliminary designs.
      (3) Use Fair's flooding correlation, and set the velocity at 60 to 85% of the flooding value.
   c. About 12% of the area is required for downcomers.
   d. Tower diameters should be specified in 6-in. increments.
   e. Use the largest diameter for the rectifying or stripping sections if they are close to the same or build a swedge tower if they are very different.
   f. Limitations.
      (1) If the tower diameter is large, i.e., greater than 12 to 15 ft, consider increasing the tray spacing to 36 in.

(2) If the tower diameter is less than 18 in., redesign the tower as a packed column.
- (*a*) Use 1-in. pall rings of a suitable metal.
- (*b*) For atmospheric operation assume a superficial vapor velocity of 3 ft/s, a pressure drop of 0.5 in.H$_2$O/ft, and $H_{OG} = 2.2$ for preliminary designs.
- (*c*) For pressures less than 100 mmHg, assume a velocity of 6 to 8 ft/s, a pressure drop of 1.0 in.H$_2$O/ft, and $H_{OG} = 3$ for preliminary designs.

(3) For towers in the range from 1.5 to 4.5 ft, use either a plate or a packed tower.

(4) Other applications for packed towers.
- (*a*) Corrosive materials.
- (*b*) Materials which foam badly.
- (*c*) If a low-pressure drop required.

D. *Sequences of Distillation Columns*

The heuristics may be contradictory.
1. Remove corrosive and hazardous materials early in the sequence.
2. Remove the most volatile component first.
3. If a component is predominant in the feed, remove it first.
4. Make 50/50 splits whenever possible.
5. Make the easiest separation first.
6. Make the cheapest separation first.
7. Remove the components one by one as column overheads.
8. Save difficult separations until last.
9. Prefer to operate as close to ambient conditions as possible, but prefer higher pressures to vaccum operation and higher temperatures to refrigeration conditions.
10. Prefer not to introduce a component absent from the original mixture (absorption or extraction); but if we introduce a foreign species, we desire to recover it as soon as possible.

III. *Heat Exchangers*

Heat exchangers also are very common pieces of process equipment, and the total heat-exchange costs normally are a large fraction of the total processing costs. Some heuristics are given below for approach temperatures, but better results can be obtained by using the energy integration procedure described in Chap 8.

A. *Heat-Transfer Coefficients*

Use the overall heat-transfer coefficients given in Table A.7-1 or A.7-2 or similar values.

B. *Design Conditions*
1. Water coolers.
- *a.* Assume cooling water is available at 90°F (on a hot summer day) and must be returned to the cooling towers at less than 120°F to prevent scaling of exchanger surfaces. Solutions of optimization

problems that estimate the effluent cooling-water temperature normally give values greater than 120°F, because scaling problems are
not considered.

b. Assume an approach temperature between the stream being cooled
and the inlet cooling water of 10°F (for very clean materials) or
20°F.

2. Condensers.
   a. Distillation columns.
      (1) Use a total condenser if possible.
      (2) Pure components condense at constant temperatures in a total
          condenser.
      (3) For mixtures calculate the dew points and bubble points, and
          use these to estimate the log-mean ΔT for a total condenser.
      (4) For partial condensers make certain that there is an adequate
          amount of reflux.
   b. Cooler condensers in processes.
      (1) Use the same approach temperature as for water coolers.
      (2) Cool the vapor to its dew point, condense and cool to its bubble
          point, and then subcool the liquid.
          (a) Base the log-mean ΔT on the inlet and outlet temperatures if
              the latent heat effects are small compared to the sensible
              heat changes.
          (b) Base the log-mean ΔT on the dew point and bubble point if
              the latent heat effects are large compared to the sensible heat
              changes.

3. Reboilers—to prevent film boiling assume $\Delta T = 45°F$ and $U = 250$
   Btu/(hr·ft²·°F), so that $U \Delta T = 11,250$ Btu/ft² (up to 13,000 is
   satisfactory).

4. Waste-heat boilers—use an approach temperature of 60°F for first
   estimates.

5. Air-cooled exchangers and condensers.
   a. Assume $U = 70$ Btu/(hr·ft²·°F) based on nonfinned surface.
   b. Use an approach temperature of 40°F based on the inlet high-
      temperature fluid.
   c. Assume the air temperature is 90°F (on a hot summer day) to ensure
      that the area is adequate.
   d. Assume 20-hp power input per 1000 ft² of base surface area.

6. Countercurrent exchangers.
   a. Assume an approach temperature of 30°F between the entering hot
      stream and the exiting cold stream for multipass exchangers.
      (1) The optimum value will depend on the thermal capacitances of
          the hot and cold streams and the number of tube-and-shell
          passes.
      (2) The minimum value is about 10°F.

   b. Multipass exchangers.
      (1) Use one of the published charts to correct the log-mean ΔT.
      (2) If the correction factor is less than 0.8, redesign the exchanger to obtain a higher correction factor.
   c. If the area is less than 200 ft^2, consider the use of finned tubes.
   7. Feed-effluent exchangers.
      a. Recover about 75% of the available heat for liquid-liquid exchangers.
      b. Recover about 50% of the available heat for gas-gas or gas-liquid exchangers.
      c. The heuristics above often give poor results. Use the energy integration procedure presented in Chap. 8.
C. *Standard Designs*
   1. Tube diameter.
      a. In a given shell, the smaller the tubes, the greater the surface area, but the greater the difficulty keeping the tubes clean.
      b. Use $\frac{3}{4}$-in. tubes for clean materials, 1-in. for most fluids, and 1.5-in. for very dirty fluids.
   2. Tube lengths.
      a. Most chemical companies use 8-ft-long exchangers.
      b. Most petroleum companies use 16-ft-long exchangers, although in some cases 20-ft-long exchangers are used.
      c. For 16-ft-long exchangers, the maximum surface area space shell is
         (1) 3500 ft^2 for $\frac{3}{4}$-in. tubes.
         (2) 3200 ft^2 for 1-in. tubes.
         (3) 2000 ft^2 for 1.5-in. tubes.
   3. Pitch.
      a. Triangular pitch gives smallest shell size for a given area.
      b. Square pitch is much easier to clean.
   4. Pressure drop—assume a 5-psi pressure drop for both the tube side and the shell side of the exchanger for first estimates.
IV. *Furnaces*
   A. Assume that the fluid velocity in the tubes for a nonvaporizing liquid is 5 ft/s.
   B. Assume an 80% efficiency based on the net heating value.
V. *Compressors*
   A. Design equation: $hp = 3.03 \times 10^{-5}$ or $(3.03 \times 10^{-5}/K)$ $P_{in}Q_{in}[(P_{out}/P_{in})^K - 1]/K$ and $T_{out}/T_{in} = (P_{out}/P_{in})^K$, where $hp = $ horsepower, $P_{in} = $ lbf/ft^2, Q is in cfm, and $K = (C_p/C_v - 1)(C_p/C_v)$.
      1. For monatomic gases use $K = 0.4$.
      2. For diatomic gases use $K = 0.29$.
      3. For more complex gases (CO_2, CH_4) use $K = 0.23$.
      4. For other gases use $K = R/C_p$, where R is the gas constant.

B. Efficiency.
1. Assume a compressor efficiency of 90% to account for fluid friction in suction and discharge valves and ports, friction of moving metal surfaces, fluid turbulence, etc.
2. Assume a driver efficiency of 90% to account for the conversion of the input energy into shaft work.

C. Spares.
1. Compressors are very expensive, so spares are seldom provided.
2. A common practice is to provide two compressors with each handling 60% of the load, so that partial operation of the plant can be maintained if one compressor fails.

VI. *Gas Absorbers*
A. Plate columns—similar to distillation.
1. Assume 99% or greater recoveries.
2. Use $L/(mG) = 1.4$.
3. Use the Kremser equation for design for dilute concentrations.
4. If heat effects are large, use sidestream coolers.
5. Assume vapor velocity is 3 ft/s, or estimate it from F factor, $F = 1.5$.
6. Assume 2-ft tray spacing—use sieve trays.
7. Assume pressure drop is 3 in.H_2O per tray.
8. Use O'Connell's correlation for tray efficiencies—the efficiencies in absorbers are lower than in distillation columns because the liquid is not boiling.

B. Packed columns.
1. Applications.
   a. Corrosive materials.
   b. Foaming materials.
   c. If low-pressure drop required.
   d. If tower diameter is less than 2 ft.
2. Design.
   a. Assume 99.9% recovery.
   b. Use $L/(mG) = 1.4$.
   c. Kremser equation gives $N_{OG} = 20$ for 99% recovery.
   d. Assume $H_{OG} = 2$ ft.
   e. Packed height $= 40$ ft.
   f. Assume vapor velocity is 3 ft/s.
   g. Use 1-in. Pall rings in metal or plastic.
   h. Assume pressure drop is 0.5 in.H_2O/ft.

VII. *Other Equipment*
A more complete collection of shortcut design procedures is available in J. Happel and D. G. Jordan, *Chemical Process Economics*, 2d ed., Dekker, New York, 1975, appendix C. Another set is given in W. D. Baasel, *Preliminary Chemical Engineering Plant Design*, Elsevier, New York, 1976, pp. 114, 211. In addition, numerous textbooks on unit operations list values.

An excellent recent collection of shortcut equipment-design procedures has been published by F. Aerstin and G. Street, *Applied Chemical Process Design*, Plenum, New York, 1978. Also the *Manual of Economic Analysis of Chemical Processes*, published by the Institut François du Petrole, McGraw-Hill, 1981, contains much useful information.

NOMECLATURE

a, b, c	Coefficients in Smoker's equation solution
A	$= L/(mG)$, absorption factor
A_C, A_R	Heat-exchanger area of condenser and reboiler
A_i, A_0	Inside and outside areas
A_T	Tower cross-sectional area (ft^2)
A_{12}, A_{21}	Binary interaction parameters in Margules equations
b	Intercept of operating line
B	Bottoms flow (mol/hr)
c	Design parameter
C_D	Drag coefficient
C_p	Heat capacity [Btu/(mol · °F)]
C_S	Cost of steam ($/Klb)
C_W	Cost of cooling water, $/Kgal
D	Distillate flow (mol/hr)
D	Tube diameter
D_T	Tower diameter (ft)
E_0	Overall plate efficiency
F	F factor
F_{RF}	Refrigerant flow (mol/hr)
F_C	Correction factors in cost correlations; see Appendix C
g	Gravitational acceleration
G	Total gas rate (mol/hr)
G	$= \rho v$, mass flow rate
G_S	Flow of carrier gas (mol/hr)
H	Height of a plate column
H_F	Enthalpy of feed (Btu/mol)
H_G	Enthalpy of vapor (Btu/mol)
h_i, h_0	Film heat-transfer coefficients
H_L	Enthalpy of liquid (Btu/mol)
H_0	Disengagement plus sump heights
H_{OG}	Height of a transfer unit (ft)
k	Parameter
k	Thermal conductivity
K_0	Constant

L	Total liquid flow (mol/hr)
L_S	Flow of solvent (mol/hr)
m	Slope of equilibrium line
M	Slope of operating line
M_L	Molecular weight of liquid
M&S	Marshall and Swift index
n	Plate number
N	Total number of theoretical plates
N_{act}	Actual number of plates
N_m	Minimum number of trays
N_R, N_S	Theoretical trays in rectifying and stripping sections
$P°$	Vapor pressure (atm or psia)
P, P_T	Total pressure (atm or psia)
q	Feed quality
Q_C, Q_R	Heat load of condenser and reboiler
Q_i	Heat loads (Btu/hr)
Q_{in}	Volumeric flow rate
R	Gas constant
R	Drop radius
r_i	Fractional recovery
R_m	Minimum reflux ratio
R_S	Term in Smoker's equation solution
S	Reboil ratio
SF	Separation factor
T_b	Bubble point (°F)
T_C	Condensate temperature (°F)
T_L	Inlet liquid temperature (°F)
U, U_c, U_R	Heat-transfer coefficients [Btu/(hr·ft^2·°F)]
v	Gas velocity (ft/s)
v_i	Weight fraction of solute in solvent phase
$\bar{V}$	Vapor flow in stripping section
V, V'	Vapor rate in rectifying and stripping sections (mol/hr)
w_c	Flow of cooling water
w_i	Weight fraction of solute in nonsolvent phase
W_S	Steam flow rate (lb/hr)
x, x_n	Mole fraction of solute in liquid
x_B	Bottoms mole fraction
x_D	Distillate mole fraction
x_F	Feed mole fraction—liquid
X_{in}, X_{out}	Molar ratio of solute in the liquid phase (mol/mol)
x_0, x'_0	Composition variables
x_W	Bottoms mole fraction

y, y_n	Mole fraction of solute in gas
y_F	Feed mole fraction—vapor
y_{in}, y_{out}	Inlet and outlet molar ratio of solute in the gas phase (mol/mol)
Y_i	Molar ratio in the gas phase
z_f	Feed composition

Greek

α	Relative volatility
ΔH_V	Heat of vaporization (Btu/mol)
ΔT_m	Log-mean temperature driving force (°F)
$\Delta H_{\bar{V}}$	Heat of vaporization of steam (Btu/lb)
γ	$= (C_p/C_v - 1)/(C_p/C_v)$
γ	Liquid-phase activity coefficient
μ_F, μ_L	Viscosity of feed or solute (C_p)
ρ_i	Density (lb/ft^3)
ρ_m	Molar density (mol/ft^3)
θ, θ_i	Root of Underwood's equation
ϵ	Small parameter

APPENDIX
B

HDA
CASE
STUDY

The design of a process to produce benzene by the hydroalkylation of toluene has been discussed in numerous chapters in this text. The purpose of this appendix is simply to collect and present a set of sample calculations for this case study. This problem is a modified version of the 1967 AIChE Student Contest problem.*

LEVEL 0: INPUT INFORMATION

The definition of the problem was given in Example 4.1-1. The information of importance is as follows:

Reactions:

$$\text{Toluene} + \text{H}_2 \rightarrow \text{Benzene} + \text{CH}_4 \qquad \text{(B-1)}$$
$$2\,\text{Benzene} \rightleftharpoons \text{Diphenyl} + \text{H}_2 \qquad \text{(B-2)}$$

Reaction conditions: Reactor inlet temperature = 1150°F (to get a high enough rate) and reactor pressure = 500 psia.

Selectivity: Moles being produced per mole of toluene converted = S

---

* See J. J. McKetta, "Encyclopedia of Chemical Processing and Design," vol. 4, Dekker, New York, 1977, p. 182.

TABLE B-1
Selectivity of HDA process

S	0.99	0.985	0.977	0.97	0.93
x	0.5	0.60	0.70	0.75	0.85

From the 1967 AIChE Student Contest Problem. See
J. J. McKetta, "Encyclopedia of Chemical Processing and
Design," vol. 4, Dekker, New York, 1977, p. 182.

The 1967 AIChE Student Contest problem presents the selectivity data given in Table B-1. If we plot the data as $\ln(1 - S)$ versus $\ln(1 - x)$, so that we make the data as sensitive as possible, we obtain the results shown in Fig. B-1.

1. Now, if we fit an equation to the data and rearrange the results, we obtain

$$S = 1 - \frac{0.0036}{(1 - x)^{1.544}} \tag{B-3}$$

Conditions: Gas phase, no catalyst.

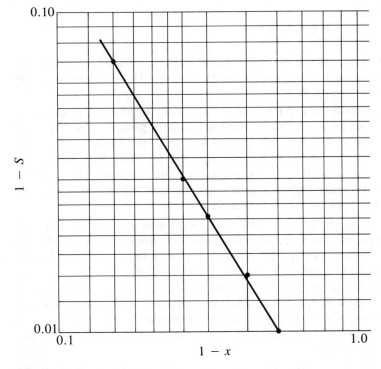

FIGURE B-1
Selectivity of the HDA process.

2. Production rate = 265 mol/hr of benzene.

3. Product purity:

$$x_D \geq 0.9997$$

4. Raw materials: Pure toluene at ambient conditions; 95 % H_2, 5 % CH_4 at 100°F, 550 psia.

5. Constraints: H_2/aromatics = 5 at reactor inlet (to prevent coking); reactor outlet temperature $\leq$ 1300°F (to prevent hydrocracking reactions); quench reactor effluent to 1150°F (to prevent coking).

6. Other plant and site data are given where needed.

LEVEL-1 DECISION: BATCH VERSUS CONTINUOUS

Choose a continuous process.

LEVEL-2 DECISIONS: INPUT-OUTPUT STRUCTURE

See Examples 5.1-2, 5.2-1, and 5.3-1.

1. Purify feed streams: The toluene feed stream is pure; do not purify the hydrogen feed stream because the methane impurity is small. Also, methane is a by-product of the reactions, and the separation of gases is expensive.

2. Reversible by-products: Diphenyl is a by-product formed by a reversible reaction. Thus, we can either recover diphenyl from the process or let it build up to its equilibrium level in a recycle loop. If we remove the diphenyl, we have a selectivity loss of toluene to produce the diphenyl. However, if we recycle the diphenyl, we can avoid the selectivity loss, but we must oversize all of the equipment in the recycle loop to accommodate the recycle flow of diphenyl. We guess that it is cheaper to recover the diphenyl.

3. Recycle and purge: Since a reactant (hydrogen) and both a feed impurity (methane) and a reaction by-product (methane) boil lower than propylene, we will need a gas recycle and a purge stream; i.e., we want to recycle the hydrogen, but the methane will build up in the recycle loop so that it must be purged. A membrane separation process, which could be used to separate the hydrogen and methane, might be less expensive than the loss of hydrogen in a purge stream. Unfortunately, however, no design procedure or cost correlation is available for membrane separators, and so we base our first design on a gas recycle and purge stream.

4. Excess reactants: Since neither O_2 from air nor H_2O is a reactant there are no excess reactants.

TABLE B-2
Component destinations

Component	NBP,°F	Destination
H_2	-423	Gas recycle and purge
CH_4	-259	Gas recycle and purge
Benzene	776.2	Primary product
Toluene	231.1	Recycle
Diphenyl	496.4	By-product

5. Number of product streams: The component boiling points and destinations are given in Table B-2. The H_2 and CH_4 are lumped as one product stream, and the benzene and diphenyl are two additional product streams, Thus, the flowsheet is given in Fig. 5.1-2.

6. Material balances and stream costs: Let

$$P_B = 265 \text{ mol/hr benzene} \qquad x = 0.75 \qquad y_{PH} = 0.4 \qquad \text{(B-4)}$$

$$S = 1 - \frac{0.0036}{(1-x)^{1.544}} = 0.9694 \qquad \text{(B-3)}$$

Fresh feed toluene:

$$F_F = \frac{P_B}{S} = \frac{265}{0.9694} = 273.4 \text{ mol/hr} \qquad \text{(B-5)}$$

Diphenyl produced:

$$P_D = P_B(1-S)/2S \qquad \text{(B-6)}$$
$$= 265(1 - 0.9694)/2(0.9694) = 4.18$$

Extents:

$$\text{Extent 1} = 273.4 \qquad \text{Extent 2} = 4.18 \qquad \text{(B-7)}$$

Makeup gas:

H_2:
$$y_{FH}F_G - \frac{P_B}{S} + \frac{P_B(1-S)}{2S} = y_{PH}P_G \qquad \text{(B-8)}$$

CH_4:
$$(1 - y_{FH})F_G + \frac{P_B}{S} = (1 - y_{PH})P_G \qquad \text{(B-9)}$$

Add these expressions to obtain

$$P_G = F_G + P_B \frac{1-S}{2S} \qquad \text{(B-10)}$$

Combine Eqs. B-8 and B-10:

$$F_G = \frac{P_B[1 - (1 - y_{PH})(1 - S)/2]}{S(y_{FH} - y_{PH})} \tag{B-11}$$

$$= \frac{265[1 - (1 - 0.4)(1 - 0.9694)/2]}{0.9694(0.95 - 0.4)} = 492.5$$

$$P_G = 492.5 + \frac{265(1 - 0.9694)}{2(0.9694)} = 496.7 \tag{B-12}$$

Excess H_2:

$$F_E = y_{PH}F_G = 0.4(492.5) = 197.0$$

$$H_2 \text{ Reacted} = \frac{P_B}{S} - \frac{P_B}{2S}(1 - S) \tag{B-13}$$

$$= 273.4 - 4.18 = 269.2$$

Stream costs:

$$\text{Value of Benzene} = \$0.85/\text{gal} = \$9.04/\text{mol} = C_B \tag{B-14}$$

$$\text{Toluene} = \$0.50/\text{gal} = \$6.40/\text{mol} = C_T \tag{B-15}$$

(assuming a captive internal price)

$$H_2 \text{ Feed} = \$1.32/\text{mol} = C_{H_2} \tag{B-16}$$

Heat values:

$$H_2 = 0.123 \times 10^6 \text{ Btu/mol} = \Delta H_{C,H} \tag{B-17}$$

$$CH_4 = 0.383 \times 10^6 \text{ Btu/mol} = \Delta H_{C,M} \tag{B-18}$$

$$\text{Benzene} = 1.41 \times 10^6 \text{ Btu/mol} = \Delta H_{C,B} \tag{B-19}$$

$$\text{Toluene} = 1.68 \times 10^6 \text{ Btu/mol} = \Delta H_{C,T} \tag{B-20}$$

$$\text{Diphenyl} = 2.69 \times 10^6 \text{ Btu/mol} = \Delta H_{C,D} \tag{B-21}$$

(We assume that the fuel value of diphenyl is $5.38/mol.)
Economic potential:

$$EP = C_B P_B - C_T F_F + C_F\{\Delta H_{C,D}P_D + [\Delta H_{C,H}\, y_{PH} + \Delta H_{CM}(1 - y_{PH})]P_G\} \tag{B-22}$$

We can use this relationship and the expressions above to prepare Fig. 5.3-1.

Level-2 Alternatives

We made several decisions concerning the structure of the flowsheet, and if we had made different decisions, we would have changed the flowsheet. If we list the process alternatives as we develop a design, it is easier to review these decisions after we complete a base-case design. The alternatives at level 2 are as follows.

1. Remove the CH_4 from the H_2 feed stream—This is probably not desirable because we produce CH_4 as a by-product anyway.
2. Recycle the diphenyl rather than removing it—With this approach we avoid any selectivity loss of toluene to diphenyl, but we must oversize all the equipment in the diphenyl-recycle loop to accommodate the equilibrium flow of diphenyl.
3. Recover some H_2 from the purge stream — We must determine whether the H_2 recovery is justified by determining the cost of the recovery system.

LEVEL-3 DECISIONS: RECYCLE STRUCTURE OF THE FLOWSHEET

Design Decisions

The design decisions for the recycle structure were discussed in Example 6.1-2. These are briefly reviewed now:

1. Only one reactor is required (the reactions take place at the same temperature and pressure).
2. There are two recycle streams—a gas recyle (and purge) of $H_2 + CH_4$ and a liquid toluene-recycle stream.
3. We must use a 5/1 hydrogen-to-aromatics ratio at the reactor inlet according to the problem statement.
4. A gas-recycle compressor is required.

Before we can decide on the reactor heat effects, we must calculate the recycle material balances. The recycle flowsheet is shown in Fig. 6.2-1.

Recycle Material Balances

The recycle material balances were developed as Eqs. 6.2-2 and 6.2-10.
Toluene to reactor:

$$F_T = \frac{F_{FT}}{x} = \frac{P_B}{Sx} \tag{B-23}$$

$$= \frac{265}{0.9694(0.75)} = 365$$

Recycle gas:

$$R_G = \frac{P_B}{Sy_{PH}}\left(\frac{M}{x} - \frac{y_{FH}}{y_{FH} - y_{PH}}\right) \tag{B-24}$$

$$= \frac{265}{0.9694(0.4)}\left(\frac{5}{0.75} - \frac{0.95}{0.95 - 0.4}\right) = 3376$$

Reactor Heat Effects

The reactor heat effects were discussed in Examples 6.3-1 and 6.3-3. The reactor heat load is

$$Q_R = \Delta H_R F_{FT} = \Delta H_R P_B / S$$

$$= -21,530\left(\frac{265}{0.9694}\right) = 5.836 \times 10^6 \text{ Btu/hr} \tag{B-25}$$

and the adiabatic exit temperature is given by

$$Q_R = \Delta H_R \frac{P_B}{S}$$

$$= \{7.16F_G + [7y_{PH} + 10.1(1 - y_{PH})]R_G + 48.7F_T\}(T_{R,\text{out}} - 1150)$$

$$T_{R,\text{out}} = 1150 + \frac{21,530P_B/S}{7.16F_G + (10.1 - 3.1y_{PH})R_G + 48.7P_B/Sx}$$

$$= 1150 + \frac{21,530(265/0.9694)}{7.16(492.5) + [10.1 - 3.1(0.4)]3376 + 48.7(265)/[0.9694(0.75)]}$$

$$= 1265 \tag{B-26}$$

These results are shown in Fig. 6.3-1, and we find that we expect that an adiabatic reactor will be acceptable.

Recycle Compressor Costs

The reactor pressure is given as 500 psia. We guess that the pressure at the phase splitter is 465 psia and that the recycle compressor must increase the pressure to 555 psia. This allows a pressure drop of 90 psi though the gas-recycle loop. We need to assess the sensitivity of our design to changes in this guess.

The design equation for a gas compressor is

$$\text{hp} = \left(\frac{3.03 \times 10^{-5}}{\gamma}\right)P_{\text{in}}Q_{\text{in}}\left[\left(\frac{P_{\text{out}}}{P_{\text{in}}}\right)^{\gamma} - 1\right] \tag{B-27}$$

Using Table 6.5-1, we can write

$$\gamma = 0.29y_{PH} + 0.23(1 - y_{PH}) = 0.254 \tag{B-28}$$

Also,
$$\left(\frac{P_{out}}{P_{in}}\right)^{\gamma} = \left(\frac{555}{465}\right)^{0.254} = 1.046 \left(= \frac{T_2}{T_1}\right) \tag{B-29}$$

and for our first design we assume that this value is constant. The gas density at $100°F$ and 465 psia, given an ideal gas, is

$$\rho_m = \left(\frac{1}{359} \frac{mol}{ft^3}\right)\left(\frac{460 + 32}{460 + 100}\right)\frac{465}{14.7} = 0.0774 \text{ mol/ft}^3 \tag{B-30}$$

and the inlet pressure is

$$P_{in} = 465 \text{ lb/in.}^2 \ (144 \text{ in.}^2/ft^2) = 6.70 \times 10^4 \text{ lb/ft}^2 \tag{B-31}$$

The volumetric flow rate is

$$Q_{in} = \frac{R_G(\text{mol/hr})}{(\rho_m \text{ mol/ft}^2)(60 \text{ min/hr})} = \frac{R_G}{60\rho_m} \tag{B-32}$$

Then since

$$R_G = \frac{273}{0.4}\left(\frac{5}{0.75} - \frac{0.95}{0.95 - 0.41}\right) = 3371 \text{ mol/hr} \tag{B-33}$$

$$Q_{in} = \frac{3371}{60(0.0774)} = 727 \text{ ft}^3/\text{min} \tag{B-34}$$

Also,
$$hp = \frac{3.03 \times 10^{-5}}{0.254} (6.70 \times 10^4)(727)(1.046 - 1) = 267 \tag{B-35}$$

and the brake horsepower is

$$bhp = \frac{267}{0.8} = 334 \tag{B-36}$$

Guthrie's correlation (Appendix E.1) gives

$$\text{Install. Cost} = \frac{M\&S}{280} (517.5)(bhp)^{0.82}(2.11 + F_c) \tag{B-37}$$

If $M\&S = 792$ and $F_c = 1.0$ (a centrifugal compressor), we find that

$$\text{Compressor Cost} = \left(\frac{792}{280}\right)(517.5)(334)^{0.82}(3.11) = \$533,400/\text{yr} \tag{B-38}$$

For optimization calculations, we can write

$$\text{Comp. Cost} = \left(\frac{R_G}{3371}\right)^{0.82}\left[\frac{(555/465)^{\gamma} - 1}{1.046}\right]^{0.82} \tag{B-39}$$

The operating cost is based on a motor efficiency (compressor plus motor) of 0.8 and a power cost of $0.045/kwhr:

$$\text{Ann. Cost} = \left(\frac{334 \text{ hp}}{0.8}\right)\left(\frac{1 \text{ kw}}{1.341 \text{ hp}}\right)\left(\frac{0.045}{\text{kw hr}}\right)\left(8150 \frac{\text{hr}}{\text{yr}}\right) = \$114{,}000/\text{yr} \quad \text{(B-40)}$$

We can also write this expression as

$$\text{Power Cost} = 114{,}000\left(\frac{R_G}{3371}\right) = \$33.81 R_G/\text{yr} \quad \text{(B-41)}$$

Reactor Cost

The kinetics for the primary reaction have been discussed by Silsby et al.[*] and by Zimmerman and York.[†] Since the amount of diphenyl produced is very small, we base the reactor design only on the primary reaction. (However, a kinetic model for the by-product reaction is available in Hougen and Watson.[‡]) So

$$r = -k(T)(H)^{1/2} \quad \text{(B-42)}$$

$$\text{where} \quad k = 6.3 \times 10^{10} \ (\text{g} \cdot \text{mol/L})^{-1/2}(\text{s}^{-1}) \exp\left[\frac{-52{,}000 \text{ cal}/(\text{g} \cdot \text{mol})}{R T_R}\right] \quad \text{(B-43)}$$

Since there is a large excess of hydrogen, we assume that $H^{1/2}$ is a constant. Also, we assume that we can estimate the reactor volume based on isothermal operation, although we will base the isothermal reactor temperature on a mean value between the reactor inlet and outlet temperatures $T_R = (1150 + 1265)/2 = 1208$. Thus, we write the reactor volume V_R as

$$V_R = \frac{F \ln [1/(1-x)]}{k\rho_m} \quad \text{(B-44)}$$

where the molar density at the reactor conditions (assuming an ideal gas) is

$$\rho_m = \left(\frac{1}{359} \frac{\text{mol}}{\text{ft}^3}\right)\left(\frac{460 + 32}{460 + T_R}\right)\left(\frac{500}{14.7}\right) = 0.0279 \text{ mol/ft}^3 \quad \text{(B-45)}$$

---

[*] R. I. Silsby and E. W. Sawyer; *J. Appl. Chem.*, **6**: 347 (August 1956); W. D. Betts, F. Popper, and R. I. Silsby, *J. Appl. Chem.*, **7**: 497 (September 1957).

[†] C. C. Zimmerman and R. York, *I&EC Proc. Des. Dev.*, **3**: 254 (July 1964).

[‡] O. A. Hougen and K. W. Watson, *Chemical Process Principles: Part III, Kinetics and Catalysis*, Wiley, New York, 1947, p. 875.

where $T_R = (1150 + 1265)/2 = 1207$. From Eqs. B-42 and B-43

$$kH^{1/2} = 6.3 \times 10^{10} \left(\frac{\text{g} \cdot \text{mol}}{\text{L}}\right)^{-1/2} \left[\left(\frac{1 \text{ lb} \cdot \text{mol}}{454 \text{ g} \cdot \text{mol}}\right)\left(\frac{28.32 \text{ L}}{\text{ft}^3}\right)\right]^{-1/2}$$

$$\times \left[0.4(0.0279) \frac{\text{mol}}{\text{ft}^3}\right]^{1/2} \exp\left[\frac{-52,000(1.8)}{1.987(1685)}\right] \qquad \text{(B-46)}$$

$$= 2.522 \times 10^{11} [0.4(0.0279)^{0.5}] \exp\left[\frac{-93,600}{1.987(1667)}\right] = 0.0142 \text{ s}^{-1}$$

Then, from Eq. B-44,

$$V_R = \frac{(492.5 + 2048 + 364)(\text{mol/hr}) \ln [1/(1 - 0.75)]}{(0.0142 \text{ s}^{-1})(3600 \text{ s/hr})(0.0279 \text{ mol/ft}^3)} = 4090 \text{ ft}^3 \qquad \text{(B-47)}$$

If we assume a cylindrical reactor with $L_R/D_R = 6$, then

$$V_R = \frac{(\pi D_R^2 L_R)}{4} = \frac{\pi}{4}(6D_R^3) \qquad \text{(B-48)}$$

and thus, $\qquad\qquad D_R = 9.52 \text{ ft} \qquad L_R = 56.96 \text{ ft} \qquad\qquad \text{(B-49)}$

The inlet and outlet temperatures for the reactor are $1150°F$ and $1265°F$, which is very high, and therefore we will add 6 in. of insulation on the inside of the reactor to try to keep the inside wall temperature below $900°F$. The addition of this insulation requires that we add another foot to the diameter of our reactor shell, and to be somewhat conservative we let

$$D_R = 10 \text{ ft} \qquad L_R = 60 \text{ ft} \qquad\qquad \text{(B-50)}$$

Then, from Guthrie's correlation for pressure vessels,

$$\text{Ann. Cost} = \left(\frac{792}{282}\right)(101.9)(10^{1.066})(60^{0.82}) \frac{2.18 + 3.67 + 1.6}{3}$$

$$= \$717,900/\text{yr} \qquad\qquad \text{(B-51)}$$

We neglect the cost of the insulation in our first estimate of the reactor cost, hoping that the additional cost associated with oversizing the reactor will compensate for the insulation.

Economic Potential

By subtracting the annualized reactor cost and both the annualized capital and the operating cost of the compressor from the stream costs (i.e., the level-2 economic potential, Eq. B-22), we can calculate the economic potential at level 3 as a function of the design variables (conversion and purge composition); see Fig. 6.7-1. We note that there is an optimum value for both conversion and purge composition. Of course, these optima are not the true optima because we have not considered the separation costs or the heat exchanger costs yet. However, we do note that the range of the design variables where profitable operation is possible has been significantly reduced. Since the process is still profitable, we continue to the next level.

LEVEL 4: SEPARATION SYSTEM

Now we add the details of the separation system. To determine the general structure of the separation system, we must decide which of the flowsheets shown in Figs. 7.1-2 through 7.1-4 is applicable. The reactor effluent is at 1265°F and therefore is all vapor. Then, providing we obtain a phase split at 100°F and 465 psia, the flowsheet given in Fig. 7.1-4 will be correct.

We can use either the Hadden and Grayson method given in Sec. C.1 or a FLOWTRAN program to find the K_i values for the flash drum. The flash calculations were discussed in Sec. 7.1, and as shortcut expressions we use these:
Light components, $K_i > 10$:

$$l_i = \frac{f_i \sum f_j}{K_i \sum f_i} \qquad v_i = f_i\left(1 - \frac{\sum f_j}{K_i \sum f_i}\right) \qquad \text{(B-52)}$$

Heavy components, $K_j < 0.1$:

$$l_j = f_j\left(1 - \frac{K_j \sum f_i}{\sum f_j}\right) \qquad v_j = \frac{K_j f_j \sum f_i}{\sum f_j} \qquad \text{(B-53)}$$

With these expressions, we calculate the vapor and liquid flows leaving the flash drum (see Table B-3)

$$\sum f_i = 1549 + 2323 = 3872 \qquad \sum f_j = 265 + 91 + 4 = 360 \qquad \text{(B-54)}$$

From these results, we see that we obtain a reasonable phase split. However, a significant amount of benzene leaves with the flash vapor. Some of this benzene will be lost in the purge stream, and the remainder will be recycled with the gas stream. From the reactions in Eqs. B-1 and B-2, we would expect that some of this recycled benzene will be converted to diphenyl. However, our selectivity correlation, Eq. B-3, does not indicate that there will be any loss.

We suspect that the data given in Table B-1 were for a pure toluene feed, so that no loss would be apparent. Hence, from the information available it is not possible to estimate the benzene loss, and we should ask the chemist or the pilot

TABLE B-3
Flash calculations

Component	f_i or f_j	K_i or K_j	v	l
H_2	1549	99.07	1547	2
CH_4	2323	20.00	2312	11
Benzene	265	0.0104	29.6	235.4
Toluene	91	0.00363	3.6	87.4
Diphenyl	4	0.000008	0	4

plant group to run some additional experiments (which is why we want to undertake the conceptual design study very early in the life of a project). The value of the benzene in the flash vapor stream, assuming that all of it is lost, is

$$\text{Potential Benzene Loss} = (29.6 \text{ mol/hr})(\$9.04/\text{mol})(8150 \text{ hr/yr})$$

$$= \$2.18 \times 10^6/\text{yr} \tag{B-55}$$

which is very large. Thus, there is a great incentive for undertaking the experiments.

As an alternative, we could attempt to increase the pressure of the flash drum. Since $P_T y_i = \gamma_i P_i^\circ x_i$, we often can obtain a reasonable estimate of the effect of pressure by using the expression

$$P_T K_i = \text{Constant} \tag{B-56}$$

so that increasing the pressure will decrease the K value for benzene and will decrease the amount of benzene in the flash vapor. Since the number of moles is conserved in the reactions, Eqs. B-1 and B-2, pressure does not appear to affect the reaction rate or the equilibrium. However, the possibility of coke formation or hydrocracking reactions (both of which were mentioned as the causes of some constraints in the problem formulation) again makes it essential to undertake additional experiments before we undertake a design at a higher pressure.

Rather than get bogged down in details about the benzene-recycle loss or pressure effects, we go ahead and complete a first design. In particular, if the process is not profitable in any event, so that we decide not to proceed to a final design, then we do not want to waste money on experiments developing a data base for a plant that we are not going to build.

LEVEL 4a: VAPOR RECOVERY SYSTEM

If we are going to include a vapor recovery system, we must decide on the location (flash vapor, purge stream, or gas recycle) and the type of recovery system (absorption, condensation, adsorption, or membrane system). In the discussion above, we estimated the value of the total amount of benzene leaving in the flash vapor. If it should be acceptable to recycle benzene, we still must estimate the benzene and toluene loss in the purge stream, to see whether we should place a recovery system on this purge stream. The flash vapor flows of benzene and toluene are given in Table B-3, the purge flow is given by Eq. B-10, and the gas-recycle flow is given by Eq. B-33. Thus, the fraction of the flash vapor that leaves as purge is

$$\text{Fraction Purged} = \frac{P_G}{P_G + R_G} = \frac{496}{496 + 3376} = 0.128 \tag{B-57}$$

The benzene and toluene losses are then

$$\text{Benzene in Purge} = 0.128(29.6) = 3.79 \text{ mol/hr}$$

$$\text{Toluene in Purge} = 0.128(3.6) = 0.461 \tag{B-58}$$

and the values of these streams are (see Eq. B-54)

Benzene Loss = ($9.04/mol)(3.79 mol/hr)(8150 hr/yr) = $279,200/yr

Toluene Loss = ($6.40/mol)(0.461 mol/hr)(8150 hr/yr) = $24,000/yr

$$(B-59)$$

Now we must guess whether the addition of a benzene and toluene recovery system on the purge stream can be justified. However, if the costs of the liquid separation system and the heat-exchanger network are sufficiently large that the process loses money (and we decide not to build it), we do not want to spend much effort on the design of a vapor recovery system. Thus, tentatively, we accept this purge loss, although we might return to this problem later. (For example, we could use the toluene feed as the solvent in an absorber.)

LEVEL 4b: LIQUID SEPARATION SYSTEM

From the results of the flash calculation (see Table B-2), we have an estimate of the amount of H_2 and CH_4 dissolved in the flash liquid. If we assume that we recover all the benzene, then the feed to the liquid separation system is

Feed: $H_2 = 2$ $CH_4 = 11$ Benzene = 235.4   Toulene = 87.4
       Diphenyl = 4 mol/hr $$(B-60)$$

Level-4b Decisions

The decisions we must make concerning the liquid separation system include

1. How should the light ends be removed if they might contaminate the product?
2. What is the best destination of the light ends?
3. Do we recycle components that form azeotropes with the reactants?
4. What separations can be made by distillation?
5. What sequence of columns should we use?
6. How should we accomplish separations if distillation is not feasible?

We discuss most of these decisions below. However, no azeotropes are formed with the reactants and distillation separations are easy for all the components, so that items 2 and 6 are not considered.

LIGHT ENDS. If we recovered all the benzene as well as the H_2 and CH_4 overhead in a product column, the product purity would be

$$x_D = \frac{235.4}{2 + 11 + 235.4} = 0.948 \qquad (B-61)$$

which is well below our product purity requirement of $x_D = 0.9997$. Of course, some toluene must leave in this stream.

If we attempt to drop the pressure to 50 psia and flash off the H_2 and CH_4, we can obtain a rough estimate of the K values by using Eq. B-56 (it would be better to

TABLE B-4
Low-pressure flash

Component	f_i or f_j	K_i or K_j	v	l
H_2	2	921	1.95	0.05
CH_4	11	186	9.51	1.49
Benzene	235.4	0.0906	0.85	234.55
Toluene	87.4	0.0337	0.11	82.29
Diphenyl	4	0.000074	0	0

use the Hadden and Grayson correlations given in Sec. C.1 for such a large pressure change). Thus, we obtain the K values and the component flows from Table B-4:

$$\sum f_i = 2 + 11 = 13 \qquad \sum f_j = 235.4 + 87.4 + 4.0 = 326.8 \qquad \text{(B-62)}$$

Now if we recover all the H_2, CH_4, and benzene, we find that

$$x_D = \frac{234.55}{234.55 + 0.05 + 1.49} = 0.9935 \qquad \text{(B-63)}$$

which is still less than our desired purity of $x_D = 0.9997$ (even if we neglect the toluene in this stream). Moreover, there is a fairly large benzene loss from this low-pressure flash; i.e., from Eq. B-59

$$\text{Benzene Loss} = (\$483,000/\text{yr})\left(\frac{0.85}{3.76}\right) = \$109,200/\text{yr} \qquad \text{(B-64)}$$

Hence, our estimates indicate that we probably will need a stabilizer to obtain a product stream with the required purity (a pasteurization section on the product column might be acceptable, and we list this as an alternative).

If we recover the light ends (H_2 and CH_4) in a stabilizer, then we would normally send these light ends to the vapor recovery system to recover any benzene or toluene that leaves with this stream (or to any unit that recovers and recycles some of the hydrogen from the purge stream). However, since we have not included any units of this type (at least at this time), we would probably send the light ends to the fuel supply.

COLUMN SEQUENCING. To use a stabilizer column to remove the H_2 and CH_4 from the benzene product, we normally pressurize this column to make it easier to condense the overhead and thereby to obtain an adequate amount of reflux. That is, the H_2 and CH_4 are removed as a vapor stream from the reflux drum after a partial condenser, but it is necessary to take some benzene overhead to provide an adequate amount of liquid reflux. From Table B-3 the K value of benzene is fairly high at 50 psia, so that we might set the stabilizer pressure at 150 psia, or so.

If we use a column sequence that does not remove the light ends in the first column, then every column that totally condensed the H_2 and CH_4 in the overhead would have to be operated at high pressure. Thus, the capital cost probably would be greater than if we had removed the light ends in the first column. Similarly, operation of several columns at high pressure increases the bubble points of the bottoms streams, so that a higher-pressure steam might be required to drive the reboilers. For these reasons, we assume that it is cheapest to remove the light ends in the first column.

When we then consider the separation of benzene (234.5 mol/hr), toluene (87.4 mol/hr), and diphenyl (4 mol/hr), almost all the heuristics (i.e., lightest first, most plentiful first, favor equimolar splits—but not the easiest first or save difficult, high-purity splits until last), favor the direct sequence. Hence, for our base-case design we will choose the lightest first sequence to evaluate. However, complex columns might provide a cheaper separation system.

Now we want to estimate the sizes and the costs of the columns. Since the recycle column, i.e., the toluene-diphenyl split, involves only a binary mixture, we consider the design of this column first. Similarly, we consider the design of the product column before we consider that of the stabilizer. We design both of the recycle and product columns to operate at slightly above ambient pressure because it is easy to condense both toluene and benzene with cooling water at this pressure.

TOLUENE COLUMN. From a plot of the vapor pressures of toluene and diphenyl, we find that the slopes of the vapor-pressure curves for these two components are somewhat different. This result, according to the Clausius-Clapeyron equation, implies that the heats of vaporization of the two compounds are different, which in turn implies that the common assumption of equal molal overflow in the column will not be correct. We could correct for this difference in latent heats by introducing a fictitious molecular weight for one of the components and then using the McCabe-Thiele procedure to design the column. However, for our preliminary calculations we ignore this potential difficulty.

From the vapor-pressure data, we find that

$$\alpha_{top} = \frac{760}{7.4} = 102.7 \qquad \alpha_{bottom} = \frac{10,000}{405} = 24.7 \qquad \text{(B-65)}$$

This is a large variation in α, as well as a very large temperature gradient across the column, that is, $110.6°C = 231°F$ at the top versus $254.9°C = 492°F$ at the bottom. Thus, we expect that some of the simplified design procedures, such as Fenske's equation, Gilliland's correlation, or Smoker's equation, which we often use to estimate column designs, may give misleading predictions. Nevertheless, we might be able to get some idea of the column design with these shortcut design procedures if we choose a conservative estimate of α_{av}. For this reason we let $\alpha = 25$.

Given no losses of aromatics anywhere in the process (which, of course, is not really consistent with our other calculations, but the error is small), the feed rate to the toluene column is 87.4 mol/hr of toluene and 4 mol/hr of diphenyl, so that

$$x_F = 87.4/(87.4 + 4) = 0.956 \qquad \text{(B-66)}$$

Thus, the feed composition of toluene is quite high.

If we recover 99.5% of the toluene overhead and 99.5% of the diphenyl in the bottoms, then we find that

$$d_T = 0.995(87.4) = 87.1 \qquad d_D = 0.005(4) = 0.021 \qquad \text{(B-67)}$$

and

$$x_D = \frac{87.1}{87.1 + 0.021} = 0.9996 \qquad \text{(B-68)}$$

Also,

$$w_T = 0.005(87.4) = 0.438 \qquad w_D = 0.995(4) = 0.398 \qquad \text{(B-69)}$$

and

$$x_B = \frac{0.438}{0.438 + 3.98} = 0.095 \qquad \text{(B-70)}$$

Assuming a saturated-liquid feed, Underwood's equation for minimum reflux

$$R_m = \frac{1}{\alpha - 1}\left[\frac{x_D}{x_F} - \frac{\alpha(1 - x_D)}{1 - x_F}\right] \qquad \text{(B-71)}$$

gives

$$R_m = \frac{1}{25 - 1}\left[\frac{0.9997}{0.956} - \frac{25(1 - 0.9997)}{1 - 0.956}\right] \qquad \text{(B-72)}$$

$$= \tfrac{1}{24}(1.0456 - 0.227) = 0.0347 \qquad \text{(B-73)}$$

which is very low; the feed composition is very high, and α is very large. With a very low value of reflux such as this, we should also consider the use of only a stripping column as an alternative. However, we continue with the design, and we let

$$R \approx 1.5R_m = 1.5(0.0347) = 0.05 \qquad \text{(B-74)}$$

According to Fenske's equation, the minimum number of theoretical trays at total reflux needed for the separation is about

$$N_m = \frac{\ln\left[x_D/(1 - x_D)\right]\left[(1 - x_w)/x_w\right]}{\ln \alpha} = \frac{\ln[(0.9996/0.0004)(0.905/0.095)]}{\ln 25} = 3.13 \qquad \text{(B-75)}$$

We can obtain an estimate of the number of theoretical trays required at the operating reflux ratio by using Gilliland's approximation

$$N_T = 2N_m = 6.2 \qquad \text{(B-76)}$$

The overall plate efficiency is given by O'Connell's correlation, Eq. A2-72. For a quick estimate we assume that $\mu_F = 0.3$, and we write that

$$E_0 \approx \frac{0.5}{(0.3\alpha)^{0.25}} = 0.302 \qquad \text{(B-77)}$$

Then the actual number of required trays is

$$N = \frac{6.2}{0.30} = 21.6 = 22 \qquad \text{(B-78)}$$

For a 2-ft tray spacing and an additional 15 ft at the ends, the tower height is

$$H = 2(22) + 15 = 59 \approx 60 \text{ ft} \tag{B-79}$$

The tower cross-sectional area can be estimated by using Eq. A.3-12

$$A = 2.124 \times 10^{-4}\sqrt{M(T_b + 460)V} \tag{B-80}$$

and we want to base the design on the bottom of the tower, i.e., the diphenyl, where $M = 154$ and $T_b = 492°F$. The vapor rate is written as

$$V = L + D = (R + 1)D = 1.05(87.1 + 0.021) = 91.73 \tag{B-81}$$

and the area becomes

$$A = 2.124 \times 10^{-4}\sqrt{154(492 + 460)}\,(91.7) = 7.5 \text{ ft}^2 \tag{B-82}$$

The column diameter is

$$D = \sqrt{\frac{4A}{\pi}} = 2.7 \approx 3 \text{ ft} \tag{B-83}$$

Now we can use Guthrie's correlation to find the cost

$$\text{Ann. Cost} = \left(\frac{792}{280}\right)(101.9)(3^{1.066})(60^{0.802})\left(\frac{3.18}{3}\right) = \$78,900/\text{yr} \tag{B-84}$$

TOLUENE COLUMN CONDENSER AND COOLING WATER. A condenser heat balance gives

$$Q_C = \Delta H_V\,V = U_C A_C\,\Delta T_m = w_c C_p(120 - 90) \tag{B-85}$$

and if $\Delta H_v = 14,400$ Btu/mol, $U_C = 100$ Btu/(hr·ft²·°F), and $T_b = 231°F$, then

$$\Delta T_m = \frac{120 - 90}{\ln\left[(231 - 90)/(231 - 120)\right]} = 125°F \tag{B-86}$$

Thus,

$$A_C = \frac{14,400(91.7)}{100(125)} = 105 \text{ ft}^2 \tag{B-87}$$

and the annual cost is

$$\text{Ann. Cost} = \left(\frac{792}{280}\right)(101.9)(105^{0.65})\left(\frac{3.29}{3}\right) = \$19,500/\text{yr} \tag{B-88}$$

Also, the cooling-water costs are

$$\text{Ann. Cost} = \left(\frac{\$0.06}{1000 \text{ gal}}\right)\left(\frac{1 \text{ gal}}{8.34 \text{ lb}}\right)\left[\frac{14,400(91.7)}{30}\frac{\text{lb}}{\text{hr}}\right]\left(8150\,\frac{\text{hr}}{\text{yr}}\right) = \$2600/\text{yr} \tag{B-89}$$

TOLUENE COLUMN REBOILER AND STEAM. A heat balance for the reboiler gives

$$Q_R = \Delta H_V\,\bar{V} = U_R A_R\,\Delta T_m = W_s\,\Delta H_s \tag{B-90}$$

The area then becomes

$$A_R = \frac{19,600\ (91.7)}{11,250} = 160\ \text{ft}^2 \tag{B-91}$$

and the cost is

$$\text{Ann. Cost} = \left(\frac{792}{280}\right)(101.9)(160^{0.65})\left(\frac{3.29}{3}\right) = \$25,500/\text{yr} \tag{B-92}$$

The boiling point of diphenyl is 492°F, and so we must use 1000-psi steam (or some high-pressure level) in the reboiler. Of course, we could use 420-psi steam if we operated the tower under a vacuum, so that the boiling point of diphenyl was reduced to about 420°F (which would allow a ΔT of 30°F). However, since the costs associated with this column are reasonably small up to this point, we use 1000-psi steam. Then the steam costs are

$$\text{Ann. Cost} = \left(\frac{\$2.25}{1000\ \text{lb}}\right)\left[\frac{19,600(91.7)}{667.5}\frac{\text{lb}}{\text{hr}}\right]\left(8150\ \frac{\text{hr}}{\text{yr}}\right) = \$49,400/\text{yr} \tag{B-93}$$

This cost is fairly high, so that we might want to examine some alternatives later.

BENZENE COLUMN. Again assuming perfect separations and no losses, we see that the flow rate to the benzene column contains 235.4 mol/hr of benzene, 87.4 mol/hr of toluene, and 4 mol/hr of diphenyl. Since the diphenyl flow rate is so small, we assume that we could obtain a reasonable estimate of the column design if we lump it together with the toluene. Then

$$x_F = \frac{235.4}{235.4 + 91.4} = 0.720 \tag{B-94}$$

In addition, our product specification for benzene requires that the purity be 99.97%, and we want to recover 99.5% of the benzene overhead. With these restrictions

$$d_B = 0.995(235.4) = 234.2 \qquad d_T = 234.2\left(\frac{1 - 0.9997}{0.9997}\right) = 0.07 \tag{B-95}$$

and $\qquad w_B = 235.4 - 234.2 = 1.2 \qquad w_T = (87.4 + 4 - 0.07) = 92.33$

so

$$x_w = 0.013 \tag{B-96}$$

From the vapor-pressure data we find that $\alpha_{av} \approx 2.5$, and then Eq. B-71 gives

$$R_m \approx \frac{1}{(\alpha - 1)x_F} = \frac{1}{(2.5 - 1)0.72} = 0.926 \tag{B-97}$$

Hence $\qquad R \approx 1.2R_m = 1.2(0.926) = 1.11 \tag{B-98}$

Also, from Eq. B-75,

$$N_m = \frac{\ln\left[(0.9997/0.0003)(0.987/0.013)\right]}{\ln 2.5} = 13.6 \qquad \text{(B-99)}$$

Then, by Gilliland's approximation

$$N_T \approx 2N_m = 2(13.6) = 27.2 \qquad \text{(B-100)}$$

For α values in the range of 2 to 3, E_0 is insensitive to α (see Eq. A.2-72), and therefore we can guess that $E_0 = 0.5$. Hence, the total number of trays required is

$$N \approx \frac{27.2}{0.5} = 54.4 \approx 55 \qquad \text{(B-101)}$$

and the column height (allowing 15 ft at the ends) is

$$H = 2(55) + 15 = 125 \text{ ft} \qquad \text{(B-102)}$$

The cross-sectional area of the column is given by Eq. B-80, where we evaluate the area at the bottom of the colum and V is given by Eq. B-81. Hence,

$$A = (2.124 \times 10^{-4})\sqrt{92(231 + 460)}\,[(1.11 + 1)(234.2 + 0.07)$$
$$= 19.73 \qquad \text{(B-103)}$$

and the diameter is

$$D = \sqrt{\frac{4A}{\pi}} = \sqrt{4\left(\frac{19.73}{\pi}\right)} = 5.01 \text{ ft} \qquad \text{(B-104)}$$

Then, from Guthrie, the cost is

$$\text{Ann. Cost} = \left(\frac{792}{280}\right)(101.9)(125^{1.066})(5.01^{0.802})\left(\frac{3.18}{3}\right)$$
$$= \$243,900/\text{yr} \qquad \text{(B-105)}$$

BENZENE COLUMN CONDENSER AND COOLING WATER. A heat balance on the condenser gives

$$Q_C = \Delta H_V\, V = U_C A_C\, \Delta T_m = w_c C_p(120 - 90) \qquad \text{(B-106)}$$

With $\Delta H_V = 13,300$, $U_C = 100$, and

$$\Delta T_m = \frac{120 - 90}{\ln[(177 - 90)/(177 - 120)]} = 70.9 \qquad \text{(B-107)}$$

we find that

$$A_C = \frac{13,300(1.11 + 1)(234.2 + 0.07)}{100(70.9)} = 928 \text{ ft}^2 \qquad \text{(B-108)}$$

and the cost is

$$\text{Ann. Cost} = \left(\frac{792}{280}\right)(101.3)(928^{0.65})\left(\frac{3.29}{3}\right)$$

$$= \$80,100/\text{yr} \tag{B-109}$$

Also, the cooling-water cost is

$$\text{Ann. Cost} = \left(\frac{\$0.06}{1000\ \text{gal}}\right)\left(\frac{1\ \text{gal}}{8.34\ \text{lb}}\right)\left[\frac{13,300(494.7)}{30}\ \frac{\text{lb}}{\text{hr}}\right]\left(8150\ \frac{\text{hr}}{\text{yr}}\right)$$

$$= \$12,900/\text{yr} \tag{B-110}$$

BENZENE COLUMN REBOILER AND STEAM COSTS. A reboiler heat balance gives

$$Q_R = \Delta H_{\bar{V}}\ \bar{V} = U_R A_R\ \Delta T_R = W_S\ \Delta H_S \tag{B-111}$$

so that with $\Delta H_{\bar{V}} = 14,400$, $\bar{V} = V$, and $U_R\ \Delta T_R = 11,250$, then

$$A_R = \frac{14,400(494.7)}{11,250} = 708\ \text{ft}^2 \tag{B-112}$$

and the cost is

$$\text{Ann. Cost} = \left(\frac{792}{280}\right)(101.3)(708^{0.65})\left(\frac{3.29}{3}\right)$$

$$= \$62,400/\text{yr} \tag{B-113}$$

Moreover, the steam cost is given by

$$\text{Ann. Cost} = \left(\frac{\$1.65}{1000\ \text{lb}}\right)\left[\frac{14,400(494.7)}{933.7}\ \frac{\text{lb}}{\text{hr}}\right]\left(8150\ \frac{\text{hr}}{\text{yr}}\right)$$

$$= \$102,600/\text{yr} \tag{B-114}$$

STABILIZER. The design of the stabilizer is not as simple as that for the other columns. First, it is very difficult to find reliable thermodynamic data for mixtures of hydrogen and methane with aromatics. The variety of points that could be used on the Hadden and Grayson charts (see Sec. C.1) illustrates this difficulty.

In addition, the mixture of hydrogen and methane that we desire to remove from the top of the column is essentially noncondensible, so to obtain a sufficient amount of liquid for reflux, we must allow some benzene to go overhead. However, as we saw in the purge-loss calculation (Eq. B-59), even small flows of benzene are quite valuable. Therefore, we want to pressurize the stabilizer column, both to minimize the benzene loss from the partial condenser and to be able to use normal cooling water in this condenser. Of course, as we increase the operating pressure, we increase the capital costs of the column, of the condenser, and of the reboiler because the wall thickness must be increased. Hence, we anticipate that there will be an optimum operating pressure for the column.

To fix the operating pressure of the stabilizer, we might start by examining the behavior of the flash drum when $K_i < 0.1$ for benzene. For example, by considering the value listed in Table B-3 for benzene, we might initially guess that we want $K_B = 0.02$. Then, from the Hadden and Grayson correlation (Sec. C.1), the pressure of the flash drum has to be 280 psia to condense the overhead at 100°F, and it is not even possible to condense the overhead at 130°F. Similarly, if we let $K_B = 0.05$, we find that the pressure is 80 psia at 100°F and 165 psia at 130°F, so the pressure calculation is very sensitive.

Now, to determine the importance of choosing values of K_B that are very small, i.e., in the range from 0.02 to 0.05, we must estimate the value of the benzene that is lost with the vapor leaving the flash drum. We expect that the values of K_{H_2} and K_{CH_4} will be very large compared to unity, and so we expect that the liquid in the flash drum will be essentially pure benzene. Then the equilibrium relationship when $x_B \approx 1.0$ gives

$$y_B = K_B x_B \approx K_B \tag{B-115}$$

From Table B-3, we expect the hydrogen and methane flows leaving in the stabilizer overhead to be about 2 and 11 mol/hr, respectively, so

$$y_B = K_B = \frac{n_B}{2 + 11 + n_B} \tag{B-116}$$

or

$$n_B = \frac{13 K_B}{1 - K_B} \tag{B-117}$$

If $K_B = 0.05$, then the benzene loss calculated from this expression is about 0.7 mol/hr, and from Eq. B-59 the value of the benzene loss is

$$\text{Ann. Loss} = 9.04(0.7)8150 = \$51,600/\text{yr} \tag{B-118}$$

which is fairly large. Of course, this loss is compensated, at least in part, by the fuel value of the benzene in the overhead stream.

After examining the sensitivity of the benzene loss from the flash drum of the stabilizer, we decided to choose $K_B = 0.04$ with a condensing temperature of 115°F, so that for our first design the operating pressure of the stabilizer is 150 psia. At these conditions

$$K_{H_2} = 185 \qquad K_{CH_4} = 36 \tag{B-119}$$

and the benzene loss is

$$n_B = \frac{13(0.04)}{1.0 - 0.04} = 0.54 \tag{B-120}$$

and

$$\text{Ann. Loss} = 9.04(0.54)8150 = \$39,800/\text{yr} \tag{B-121}$$

The distillate flows are then

$$d_{H_2} = 2 \qquad d_T = 11 \qquad d_B = 0.54 \tag{B-122}$$

so that

$$x_{D,H_2} = 0.148 \qquad x_{D,CH_4} = 0.812 \qquad X_{D,B} = 0.040 \tag{B-123}$$

We can simplify the design calculations if we lump all the aromatics and consider them to be benzene, so that the benzene feed rate to the column becomes (see Table B-3) $235.4 + 87.4 + 4 = 326.8$ mol/hr. Hence, the feed rate is

$$F = 2 + 11 + 326.8 = 379.8 \text{ mol/hr} \tag{B-124}$$

and the feed composition is

$$x_{H_2} = 0.0055 \qquad x_{CH_4} = 0.0324 \qquad x_B = 0.961 \tag{B-125}$$

With this assumption, the effluent from the bottom of the tower will be essentially pure benzene boiling at 150 psia. Then from the Hadden and Grayson correlations for $K_B = 1.0$ and $P = 150$ psia, we find that the temperature is 360°F. However, we expect that the presence of the toluene and diphenyl will increase this boiling point, thus we guess that the bottoms temperature is about 400°F. Also, at 150 psia and 400°F, we find that

$$K_{H_2} = 70 \qquad K_{CH_4} = 36 \tag{B-126}$$

Now, from Eqs. B-119 and B-125 we can estimate the α values:

Top:
$$\alpha_{H_2} = \frac{185}{0.04} = 4600 \qquad \alpha_{CH_4} = \frac{36}{0.04} = 900 \tag{B-127}$$

Bottom:
$$\alpha_{H_2} = \frac{70}{1} = 70 \qquad \alpha_{CH_4} = \frac{36}{1} = 36$$

Clearly to introduce an assumption of constant relative volatility would be rather "shaky" but all the shortcut design procedures rely on this assumption. Hence, for our first design we assume that α_i is constant, but we use the smallest possible value, Eq. B-127 at the bottom, in our calculations.

Underwood's expression for the minimum reflux rates for binary separations should give us a conservative estimate for multicomponent separations, so

$$R_m = \frac{1}{\alpha - 1}\left[\frac{x_{D,C_1}}{x_{F,C_1}} - \alpha_{C_1 - B}\frac{x_{D,B}}{x_{F,B}}\right] \tag{B-128}$$

$$= \frac{1}{36 - 1}\left[\frac{0.812}{0.0324} - 36\left(\frac{0.04}{0.961}\right)\right] = 0.689 \tag{B-129}$$

Then, for an operating reflux, we choose

$$R \approx 1.5(0.689) \approx 1.033 \tag{B-130}$$

We estimate the minimum number of trays by using Fenske's equation, which we write as

$$N_m = \frac{\ln\left[(x_{C_1}/x_B)_D\,(x_B/x_{C_1})_w\right]}{\ln \alpha_{C_1 - B}} \tag{B-131}$$

If $(x_B/x_{C_1})_w = 10^{-6}$, then

$$N_m = \frac{\ln\left[(0.812/0.04)\,0.719/(6.62 \times 10^{-5})\right]}{\ln 36} = 3.4 \tag{B-132}$$

Then, from Gilliland's approximation

$$N_T = 2N_m = 6.8 \tag{B-133}$$

Using Eq. B-79 to estimate the overall plate efficiency (note that we are not certain that $\mu_F = 0.3$ centipoise at high pressures) gives

$$E_0 = \frac{0.5}{[0.3(36)]^{0.25}} = 0.276 \tag{B-134}$$

so that the total number of required trays is about

$$N = \frac{6.86}{0.276} = 24.6 \tag{B-135}$$

and the column height is

$$H = 2(25) + 15 = 65 \text{ ft} \tag{B-136}$$

The vapor rate is

$$V = (R + 1)D \approx (1.033 + 1)(2 + 11 + 0.54) \approx 27.5 \tag{B-137}$$

and the tower diameter is (for an ideal gas)

$$A = 0.000214 \left[78(400 + 460)\left(\frac{15}{150}\right) \right]^{1/2} (27.5) = 0.478 \tag{B-138}$$

and the column diameter is

$$D = \sqrt{\frac{4A}{\pi}} = 0.8 \text{ ft} \approx 1 \text{ ft} \tag{B-139}$$

Unfortunately, this diameter is less than the allowable minimum value of 1.5 ft, so we should consider the possibility of using a packed tower. However, clearly we have based our design on a number of questionable assumptions, so at some point we must undertake a more careful analysis. Nevertheless, our primary focus at this time should be the processing costs of the total plant, rather than an accurate design of the stabilizer, and since we do not expect that the stabilizer will be very expensive, we estimate the cost

$$\text{Ann. Cost} = \left(\frac{792}{280}\right)(101.3)(65^{1.066})(1^{0.802})\left(\frac{2.18 + 1.15}{3}\right) = \$16{,}200/\text{yr} \tag{B-140}$$

STABILIZER CONDENSER AND COOLING WATER. A heat balance for the condenser gives

$$Q_C = \Delta H_V V = U_C A_C \, \Delta T_m = w_c C_p(t_{\text{out}} - 90) \tag{B-141}$$

Since we are condensing the overhead at 115°F, we cannot allow the temperature of the cooling water to increase form 90 to 120°F. Thus, for our first design we assume that $t_{\text{out}} = 100$°F. Then

$$\Delta T_m = \frac{100 - 90}{\ln{[(115 - 90)/(115 - 100)]}} = 19.6 \approx 20°F \tag{B-142}$$

We let $U_C = 100$ Btu/(hr·ft²·°F), and we assume that $\Delta H_{H_2} = 389$ and $\Delta H_{CH_4} = 3520$ Btu/mol. Hence

$$A_c = \frac{[389(\frac{2}{13}) + 3520(\frac{11}{13})]27.5}{(100)(20)} = 41.8 \text{ ft}^2 \tag{B-143}$$

and the cost is

$$\text{Ann. Cost} = \left(\frac{792}{280}\right)(101.3)(41.8^{0.65})\left(\frac{3.29}{3}\right) = \$10,800/\text{yr} \tag{B-144}$$

Also, the cost of cooling water is

$$\text{Ann. Cost} = \left(\frac{\$0.06}{1000 \text{ gal}}\right)\left(\frac{1 \text{ gal}}{8.34 \text{ lb}}\right)\left[\frac{83,600}{1(100-90)} \frac{\text{lb}}{\text{hr}}\right]\left(8150 \frac{\text{hr}}{\text{yr}}\right) = \$500/\text{yr} \tag{B-145}$$

STABILIZER REBOILER AND STEAM. The reboiler heat balance is

$$Q_R = \Delta H_{\bar{V}} \bar{V} = U_R A_R \Delta T_R = W_S \Delta_{H_S} \tag{B-146}$$

We assume that $\bar{V} = V$ and $\Delta H_{\bar{V}} = \Delta H_v$, so that

$$A_R = \frac{13,300(27.5)}{11,250} = 32.5 \text{ ft}^2 \tag{B-147}$$

and the cost is

$$\text{Ann. Cost} = \left(\frac{792}{280}\right)(101.3)(32.5^{0.65})\left(\frac{3.29}{3}\right) = \$9000/\text{yr} \tag{B-148}$$

The boiling point of the bottoms is about 400°F, so we use 420-psia steam in the reboiler. Then, using the values in Sec. E.1, we find that the cost is

$$\text{Ann. Cost} = \left(\frac{\$2}{1000 \text{ lb}}\right)\left[\frac{13,300(27.5)}{775.6} \frac{\text{lb}}{\text{hr}}\right]\left(8150 \frac{\text{hr}}{\text{yr}}\right) = \$7700/\text{yr} \tag{B-149}$$

LEVEL-4 ECONOMIC POTENTIAL. To calculate the economic potential at level 4, we subtract the annualized capital costs of the columns, condensers, and reboilers, as well as the steam and cooling-water costs, from the values of the level-3 economic potential. In addition, we must subtract the purge losses of benzene and toluene (as well as any recycle losses, but we may take fuel credit for the benzene and toluene in the purge and the stabilizer overhead). The results are shown in Fig. 7.3-13, and we see that the range of the design variables corresponding to profitable operation has been dramatically reduced.

Of course, with a different process alternative, the profitability might be increased. However, we want to complete the design before we examine any alternatives.

LEVEL-4 ALTERNATIVES. There are numerous alternatives that we could consider at level 4. A vapor recovery system might prove to be profitable on either the flash vapor (if we incur recycle losses) or the purge stream. Absorption, condensation, adsorption, or a membrane recovery system could be used as this vapor recovery process.

There are also numerous alternatives for the distillation system. Since the reflux ratios are so small for both the stabilizer and the recycle column, we might be able to use just stripping columns, i.e., no rectifying sections. Similarly, since both the CH_4-benzene and the toluene-diphenyl splits are very easy, we should consider the possiblity of using pasteurization sections to decrease the number of columns. However, we defer an attempt to evaluate the costs of these alternatives until we have evaluated the heat-exchanger network for our base-case design.

LEVEL 5:   HEAT-EXCHANGER NETWORK

The procedure for designing a heat-exchanger network was presented in Chap. 8. To develop the data we need for this procedure, we must calculate the temperature-enthalpy curves for each process stream. We would recommend that normally this information be developed with the use of a CAD package such as FLOWTRAN, particularly for processes where mixtures exhibit a phase change (i.e., a set of flash calculations is needed along the length of an exchanger to calculate the *T-H* profile). Similarly, for cases where a phase change takes place at high pressures, it is usually simpler to use a CAD package to calculate the physical properties required than it is to use handbooks and empirical correlations. The use of FLOWTRAN for these calculations is discussed in Chap. 12.

APPENDIX
C

DESIGN
DATA

C.1 HYDROCARBON VAPOR-LIQUID EQUILIBRIA*

Use Figs. C.1-1 and C.1-2 for compounds with lower boiling points than heptane, and use Table C.1-1 with the graphs for compounds or cuts with higher boiling points than heptane.

---

*Taken from S. T. Hadden and H. G. Grayson, "New Charts for Hydrocarbon Vapor-Liquid Equilibria," *Hydrocarb. Proc. and Petrol. Refiner*, **40**: 91, 207 (September 1961).

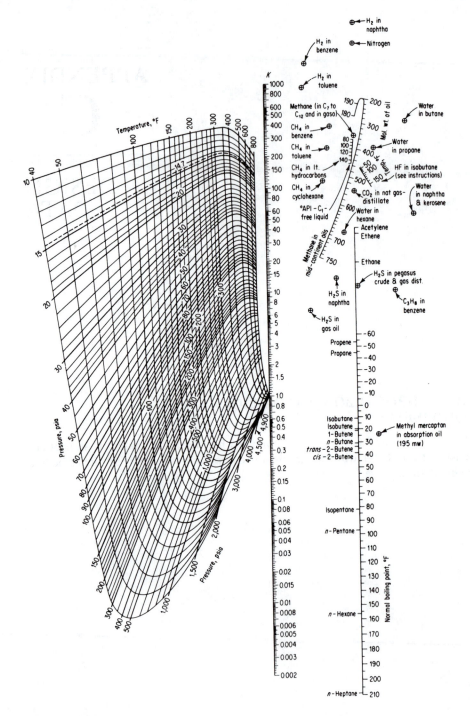

FIGURE C.1-1
Vapor-liquid equilibria, 40 to 800°F. [*From S. T. Hadden and H. G. Grayson, Hydrocarb. Proc. and Petrol. Refiner,* **40**: *207 (September 1961).*]

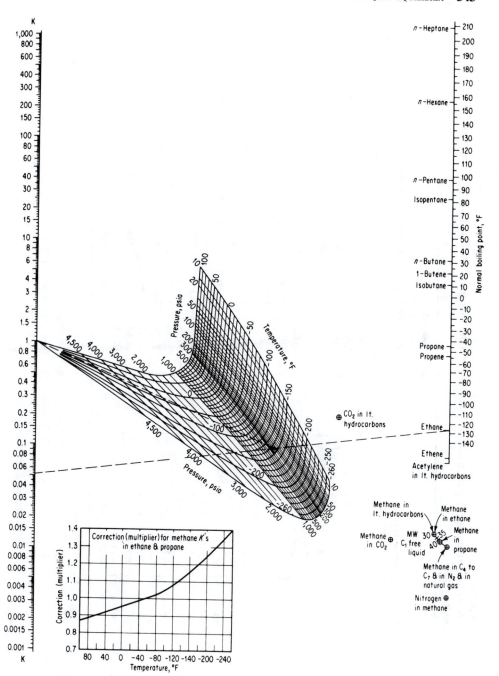

FIGURE C.1-2

Vapor-liquid equilibria, −260 to +100°F. [*From S. T. Hadden and H. G. Grayson, Hydrocarb. Proc. and Petrol. Refiner,* **40**: *207 (September 1961).*]

TABLE C1-1
Volatility exponent for pure hydrocarbons and true boiling point 50°F cuts at 5°F intervals on boiling point

Volatility exponent b

Temp. interval, °F	Pure hydrocarbons, NBP, °F			50°F Cuts, TBP mid-boiling-point temperatures, °F						
	200	300	400	100	200	300	400	500	600	700
00		0.360	0.834		−0.081	0.326	0.792	1.348	2.016	2.801
05		0.381	0.861		−0.061	0.348	0.817	1.379	2.053	2.843
10	0.003	0.403	0.888		−0.041	0.370	0.842	1.410	2.090	2.886
15	0.021	0.425	0.916		−0.021	0.392	0.868	1.441	2.126	2.928
20	0.040	0.447	0.944		−0.002	0.414	0.894	1.472	2.164	2.972
25	0.060	0.469		−0.376	0.018	0.436	0.921	1.504	2.202	3.014
30	0.079	0.491		−0.356	0.038	0.458	0.948	1.536	2.240	3.058
35	0.098	0.514		−0.336	0.058	0.481	0.974	1.569	2.278	3.102
40	0.117	0.537		−0.316	0.078	0.504	1.002	1.602	2.316	3.146
45	0.137	0.560		−0.296	0.098	0.526	1.029	1.634	2.355	3.190
50	0.156	0.584		−0.276	0.118	0.550	1.057	1.668	2.394	3.234
55	0.176	0.607		−0.257	0.138	0.572	1.085	1.701	2.434	3.280
60	0.196	0.631		−0.237	0.159	0.596	1.113	1.735	2.474	3.324
65	0.216	0.656		−0.218	0.180	0.620	1.142	1.769	2.514	3.370
70	0.236	0.680		−0.198	0.200	0.644	1.170	1.804	2.554	3.415
75	0.256	0.705		−0.178	0.221	0.668	1.199	1.838	2.594	3.461
80	0.277	0.730		−0.158	0.242	0.692	1.228	1.874	2.635	3.507
85	0.297	0.756		−0.139	0.262	0.716	1.258	1.908	2.676	3.553
90	0.318	0.781		−0.120	0.284	0.742	1.288	1.944	2.718	3.600
95	0.339	0.808		−0.100	0.305	0.766	1.318	1.980	2.759	3.616

$$K_H = K_7/(K_2/K_7)^b$$

where K_H = K value of the cut or high-boiling pure hydrocarbon
K_2 = K value of ethane at the temperature, pressure and convergence pressure of the system
K_7 = K value of n-heptane at the temperature, pressure and convergence pressure of the system
b = Volatility exponent from Table C.1-1

C.2   TEMPERATURE RANGES FOR SOME MATERIALS

See Table C.2-1.

TABLE C.2-1
Temperature ranges for some materials

Material	Safe operating range, °F
Mild-carbon steel	−20 to 650
Low-nickel stainless steel	−150 to 700
18/8 Stainless steel	−325 to 1400
Aluminum	−325 to 700
Incoloy (35/25)	−325 to 2100

APPENDIX
D

FORTRAN INPUT FORMS

The tables in this appendix are taken from J. D. Seader, W. D. Seider, and A. C. Pauls, "Flowtran Simulation—An Introduction," 2d ed., CAChE Corp., Cambridge, Mass., 1977. Copies are available from Ulrich's Book Store, 549 E. University Ave., Ann Arbor, Mich., 48104.

D.1 CHEMICALS IN THE PUBLIC DATA FILE

Physicial properties are available for the components in Table D.1-1.

TABLE D.1-1
Chemicals in the Public Data File

Empirical formula	Component name	Empirical formula	Component name
		Inorganic chemicals	
Ar	Argon	Cl_2	Chlorine
Br_2	Bromine	HI	Hydrogen iodide
CCl_4	Carbon tetrachloride	H_2	Hydrogen
CO	Carbon monoxide	H_2O	Water
$COCl_2$	Phosgene	H_2S	Hydrogen sulfide
CO_2	Carbon dioxide	H_3N	Ammonia
CS_2	Carbon disulfide	Ne	Neon
C_2OCl_4	Trichloroacetyl-CL	NO	Nitric oxide
ClH	Hydrogen chloride	NO_2	Nitrogen dioxide

548

TABLE D.1-1 (*Continued*)

Empirical formula	Component name	Empirical formula	Component name
		Inorganic chemicals Continued	
N_2	Nitrogen	O_2S	Sulfur dioxide
N_2O	Nitrous oxide	O_3S	Sulfur trioxide
O_2	Oxygen		
		Organic chemicals	
$CHCl_3$	Chloroform	C_3H_8O	*N*-Propanol
CHN	Hydrogen cyanide	C_3H_9N	Trimethylamine
CH_2O	Formaldehyde	C_4H_4	Vinylacetylene
CH_3Cl	Methyl chloride	C_4H_4S	Thiophene
CH_3I	Methyl iodide	C_4H_5N	Methacrylonitrile
CH_4	Methane	C_4H_6	Butadienes
CH_4O	Methanol	C_4H_6	Dimethylacetylene
CH_5N	Methyl amine	C_4H_6	Ethylacetylene
C_2HCl_3	Trichloroethylene	C_4H_6	1,2-Butadiene
C_2HCl_3O	Dichloroacetyl-CL	C_4H_6	1,3-Butadiene
C_2H_2	Acetylene	C_4H_8	Butylenes
$C_2H_2Cl_2O$	Chloroacetyl-CL	C_4H_8	1-Butene
C_2H_3Cl	Vinyl chloride	C_4H_8	*Cis*-2-butene
C_2H_3ClO	Acetyl chloride	C_4H_8	Isobutene
$C_2H_3Cl_3$	112Tri-CL-ethane	C_4H_8	*Trans*-2-butene
C_2H_3N	Acetonitrile	C_4H_8O	Isobutyraldehyde
C_2H_4	Ethylene	C_4H_8O	MEK
$C_2H_4Cl_2$	1,1-Dichloroethane	$C_4H_8O_2$	*N*-Butyric acid
$C_2H_4Cl_2$	1,2-Dichloroethane	$C_4H_8O_2$	Ethyl acetate
C_2H_4O	Acetaldehyde	$C_4H_8O_2$	Methyl propionate
C_2H_4O	Ethylene oxide	$C_4H_8O_2$	Propyl formate
$C_2H_4O_2$	Acetic acid	C_4H_9NO	Dimethyl acetamide
$C_2H_4O_2$	Methyl formate	C_4H_{10}	Isobutane
C_2H_5Cl	Ethyl chloride	C_4H_{10}	*N*-Butane
C_2H_6	Ethane	$C_4H_{10}O$	Isobutanol
C_2H_6O	Dimethyl ether	$C_4H_{10}O$	*N*-Butanol
C_2H_6O	Ethanol	$C_4H_{10}O$	*T*-Butyl alcohol
$C_2H_6O_2$	Ethylene glycol	$C_4H_{10}O$	Diethyl ether
C_2H_6S	Dimethyl sulfide	$C_4H_{10}O_3$	Diethylene glycol
C_2H_6S	Ethyl mercaptan	$C_5H_4O_2$	Furfural
C_2H_7N	Ethylamine	C_5H_8	Isoprene and C5
C_3H_3N	Acrylonitrile	C_5H_{10}	2-Me-1-butene
C_3H_4	Methylacetylene	C_5H_{10}	2-Me-2-butene
C_3H_4	Propadiene	C_5H_{10}	3-Me-1-butene
C_3H_6	Propylene	C_5H_{10}	Cyclopentane
C_3H_6O	Acetone	C_5H_{10}	1-Pentene
$C_3H_6O_2$	Ethyl formate	C_5H_{10}	*Cis*-2-pentene
$C_3H_6O_2$	Methyl acetate	C_5H_{10}	*Trans*-2-pentene
$C_3H_6O_2$	Propionic acid	$C_5H_{10}O$	Diethyl ketone
C_3H_7NO	Dimethylformamide	$C_5H_{10}O_2$	*N*-Propyl acetate
C_3H_8	Propane	C_5H_{12}	Isopentane
C_3H_8O	Iso-propanol	C_5H_{12}	*N*-Pentane

Continued

TABLE D.1-1 (*Continued*)

Empirical formula	Component name	Empirical formula	Component name
	Inorganic chemicals *Continued*		
C_5H_{12}	Neo-pentane	C_9H_{10}	Methylstyrene
$C_6H_3Cl_3$	1,2,4-Tri-CL-BZ	C_9H_{12}	C,3-Alkylbenzene
$C_6H_4Cl_2$	*M*-Dichlorobenzene	C_9H_{12}	1-ET-2-Me-benzene
$C_6H_4Cl_2$	*O*-Dichlorobenzene	C_9H_{18}	*N*-Propylcyclohexan
$C_6H_4Cl_2$	*P*-Dichlorobenzene	C_9H_{20}	*N*-Nonane
C_6H_5Br	Bromobenzene	$C_{10}H_8$	Naphthalene
C_6H_5Cl	Chlorobenzene	$C_{10}H_{10}$	1-Methylindene
C_6H_5I	Iodobenzene	$C_{10}H_{10}$	2-Methylindene
C_6H_6	Benzene	$C_{10}H_{12}$	Dicyclopentadiene
C_6H_6O	Phenol	$C_{10}H_{14}$	*N*-Butylbenzene
C_6H_7N	Aniline	$C_{10}H_{14}$	1,2-DIME-3-ETHBZ
C_6H_{12}	Cyclohexane	$C_{10}H_{20}$	*N*-Butylcyclohexane
C_6H_{12}	Methylcyclopentane	$C_{10}H_{22}$	*N*-Decane
C_6H_{12}	1-Hexene	$C_{11}H_{10}$	1-Methylnaphthalen
C_6H_{14}	2,2-Dimethylbutane	$C_{11}H_{10}$	2-Methylnaphthalen
C_6H_{14}	2,3-Dimethylbutane	$C_{11}H_{24}$	*N*-Undecane
C_6H_{14}	*N*-Hexane	$C_{12}H_8$	Acenaphthylene
C_6H_{14}	2-Methylpentane	$C_{12}H_{10}$	Biphenyl
C_6H_{14}	3-Methylpentane	$C_{12}H_{12}$	2,7-Dimethylnaphth
$C_6H_{14}O_4$	Triethylene glycol	$C_{12}H_{13}$	1,2,3-TRIME-Indene
C_7H_8	Toluene	$C_{12}H_{26}$	*N*-Dodecane
C_7H_8O	*O*-Cresol	$C_{13}H_{10}$	Fluorene
C_7H_{14}	Methylcyclohexane	$C_{13}H_{14}$	C,3-Alkylnaphtha-
C_7H_{14}	Ethylcyclopentane		lene
C_7H_{14}	1-Heptene	$C_{13}H_{14}$	1-Me-4-ETH-Naph-
C_7H_{16}	*N*-Heptane		tha
C_8H_8	Styrene	$C_{13}H_{14}$	2,3,5-TRIME-Naph-
C_8H_{10}	Ethylbenzene		tha
C_8H_{10}	Xylenes and ETB	$C_{13}H_{28}$	*N*-Tridecane
C_8H_{10}	*M*-Xylene	$C_{14}H_{10}$	Phenanthrene
C_8H_{10}	*O*-Xylene	$C_{14}H_{30}$	*N*-Tetradecane
C_8H_{10}	*P*-Xylene	$C_{15}H_{12}$	1-Phenylindene
C_8H_{12}	*N*-Propylbenzene	$C_{15}H_{14}$	2-Ethylfluorene
C_8H_{16}	Ethylcyclohexane	$C_{15}H_{32}$	*N*-Pentadecane
C_8H_{16}	*N*-Propylcyclopenta	$C_{16}H_{10}$	Fluoranthene
C_8H_{18}	*N*-Octane	$C_{16}H_{10}$	Pyrene
$C_8H_{18}O_5$	Tetraethene glycol	$C_{16}H_{12}$	1-Phenylnaphthalen
C_9H_8	Indene	$C_{16}H_{34}$	*N*-Hexadecane
C_9H_{10}	Indan	$C_{18}H_{12}$	Chrysene

D.2 IFLSH

Description

IFLSH (Isothermal flash) determines the quantity and composition of liquid and vapor streams resulting when a feed stream is flashed at a specified temperature and pressure. If the flash conditions are such that only a single phase product occurs, then the appropriate product composition is set equal to the feed and the other stream is set to zero. The block also calculates the rate of heat addition (positive) or removal (negative) from the flash vessel in order to maintain the specified temperature and pressure.

Output

The output gives the unit name, the feed and product stream names, the flash temperature and pressure, the heat added to the vessel, and the fraction of feed which leaves as vapor. An option is provided for printing stream flows and physical properties including equilibrium K-values.

Physical Properties

Vapor-liquid equilibria and enthalpies are required.

Block List

List type	BLOCK
Unit name	_____
Unit type	IFLSH
Name of feed stream	_____
Name of liquid product stream	_____
Name of vapor product stream	_____

Parameter List

List type	PARAM
Unit name	_____
Index of first entry	1
1. Flash temperature, °F	_____
2. Flash pressure, psia	_____
3. Print stream flows, physical properties, and K-values: 0 = no, 1 = yes.	_____

D.3   AFLSH

Description

AFLSH (Adiabatic flash) determines the quantity and composition of liquid and vapor streams resulting when up to seven feed streams are mixed and flashed adiabatically. The number of product streams may be 1 or 2. The block can be used to simulate a pressure drop across a valve or through a pipeline. If two product streams are specified and the flash conditions result in a single phase, the appropriate product is set equal to the sum of the feed streams and the other stream is set to zero. You can also specify heat addition to or removal from the flash unit.

Output

The output gives the unit name, the feed and product stream names, the flash temperature and pressure, the heat added to or removed from the system, and the fraction of feed which leaves as vapor. An option is provided for printing stream flows and physical properties including equilibrium K-values.

Properties Used

Vapor-liquid equilibria and enthalpies are required.

Block List

List type	BLOCK
Unit name	_____
Unit type	AFLSH
Name of 1st feed stream	_____
Name of 2d feed stream or 0	_____
Name of 3d feed stream or 0	_____
Name of 4th feed stream or 0	_____
Name of 5th feed stream or 0	_____
Name of 6th feed stream or 0	_____
Name of 7th feed stream or 0	_____
Name of liquid product stream*	_____
Name of vapor product stream or 0*	_____

Parameter List

List type	PARAM
Unit name	_____
Index of first entry	1

1. Flash pressure, psia, if positive.   _____
   Flash pressure minus the minimum feed
   pressure psia, if 0 or negative.
2. Heat added, Btu/hr (if negative, heat removed).   _____
3. Print stream flows, physical properties,   _____
   K-values: 0 = no, 1 = yes.

---

* If only one product stream is specified, its phase condition will be determined.

D.4   SEPR

Description

SEPR (Constant split-fraction separation) can be used to simulate distillation or other separation processes when details of the process are unknown or irrelevant. The basis of the model is that for each component the ratio of the moles in the overhead to the moles in the feed is constant. These split fractions, one for each component, are input values. The bottoms product may be assumed to be a saturated liquid, and the overhead product may be either saturated liquid or saturated vapor. Product temperatures are then determined from the saturation requirement. The product temperatures may be alternatively set equal to the feed temperature.

Output

The output gives the unit name, the type of condenser (total or partial) if one is simulated, the stream names, the total fraction of feed taken as overhead, and the split fraction for each component.

Properties Used

Vapor-liquid equilibria and enthalpies are required.

Block List

List type	BLOCK
Unit name	
Unit type	SEPR
Name of feed stream	
Name of bottoms product stream	
Name of overhead product stream	

Parameter List

List type	PARAM
Unit name	
Index of first entry	1
Split fractions, moles i in overhead/moles i in feed*	

---

* Must be in range from 0.0 to 1.0.

1. Component 1 _____

2. Component 2 _____

3. Component 3 _____

4. Component 4 _____

5. Component 5 _____

D.5 ADD

Description

ADD (stream addition) adds up to seven process streams and determines the product stream temperature and phase condition which satisfies the condition of zero enthalpy change. The pressure of the outlet stream is taken as the minimum of the nonzero pressures of the inlet streams less a specified pressure drop. This block is similar to the adiabatic flash block AFLSH.

Output

The output gives the unit name, the feed stream names and product stream name, temperature, and pressure. An option is provided for printing stream flows and physical properties including equilibrium K-values.

Properties Used

Vapor-liquid equilibria and enthalpies are used.

Block List

List type	**BLOCK**
Unit name	_____
Unit type	ADD
Name of 1st feed stream	_____
Name of 2d feed stream	_____
Name of 3d feed stream or 0	_____
Name of 4th feed stream or 0	_____
Name of 5th feed stream or 0	_____
Name of 6th feed stream or 0	_____
Name of 7th feed stream or 0	_____
Name of product stream	_____

Parameter List

List type	**PARAM**
Unit name	_____
Index of first entry	1

1. ΔP, pressure drop, psi (subtracted from minimum inlet pressure). _____
2. Print stream flows, properties, and K-values: 0 = no, 1 = yes. _____

D.6   SPLIT

Description

SPLIT (stream split) separates an input stream into as many as seven output streams, each having the same composition, temperature, and pressure as the input. You must specify the amount of each product stream as a fraction of the input stream. If the sum of the fractions does not equal unity, the fractions will be normalized. This makes it possible for a given fraction to serve as a manipulated variable in a control loop.

Output

The output gives the unit name, the input and output stream names, and the normalized split fractions.

Properties Used

None are used.

Block List

List type	BLOCK
Unit name	_____
Unit type	SPLIT
Name of input stream	_____
Name of 1st output stream	_____
Name of 2d output stream	_____
Name of 3d output stream or 0	_____
Name of 4th output stream or 0	_____
Name of 5th output stream or 0	_____
Name of 6th output stream or 0	_____
Name of 7th output stream or 0	_____

Parameter List

List type	PARAM
Unit name	_____
Index of first entry	1

1. Number of output streams (2 to 7)   —————

2. Split fraction for 1st output stream   —————

3. Split fraction for 2d output stream   —————

4. Split fraction for 3d output stream   —————

5. Split fraction for 4th output stream   —————

6. Split fraction for 5th output stream   —————

7. Split fraction for 6th output stream   —————

8. Split fraction for 7th output stream   —————

D.7   PUMP

Description

PUMP (centrifugal pump size and power) raises the pressure of a stream to a specified value. It should be used only for a totally liquid stream. Typical curves from the literature are used to estimate pump and driver efficiencies so that driver size and electric power requirements can be calculated. These curves are based on water; lower pump efficiency can be expected for more viscous fluids.

Output

The output gives the unit name, the inlet and outlet stream names, the actual flow in gpm, pressure change in psia, fluid horsepower, pump efficiency, brake horsepower, driver efficiency, and kilowatts.

Properties Used

Liquid molal volume is used.

Reference

M. S. Peters, *Plant Design and Economics for Chemical Engineers*, McGraw-Hill, New York, 1958, p. 291.

Block List

List type	<u>BLOCK</u>
Unit name	—————
Unit type	<u>PUMP</u>
Name of feed stream	—————
Name of product stream	—————

Parameter List

List type   **PARAM**
Unit name
Index of first entry   1

1. Outlet pressure, psia

D.8   GCOMP

Description

GCOMP (centrifugal compressor, positive-displacement compressor, and turbine) computes the work required for compression or the work yielded by expansion and the outlet stream temperature, phase condition, and enthalpy.

The block will perform four different types of calculations:

1. Centrifugal compressor using the polytropic efficiency, sometimes called the *N method*
2. Positive-displacement compressor using the polytropic efficiency
3. Turbine or expander using the isentropic efficiency, also called the *adiabatic efficiency* and the *Mollier method*
4. Centrifugal compressor using the isentropic efficiency

With each method, you may specify the appropriate efficiency; all outlet stream conditions and shaft work will be computed. However, if the outlet temperature is known, it may be entered for the second parameter, and the program will compute the efficiency. This option is useful for characterizing an existing compressor.

The polytropic calculation must not be used for streams that are partially liquefied at inlet or outlet conditions.

The equations used are as follows where the subscript 1 refers to inlet conditions and subscript 2 refers to outlet conditions:

1. Polytropic compression

$$K = \frac{C_{p_1}}{C_{v_1}}$$ Heat-capacity ratio

$$\eta_p = \frac{(K-1)/K}{(n-1)/n}$$ Polytropic efficiency

$$\Delta h = \frac{P_1 v_1}{\eta_p^{n-1/n}}\left[\left(\frac{P_2}{P_1}\right)^{n-1/n} - 1\right]$$ Enthalpy change per mole (with appropriate units)

$$\text{ihp} = w\,\Delta h$$ Indicated horsepower

$$\text{bhp} = \frac{\text{ihp}}{\eta_m}$$ Brake horsepower

2. Isentropic compression

$$\Delta h = h_2 - h_1 = \frac{h_s - h_l}{\eta_s}$$

3. Isentropic expansion

$$\Delta h = h_2 - h_1 = (h_s - h_l)\eta_s$$

4. Volumetric efficiency

$$\eta_v = 1.0 - 0.01\frac{P_2}{P_1} + c\left(1 - \frac{V_1}{V_2}\right)$$

5. Outlet temperature is computed from the outlet enthalpy:

$$h_2 = h_1 + \Delta h$$

Output

The output consists of the inlet and outlet stream names, the discharge pressure, actual and constant-entropy discharge temperatures, constant-entropy horsepower, indicated horsepower, and brake horsepower. Work done by the fluid is negative. If the positive-displacement-compression option is chosen, the output also includes the volumetric efficiency and the required displacement in ft^3/hr.

Properties Used

Entropies, enthalpies, specific heats, specific volumes, and vapor-liquid equilibria are used.

Nomenclature

C	Clearance fraction
C_p	Heat capacity at constant pressure
C_v	Heat capacity at constant volume
h_s	Enthalpy on constant-entropy path
n	Polytropic exponent
w	Stream flow
η_m	Mechanical efficiency
η_p	Polytropic efficiency
η_s	Isentropic efficiency
η_v	Volumetric efficiency

References

Dresser Industries, Inc., *Clark Multistage Centrifugal Compressors*, Franklin Park, Ill., 1969.

Elliott Company, *Elliott Compressor Refresher*, Jeannette, Pa.

Block List

List type	BLOCK
Unit name	
Unit type	GCOMP
Name of inlet stream	
Name of outlet stream	

Parameter List

List type	PARAM
Unit name	
Index of first entry	1

  I. Outlet pressure, psia   _____

 II. Outlet temperature, °F (computed if 0)   _____

III. Type of unit   _____
    1 Centrifugal compressor by polytropic method
    2 Positive-displacement compressor by polytropic method
    3 Turbine by isentropic method
    4 Centrifugal compressor by isentropic method

IV. Clearance fraction for positive-displacement compressor   _____

 V. Mechanical efficiency (1 if 0)   _____

VI. Polytropic efficiency for options 1 and 2;
    isentropic efficiency for options 3 and 4   _____
    (0.72 if 0)

D.9   SCVW

Description

SCVW (bounded Wegstein stream convergence) is a stream convergence block which is capable of converging one, two, or three streams simultaneously. The accelerated convergence method of Wegstein with a modification by Kleisch and Sullivan is used (see References).

The following options are included:

1. The convergence tolerance may be specified. The same value is used in each stream for every variable except enthalpy and fraction vapor. If a value is not specified, 0.0005 is chosen (see parameter 2). For each variable in a stream, the error is defined as the smaller of either the difference between the calculated (input) and estimated (output) values or the differences divided by the estimated value. Errors which exceed the convergence tolerance are indicated in the history by an asterisk. The first N variables in the stream are the molar flow rates of each of the N components, $N + 1$ is the total molar flow rate, $N + 2$ is temperature, $N + 3$ is pressure, $N + 4$ is the enthalpy flow, and $N + 5$ is the fraction vapor.

2. The maximum number of iterations may be specified. If the number is not specified, 30 iterations are automatically chosen (see parameter 3). The block may operate as a direct iteration block by specifying the number of direct iterations (parameter 6) to be equal to the maximum number of iterations. The direct iterations may be damped by a constant factor (parameter 7), or variable damping may be applied, as explained below.

3. You can continue the simulation even though the maximum number of iterations has been reached. If you do, the last estimate of the output streams and a message indicating that the simulation continued with unconverged output are printed in the history. See parameter 4.

4. You may specify whether the output stream is to be flashed. You may not wish to flash the stream if there is a possibility of three phases existing or if a heat balance is not of interest. See parameter 5.

In the convergence method, an acceleration parameter q_n is calculated and used to weight the previous stream estimates to obtain a new estimate. The acceleration factor is defined as the slope divided by the slope minus 1. The slope is determined for each variable in each stream from

$$\text{Slope} = \frac{f(x_n) - f(x_{n-1})}{x_n - x_{n-1}}$$

where x_{n-1} is the value given to the process for a particular stream variable (block output) and $f(x_{n-1})$ is the corresponding value calculated from the process (block input). The values x_n and $f(x_n)$ are the next pair of input and response stream values. Since two values of $f(x)$ are needed to calculate q_n, at least one direct iteration must be made before acceleration can be applied.

Following the initial direct iterations, the new estimate for each stream variable is calculated from

$$x_{n+1} = q_n x_n + (1 - q_n) f(x_n)$$

If $q_n = 0$, the calculation is direct iteration. If $0 < q_n < 1$, it is direct iteration with damping, while $q_n < 0$ applies acceleration. Positive values of q_n give slow, stable convergence; negative values can considerably speed up convergence but increase the likelihood of instability.

To obtain a favorable trade-off between speed and stability, each q_n is limited to the range

$$Q_{\min} \leq q_n \leq Q_{\max}$$

See parameters 8 to 11. Unless you believe your system is exceptionally stable (or unstable), you should use the default values of $Q_{min} = -5.0$ and $Q_{max} = 0.0$. If the solution is found to oscillate or diverge, one or both of these values should be algebraically increased. However, if the system is stable, decreasing Q_{min} and/or Q_{max} may give convergence in fewer iterations.

If Q_{min} and Q_{max} are both nonnegative, there will be no acceleration, convergence being equivalent to direct iteration with damping in the range between Q_{min} and Q_{max}. Further, unless q_n is always somewhat less than 1.0, the block will repeatedly try the same value of x_{n+1} (and thus appear to have converged). Setting $Q_{max} < 1.0$ forces the block to always take at least a $1.0 - Q_{max}$ fraction of the direct iteration step.

In addition to changing Q_{min} and Q_{max}, stability can be influenced by changing the initial number of direct iterations, parameter 6, and the amount of damping, parameter 7. In general, this will also decrease the rate of convergence.

Output

There is none.

Properties Used

Enthalpies and vapor-equilibria data are needed if the output stream is flashed.

References

H. C. Kleisch, "A Study of Convergence Accelerator Algorithms Used in Steady State Process Simulation," Master's Thesis, Tulane University, New Orleans, 1967.
J. H. Wegstein, "Accelerating Convergence of Iteration Processes," *Comm. ACM*, 1: 9 (1958).

Block List

List type	BLOCK
Unit name	
Unit type	SCVW
Name of first input stream	
Name of second input stream or 0	
Name of third input stream or 0	
Transfer point (unit name of the first block in the recycle system)	
Name of first output stream	
Name of second output stream or 0	
Name of third output stream or 0	

Parameter List

List type	PARAM
Unit name	
Index of first entry*	1

1. Print the q vector after each iteration: $0 = $ no, $1 = $ yes. _____
2. Convergence tolerance (0.0005 if 0). _____
3. Maximum number of iterations (30 if 0). _____
4. Action if unconverged after the maximum number of iterations: $0 = $ terminate simulation, $1 = $ continue with last estimate of output streams. _____
5. Flash output streams: $0 = $ yes, $1 = $ no. _____
6. Number of direct iterations to be performed before acceleration is applied (1 if 0). _____
7. Damping factor for direct iterations ($0 = $ no damping). _____
8. Code for determining upper limit on q, Q_{max}. _____
   a. If 0, Q_{max} is set to 0.
   b. If 1, Q_{max} is set by parameter 9.
9. Maximum value of q, Q_{max} (see parameter 8). _____
10. Code for determining lower limit on q, Q_{min}. _____
   a. If 0, Q_{min} is set to -5.
   b. If 1, Q_{min} is set by parameter 11.
11. Minimum value of q, Q_{min} (see parameter 10). _____

D.10   DSTWU

Description

DSTWU (shortcut distillation) uses the Winn-Underwood method (see References) to calculate the overhead and bottoms streams for a single feed distillation column with either a total or a partial condenser; that is, the overhead product is either saturated liquid or saturated vapor. The desired column performance is specified as splits for light- and heavy-key components. The Gilliland correlation (see References) is used to calculate either the actual reflux ratio for a given number of trays or the actual number of trays for a specified reflux ratio. The reflux ratio may be given as either a fixed value or a multiple of the minimum reflux. If the

---

* All parameters for this block are optional. Default values are assigned to zero and unspecified parameters.

specified reflux ratio or number of trays is less than the minimum that will produce the desired separation, twice the minimum is used.

Output

Minimum reflux ratio, minimum number of trays, actual reflux ratio, actual number of trays, number of trays in rectifying section for optimum feed tray location, and condenser and reboiler temperatures are output.

Properties Used

Vapor-liquid equilibria and enthalpies are employed.

References

F. W. Winn, *Petrol. Refiner,* **37** (5): 216 (1958).

A. J. V. Underwood, *Chem. Eng. Progr.,* **44** (8): 603 (1948).

E. R. Gilliland and C. S. Robinson, *Elements of Fractional Distillation,* 4th ed, McGraw-Hill, New York, 1950, p. 347.

Block List

List type	BLOCK
Unit name	
Unit type	DSTWU
Name of feed	
Name of bottoms product	
Name of overhead product	

Parameter List

List type	PARAM
Unit name	
Index of first entry	1

1. Heavy-key component number
2. Light-key component number
3. Split for light key, distillate mole fraction/ bottoms mole fraction
4. Split for heavy key, bottoms mole fraction/ distillate mole fraction

5. Quality of feed* (computed if 0) _____

6. Desired reflux ratio if positive; if less than -1, the reflux ratio is $|p(6)|$ times the minimum reflux _____

7. Desired number of trays[†] _____

8. Top pressure, psia _____

9. Bottom pressure, psia _____

10. Condenser type: $0 = $ total, $1 = $ partial _____

D.11 REACT

Description

REACT (chemical reactor) is a block which computes the composition of an effluent stream from a reactor. The first reaction uses the component flows found in the feed stream. A second reaction would use the final reaction mass of the first reaction as a feed stream. This procedure is applied in a similar manner to the third and fourth reactions. The exact computation is as follows:

$$ROUT_i = FEED_i - (FEED_{key} \times CONV)(COEF_i/COEF_{key})$$

where i is the ith component in the reaction and

$FEED_i$ = amount of ith component in feed stream
$ROUT_i$ = amount of ith component in effluent stream
$FEED_{key}$ = amount of key component in feed stream
key = key component (a reactant)
CONV = fractional conversion of key component
$COEF_i$ = molar stoichiometric coefficient of component i for the reaction; positive for products, negative for reactants
$COEF_{key}$ = Molar stoichiometric coefficient of key component for reaction (always negative, usually -1)

---

* The feed quality is the amount of heat needed to completely vaporize the feed divided by the "latent heat" (difference between dew- and bubble-point enthalpies) of the feed.

† If both the desired reflux and the desired number of trays are nonzero, then the desired reflux is ignored. If they are both zero, the program gives the reflux for twice the minimum number of trays.

The maximum number of reactions that can be specified is 4. More than four reactions can be handled by placing reactor blocks in series. Up to seven feeds can be handled by REACT which adds the component flow rates in one feed to the reactor. No heat of reaction is calculated for the reactor; the reactor temperature and pressure must be specified. The effluent is flashed into liquid and vapor products at the specified temperature and pressure if two products are specified. The output can be a single stream.

Parameter 3 is used to determine the block action when the specified conversion is greater than the maximum possible conversion (when all the reactant is depleted). A value of 0 causes the block to reset the conversion to the maximum possible and continue; a value of 1 causes the simulation to terminate with an error message.

The parameter lists allow 25 components for each of the four reactions. Stoichiometric coefficients for components not in a reaction should be zero. Components are numbered sequentially from 1 to 25 in the order in which they occur in the component list.

Output

Error messages printed in the history are as follows:

WARNING REACTION X CONVERSION RESET TO = Y
If parameter 3 is set to 0 and you specify a conversion for reaction X higher than the maximum possible, the block uses the maximum Y and continues.

REACTION X EXCEEDS MAX POSSIBLE CONV OF Y
If parameter 3 is set to 1 and you specify a conversion for reaction X higher than the maximum possible, Y, the simulation is terminated.

REACTION X KEY COMP FLOW = 0
For reaction X the flow rate for the specified key component is zero. The simulation continues on to the next reaction in the sequence.

FEED STREAM IS ZERO
All feeds to the block are zero. The simulation continues with the next block. Output streams are set to zero.

ERROR IN MFLSH
The flash program cannot find the enthalpy of the exit streams. Simulation is terminated.

MAX NO. REACTIONS EQUAL 4
Simulation is terminated.

KEY COMP COEF POSITIVE OR ZERO
Key component coefficient must be negative. Simulation is terminated.

Block output consists of the block name, the reaction temperature and pressure, and the names of the input and output streams.

Properties Used

Enthalpies are used.

Block List

List type	BLOCK
Unit name	_____
Unit type	REACT
Name of 1st feed stream	_____
Name of 2d feed stream or 0	_____
Name of 3d feed stream or 0	_____
Name of 4th feed stream or 0	_____
Name of 5th feed stream or 0	_____
Name of 6th feed stream or 0	_____
Name of 7th feed stream or 0	_____
Name of liquid product stream or single stream output	_____
Name of vapor product stream or 0	_____

Parameter List

List type	PARAM
Unit name	_____
Index of first entry	1

1. Temperature of reactor, °F. _____
2. Pressure of reactor, psia. _____
3. Terminate simulation if specified conversion exceeds maximum conversion: 0 = no, 1 = yes. _____
4. Number of reactions. _____

Data for 1st reaction
5. Key component number _____
6. Fractional conversion for key component _____
7. Stoichiometric coefficient 1st component* _____
8. Stoichiometric coefficient 2d component _____
9. Stoichiometric coefficient 3d component _____
10. Stoichiometric coefficient 4th component _____
11. Stoichiometric coefficient 5th component _____

---

* Component order is the order in which they occur in the component list.

12. Stoichiometric coefficient 6th component   _____

...

31. Stoichiometric coefficient 25th component   _____

Data for 2d reaction

  32. Key component number   _____

  33. Fractional conversion for key component   _____

  34. Stoichiometric coefficient 1st component   _____

...

58. Stoichiometric coefficient 25th component   _____

Data for 3d reaction

  59. Key component number   _____

  60. Fractional conversion for key component   _____

  61. Stoichiometric coefficient 1st component   _____

...

85. Stoichiometric coefficient 25th component   _____

Data for 4th reaction

  86. Key component number   _____

  87. Fractional conversion for key component   _____

  88. Stoichiometric coefficient 1st component   _____

...

112. Stoichiometric coefficient 25th component   _____

APPENDIX
E

COST
DATA

E.1 OPERATING COSTS

Chemicals

The costs of raw materials, products, and by-products can normally be found in the *Chemical Marketing Reporter*. The values listed are the current market prices, which may be significantly different from the price used in a particular company because of long-term contracts. The costs of light gases usually are not listed in the *Chemical Marketing Reporter* because these materials often are sold "over the fence" (a vendor builds a special plant to produce these materials which is located next to the site that will use them) or a long-term contract is negotiated.

Utilities

The best way to estimate the cost of utilities is to relate the costs of any utility to its equivalent fuel value by using thermodynamics and typical efficiencies of power plants, turbines, boilers, etc. Market fluctuations might occur at times which make the value of steam less than that of fuel, but large cost penalties can be encountered

568

TABLE E.1-1
Utilities costs

Utility	Factor	Price
Fuel (oil or gas)	1.0	$4.00/10^6 Btu
Steam		
600 psig at 750°F	1.30	$5.20/1000 lb
Saturated steam		
600 psig	1.13	$4.52
250 psig	0.93	3.72
150 psig	0.85	3.4
50 psig	0.70	2.8
15 psig	0.57	2.28
Electricity	1.0	$0.04/kwhr
Cooling water	0.75	$0.03/1000 gal

if a design is based on distorted prices and then the costs revert to their normal pattern.

A reasonable set of factors to use is given in Table E.1-1. Once the value of fuel has been specified, the costs of the other utilities can easily be calculated. Note that the values given in Table E.1-1 were not used throughout this text. Similarly, the costs used in different problems are sometimes different. However, the costs used in various problems are identified as the solution is developed.

E.2 SUMMARY OF COST CORRELATIONS

The 1970s have been a period of rapid cost escalation (see Fig. 2.2-11), and so very few cost correlations were published during this period. We use Guthrie's cost correlations in this text, whenever possible, to illustrate costing procedures, but note that these correlations are out of date. We update the correlations from the mid-1968 values[*] by using a ratio of the M&S indices, but this is not a recommended practice for such a long time span. Instead, if an updated set of company cost correlations is not available, a designer should consult one or more vendors early in the costing procedure to obtain more recent cost data.

For our preliminary process designs, we use a simplified version of Guthrie's correlations. The normal material (the base costs assume carbon steel) and pressure correction factors are used to estimate the purchased cost, but the most conservative base module cost factor is used to estimate the installed costs. This approximation corresponds to a conservative cost estimate. For more accurate estimates, Guthrie's book should be consulted.[†]

---

[*] K. M. Guthrie, "Capital Cost Estimating," *Chem Eng.*, **76**(6): 114 (March 24, 1969).

[†] K. M. Guthrie, *Process Plant Estimating Evaluation and Control*, Craftsman Book Co., Solana Beach, Calif., 1974.

Process Furnaces

Mid-1968 cost, box or A-frame construction with multiple tube banks, field-erected.

$$\text{Purchased Cost, \$} = \left(\frac{\text{M\&S}}{280}\right)(5.52 \times 10^3)Q^{0.85}F_c$$

where Q = adsorbed duty, 10^6 Btu/hr; $20 < Q < 300$

$$F_c = F_d + F_m + F_p$$

$$\text{Installed Cost, \$} = \left(\frac{\text{M\&S}}{280}\right)(5.52 \times 10^3)Q^{0.85}(1.27 + F_c)$$

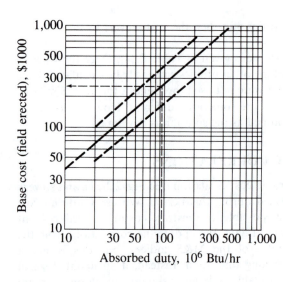

FIGURE E.2-1
Process furnaces. [*K. M. Guthrie, Chem. Eng.,* **76**(*6*): *114* (*March 24, 1969*).]

TABLE E.2-1
Correction factors F_c for process furnace

Design type	F_d	Radiant tube material	F_m	Design pressure, psi	F_p
Process heater	1.00	Carbon steel	0.0	Up to 500	0.00
Pyrolysis	1.10	Chrome/moly	0.35	1000	0.10
Reformer (no catalyst)	1.35	Stainless	0.75	1500	0.15
				2000	0.25
				2500	0.40
				3000	0.60

Direct-Fired Heaters

Mid-1968 cost, cylindrical construction, field erection.

$$\text{Purchased Cost, \$} = \left(\frac{\text{M\&S}}{280}\right)(5.07 \times 10^3)Q^{0.85}F_c$$

where Q = adsorbed duty, 10^6 Btu/hr; $2 < Q < 30$

$$F_c = F_d + F_m + F_p$$

$$\text{Installed Cost, \$} = \left(\frac{\text{M\&S}}{280}\right)(5.07 \times 10^3)Q^{0.85}(1.23 + F_c)$$

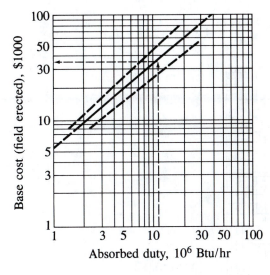

FIGURE E.2-2
Direct-fired heater. [*K. M. Guthrie, Chem. Eng.*, **76**(6): *114 (March 24, 1969).*]

TABLE E.2-2
Correction factors F_c for direct-fired heaters

Design type	F_d	Radiant tube material	F_m	Design pressure, psi	F_p
Cylindrical	1.0	Carbon steel	0.0	Up to 500	0.00
Dowtherm	1.33	Chrome/moly	0.45	1000	0.15
		Stainless	0.50	1500	0.20

Heat Exchangers

Mid-1968 cost, shell and tube, complete fabrication.

$$\text{Purchased Cost, \$} = \left(\frac{\text{M\&S}}{280}\right)(101.3A^{0.65}F_c)$$

where $A = $ area ft^2; $200 < A < 5000$

$$F_c = (F_d + F_p)F_m$$

Shell-and-Tube Material $= F_m$

Surface area, ft²	CS/ CS	CS/ Brass	CS/ MO	CS/ SS	SS/ SS	CS/ Monel	Monel/ Monel	CS/ T_i	T_i/ T_i
1000 to 5000	1.00	1.30	2.15	2.81	3.75	3.10	4.25	8.95	13.05

$$\text{Installed Cost, \$} = \left(\frac{\text{M\&S}}{280}\right)101.3A^{0.65}(2.29 + F_c)$$

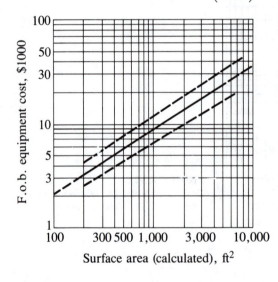

F.o.b. equipment cost, $1000

Surface area (calculated), ft²

FIGURE E.2-3
Shell-and-tube heat exchangers. [*K. M. Guthrie, Chem. Eng.,* **76**(6): *114* (*March 24, 1969*).]

TABLE E.2-3
Correction factors for heat exchangers

Design type	F_d	Design pressure, psi	F_p
Kettle, reboiler	1.35	Up to 150	0.00
Floating head	1.00	300	0.10
U-tube	0.85	400	0.25
Fixed-tube sheet	0.80	800	0.52
		1000	0.55

Gas Compressors

Mid-1968 cost, centrifugal machine, motor drive, base plate and coupling.

$$\text{Purchased Cost, \$} = \left(\frac{M\&S}{280}\right)(517.5)(\text{bhp})^{0.82}F_c$$

where bhp = brake horsepower; $30 < \text{bhp} < 10{,}000$

$$F_c = F_d$$

$$\text{Installed Cost, \$} = \left(\frac{M\&S}{280}\right)(517.5)(\text{bhp})^{0.82}(2.11 + F_c)$$

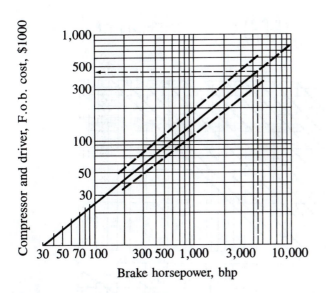

FIGURE E.2-4
Process gas compressors and drives. [*K. M. Guthrie, Chem. Eng.,* **76**(6): *114* (*March 24, 1969*).]

TABLE E.2-4
Correction factors for
Compressors

Design type F_d	Factor
Centrifugal, motor	1.00
Reciprocating, steam	1.07
Centrifugal, turbine	1.15
Reciprocating, motor	1.29
Reciprocating, gas engine	1.82

Pressure Vessels, Columns, Reactors

$$\text{Purchased Cost, \$} = \left(\frac{M\&S}{280}\right)(101.9D^{1.066}H^{0.082}F_c)$$

where D = diameter, ft

H = height, ft

$F_c = F_m + F_p$

Pressure	Up to 50	100	200	300	400	500	600	700	800	900	1000
F_p	1.00	1.05	1.15	1.20	1.35	1.45	1.60	1.80	1.90	2.30	2.50

$$\text{Installed Cost, \$} = \left(\frac{M\&S}{280}\right)101.9D^{1.066}H^{0.802}(2.18 + F_c)$$

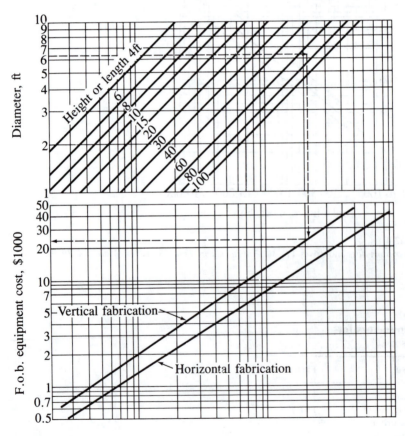

FIGURE E.2-5
Pressure vessels. [*K. M. Guthrie, Chem. Eng.,* **76**(6): *114 (March 24, 1969).*]

TABLE E.2-5
Correction factors for pressure vessels

Shell material	CS	SS	Monel	Titanium
F_m, clad	1.00	2.25	3.89	4.25
F_m, solid	1.00	3.67	6.34	7.89

Distillation Column Trays and Tower Internals

$$\text{Installed Cost, } \$ = \left(\frac{\text{M\&S}}{280}\right)4.7D^{1.55}HF_c$$

where D = diameter, ft

H = tray stack height, ft (24-in. spacing)

$F_c = F_s + F_t + F_m$

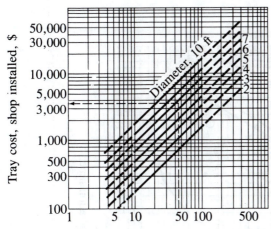

FIGURE E.2-6
Distillation column trays. [*K. M. Guthrie, Chem. Eng.,* **76**(*6*): *114* (*March 24, 1969*).]

TABLE E.2-6
Correction factors for column trays

Tray spacing, in.	24	18	12				
F_s	1.0	1.4	2.2				
Tray type		Grid (no down-comer)	Plate	Sieve	Trough or valve	Bubble cap	Koch Kascade
F_t	0.0		0.0	0.0	0.4	1.8	3.9
Tray material	CS	SS	Monel				
F_m	0.0	1.7	8.9				

TABLE E.2-7
Tower packings

Material	Materials and labor, $/ft^3
Activated carbon	14.2
Alumina	12.6
Coke	3.5
Crushed limestone	5.8
Silica gel	27.2
1-in. Raschig rings—Stoneware	5.2
Porcelain	7.0
Stainless	70.2
1-in. Berl saddles—Stoneware	14.5
Porcelain	15.9

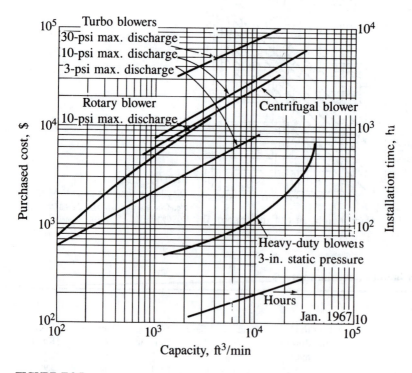

FIGURE E.2-7
Blowers (heavy-duty, industrial type). (*From M. S. Peters and K. D. Timmerhaus, Plant Design and Economics for Chemical Engineers, 3d ed., McGraw-Hill, New York, 1980, p. 562.*)

Turbo Blowers

From Peters and Timmerhaus,* January 1967 cost, see Fig. E.2-7

3-psi maximum discharge:

$$\text{Purchased Cost} = \left(\frac{\text{M\&S}}{260}\right) 39.7 Q^{0.529}$$

where Q = cfm and $100 < Q < 10{,}000$.

10-psi maximum discharge:

$$\text{Purchased Cost} = \left(\frac{\text{M\&S}}{260}\right) 126.5 Q^{0.598}$$

where Q = cfm and $1000 < Q < 30{,}000$.

30-psi maximum discharge:

$$\text{Purchased Cost} = \left(\frac{\text{M\&S}}{260}\right) 838.7 Q^{0.493}$$

where Q = cfm and $2000 < Q < 15{,}000$. Assume installation factor = 4.0.

---

* M. S. Peters and K. D. Timmerhaus, "Plant Design and Economics for Chemical Engineers," 3d ed., McGraw-Hill, New York, 1980, p. 562.

APPENDIX
F

CONVERSION FACTORS

Area

$1 \text{ ft}^2 = 0.0929 \text{ m}^2$
$= 144 \text{ in.}^2$

Density

$1 \text{ lb/ft}^3 = 16.018 \text{ kg/m}^3$
$= 1/62.4 \text{ g/cm}^3$
1 lb mole of an ideal gas, 0°C, 1 atm = 359.0 ft^3
1 lb mole of air, 0°C 1 atm = 0.0807 lb/ft^3

Energy—Also see Work

$1 \text{ Btu} = 252 \text{ cal}$
$= 1.055 \text{ kJ}$
$= 777.9 \text{ ft} \cdot \text{lbf}$
$= 3.929 \times 10^{-4} \text{ hp} \cdot \text{hr}$
$= 2.9307 \times 10^{-4} \text{ kwhr}$

Force

$$1 \text{ lbf} = 4.4482 \text{ N } (\text{kg} \cdot \text{m/s}^2)$$
$$= 32.174 \text{ lbm} \cdot \text{ft/s}^2$$
$$= 4.4482 \times 10^5 \text{ dyn } (\text{g} \cdot \text{cm/s}^2)$$

Heat Load—Also see Power

$$1 \text{ Btu/hr} = 0.29307 \text{ w}$$

Heat-Transfer Coefficient

$$1 \text{ Btu/(hr} \cdot \text{ft}^2 \cdot {}^\circ\text{F}) = 5.6782 \text{ w/(m}^2 \cdot {}^\circ\text{C})$$
$$= 1.3571 \times 10^{-4} \text{ cal/(cm}^2 \cdot \text{s} \cdot {}^\circ\text{C})$$

Length

$$1 \text{ ft} = 0.3048 \text{ m}$$

Mass

$$1 \text{ lbm} = 0.45359 \text{ kg}$$
$$1 \text{ ton (short)} = 2000 \text{ lbm}$$

Pressure

$$1 \text{ atm} = 14.7 \text{ psi}$$
$$1 \text{ psi} = 6894.76 \text{ N/m}^2 (\text{dyn/cm}^2)$$

Power—Also see Heat Load

$$1 \text{ hp} = 550 \text{ ft} \cdot \text{lbf/s}$$
$$= 0.7457 \text{ kw}$$
$$= 2546.7 \text{ Btu/hr}$$

Specific Heat

$$1 \text{ Btu/(lbm} \cdot {}^\circ\text{F}) = 4.1869 \text{ kJ/(kg} \cdot {}^\circ\text{C})$$

Work—Also see Energy

$$1 \text{ ft} \cdot \text{lbf} = 1.2851 \times 10^{-3} \text{ Btu}$$
$$= 3.7662 \times 10^{-7} \text{ kwhr}$$

Velocity

$1 \text{ ft/s} = 0.3048 \text{ m/s}$

Viscosity

$1 \text{ lbm/(ft} \cdot \text{s)} = 1.4881 \text{ kg/(m} \cdot \text{s)}$

$1 \text{ lbm/(ft} \cdot \text{hr)} = 4.1338 \times 10^{-3} \text{g/(cm} \cdot \text{s)}$

Volume

$$1 \text{ ft}^3 = 0.028317 \text{ m}^3$$
$$= 28.32 \text{ L}$$
$$= 7.481 \text{ gal}$$

INDEXES

AUTHOR
INDEX

SUBJECT
INDEX